THE CONCISE HUMAN BODY BOOK

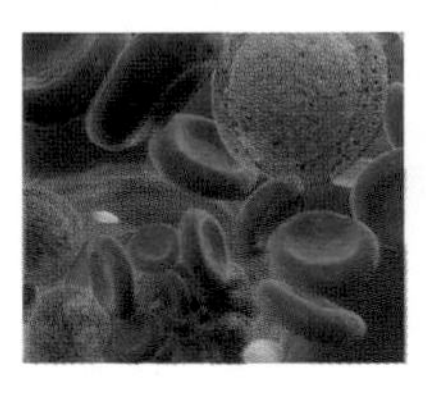
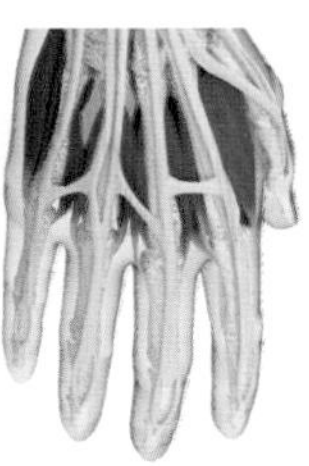

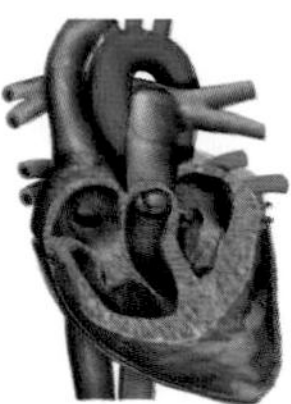
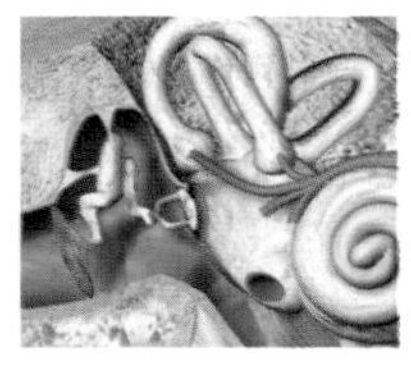

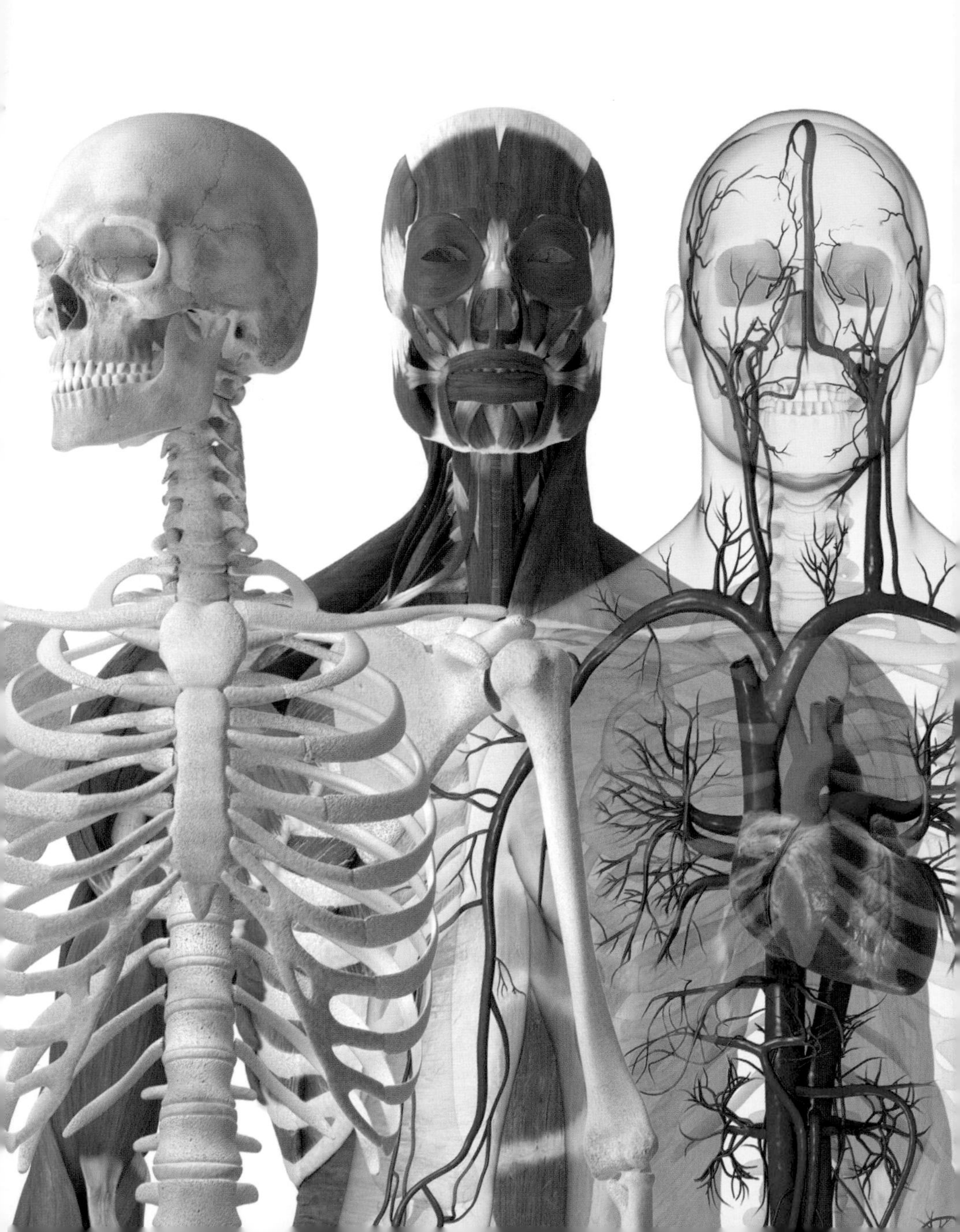

THE CONCISE HUMAN BODY BOOK

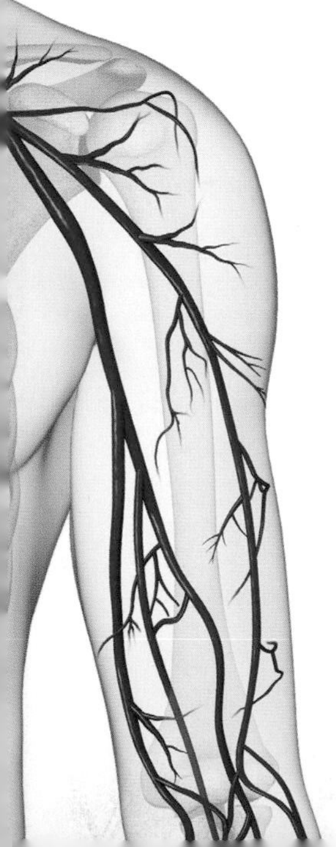

STEVE PARKER

PROJECT EDITORS Ann Baggaley, Philip Morgan, Martyn Page
PROJECT ART EDITORS Mandy Earey, Ted Kinsey
SENIOR EDITOR Simon Tuite
SENIOR ART EDITOR Vicky Short
MANAGING EDITOR Julie Oughton
MANAGING ART EDITOR Louise Dick
ASSOCIATE PUBLISHER Liz Wheeler
PUBLISHER Jonathan Metcalf
ART DIRECTOR Bryn Walls

JACKET DESIGNER Duncan Turner
PRODUCTION EDITORS Joanna Byrne, Maria Elia
PRODUCTION CONTROLLER Sophie Argyris
US EDITORS Jill Hamilton, Rebecca Warren

MEDICAL CONSULTANT Dr Penny Preston

The Concise Human Body Book provides information on a wide range of medical topics, and every effort has been made to ensure that the information in this book is accurate. The book is not a substitute for medical advice, however, and you are advised always to consult a doctor or other health professional on personal health matters

First Canadian edition 2009

Dorling Kindersley is represented in Canada by
Tourmaline Editions Inc.
662 King Street West, Suite 304
Toronto, Ontario M5V 1M7

9 10 11 12 10 9 8 7 6 5 4 3

010-TD422-June/09

A CIP catalogue record for this book is available from Library and Archives Canada

ISBN 978-1-55363-109-5

The Concise Human Body Book has been adapted from *The Human Body Book*, published in Great Britain in 2007 by Dorling Kindersley Limited

PROJECT EDITOR Rob Houston
PROJECT ART EDITOR Maxine Lea
EDITORS Ruth O'Rourke, Rebecca Warren, Mary Allen, Kim Bryan, Tarda Davidson-Aitkins, Jane de Burgh, Salima Hirani, Miezan van Zyl
DESIGNERS Matt Schofield, Kenny Grant, Francis Wong, Anna Plucinska
MANAGING EDITOR Sarah Larter
MANAGING ART EDITOR Philip Ormerod
PUBLISHING MANAGER Liz Wheeler

PICTURE RESEARCHER Louise Thomas
JACKET DESIGNER Lee Ellwood
DTP DESIGNER Laragh Kedwell
PRODUCTION CONTROLLER Tony Phipps
EDITORIAL ASSISTANTS Tamlyn Calitz, Manisha Thakkar
INDEXER Hilary Bird
PROOFREADER Andrea Bagg

CONTRIBUTORS Mary Allen, Andrea Bagg, Jill Hamilton, Katie John, Janet Fricker, Jane de Burgh, Claire Cross
MEDICAL CONSULTANTS Dr Sue Davidson, Dr Penny Preston, Dr Ian Guinan

ILLUSTRATORS

Medi-Mation
Medical & Scientific Visualization

CREATIVE DIRECTOR Rajeev Doshi
3D ARTISTS Olaf Louwinger, Gavin Whelan, Monica Taddei

ADDITIONAL ILLUSTRATORS Peter Bull Art Studio, Kevin Jones Associates, Adam Howard

Color reproduction by GRB Editrice s.r.l. in London, UK

Printed and bound in China by Hung Hing Offset Printing Company Ltd

Discover more at
www.dk.com

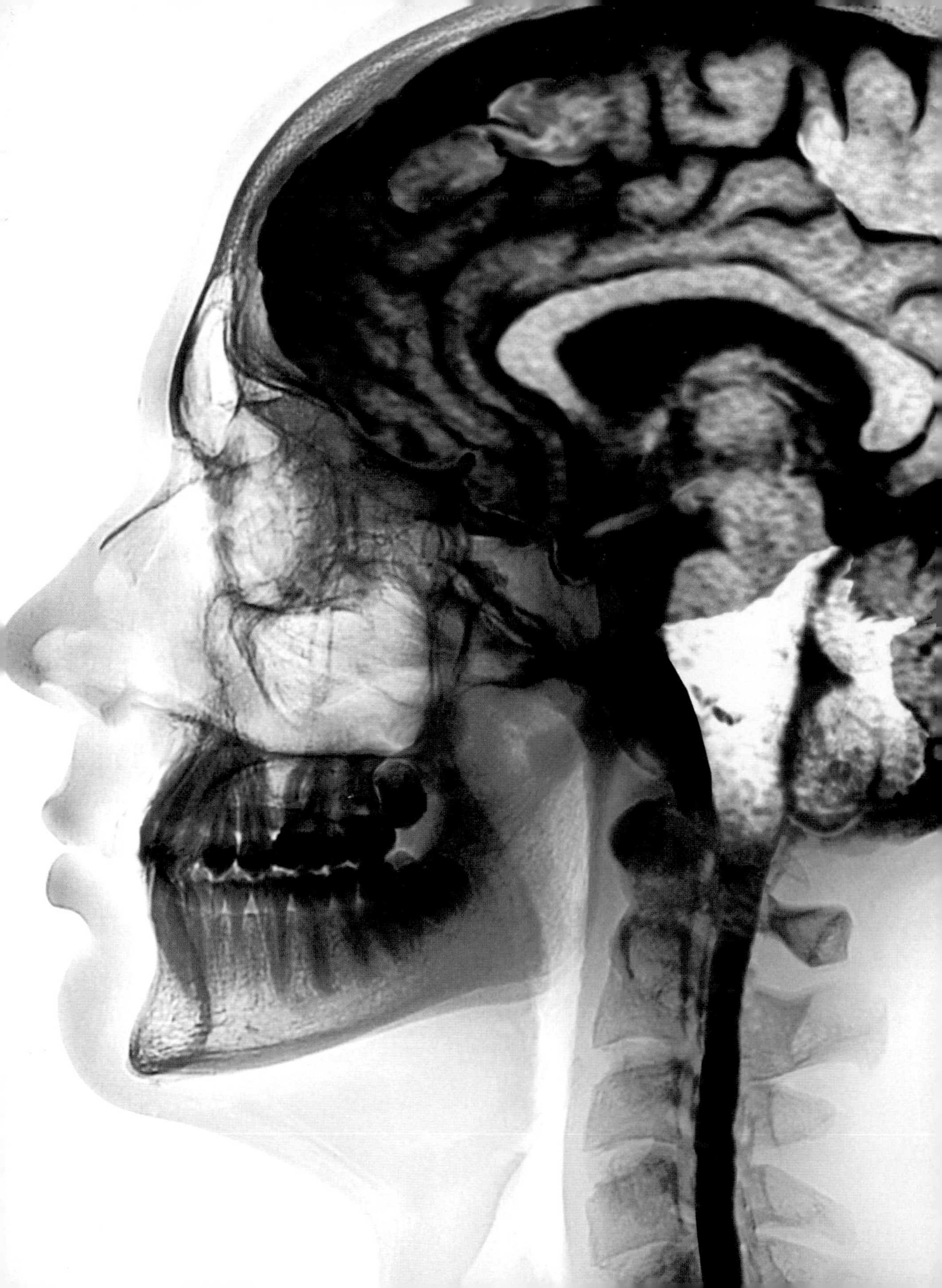

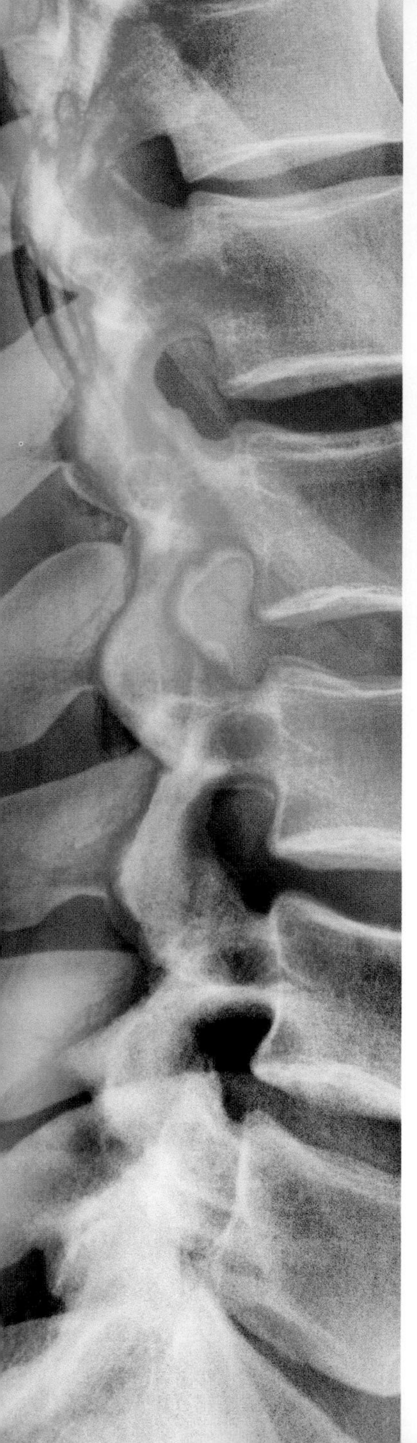

CONTENTS

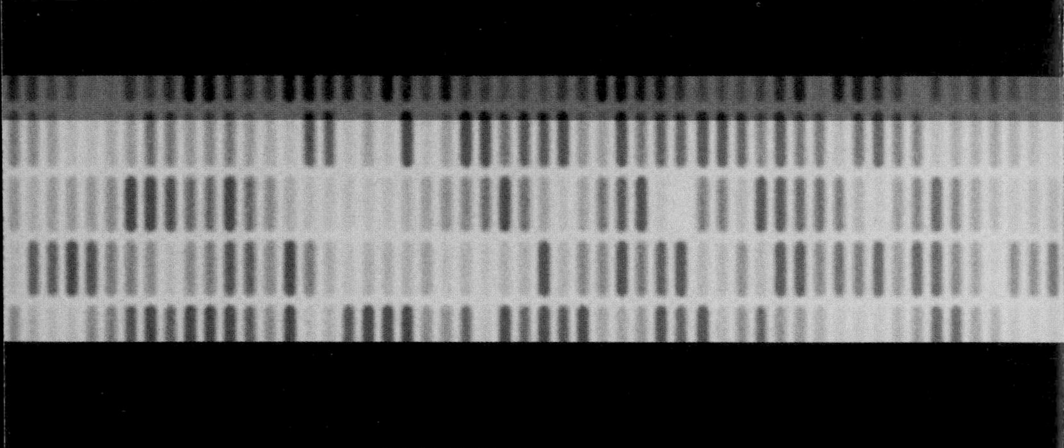

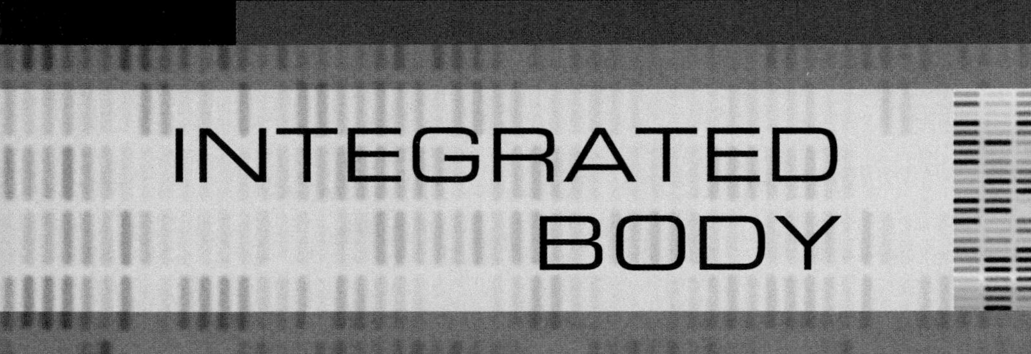
INTEGRATED
BODY

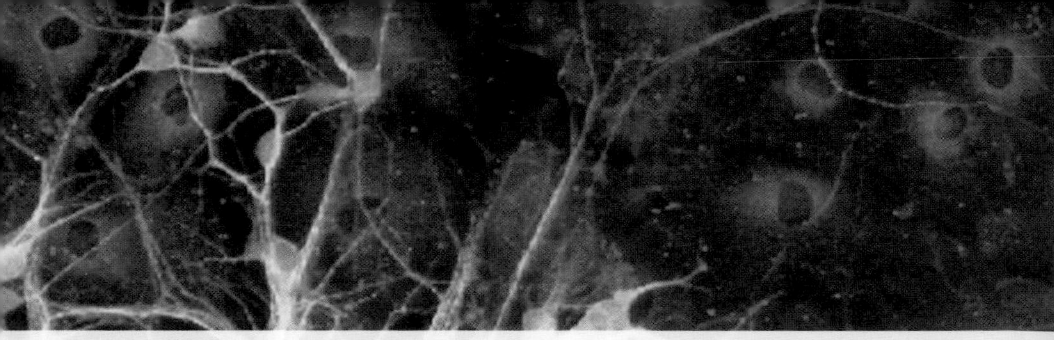

INTRODUCTION

The number of humans in the world is approaching seven billion (7,000,000,000). More than 250 babies are born every minute, while 150,000 people die daily, with the population increasing by almost three humans per second. Each of these people lives and thinks with, and within, that most complex and marvelous of possessions–a human body.

LEVELS OF ORGANIZATION

To understand the inner structure and workings of the human body, this book takes the "living machine" approach, borrowed from sciences such as engineering. This views the body as a series of integrated systems. Each system carries out one major task. In the cardiovascular system, for example, the heart pumps blood through vessels, to supply every body part with essential oxygen and nutrients. The systems are, in turn, composed of main parts known as organs. For example, the stomach, intestines, and liver are organs of the digestive system. Moving down through the anatomical hierarchy, organs consist of tissues, and tissues are made up of cells.

Cells are often called the building blocks of the body. Active and dynamic, they continually grow and specialize, function, die, and replenish themselves, by the millions every second. The whole body contains about 100 million million cells, of at least 200 different kinds. Science is increasingly able to delve deeper than cells, to the organelles within them, and onward, to the ultimate components of ordinary matter–molecules and atoms.

ANATOMY

The study of the body's structure, and how its cells, tissues, and organs are assembled, is known as human anatomy. For clarity, its elements are often shown in isolation, because the inside of the body is

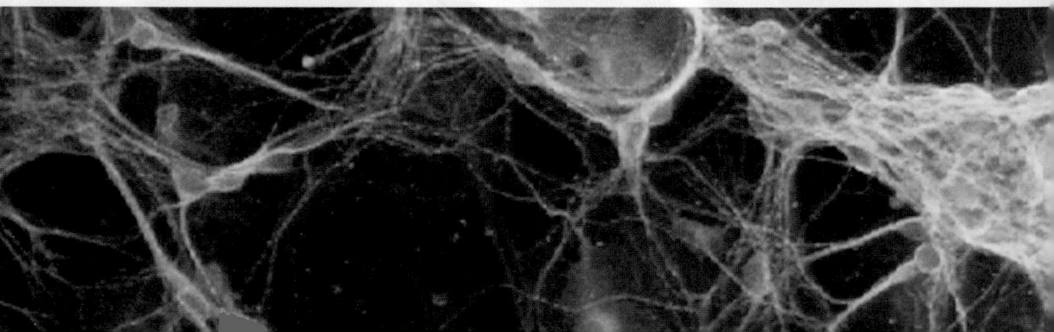

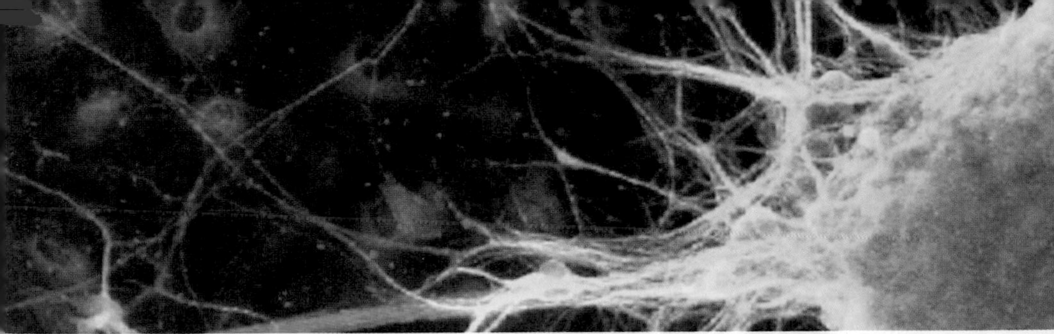

a crowded place. Tissues and organs press against one another. Body parts shift continually as we move, breathe, pump blood, and digest food. For example, swallowed food does not simply fall down the gullet into the stomach; it is forced down by waves of muscular contraction.

PHYSIOLOGY

For a rounded understanding of the body, we need to see human anatomy in combination with physiology–the study of how the body functions. Physiology focuses on the dynamic chemical minutiae at atomic, ionic, and molecular levels. It investigates the workings of such processes as enzyme action, hormone stimulation, DNA synthesis, and how the body stores and uses energy from food. As researchers look closer, and unravel more biochemical pathways, more physiological secrets are unlocked. Much of this research work is aimed at preventing or treating disease.

HEALTH AND ILLNESS

Medical science amasses mountains of evidence every year for the best ways to stay healthy. At present, an individual's genetic inheritance, which is a matter of chance, is the given starting point for maintaining health and well-being. In coming years, treatments such as gene therapy could remove or negate some of these chance elements. Many aspects of upbringing have a major impact on health, including factors such as diet–whether it is too rich or too poor. The body can also be affected by many different types of disorders, such as infection by a virus or bacteria, injury, inherited faulty genes, or exposure to toxins in the environment.

COMMUNICATION NETWORK
This microscopic image of nerve cells (neurons) shows the fibers that connect the cell bodies. Neurons transmit electrical signals around the body; each one links with hundreds of others, forming a dense web.

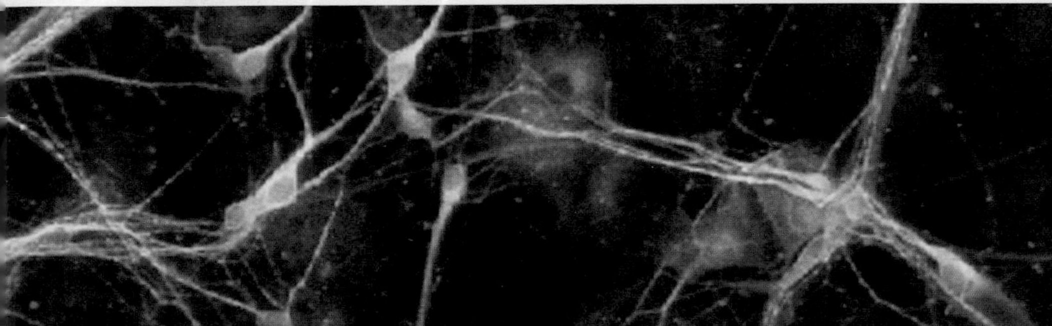

IMAGING THE BODY

IMAGING IS A VITAL PART OF DIAGNOSING ILLNESS, UNDERSTANDING DISEASE, AND EVALUATING TREATMENTS. MODERN TECHNIQUES PROVIDING HIGHLY DETAILED INFORMATION HAVE LARGELY REPLACED SURGERY AS A METHOD OF INVESTIGATION.

The invention of the X-ray made the development of noninvasive medicine possible. Without the ability to see inside the body, many disorders could be found only after major surgery. Computerized imaging now helps doctors make early diagnoses, often greatly increasing the chances of recovery. Computers process and enhance raw data, for example reinterpreting shades of gray from an X-ray or scan into colors. However, sometimes direct observation is essential. Viewing techniques have also become less invasive with the development of instruments such as the endoscope (see opposite). This book makes extensive use of internal images from real bodies.

MICROSCOPY

In light microscopy (LM), light is passed through a section of material and lenses magnify the view up to 2,000 times. Even higher magnifications are possible with scanning electron microscopy (SEM), in which light runs across a specimen coated with gold film. Electrons bounce off the surface, creating a three-dimensional image.

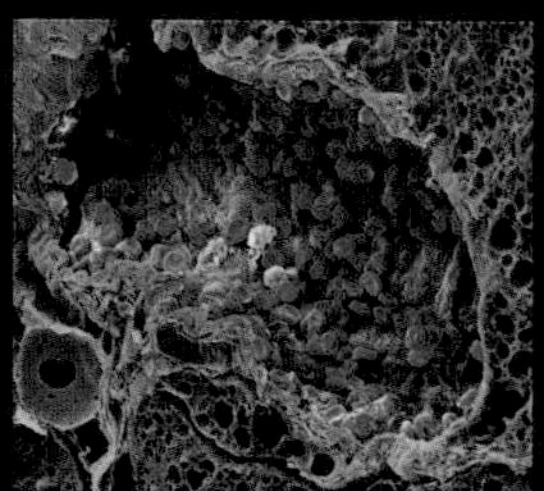

SEM OF TUMOR BLOOD SUPPLY
This image, in which the specimen has been frozen and split open, shows a blood vessel with blood cells growing into a melanoma (skin tumor).

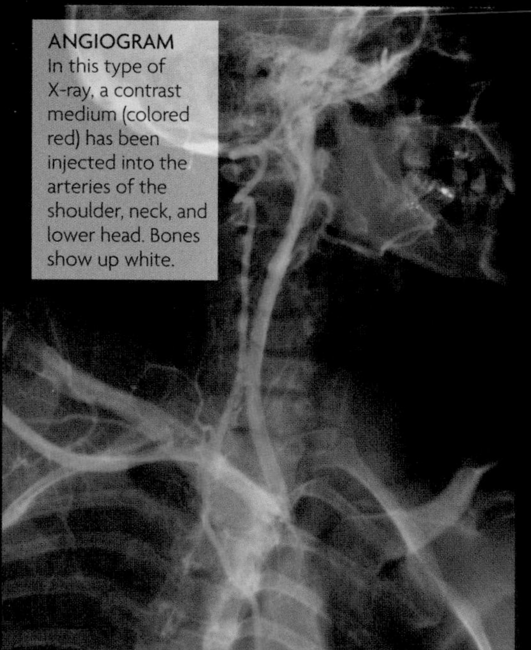

ANGIOGRAM
In this type of X-ray, a contrast medium (colored red) has been injected into the arteries of the shoulder, neck, and lower head. Bones show up white.

X-RAY

X-rays are similar to light waves, but of very short wavelength. When passed through the body they create shadow images on photographic film. Dense structures such as bone show up white; soft tissues appear as shades of gray. To show hollow or fluid-filled structures, these are filled with a substance that absorbs X-rays (a contrast medium).

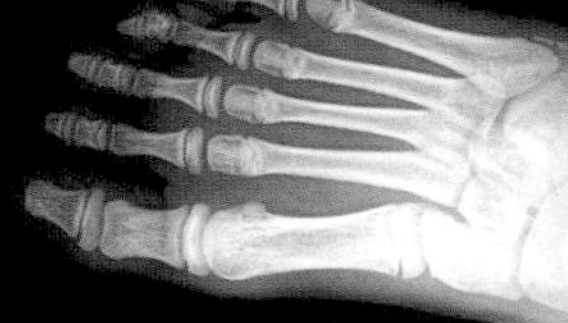

PLAIN X-RAY OF FOOT
A plain X-ray is especially useful for viewing dense tissue, such as bone. This image shows the foot bones of a nine-year-old child.

MRI AND CT SCANNING

In computerized tomography (CT) an X-ray scanner is used with a computer to build up cross-sectional images of tissues of different density. In magnetic resonance imaging (MRI), magnets are used to line up atoms in the body, then radio waves throw the atoms out of alignment. As they realign, the atoms emit signals that are used to create an image.

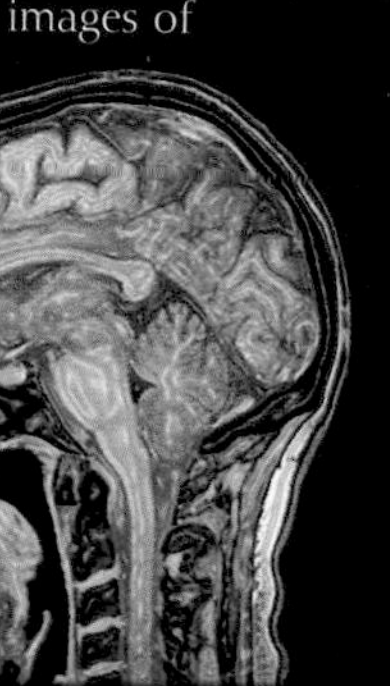

MRI SCAN OF HEAD

This digitally enhanced MRI scan shows the brain and spinal cord in orange and yellow, and muscle and bone in blue.

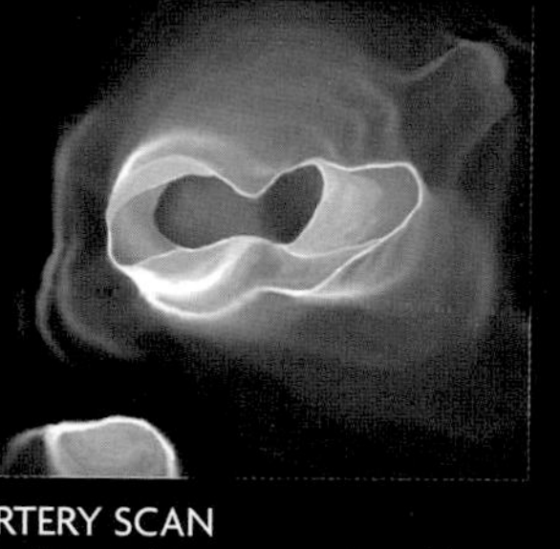

ARTERY SCAN

The layered images produced by a CT scan can be built into a three-dimensional image on a computer. The inside of a narrowed artery is seen here.

RADIONUCLIDE AND PET SCANNING

In radionuclide imaging, a radioactive chemical is injected into the body and is absorbed by the tissues. As the substance decays it emits gamma rays, which a computer forms into an image. Positron emission tomography (PET) is a type of radionuclide scanning that uses injected chemicals, which emit radioactive particles called positrons.

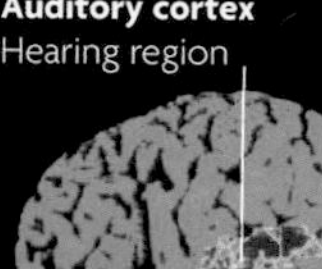

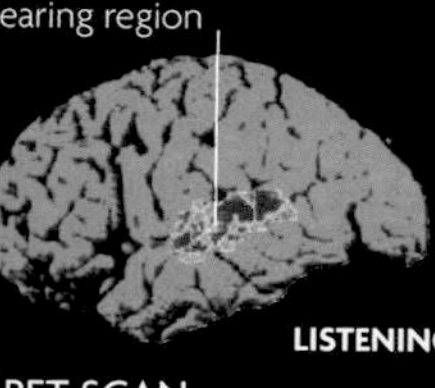
Auditory cortex
Hearing region
LISTENING

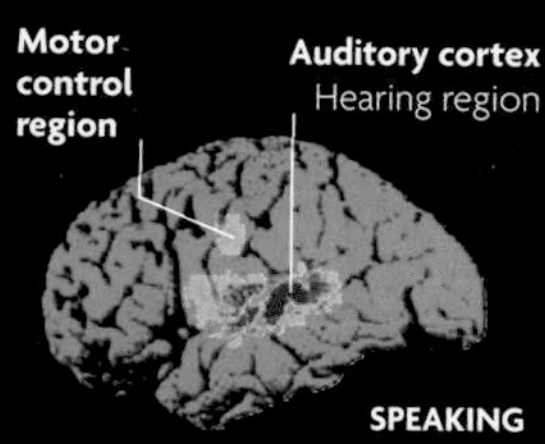
Motor control region
Auditory cortex
Hearing region
SPEAKING

PET SCAN

PET scans show function rather than anatomy. These images reveal the brain's activity as the subject listened to spoken words and then both listened to and repeated the words.

ULTRASOUND

High-frequency sound waves emitted by a device called a transducer pass into the body and echo back as electrical signals. A computer processes the signals to create images.

FETAL ULTRASOUND

Ultrasound is a very safe technique, commonly used to monitor fetal development in the uterus.

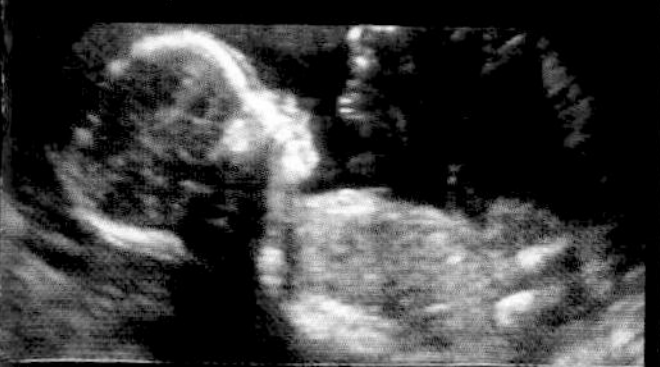

ENDOSCOPY

Endoscopes are flexible or rigid tubes inserted into the body to view its interior, perform surgical procedures, or both. They carry a light source and instruments may be passed down them.

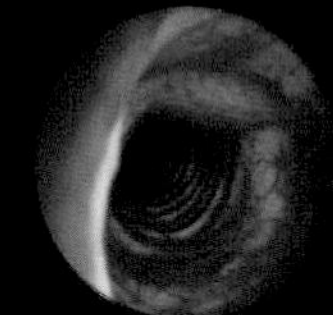

TRACHEA

An endoscopic view of the trachea (windpipe) shows the hoops of cartilage that maintain its shape.

ELECTRICAL ACTIVITY

Sensor pads applied to the skin detect electrical activity in muscles and nerves. The signals are displayed as a trace line. This technique includes electrocardiography (ECG) of the heart (see below).

Upper chambers contract

Lower chambers contract

Heart muscle relaxes

BODY SYSTEMS

THE HUMAN BODY'S SYSTEMS WORK TOGETHER AS A TRUE COOPERATIVE. EACH SYSTEM FULFILLS ITS OWN VITAL FUNCTION, BUT ALL WORK TOGETHER TO MAINTAIN THE HEALTH AND EFFICIENCY OF THE BODY AS A WHOLE.

The exact number and extent of the body's systems is debated–the muscles, bones, and joints are sometimes combined as the musculoskeletal system, for instance. Although these systems can be described as separate entities, each depends on all of the others for physical and physiological support. Most systems have some "general" body tissues, such as the connective tissues, which delineate, support, and cushion many organs. All the systems–except, and somewhat ironically, the reproductive system–are essential for our basic survival.

SKELETAL

EXPLORED ON PAGES **38–61**

The skeleton is a solid framework that supports the body. Its bones work as levers and anchor plates to allow for movement. Bones also have a role in other body systems—blood cells develop in their fatty inner tissue (red marrow), for example. The body draws from mineral stores in bones during times of shortage, such as when calcium is needed for healthy nerve function.

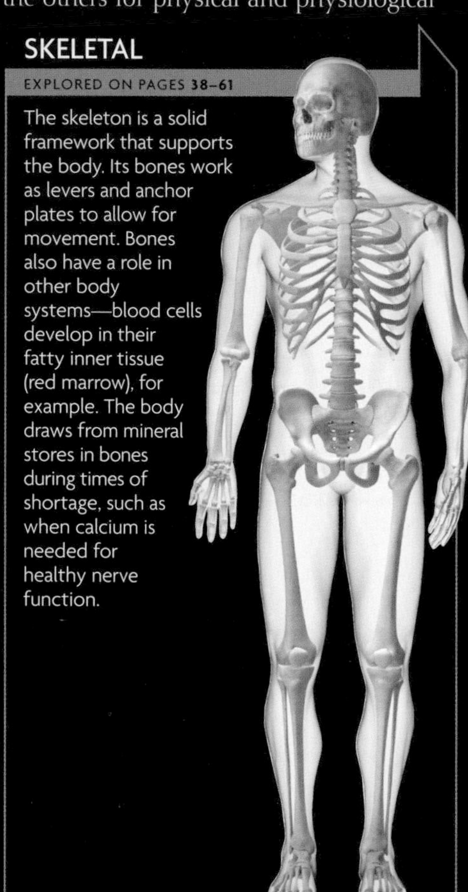

MUSCULAR

EXPLORED ON PAGES **62–75**

Muscles work in conjunction with the skeleton, providing the pulling force for varying degrees of movement, from powerful to finely tuned. Involuntary muscles work largely automatically to control internal processes, such as blood distribution and digestion. Muscles rely on nerves to control them and blood to keep them supplied with oxygen and energy.

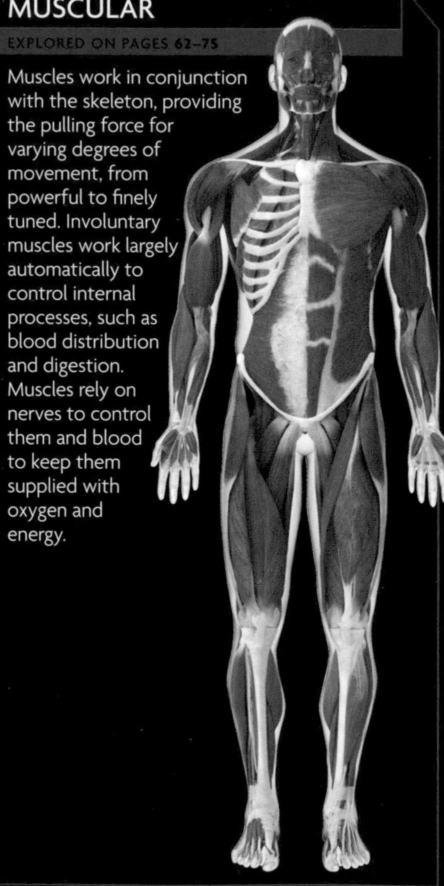

NERVOUS

EXPLORED ON PAGES **76–129**

The brain is the seat of both consciousness and creativity and, through the spinal cord and nerve branches, it controls all body movements with its motor output. The brain also receives sensory information from outside and within the body. Much of the brain's activity occurs unconsciously as it works with endocrine glands to monitor and maintain other body systems.

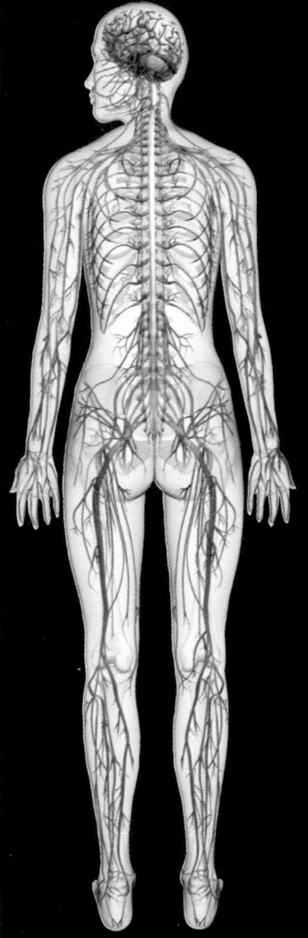

ENDOCRINE

EXPLORED ON PAGES **130–143**

The glands and cells of the endocrine system produce chemical messengers called hormones, which circulate in blood and other fluids. These maintain an optimal internal environment. Hormones also govern long-term processes such as growth, the changes that take place during puberty, and reproductive activity. The endocrine system has close links to the nervous system.

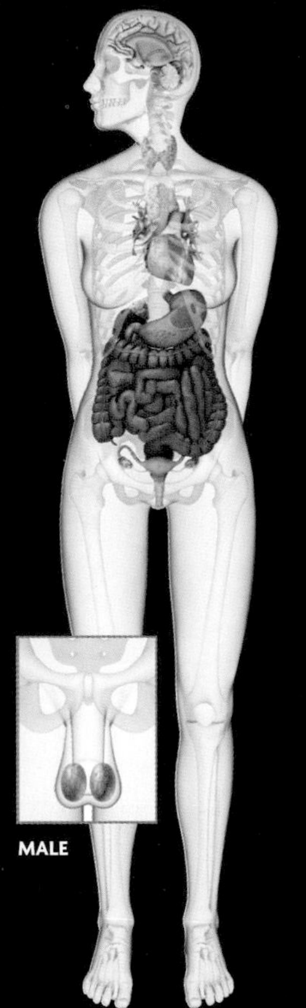
MALE

CARDIOVASCULAR

EXPLORED ON PAGES **144–159**

The most basic function of the cardiovascular, or circulatory, system is to pump blood around the body. It supplies all organs and tissues with freshly oxygenated, nutrient-rich blood. Any waste products of cell function are removed with the blood as it leaves. The circulatory system also transports other vital substances, such as nutrients, hormones, and immune cells.

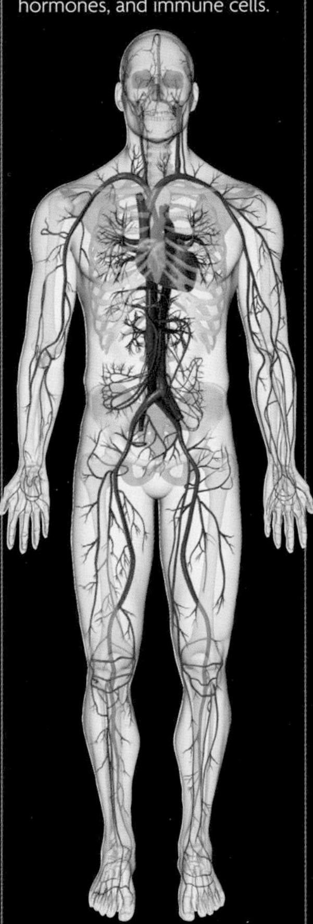

RESPIRATORY

EXPLORED ON PAGES 160–175

The respiratory tract and its movements, powered by breathing muscles, carry air into and out of the lungs. Deep inside the lungs, gases are exchanged. On inhalation, life-giving oxygen is absorbed from air, while carbon dioxide waste is passed into the air, to be expelled from the body on exhalation. A secondary function of the respiratory system is vocalization.

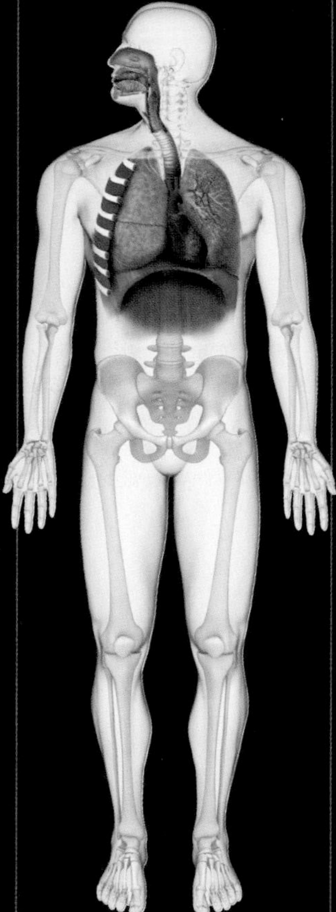

SKIN, HAIR, AND NAILS

EXPLORED ON PAGES 176–189

The skin, hair, and nails form the body's outer protective covering, and are together termed the integumentary system. They repel hazards such as physical injury, microorganisms, and radiation. The skin also regulates body temperature through sweating and hair adjustment. A layer of fat under the skin acts as an insulator, an energy store, and a shock absorber.

LYMPH AND IMMUNITY

EXPLORED ON PAGES 190–209

The immune system's intricate interrelationships of physical, cellular, and chemical defenses provide vital resistance to many threats, including infectious diseases and malfunctions of internal processes. The slowly circulating lymph fluid helps distribute nutrients and collect waste. It also delivers immunity-providing white blood cells when needed.

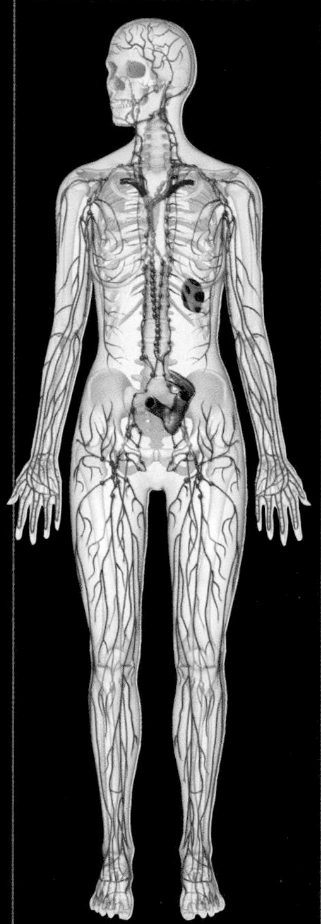

DIGESTIVE

EXPLORED ON PAGES **210–239**

The digestive tract's 30 feet of tubing, which runs from the mouth to the anus, has a complex range of functions. It chops and chews food, stores and then digests it, eliminates waste, and passes the nutrients to the liver, which processes or stores the various digestive products. Healthy digestion depends on the proper functioning of the immune and nervous systems.

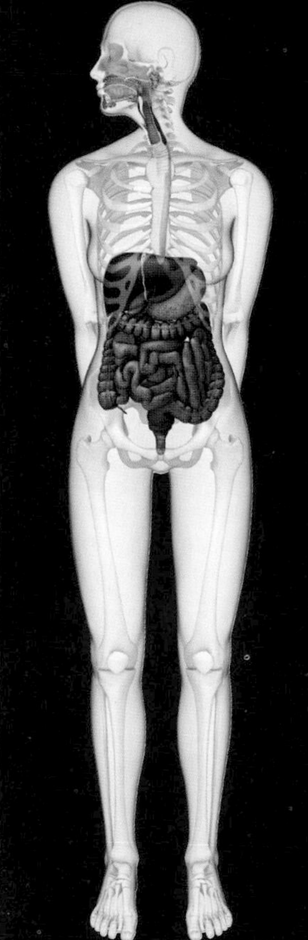

URINARY

EXPLORED ON PAGES **240–249**

The formation of urine by the kidneys eliminates unwanted substances from the blood, helping maintain the body's correct balance of fluids, salts, and minerals. Urine production is controlled by hormones and influenced by blood flow and pressure, intake of water and nutrients, fluid loss (through sweating, for instance), external temperature, and body cycles such as sleeping and waking.

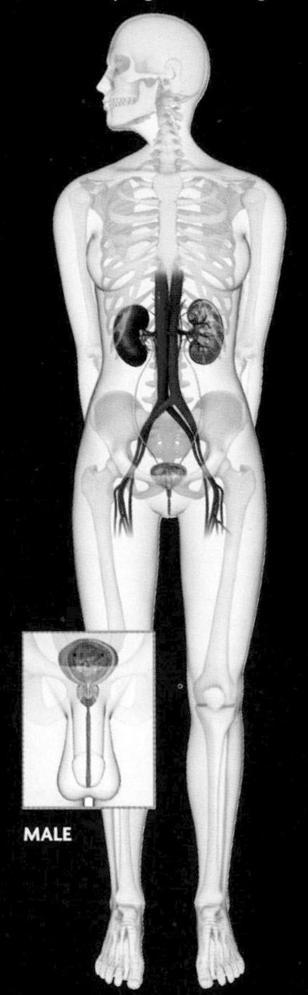

MALE

REPRODUCTIVE

EXPLORED ON PAGES **250–305**

Unlike any other system, the reproductive system differs dramatically between female and male; it functions only for part of the human lifespan and is not vital for maintaining life. The production of sperm in the male is continual while the female production of ripe eggs is cyclical. In the male, both sperm and urine use the urethra as an exit tube, but at different times.

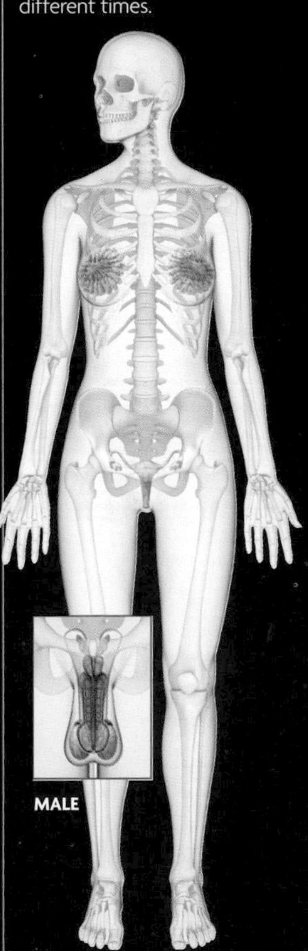

MALE

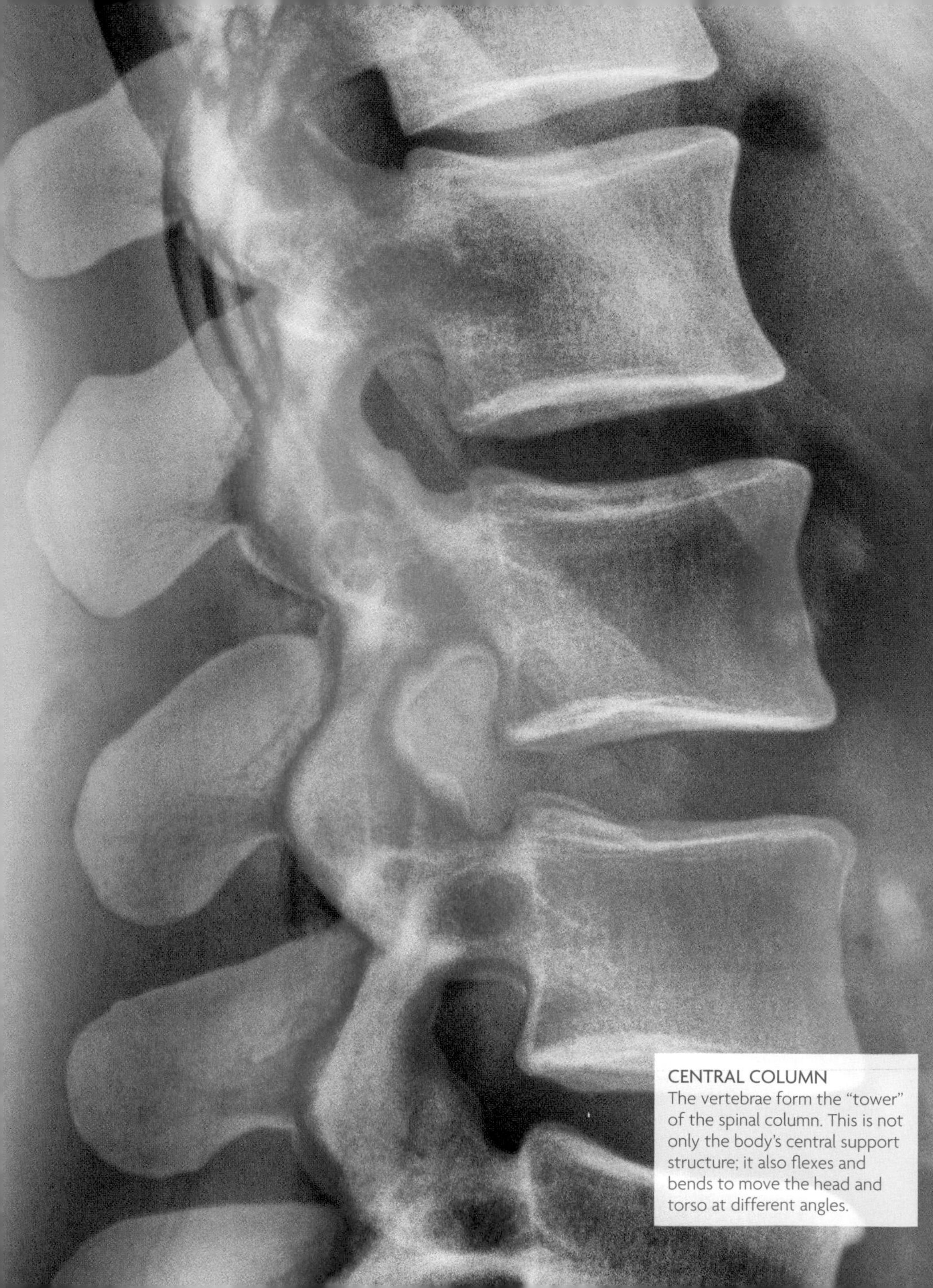

CENTRAL COLUMN

The vertebrae form the "tower" of the spinal column. This is not only the body's central support structure; it also flexes and bends to move the head and torso at different angles.

SUPPORT AND MOVEMENT

THE BODY'S MUSCLES, BONES, AND JOINTS PROVIDE A SUPPORTIVE FRAMEWORK CAPABLE OF AN ENORMOUS RANGE OF MOTION. MUSCLES AND BONES ALSO HAVE NUMEROUS INTERACTIONS WITH OTHER BODY SYSTEMS, ESPECIALLY THE NERVES.

The body's muscular system is never still. Even as the body sleeps, breathing continues, the heart beats, the intestines squirm, and skeletal muscles contract occasionally to shift the body into a new position.

SPECIALIZED CELLS AND TISSUES	36–37
SKELETAL SYSTEM	38–61
MUSCULAR SYSTEM	62–75
SKIN, HAIR, AND NAILS	176–189

MUSCLE TEAMWORK

Most movements are the result of multiple muscle contractions. A smile, for example, involves 20 facial muscles; writing utilizes more than 60 muscles in the arm, hand, and wrist. Muscles work in pairs: as one contracts to pull on a bone and initiate movement, an opposing muscle relaxes. Body action is a continuing sequence of split-second give-and-take.

POSTURE AND FEEDBACK

Sensory systems built into muscles provide the brain with information about the posture and position of the body and limbs. This is known as the proprioceptive sense, which allows us to "know," without having to look or feel, that fingers are clenched or a knee is bent. When we are learning a new motor skill, we concentrate on the movement as the brain adjusts muscle control through trial and error. With practice, the motor nerve patterns and their proprioceptive feedback become established, and eventually the movement becomes automatic. Sensory feedback also protects the muscular and skeletal systems against injury. If bones or muscles are under excessive stress, nerve messages registering discomfort or pain are sent to the brain. Awareness of the pain stimulates evasive or protective action by the body.

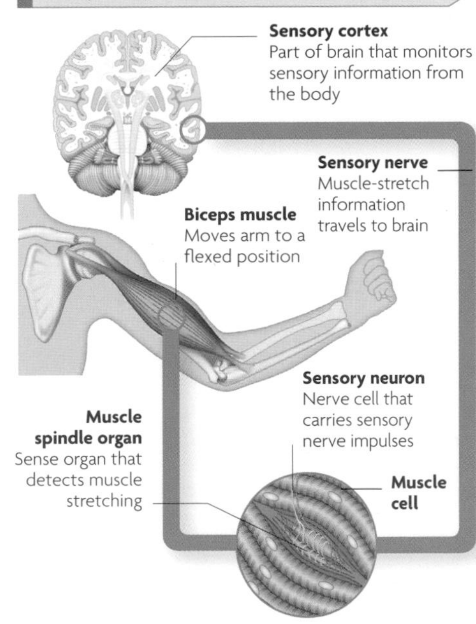

Sensory cortex
Part of brain that monitors sensory information from the body
Sensory nerve
Muscle-stretch information travels to brain
Biceps muscle
Moves arm to a flexed position
Sensory neuron
Nerve cell that carries sensory nerve impulses
Muscle spindle organ
Sense organ that detects muscle stretching
Muscle cell

SENSORY FEEDBACK
Within muscles, nerves end in sense organs (spindle organs). These respond to tension by firing signals along nerve fibres to tell the brain what is happening.

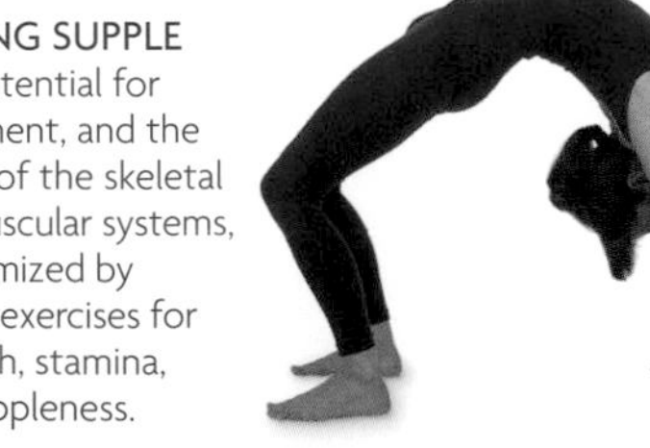

STAYING SUPPLE
Our potential for movement, and the health of the skeletal and muscular systems, is maximized by regular exercises for strength, stamina, and suppleness.

INFORMATION PROCESSING

THE BODY IS A DYNAMIC MECHANISM WHOSE INTERACTING PARTS REQUIRE CONTROL AND COORDINATION. TWO SYSTEMS ARE RESPONSIBLE FOR THESE INFORMATION-PROCESSING FUNCTIONS: THE NERVOUS AND ENDOCRINE SYSTEMS.

Information processing involves inputs, evaluation, and decision-making, followed by outputs. The body receives inputs from the senses. The brain is the central processing unit whose outputs control the physical actions of muscles and chemical responses of glands. Both nerves and hormones are involved in data management.

NERVOUS SYSTEM 76–129

ENDOCRINE SYSTEM 130–143

ELECTRICAL AND CHEMICAL PATHWAYS

The "language" of the nervous system is tiny electrical impulses. Every second, millions pass through the body's nerve network, conveying information to and from the brain. Information from the senses flows to the brain, where it is analyzed. Decisions are reached, and command messages–also in the form of electrical impulses–travel along motor nerves to the muscles to stimulate and coordinate their contractions. In addition, microreceptors monitor conditions inside the body and feed data about it to the unconscious part of the brain, which automatically evaluates the data and sends out impulses to various parts of the body to keep the internal environment at the optimum for body functioning.

SELECTIVE FOCUS
The nose sends streams of "smell" nerve signals to the brain. We can choose to ignore these or to focus on them, as part of the mind's selective awareness.

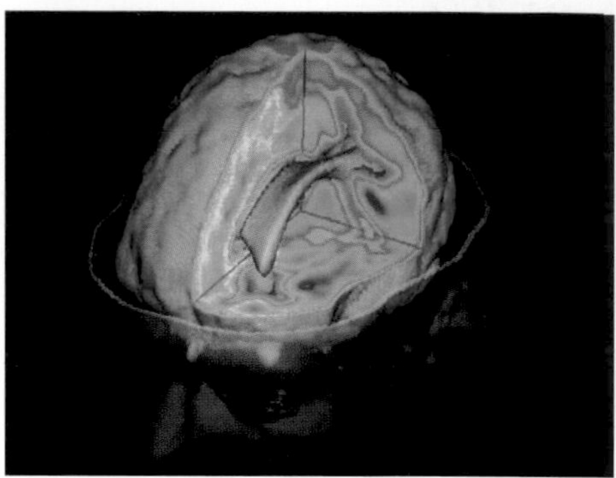

BRAIN ACTIVITY
This image is a three-dimensional functional MRI scan showing brain activity during speech. Red indicates areas of high activity, yellow indicates medium activity, and green indicates low activity.

In the endocrine system, hormones act as information carriers, instructing their target cells or tissues to perform specific functions. More than 50 hormones circulate in the bloodstream. The specific molecular structure of each hormone stimulates only those cells that have suitable receptors on their surface, instructing the cells to carry out certain processes. In general, nerves work fast–within fractions of a second. Most hormones function over longer time periods–minutes, days, or even months. Long-lasting effects, as in growth hormone for example, occur because the hormone is continuously secreted over a period of many years; an individual dose would last only a few days.

THE FLUID BODY

ROUGHLY TWO-THIRDS OF THE BODY IS COMPOSED OF WATER AND THE VARIOUS ESSENTIAL SUBSTANCES DISSOLVED IN IT. THESE FLUIDS ARE FOUND IN CELLS, AROUND THE BODY'S TISSUES, AND, MOST OBVIOUSLY, IN BLOOD AND LYMPH.

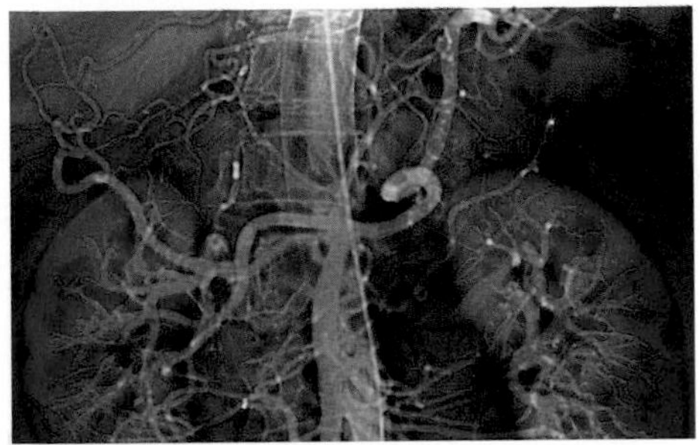

CIRCULATORY NETWORK
Blood is the fastest-circulating "fluid" in the body. Its liquid component, plasma, is constantly exchanging fluids with other body systems and structures.

There are about 70 pints (40 liters) of water in the average adult body, and it makes up the major part of most body parts. Tissues are 70–80 percent water; blood plasma is more than 90 percent; bones contain almost 25 percent; and fat is 10–15 percent water.

TYPES AND FUNCTIONS OF FLUIDS

There are two major categories of body fluids–intracellular and extracellular. Intracellular fluid (also called cytoplasm) is found inside cells. Extracellular fluid accounts for all other fluids in the body. Its subcategories are: interstitial fluid in the spaces between cells and tissues; blood plasma and lymph; the fluids in bones, joints, and dense connective tissue; and transcellular fluid, which includes saliva, mucus, sweat, and urine.

Water is an excellent solvent and the thousands of substances dissolved in it are used in the biochemical reactions that are the very basis of life. Water also distributes nutrients around the body and collects

CARDIOVASCULAR SYSTEM 144–159
LYMPH AND IMMUNE SYSTEMS 190–209

and delivers wastes. Fluids spread heat from active areas to cooler ones, and act as shock absorbers to cushion sensitive areas such as the brain. Fluids also work as lubricants, so that tissues and organs slip past each other with minimal friction.

BLOOD AND LYMPH

The blood and lymphatic circulatory systems are constantly swapping fluids (see the illustration below). Blood plasma, the fluid in which the blood cells are suspended, transports red blood cells (which carry oxygen and remove carbon dioxide) around the body. Lymph fluid transports white blood cells, which produce antibodies to fight infection and disease.

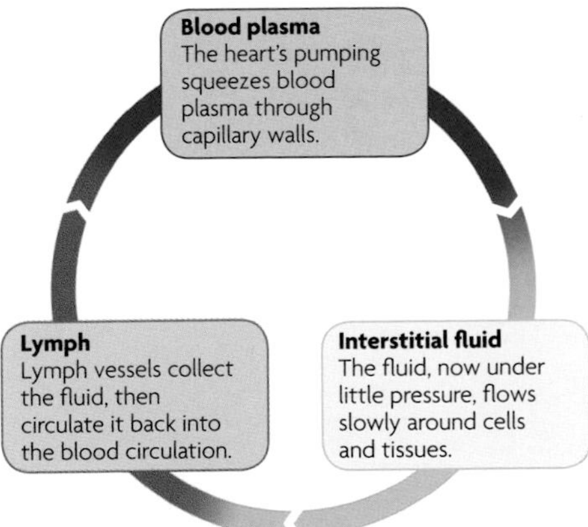
Blood plasma
The heart's pumping squeezes blood plasma through capillary walls.
Interstitial fluid
The fluid, now under little pressure, flows slowly around cells and tissues.
Lymph
Lymph vessels collect the fluid, then circulate it back into the blood circulation.

BLOOD PLASMA AND LYMPH CYCLE
Blood plasma leaks from capillaries to form interstitial fluid. Some of this drains into lymph vessels to become lymph fluid, which then returns to the blood circulation.

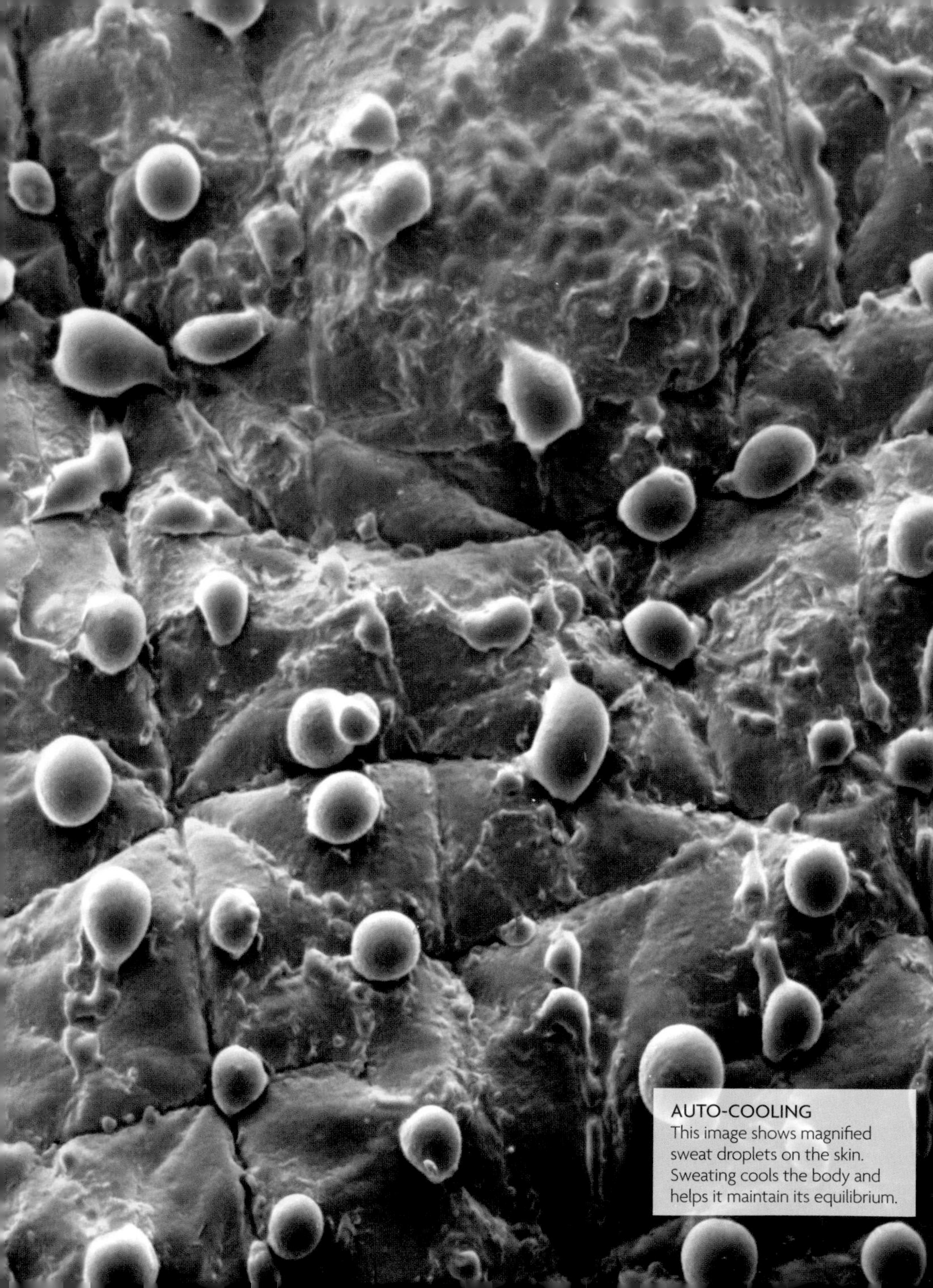

AUTO-COOLING
This image shows magnified sweat droplets on the skin. Sweating cools the body and helps it maintain its equilibrium.

EQUILIBRIUM

THE BODY'S CELLS AND TISSUES FUNCTION WELL ONLY IF ALL ASPECTS OF THEIR ENVIRONMENT ARE KEPT STABLE AND IN EQUILIBRIUM. SEVERAL SYSTEMS MAINTAIN A BALANCED INTERNAL ENVIRONMENT, A PROCESS CALLED HOMEOSTASIS.

The biochemical reactions in cells are attuned to specific conditions, such as oxygen levels, acidity, water levels, and temperature. These must be maintained within the correct limits or the reactions go awry and the body malfunctions.

CARDIOVASCULAR SYSTEM 144-159
RESPIRATORY SYSTEM 160-175
SKIN, HAIR, AND NAILS 176-189
DIGESTIVE SYSTEM 210-239
URINARY SYSTEM 240-249

HOMEOSTATIC SYSTEMS

Several systems contribute to homeostasis. For example, the respiratory system ensures that oxygen levels are maintained; the digestive system takes in and processes nutrients; and the circulatory system distributes oxygen and nutrients and gathers waste products, which are removed by the urinary and respiratory systems.

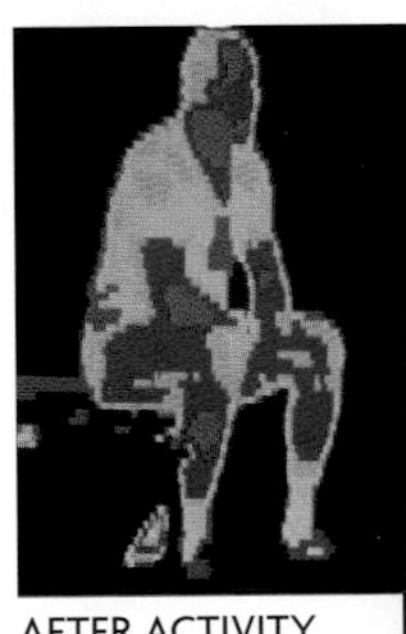

BEFORE ACTIVITY
In this thermogram, temperature is graded from blue (cooler) to red (warmer).

AFTER ACTIVITY
Following exercise, a thermogram shows that most of the exposed skin is now warmer than normal.

CONTROL AND FEEDBACK

The body's major control systems, nerves and hormones, are mainly responsible for coordinating homeostatic mechanisms using feedback loops. For example, if water levels in the tissues fall, body fluids become more concentrated. Sensors detect this and feed back information to the brain, whose homeostatic centres trigger regulating actions. Hormonal control of urinary excretion is adjusted to conserve water, and nervous activity produces thirst so that we drink. The sensors detect the changes as fluid concentrations return to normal, then they switch off until needed. again. Thermoregulation–maintaining an approximately constant body temperature–uses the same feedback principles, with mechanisms such as sweating and shivering being used to regulate heat loss, conservation, and generation. In these ways, conditions inside the body are kept relatively stable, and an ongoing equilibrium is maintained.

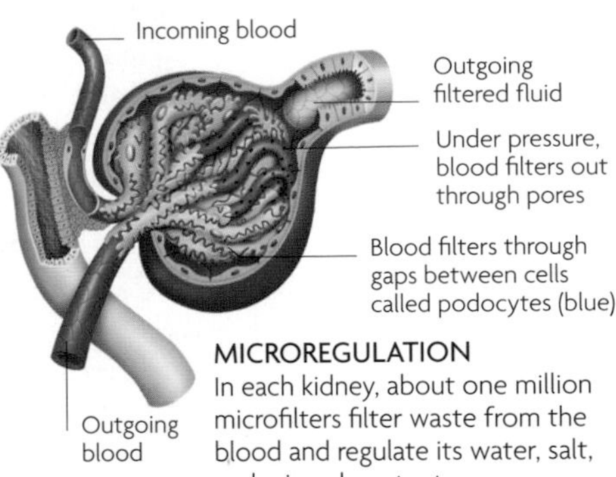
Incoming blood
Outgoing filtered fluid
Under pressure, blood filters out through pores
Blood filters through gaps between cells called podocytes (blue)
Outgoing blood

MICROREGULATION
In each kidney, about one million microfilters filter waste from the blood and regulate its water, salt, and mineral content.

BODY SYSTEMS TO CELLS

EACH SYSTEM CAN BE SEEN AS A HIERARCHY. THE SYSTEM ITSELF IS AT THE TOP OF THE HIERARCHY; NEXT ARE ITS ORGANS; THEN THE TISSUES THAT MAKE UP THE ORGANS; AND AT THE BOTTOM ARE THE CELLS FROM WHICH TISSUES ARE MADE.

A body system is usually regarded as a collection of organs and parts designed for one important task. The systems are integrated and interdependent, but each has its own identifiable components.

The main parts of a system are its organs and tissues. Most organs are composed of different tissues. The brain, for example, contains nervous, connective, and epithelial (covering or lining) tissues. A tissue is a group of cells that are similar in structure and carry out the same function.

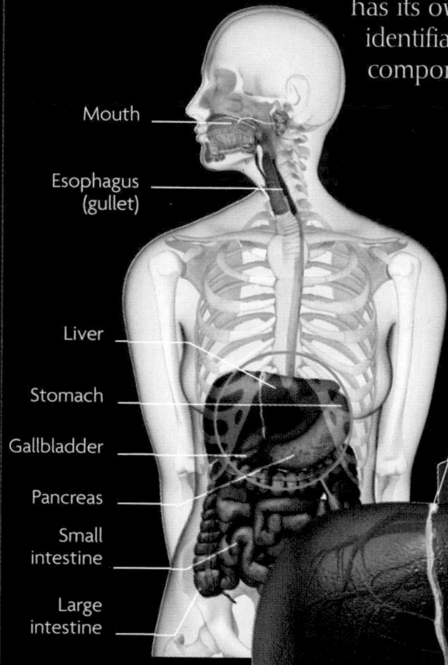

Mouth
Esophagus (gullet)
Liver
Stomach
Gallbladder
Pancreas
Small intestine
Large intestine

1 SYSTEM

The digestive system is one of the most clearly defined in the body. It consists of a long passageway—the digestive tract—and associated glands. These include the liver and pancreas, which are connected to the main tract by ducts, and empty their products, such as enzymes, into the tract.

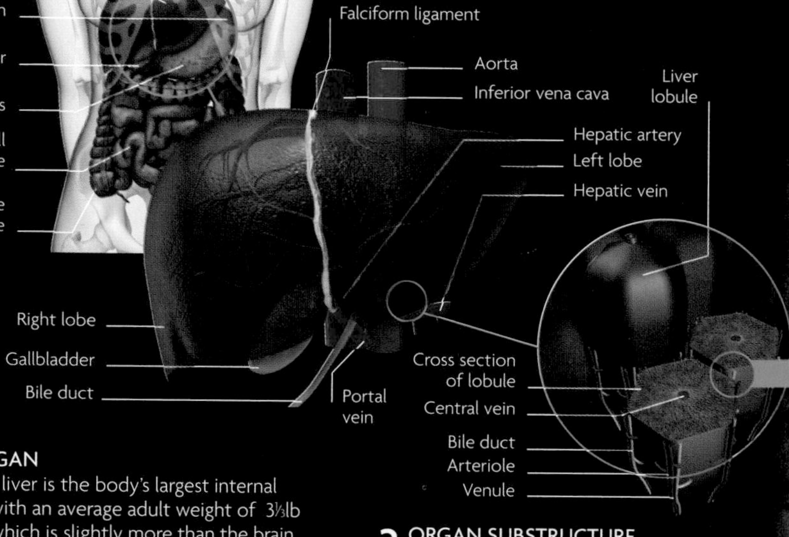

Falciform ligament
Aorta
Inferior vena cava
Liver lobule
Hepatic artery
Left lobe
Hepatic vein
Right lobe
Gallbladder
Bile duct
Portal vein
Cross section of lobule
Central vein
Bile duct
Arteriole
Venule

2 ORGAN

The liver is the body's largest internal organ, with an average adult weight of 3⅓lb (1.5kg), which is slightly more than the brain. Within the liver is a system of tubes for carrying away its digestive product, bile, which is stored in the small sac under its right end, the gallbladder.

3 ORGAN SUBSTRUCTURE

The structural–functional units of the liver are hepatic lobules. The lobules are six-sided and have blood vessels and bile ducts inside and between them.

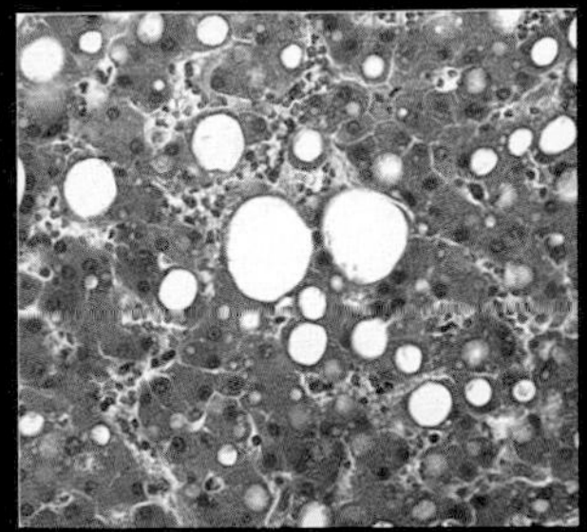

MICROSECTION OF LIVER

In this magnified section the cells (pinkish purple) and their nuclei (dark purple) are visible. The white circular areas are fatty deposits.

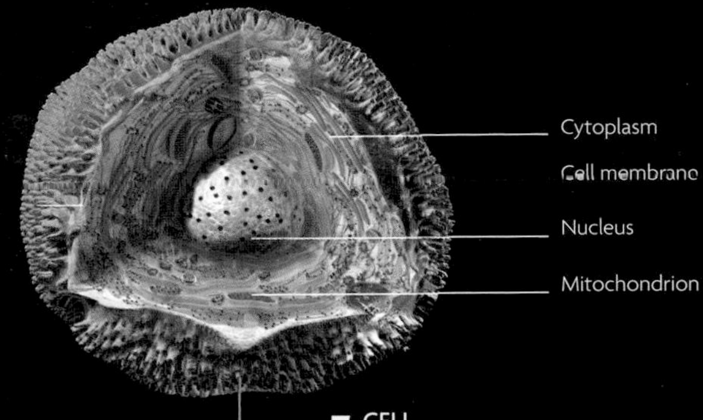
Cytoplasm
Cell membrane
Nucleus
Mitochondrion

Kupffer cell
Also known as a hepatic macrophage, a type of white blood cell specific to the liver that engulfs and digests old worn-out blood cells and other debris.

4 TISSUE

The unique tissue of the liver consists of branching sheets, or laminae, of liver cells (hepatocytes) arranged at angles. These are permeated by fluids and microscopic branches of two main kinds of tubes: blood vessels and bile ducts.

5 CELL

The fundamental living unit of all body tissues, a typical cell is capable of obtaining energy and processing nutrients. The hepatocytes of the liver are examples of body cells, containing most types of the miniature structures known as organelles inside them.

Bile canaliculus
Smallest branch of bile duct; snakes between hepatocytes

Sinusoid
A blood vessel with many pores that allow for the exchange of oxygen and nutrients

Hepatocyte

Bile duct
Collects bile fluid, made by hepatocytes, from canaliculi

Branch of hepatic portal vein

Branch of hepatic artery

Red blood cell

Lymph vessel

Central vein

White blood cell

Fat-storing cell

Vacuole
A sac that stores and transports ingested materials, wastes, and water

Nucleolus
Central region of the nucleus; plays a role in ribosome production

Nucleus
The cell's control center, containing chromatin and most of the DNA

Nuclear membrane
A two-layered membrane with pores through which substances can pass

Nucleoplasm
The fluid within the nucleus

Cytoskeleton
Internal framework of the cell, consisting of microfilaments and microtubules

Microfilament
Provides support for the cell

Cytoplasm
Jellylike fluid that contains organelles

Mitochondrion
Site of fat and sugar digestion in the cell; produces energy

Microtubule
Part of cytoskeleton; aids movement of substances in the cytoplasm

Centriole
Two cylinders of tubules; essential for reproduction

Microvilli
Projections found on some cells; they increase the cell's surface area

Released secretions
Secretions are released from the cell by exytosis; a vesicle merges with the cell membrane and releases its contents

Ribosome
Involved in protein assembly

Cell membrane
Encloses cell contents; regulates inflow and outflow of substances

Secretory vesicle
Sac that contains various substances, such as enzymes, that are produced by the cell and secreted at the cell membrane

Lysosome
Produces enzymes that aid in digestion of substances and worn-out organelles

Peroxisome
Makes enzymes that oxidize some toxic chemicals

Rough endoplasmic reticulum
Folded membranes studded with ribosomes; help transport materials through cell; site of much protein manufacture

Golgi complex
Organelle that processes and repackages proteins produced in rough endoplasmic reticulum for release at cell membrane

Smooth endoplasmic reticulum
Network of tubes and flat, curved sacs that helps transport materials through cell; site of calcium storage; main location of fat metabolism

INSIDE A CELL
This illustration shows all the tiny specialized structures (organelles) inside a generalized body cell.

THE CELL

THE CELL IS THE BASIC UNIT OF THE BODY. IT IS THE SMALLEST PART CAPABLE OF THE PROCESSES THAT DEFINE LIFE, SUCH AS REPRODUCTION, MOVEMENT, RESPIRATION, DIGESTION, AND EXCRETION—ALTHOUGH NOT ALL CELLS HAVE ALL THESE ABILITIES.

CELL ANATOMY

Most cells are microscopic–a typical cell is 20–30µm in diameter, which means 40 in a row would stretch across a period. Very specialized, long, thin cells include neurons (nerve cells) and muscle fiber cells (myofibers), which may be more than 12in (30cm) long. Most cells have an outer flexible "skin": the cell, or plasma, membrane. Inside are structures known as organelles, each with a characteristic shape, size, and function. These organelles do not float about at random. The cell is highly organized, with interior compartments linked by sheets and membranes and held in place by a flexible, latticelike "skeleton" of even tinier tubules and filaments.

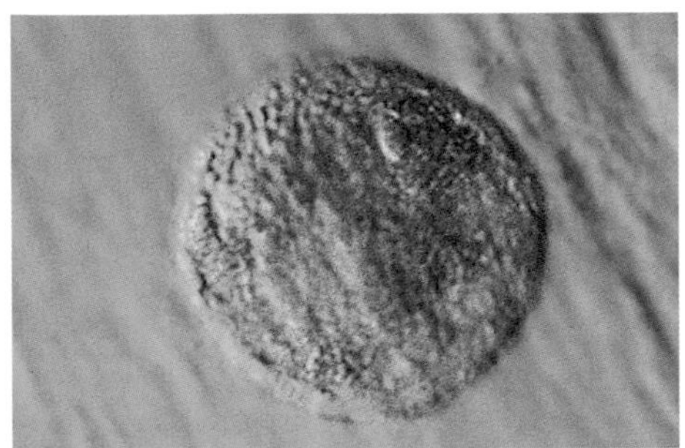

EMBRYONIC STEM CELL
Stem cells are unspecialized "beginner" cells that can develop into specialized cells. Stem cells in the embryo can develop into any of the 200-plus types of specialized cells in the body.

CELL TYPES

Cells come in many shapes and sizes, depending on their specialized functions within tissues. Speed of cell division also varies. It is most rapid in epithelial (covering and lining) cells, which are subjected to physical abrasion and wear and which must continually replace themselves. It is slow or even nonexistent in some cells that are structurally complex, such as nerve cells (neurons).

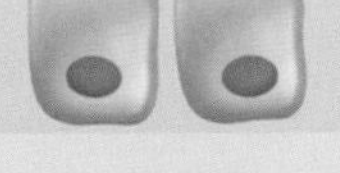

Epithelial cells
These cells form skin, cover most organs, and line hollow cavities such as the intestinal tract.

Smooth muscle cell
The large, elongated, spindlelike cells of smooth muscle contract by sliding strands of protein inside.

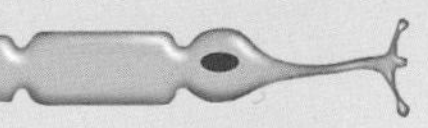

Photoreceptor cell
A cone cell is a type of light- and color-sensitive cell in the retina of the eye.

Nerve cell
Each cell has short branches (dendrites) to receive nerve signals, and a long "wire" (axon) to send them.

Red blood cell
The double-dished (biconcave) red cell (erythrocyte) is a bag of oxygen-carrying hemoglobin molecules.

Sperm cell
Each sperm has a head that carries the paternal genetic material, and a whip-like tail for propulsion.

Adipose (fat) cell
The main adipose cells, adipocytes, are bulky and crammed with droplets of fat (lipids), which store energy.

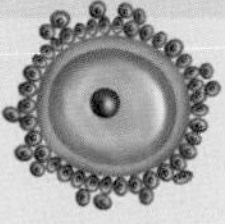

Ovum (egg) cell
These giant cells contain the maternal genetic material, and energy resources for the embryo's first cell divisions.

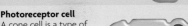

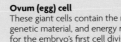

CELL MEMBRANE

Several features allow the membrane to fulfill its dual functions of protecting the cell's contents and permitting movement of materials into and out of the cell. The primary component of this membrane is a double layer of phospholipid molecules. Each phospholipid has a water-loving (hydrophilic) head group and two water-hating (hydrophobic) tails. The two layers are arranged with the heads on the outside and inside of the cell membrane, and the tails in between. The phospholipids are interspersed with protein molecules and carbohydrate chains that allow the cell to be recognized by other body cells.

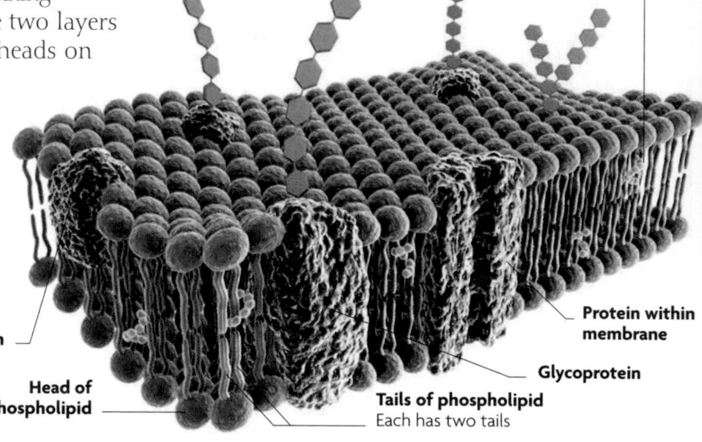
Carbohydrate chain
Cholesterol
Enhances stability
Protein within membrane
Glycoprotein
Tails of phospholipid
Each has two tails
Head of phospholipid
Protein

PERMEABLE BILAYER
The typical cell membrane is characterized by a double layer of phospholipids with embedded proteins.

SURFACE ORGANELLES

Some cells in the body have specialized structures projecting from their surface. Cells lining the small intestine have small, fingerlike projections called microvilli, which increase the surface area for absorption of nutrients. Some cells in the female reproductive tract have small, hairlike cilia that wave to move the ovum along the oviduct; similar ciliated cells in the respiratory tract move small particles out of the airways. The sperm is unique in the human body in having a long, whiplike flagellum, used for propulsion.

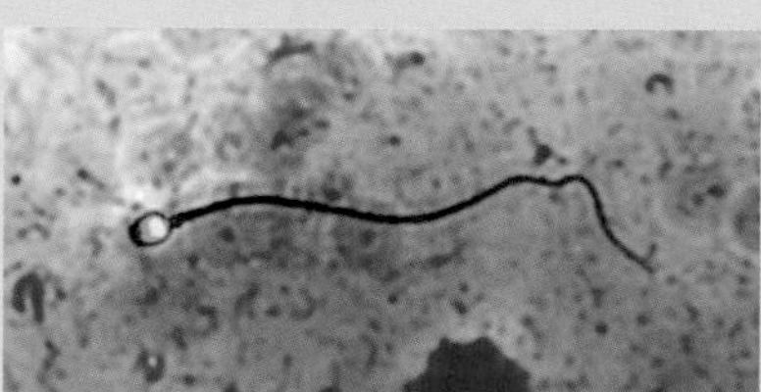

SPERM
The thin tail (flagellum) that extends from a human sperm cell is used like a propeller to help the sperm swim up the female genital tract.

CILIATED CELLS
Some of the cells lining the fallopian tubes have hairlike cilia (colored pink in this micrograph) that brush an egg along toward the uterus.

MEMBRANES OF ORGANELLES

Membranes divide the cytoplasm into sections and control the passage of materials between these regions, act as attachment points and storage areas, and shape channels along which substances move.

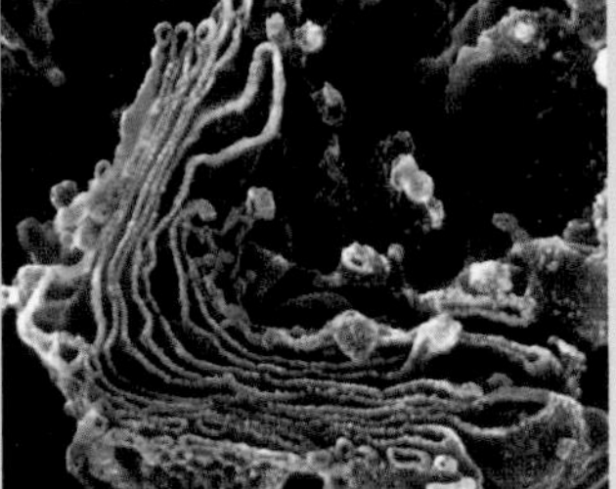

GOLGI COMPLEX
Within the membranous sacs of the Golgi complex, protein from the endoplasmic reticulum is processed.

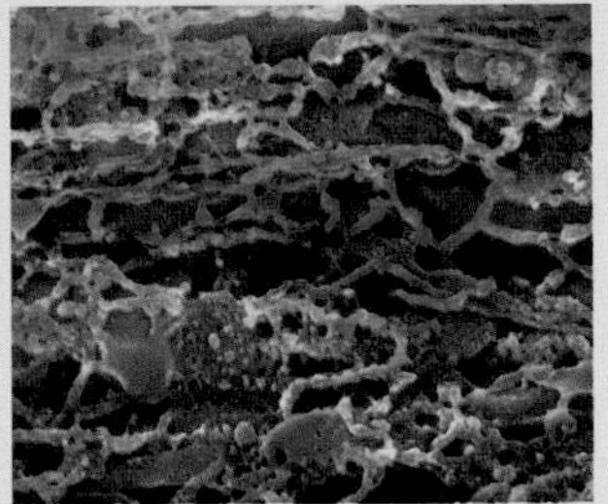

ENDOPLASMIC RETICULUM (ER)
A series of highly folded and curved ER membranes usually encloses one continuous labyrinthine space.

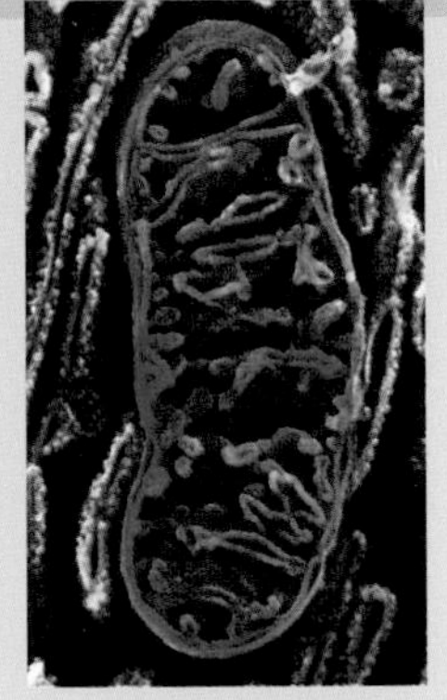

MITOCHONDRION
The inner membrane is folded to increase the area for releasing energy.

TRANSPORT

The transfer of materials through the cell membrane occurs by one of three processes. Small molecules, such as water, oxygen, and carbon dioxide, cross the membrane by diffusion. Molecules that cannot cross the phospholipid layer must cross by facilitated diffusion. When substances (such as minerals and nutrients) are at lower concentration on the outside of the cell than on the inside, they can only be conveyed into the cell by active transport, which requires energy.

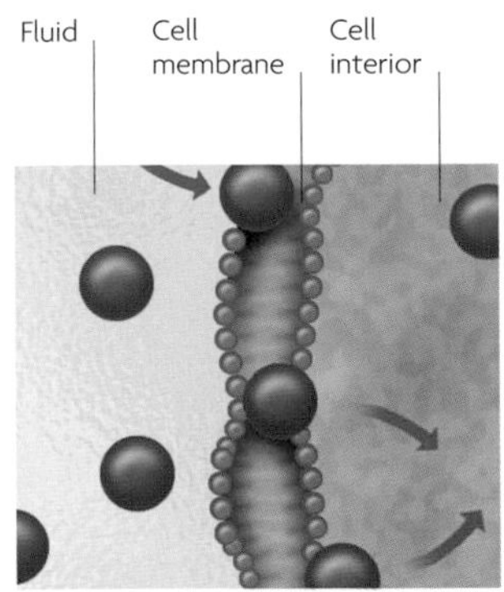
Fluid
Cell membrane
Cell interior

DIFFUSION
Many molecules naturally move from an area where they are at high concentration to one in which their concentration is lower.

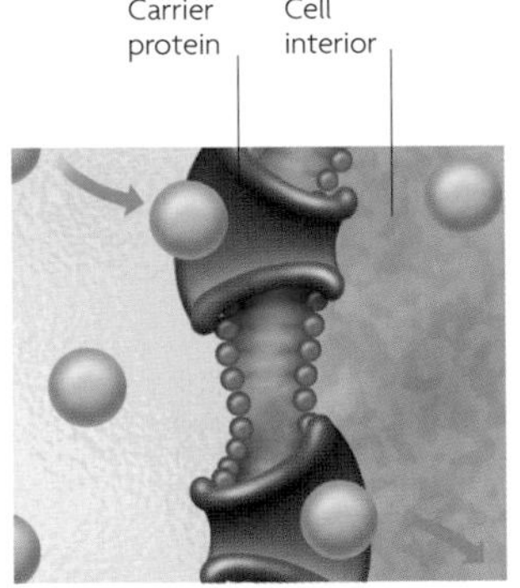
Carrier protein
Cell interior

FACILITATED DIFFUSION
A carrier protein binds with a specific molecule outside the cell, then changes shape and ejects the molecule into the cell.

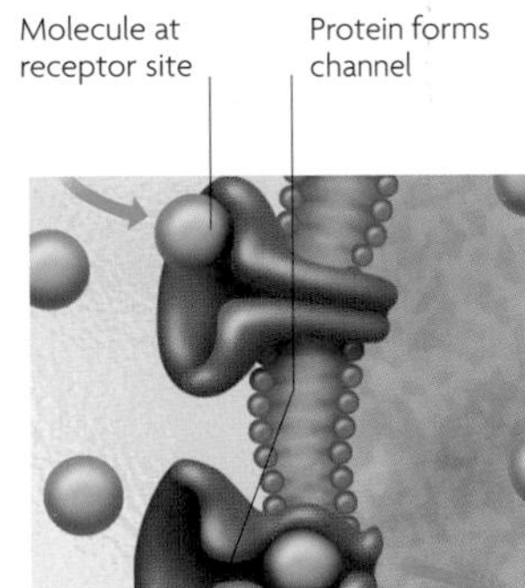
Molecule at receptor site
Protein forms channel

ACTIVE TRANSPORT
Molecules bind to a receptor site on the cell membrane, triggering a protein to change into a channel through which molecules travel.

DNA

KNOWN AS THE "MOLECULE OF LIFE," THE CHEMICAL DNA (DEOXYRIBONUCLEIC ACID) CONTAINS THE INSTRUCTIONS, KNOWN AS GENES, FOR THE BODY'S GROWTH, FUNCTION, AND REPAIR.

In nearly all human cells, DNA is packaged into 46 coiled stuctures called chromosomes, situated in the cell's nucleus. DNA's list of instructions takes the form of long, thin molecules, one per chromosome, each forming a double-helix shape. Each double-helix has two long strands that corkscrew around each other. These are linked by rungs, like a ladder. The rungs are made of pairs of chemicals called bases: adenine (A), guanine (G), thymine (T), and cytosine (C). A always pairs with T, and G with C. The order of the bases contains the chromosome's genetic code, while the way the bases link enables DNA to make copies of itself.

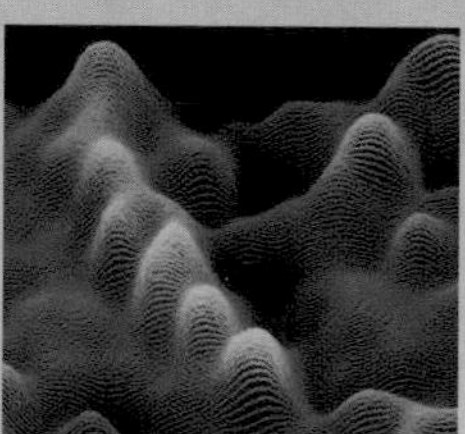

DNA UNDER THE MICROSCOPE
This scanning tunneling micrograph (STM) of DNA, magnified about one million times, shows the twists of the helix as a series of yellow peaks on the left.

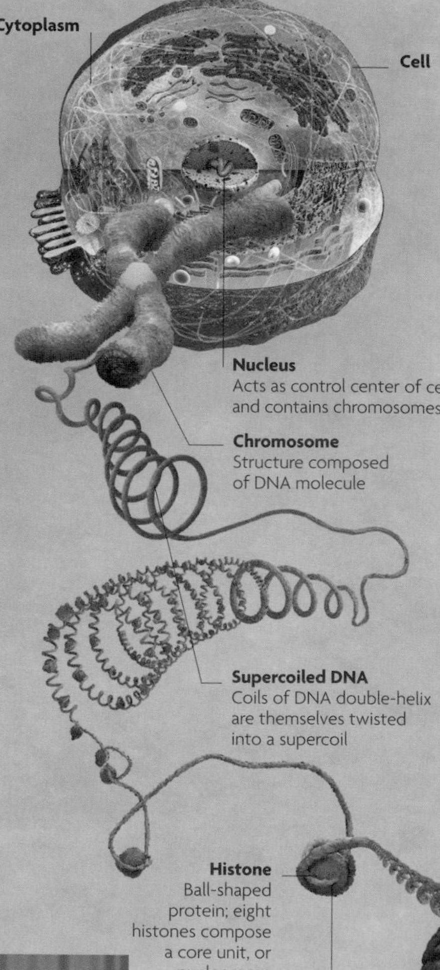
Cytoplasm
Cell
Nucleus
Acts as control center of cell and contains chromosomes
Chromosome
Structure composed of DNA molecule
Supercoiled DNA
Coils of DNA double-helix are themselves twisted into a supercoil
Histone
Ball-shaped protein; eight histones compose a core unit, or nucleosome
Core unit
Package of proteins around which 2–5 turns of DNA are wrapped; also known as a nucleosome

BASE PAIRS

The four bases can pair in only two configurations due to their chemical structures. Adenine and thymine each have two positions for forming hydrogen bonds and so fit together, while guanine and cytosine each have three hydrogen-bond locations.

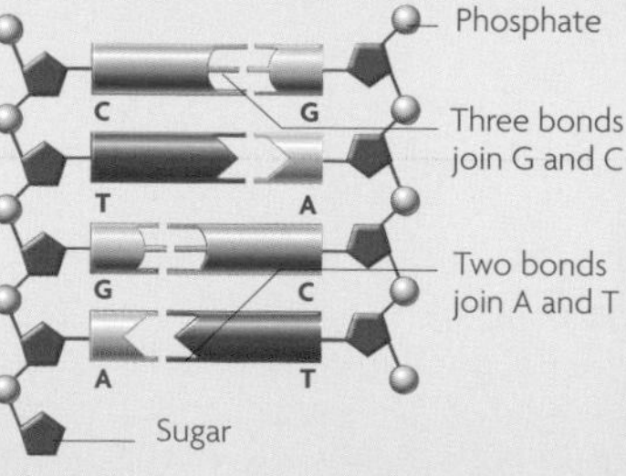
Phosphate
C
G
Three bonds join G and C
T
A
G
C
Two bonds join A and T
A
T
Sugar

COILS AND SUPERCOILS

DNA's coiled structure allows an incredible length to be packed into a tiny space. If unwound, the DNA in a chromosome would stretch about 2in (5cm). When cells are not dividing, the DNA forms a loosely tangled structure called chromatin. This allows portions to be available for protein assembly and other functions. As a cell prepares to divide, its DNA coils into supercoils, which are shorter and denser, and visible as the typical chromosome "X" shapes.

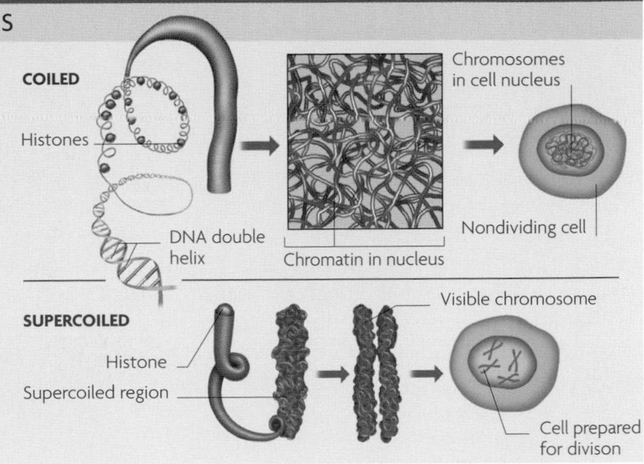
COILED
Histones
DNA double helix
Chromatin in nucleus
Chromosomes in cell nucleus
Nondividing cell
SUPERCOILED
Histone
Supercoiled region
Visible chromosome
Cell prepared for divison

Helical repeat DNA helix twists once for every 10.4 rungs of base pairs

Adenine–thymine link Adenine always forms a base pair with thymine

DOUBLE HELIX

A DNA molecule in a chromosome is coiled and supercoiled (see panel, above). The DNA molecule also loops and twists. It is accompanied by various proteins, particularly histones.

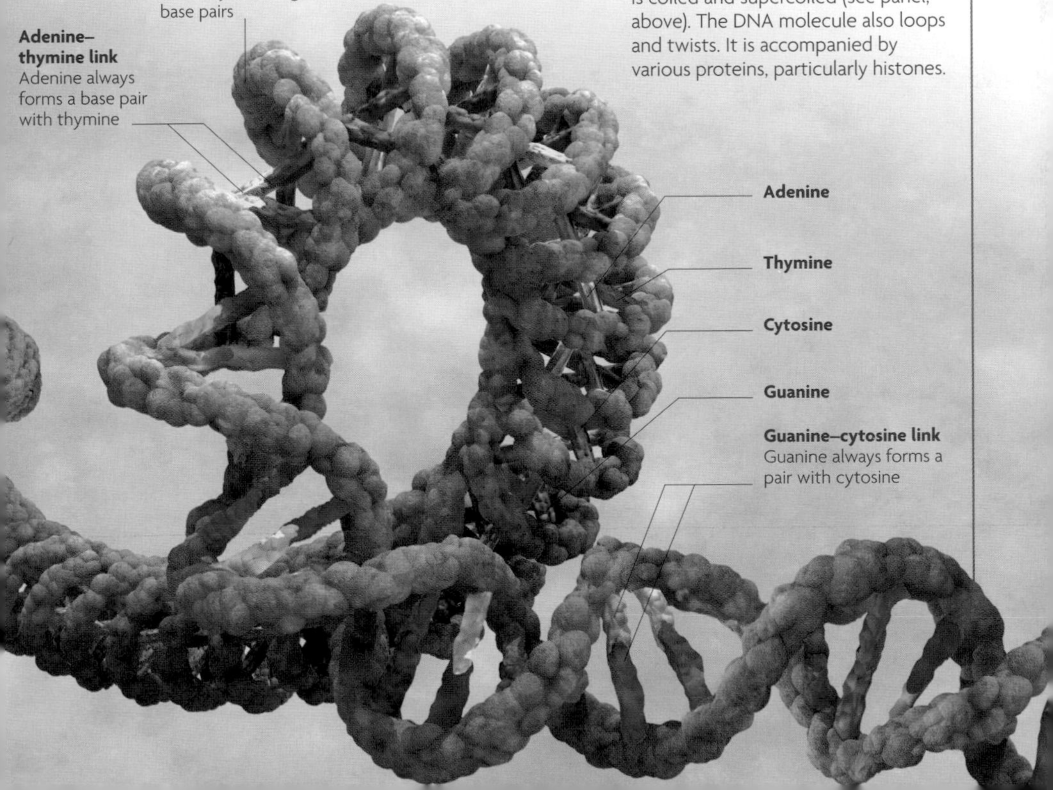

Adenine

Thymine

Cytosine

Guanine

Guanine–cytosine link Guanine always forms a pair with cytosine

HOW DNA WORKS

One of DNA's key functions is to provide the information to build proteins. Some proteins are the body's major structural molecules, while others form enzymes or hormones, which control chemical reactions within the body. Manufacture of proteins occurs in two main phases: transcription and translation. In transcription, information is taken from the DNA and copied to an intermediate type of molecule called mRNA (messenger ribonucleic acid). The mRNA moves out of the cell's nucleus to protein assembly units called ribosomes. In the translation phase, the mRNA acts as a template for the formation of units of protein, known as amino acids. There are about 20 different amino acids. Their order is specified by lengths of mRNA three bases long, called triplet codons. The order of bases in each codon is the code for a particular amino acid (hence the term "genetic code"). The mRNA carries instructions to make a specific protein from a sequence of amino acids.

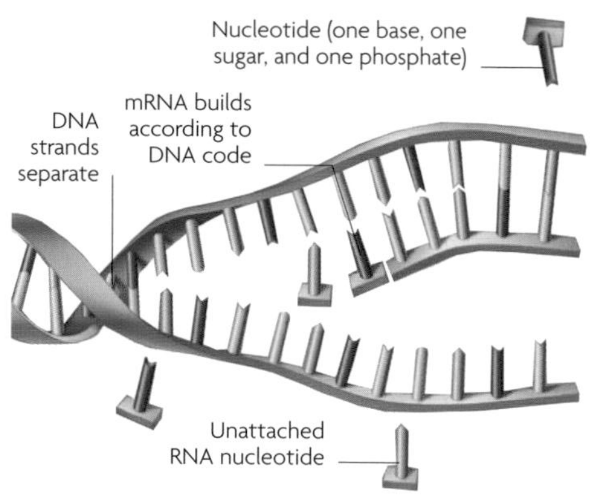
Nucleotide (one base, one sugar, and one phosphate)
DNA strands separate
mRNA builds according to DNA code
Unattached RNA nucleotide

1 TRANSCRIPTION
In the cell's nucleus, the DNA strands temporarily separate, with one acting as the template for the formation of mRNA. Separate RNA nucleotides with the correct bases lock onto the exposed DNA bases in cross-linked fashion, thereby forming a mirror image of the DNA's information.

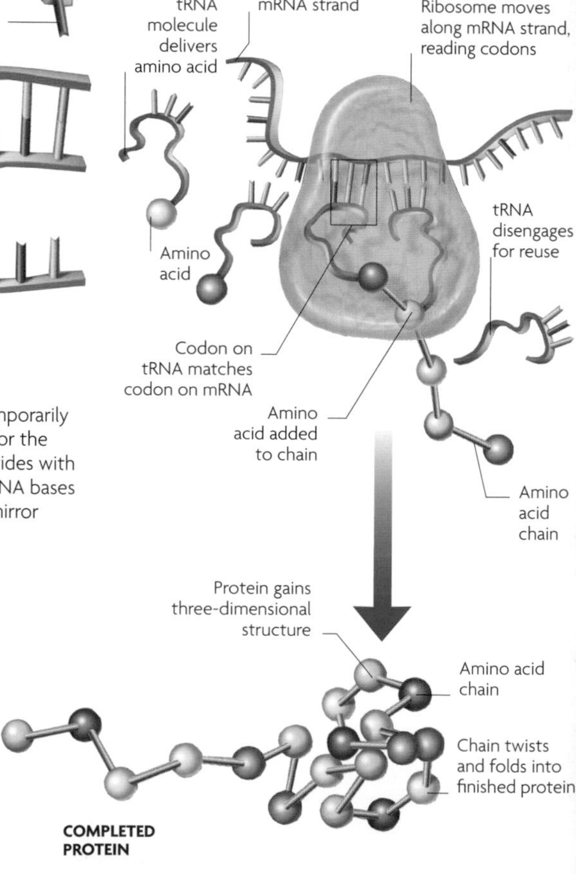
tRNA molecule delivers amino acid
mRNA strand
Ribosome moves along mRNA strand, reading codons
Amino acid
tRNA disengages for reuse
Codon on tRNA matches codon on mRNA
Amino acid added to chain
Amino acid chain
Protein gains three-dimensional structure
Amino acid chain
Chain twists and folds into finished protein
COMPLETED PROTEIN

2 TRANSLATION
In the cell's cytoplasm, the mRNA attaches to a ribosome. Individual tRNA (transfer ribonucleic acid) molecules have specific amino acids attached. They can slot onto the mRNA only if the order of their bases matches, ensuring they bring the correct amino acid. As the ribosome moves along the mRNA, the tRNAs bring the correct amino acid sequence, which fit together to construct a protein.

WHAT ARE GENES?

A gene is generally regarded as a unit of DNA needed to construct one protein. It consists of all the sections of DNA that code for all the amino acids for that protein. These sections are not necessarily on the same strand of DNA or even on the same chromosome. There may be many strands of DNA, each of them containing the code for one portion of the protein. Typically, lengths of DNA called introns and exons (see below) are both transcribed to form immature mRNA. The parts of mRNA made from the introns are then stripped out by the cell's molecular machinery, leaving mature mRNA for translation. There are also regulatory DNA sequences that code for their own proteins, affecting the rate of gene transcription.

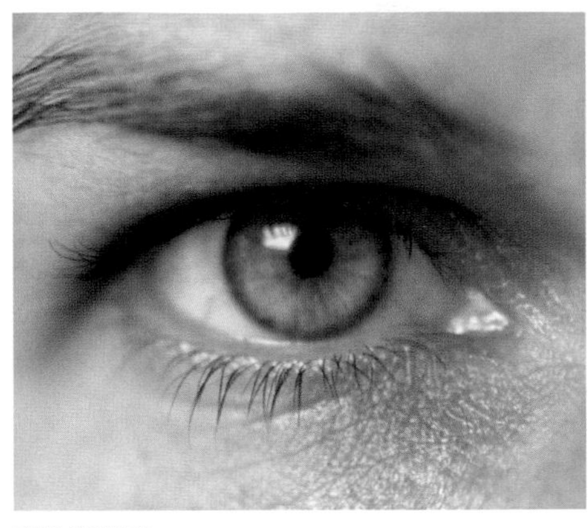

EYE COLOR

A trait such as eye color is not determined by just one gene. At least three genes, called bey 1 and bey 2 on chromosome 15, and gey on chromosome 19, influence iris color by affecting the development of pigment.

PARTS OF A GENE

Regions called introns and exons both transcribe to form mRNAs for different portions of a protein. The lengths made from introns are then spliced out chemically, to leave exon-only portions, which go on to make the protein.

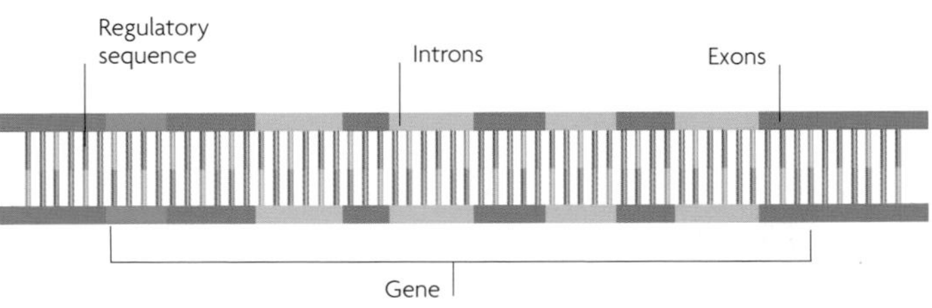
Regulatory sequence
Introns
Exons
Gene

RANGE OF GENE SIZE

Genes vary enormously in their size, which is usually measured in numbers of base pairs. Small genes may be just a few hundred base pairs long, while others are measured in millions of base pairs. The gene for beta-globin is one of the smallest. It codes for part of the hemoglobin molecule. It is compared, right, with a larger gene.

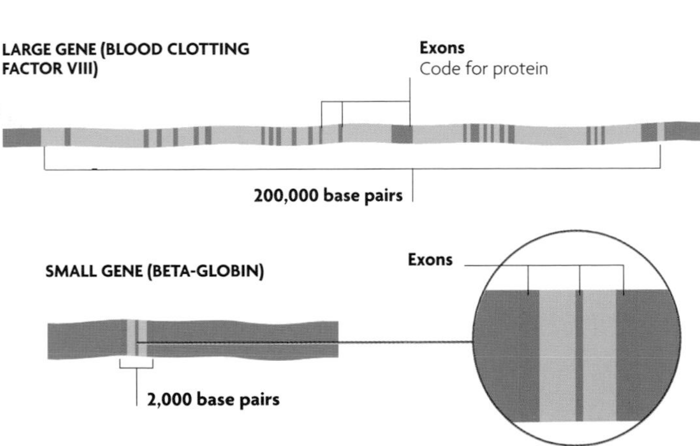
LARGE GENE (BLOOD CLOTTING FACTOR VIII)
Exons
Code for protein
200,000 base pairs
SMALL GENE (BETA-GLOBIN)
Exons
2,000 base pairs

THE GENOME

A GENOME IS THE FULL SET OF GENETIC INSTRUCTIONS FOR A LIVING THING. THE HUMAN GENOME CONSISTS OF AN ESTIMATED 20,000–25,000 GENES, CARRIED ON THE 46 CHROMOSOMES FOUND IN NEARLY EVERY KIND OF BODY CELL.

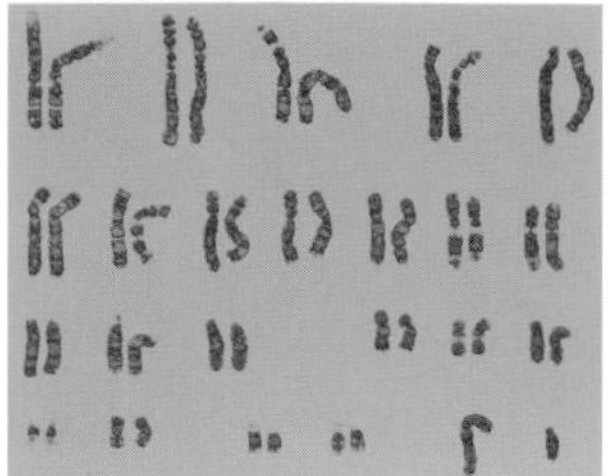

KARYOTYPE
A karyotype is a photograph of all chromosomes from a cell arranged in a standard order. This example is from a male (note the large curved "X" and small "Y" at bottom, right).

CHROMOSOMES AND DNA

The Human Genome Project, a multinational effort to map the human genome, was completed in 2003. It led to the identification of the 20,000–25,000 individual genes on the 46 human chromosomes, which collectively include about 3.2 billion base pairs. Although much of the DNA that makes up the chromosomes does not code for proteins, known as noncoding and "junk" DNA, it may still regulate gene function. Junk DNA is different from noncoding DNA in that its structure does not resemble that of genes.

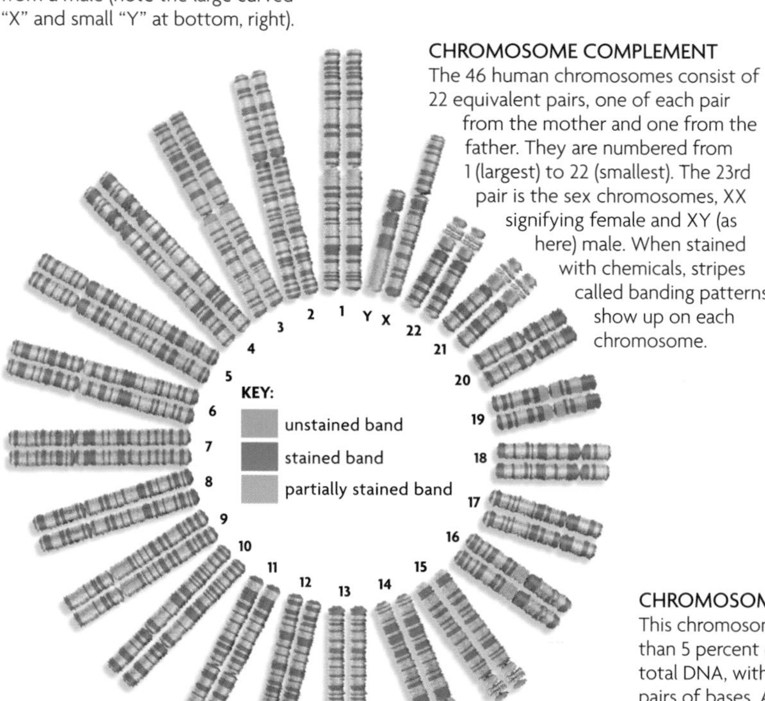
1
2
3
4
5
6
7
8
9
10
11
12
13
14
15
16
17
18
19
20
21
22
Y
X
KEY:
unstained band
stained band
partially stained band

CHROMOSOME COMPLEMENT
The 46 human chromosomes consist of 22 equivalent pairs, one of each pair from the mother and one from the father. They are numbered from 1 (largest) to 22 (smallest). The 23rd pair is the sex chromosomes, XX signifying female and XY (as here) male. When stained with chemicals, stripes called banding patterns show up on each chromosome.

p22.2
p21.3
p21.1
p15.2
p14.3
p14.1
p12.3
p12.1
q11.22
q11.23
q21.11
q21.2
q22.1
q23.3
q31.2
q31.32
q32.1
q33
q36.1
q36.3

CHROMOSOME SEVEN
This chromosome contains more than 5 percent of the genome's total DNA, with about 159 million pairs of bases. Almost 60 million are in the short arm, 7p, with the rest in the longer arm, 7q.

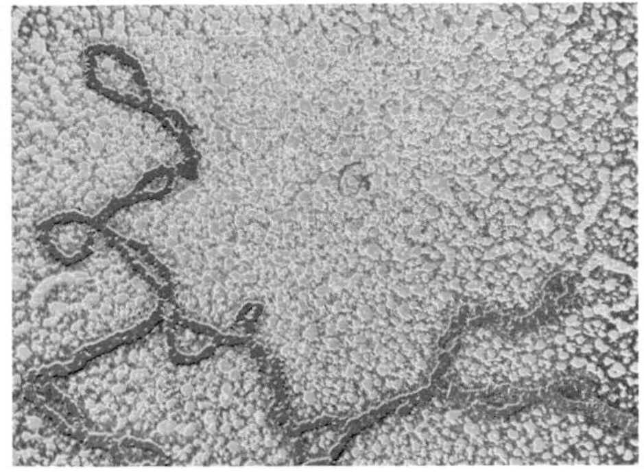

MITOCHONDRIAL NUCLEOID
Mitochondrial DNA, which occurs in structures called nucleoids, forms a closed loop, unlike other, open-ended DNA. This image shows several circular loops and reveals that the molecule is often twisted.

MITOCHONDRIAL GENES

Mitochondria have their own DNA (also known as mtDNA), RNA, and ribosomes, and therefore can make many of their own proteins. Unlike nuclear DNA, which is arranged on chromosomes, mitochondrial DNA forms circular, double-stranded chains that are 16,500 base pairs long. Each chain has 37 genes: 13 that code for proteins, 22 for transfer RNAs, and 2 for ribosomal RNAs. There are more than 1,000 identical molecules of mtDNA in humans. Mitochondrial DNA has been used to study genetic relationships because it has a high mutation rate and is inherited primarily through the mother.

GENETIC CONTROL OF CELLS

Not all genes are active in all cells. The process by which a gene is able to make its protein is called gene expression. Some genes are "switched on" and express themselves in most cells. These are concerned with basic processes such as utilizing glucose for energy. Other genes are switched off in most cells; these are for making specialized products, such as hormones. Cell specialization involves genes being switched on or off, according to exposure to chemicals such as growth factors and regulators–products of other genes.

CELLULAR DIFFERENTIATION

The first cells produced by the early divisions of a fertilized egg are generalized. As they multiply, the cells differentiate in response to preprogrammed instructions, intercellular contacts, and their chemical environment to become specialized tissues such as nerves, muscles, and skin.

PRECURSOR CELL

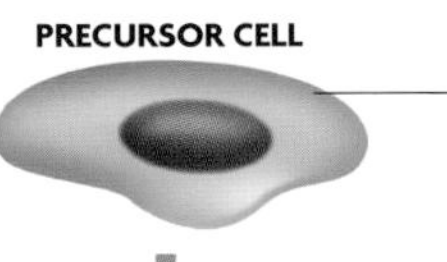

PRECURSOR CELL
This can become any of a variety of cells. Some lines of offspring cells retain the ability to generalize, while others go on to become specialists

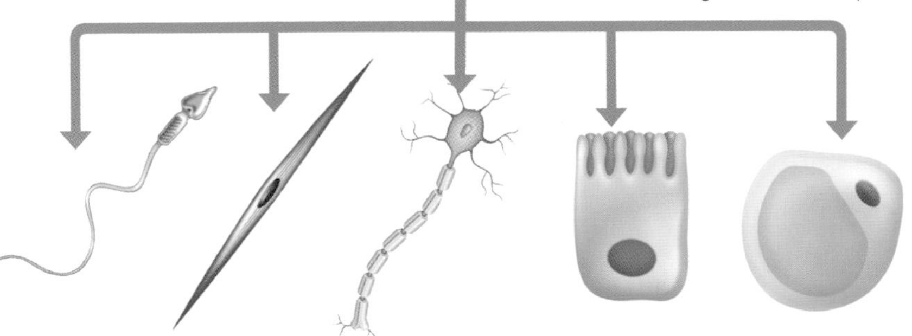

SPERM CELL
Packed with mitochondria to supply fuel

MUSCLE CELL
Long, thin cells with contractile proteins

NERVE CELL
Extreme specialization in both shape and connections

EPITHELIAL CELL
Programmed to multiply rapidly and then die

FAT CELL
Stores energy in case diet does not meet energy requirements

SPECIALIZED CELLS AND TISSUES

MORE THAN 200 TYPES OF SPECIALIZED CELL POPULATE THE HUMAN BODY. EACH TYPE FORMS CLOSELY KNIT CONFIGURATIONS, WHICH ARE RECOGNIZABLE AS SPECIFIC TISSUES. IN SOME CASES, TISSUES ARE MADE OF SEVERAL TYPES OF CELLS.

TISSUE TYPES

The cells that form tissue all have much the same structure and perform the same function. There are four primary tissue types, derived from specific cell layers in the early embryo: epithelial, connective, muscle, and nerve. Blood, bone, cartilage, tendons, and ligaments are connective tissues. The epidermis and the tissues that line almost every organ are all types of epithelial tissues. Muscle and nerve tissues, of course, form muscles and nerves.

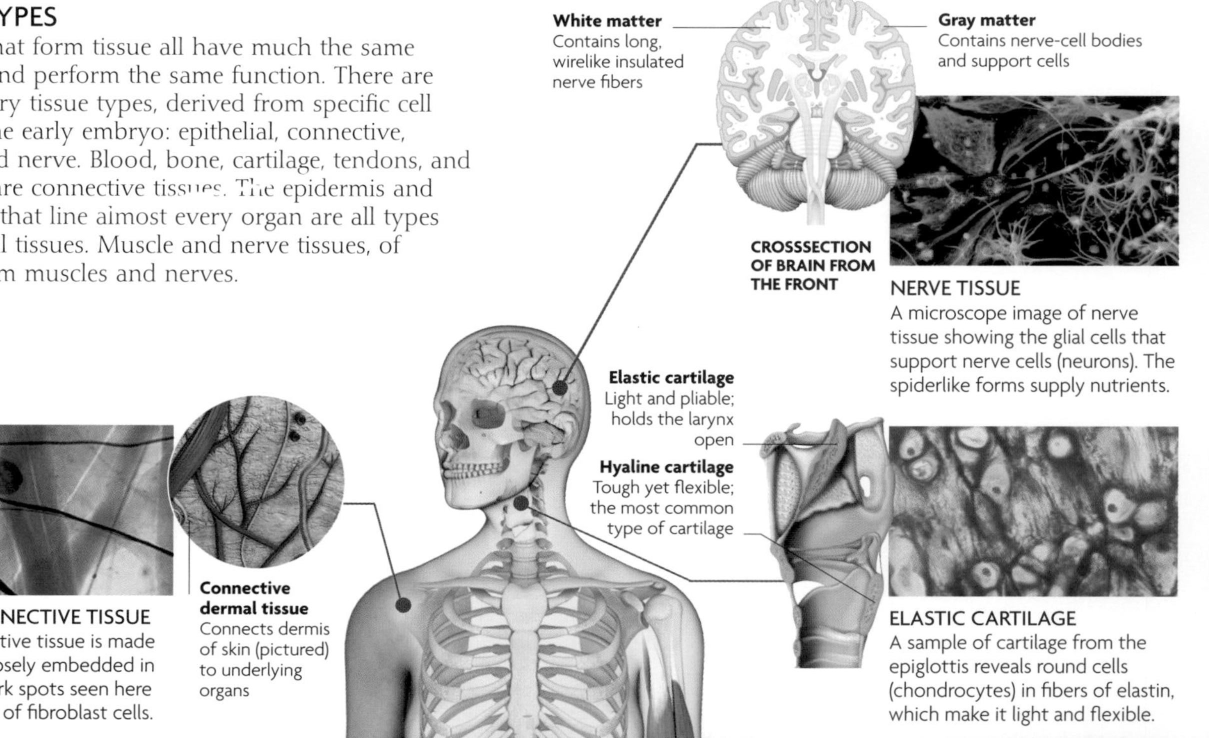
White matter
Contains long, wirelike insulated nerve fibers
Gray matter
Contains nerve-cell bodies and support cells
CROSSSECTION OF BRAIN FROM THE FRONT
Elastic cartilage
Light and pliable; holds the larynx open
Hyaline cartilage
Tough yet flexible; the most common type of cartilage
Connective dermal tissue
Connects dermis of skin (pictured) to underlying organs

NERVE TISSUE
A microscope image of nerve tissue showing the glial cells that support nerve cells (neurons). The spiderlike forms supply nutrients.

LOOSE CONNECTIVE TISSUE
Some connective tissue is made up of cells loosely embedded in fibers. The dark spots seen here are the nuclei of fibroblast cells.

ELASTIC CARTILAGE
A sample of cartilage from the epiglottis reveals round cells (chondrocytes) in fibers of elastin, which make it light and flexible.

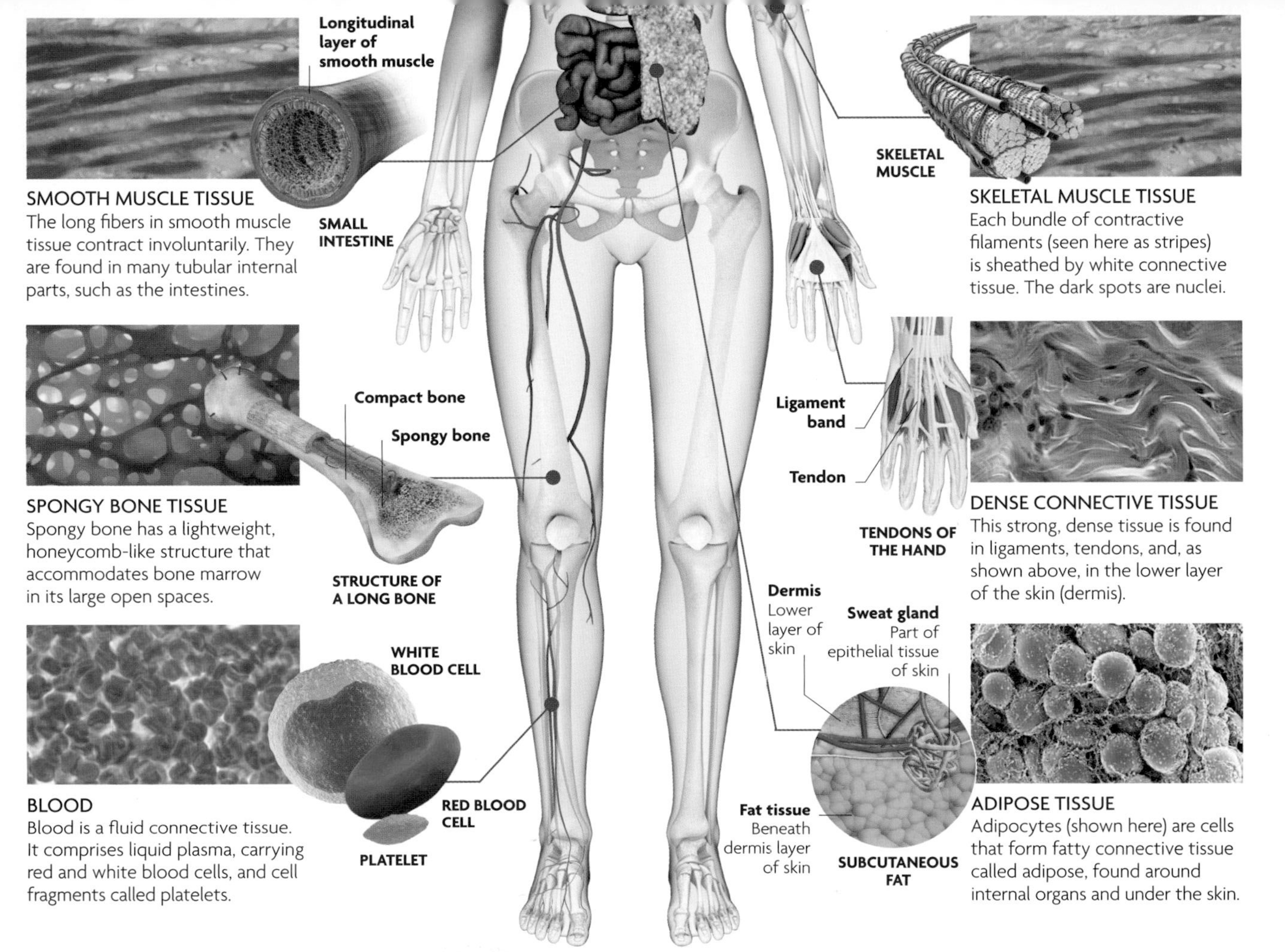
Longitudinal layer of smooth muscle
SMALL INTESTINE
Compact bone
Spongy bone
STRUCTURE OF A LONG BONE
WHITE BLOOD CELL
RED BLOOD CELL
PLATELET
SKELETAL MUSCLE
Ligament band
Tendon
TENDONS OF THE HAND
Dermis Lower layer of skin
Sweat gland Part of epithelial tissue of skin
Fat tissue Beneath dermis layer of skin
SUBCUTANEOUS FAT

SMOOTH MUSCLE TISSUE
The long fibers in smooth muscle tissue contract involuntarily. They are found in many tubular internal parts, such as the intestines.

SPONGY BONE TISSUE
Spongy bone has a lightweight, honeycomb-like structure that accommodates bone marrow in its large open spaces.

BLOOD
Blood is a fluid connective tissue. It comprises liquid plasma, carrying red and white blood cells, and cell fragments called platelets.

SKELETAL MUSCLE TISSUE
Each bundle of contractive filaments (seen here as stripes) is sheathed by white connective tissue. The dark spots are nuclei.

DENSE CONNECTIVE TISSUE
This strong, dense tissue is found in ligaments, tendons, and, as shown above, in the lower layer of the skin (dermis).

ADIPOSE TISSUE
Adipocytes (shown here) are cells that form fatty connective tissue called adipose, found around internal organs and under the skin.

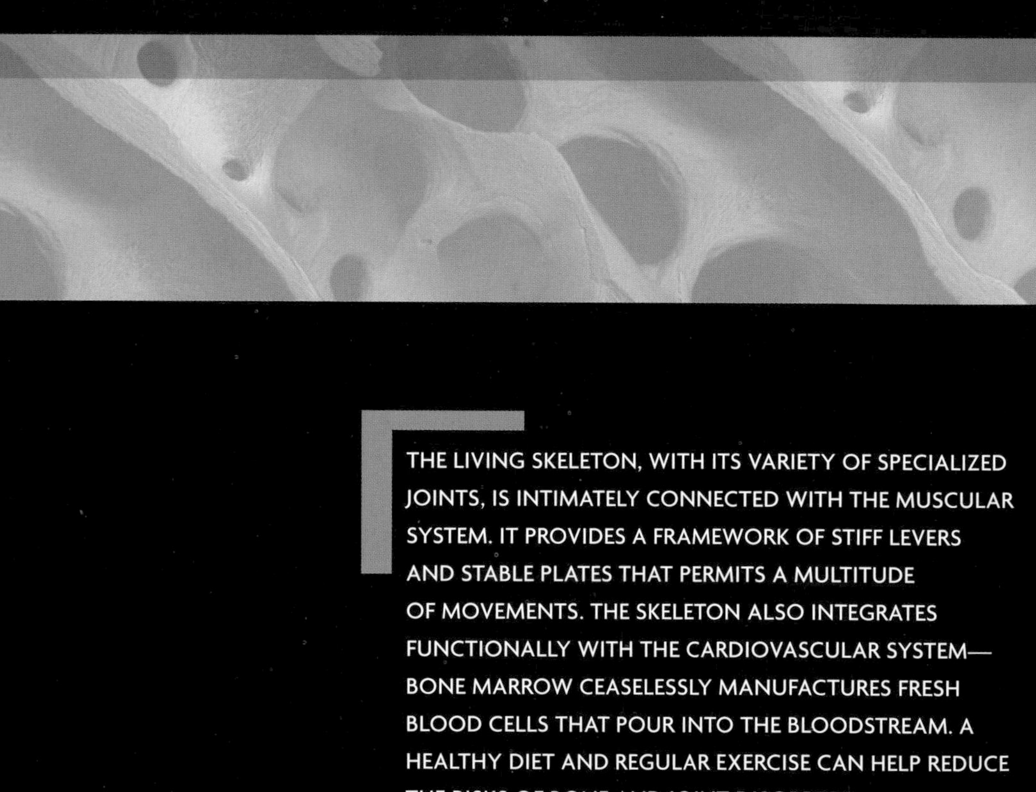

THE LIVING SKELETON, WITH ITS VARIETY OF SPECIALIZED JOINTS, IS INTIMATELY CONNECTED WITH THE MUSCULAR SYSTEM. IT PROVIDES A FRAMEWORK OF STIFF LEVERS AND STABLE PLATES THAT PERMITS A MULTITUDE OF MOVEMENTS. THE SKELETON ALSO INTEGRATES FUNCTIONALLY WITH THE CARDIOVASCULAR SYSTEM—BONE MARROW CEASELESSLY MANUFACTURES FRESH BLOOD CELLS THAT POUR INTO THE BLOODSTREAM. A HEALTHY DIET AND REGULAR EXERCISE CAN HELP REDUCE THE RISKS OF BONE AND JOINT DISORDERS.

SKELETAL
SYSTEM

SKELETON

THE SKELETON MAKES UP ALMOST ONE-FIFTH OF THE BODY'S WEIGHT. WITHOUT THIS INNER FRAMEWORK, ALL OTHER PARTS AND TISSUES WOULD COLLAPSE.

The average skeleton has 206 bones. There are natural variations: about one individual in 20 has an extra rib. Bone is an active tissue, and even though it is about 22 percent water, it has an extremely strong yet lightweight and flexible structure. The skeleton has the advantage of being able to repair itself if damaged. It can also remodel its bones to thicken and strengthen areas that experience extra stress, resulting from activities such as running and weight-lifting. The two major divisions of the skeleton are called the axial and appendicular skeletons. The axial skeleton consists of the skull, vertebral (spinal) column, ribs, and sternum. The appendicular skeleton includes the bones of the shoulder, arm, wrist, and hand, and the hips, legs, ankles, and feet. Of the 206 bones, 80 are in the axial skeleton, with 64 in the upper appendicular and 62 in the lower appendicular skeleton.

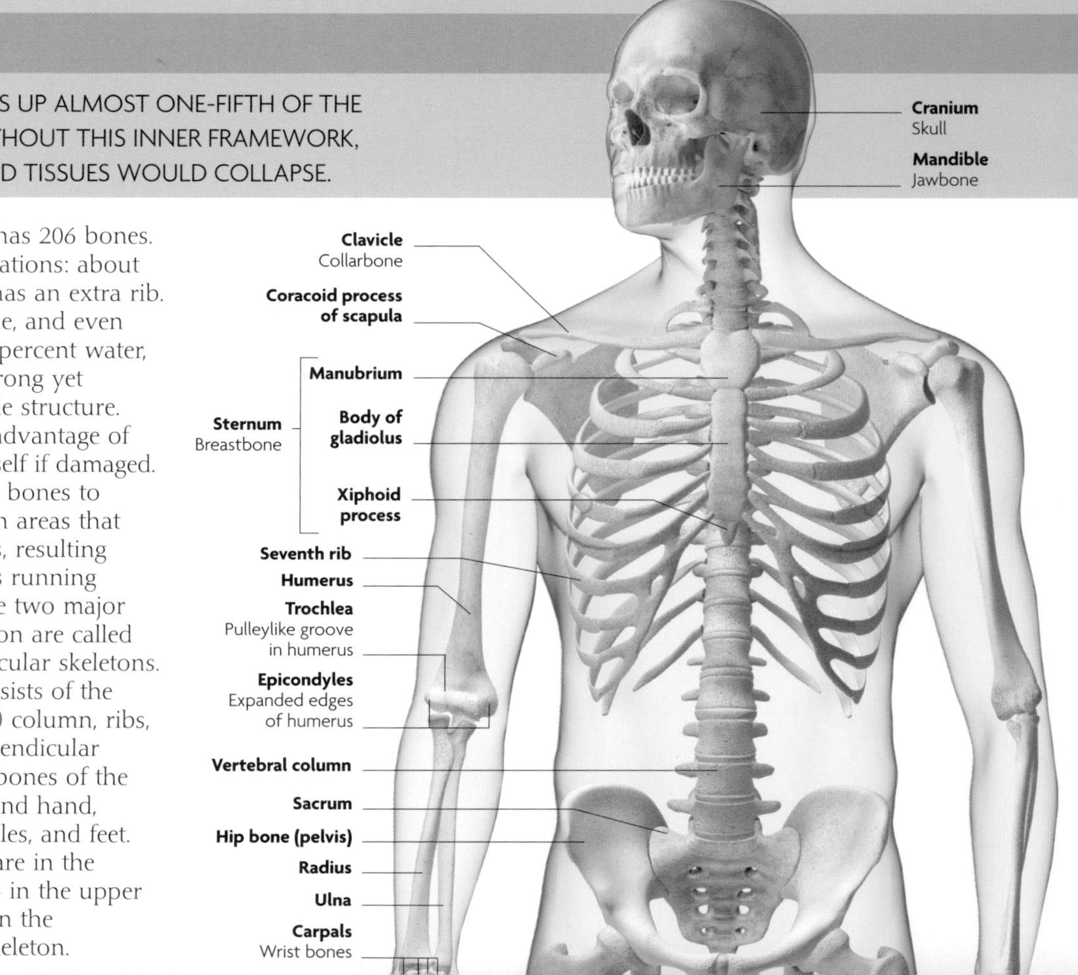
Cranium
Skull
Mandible
Jawbone
Clavicle
Collarbone
Coracoid process of scapula
Sternum
Breastbone
Manubrium
Body of gladiolus
Xiphoid process
Seventh rib
Humerus
Trochlea
Pulleylike groove in humerus
Epicondyles
Expanded edges of humerus
Vertebral column
Sacrum
Hip bone (pelvis)
Radius
Ulna
Carpals
Wrist bones

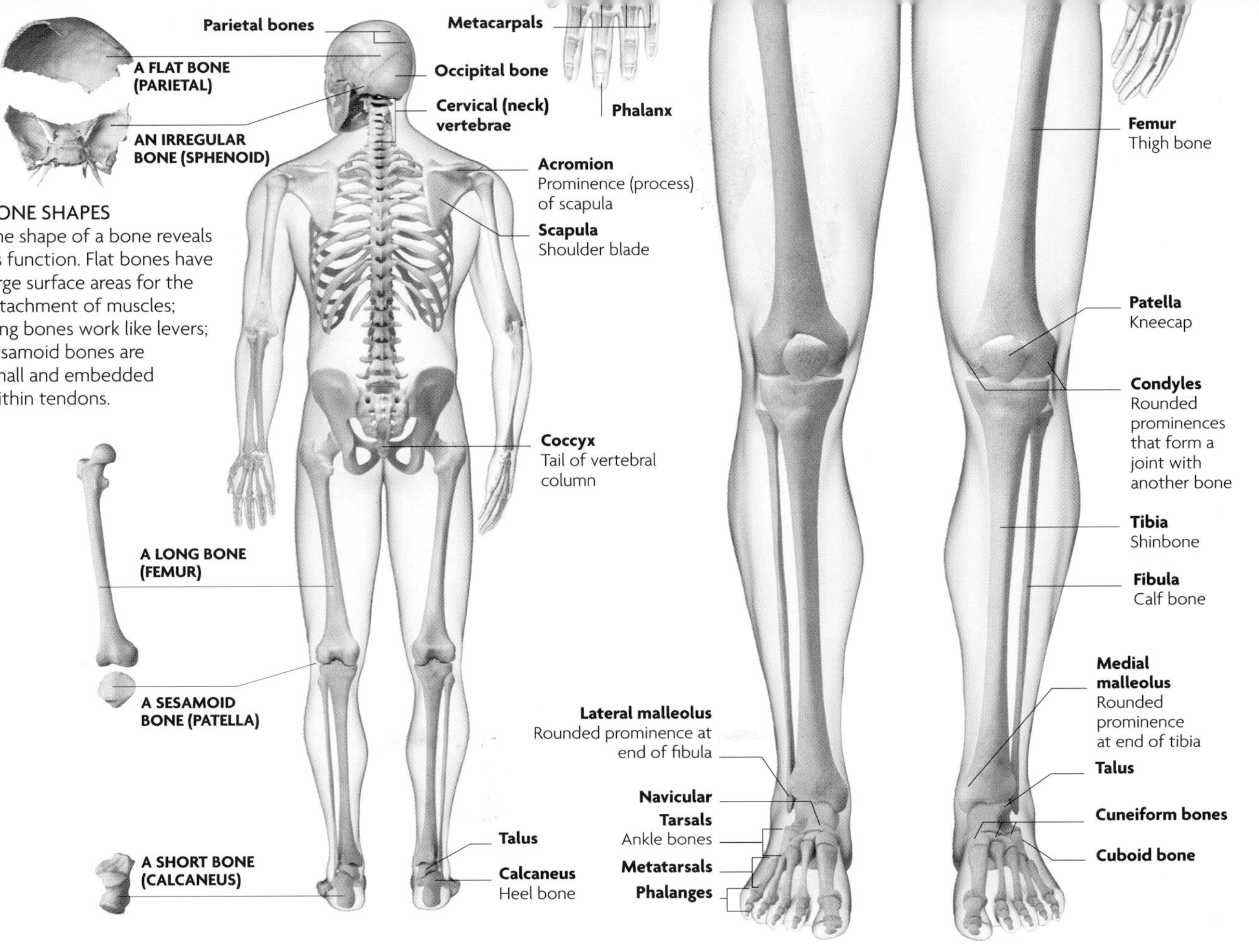
Parietal bones
A FLAT BONE (PARIETAL)
AN IRREGULAR BONE (SPHENOID)
Metacarpals
Occipital bone
Cervical (neck) vertebrae
Phalanx
Acromion
Prominence (process) of scapula
Scapula
Shoulder blade
Coccyx
Tail of vertebral column
A LONG BONE (FEMUR)
A SESAMOID BONE (PATELLA)
A SHORT BONE (CALCANEUS)
Talus
Calcaneus
Heel bone
Femur
Thigh bone
Patella
Kneecap
Condyles
Rounded prominences that form a joint with another bone
Tibia
Shinbone
Fibula
Calf bone
Lateral malleolus
Rounded prominence at end of fibula
Navicular
Tarsals
Ankle bones
Metatarsals
Phalanges
Medial malleolus
Rounded prominence at end of tibia
Talus
Cuneiform bones
Cuboid bone

BONE SHAPES

The shape of a bone reveals its function. Flat bones have large surface areas for the attachment of muscles; long bones work like levers; sesamoid bones are small and embedded within tendons.

BONE STRUCTURE

BONE IS A TYPE OF CONNECTIVE TISSUE THAT IS AS STRONG AS STEEL BUT AS LIGHT AS ALUMINUM. IT IS MADE OF SPECIALIZED CELLS AND PROTEIN FIBERS. NEITHER IMMOBILE NOR DEAD, BONE CONSTANTLY BREAKS DOWN AND REBUILDS ITSELF.

STRUCTURE OF A BONE

Along the central shaft of a long bone (such as the femur or humerus) is the medullary canal or marrow cavity. This contains red bone marrow, which produces blood cells; yellow marrow, which is mostly fatty tissue; and blood vessels. Surrounding the marrow cavity is a layer of spongy (cancellous) bone, the honeycomb-like cavities of which also contain marrow. Around this is a layer of compact (cortical) bone, which is hard, dense, and strong. Small canals connect the marrow cavity with the periosteum–a membrane covering the bone surface. Bone tissue is made of specialized cells and protein fibers, woven into a matrix of water, mineral salts, and other substances. Bone cells include osteoblasts, which calcify bone as it forms; osteocytes, which maintain healthy bone structure; and osteoclasts, which absorb bone tissue where it is degenerating or not needed.

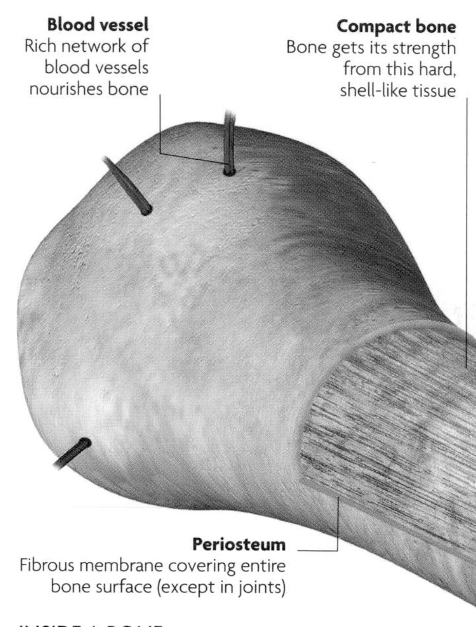
Blood vessel
Rich network of blood vessels nourishes bone
Compact bone
Bone gets its strength from this hard, shell-like tissue
Periosteum
Fibrous membrane covering entire bone surface (except in joints)

INSIDE A BONE
Long bones, for example those in the leg, comprise several different types of bone tissue.

BONE GROWTH

During development in the womb and infancy, most bones develop from structures made of cartilage. Ossification is the process by which this cartilage is converted into bone tissue by the deposition of mineral salts and crystals. Near each end of a long bone is an area called the growth plate, where lengthening and ossification occur. Cartilage cells (see right) multiply here and form columns toward the bone shaft. As the cartilage cells enlarge and die, the space they occupied is filled by new bone cells.

CARTILAGE TO BONE
Initially, ossification in a long bone occurs between the shaft and the head. Later, ossification also takes place inside the head.

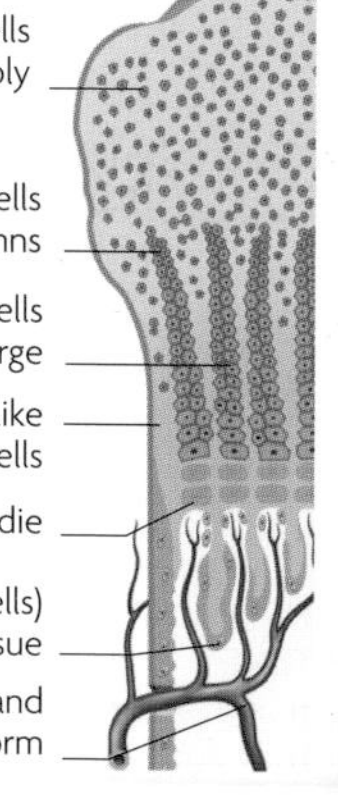
Cartilage cells multiply
Cartilage cells form columns
Cartilage cells enlarge
Calcium is deposited in gel-like matrix surrounding cartilage cells
Old cartilage cells die
Osteoblasts (specialized bone cells) attach to the calcified tissue
New blood vessels and bone tissue form

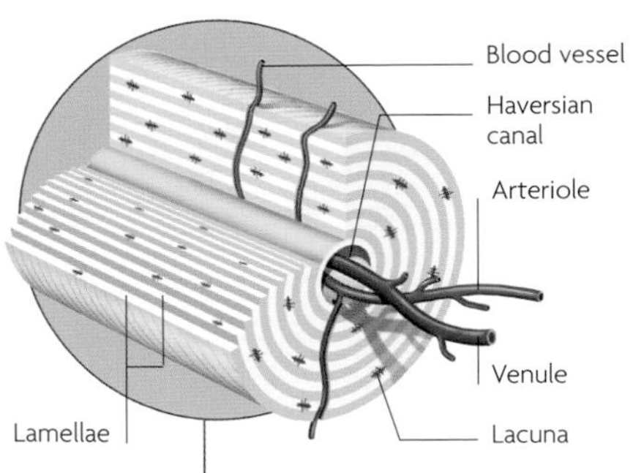
Blood vessel
Haversian canal
Arteriole
Venule
Lacuna
Lamellae

OSTEON

This rod-shaped unit is the building block of compact bone. Its central (Haversian) canal, containing blood vessels and nerves, is surrounded by concentric layers of tissue (lamellae). Gaps (lacunae) in the tissue contain osteocytes, which maintain bones.

CARTILAGE

Cartilage is a tough, adaptable form of connective tissue. It consists of a gel-like matrix containing many chemicals, such as proteins and carbohydrates. Various types of fibers are embedded in this tissue, as well as chondrocytes cells, which make and maintain the whole tissue. There are several kinds of cartilage, including hyaline cartilage, fibrocartilage, and elastic cartilage, a springy material found at sites such as the outer ear flap and larynx.

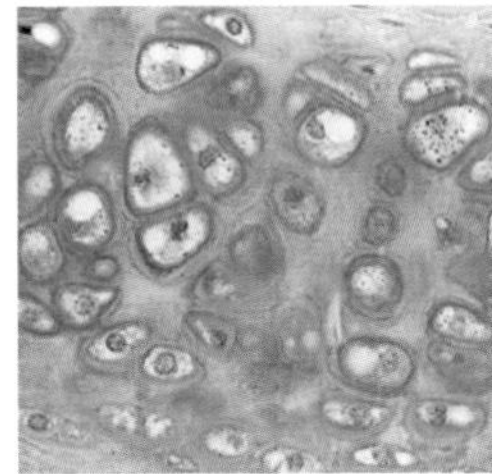

HYALINE CARTILAGE

Dense collagen fibers make this cartilage extra tough and resistant. It covers bone ends in joints and attaches ribs to the sternum, and is also found in the trachea and nose.

Bone marrow
Tissue filling a bone's central cavity; at first, long bones have red marrow—later this turns into yellow marrow

Osteon

Vein

Artery

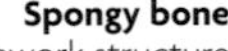

Spongy bone
Latticework structure consisting of bony spikes (trabeculae), arranged along lines of greatest stress

Epiphysis
Expanded head of bone containing mainly spongy bone tissue

Bone shaft
Mostly compact bone and marrow

FIBROCARTILAGE

This is mostly dense bundles of collagen fibres, with little gel-like matrix. It is found in the jaw, knee joints, and intervertebral disks.

JOINTS

THE SITE AT WHICH TWO BONES MEET IS CALLED A JOINT OR AN ARTICULATION. JOINTS CAN BE CLASSIFIED ACCORDING TO THEIR STRUCTURE AND BY THE TYPES OF MOVEMENT THEY ALLOW. THE BODY HAS MORE THAN 300 DIFFERENT JOINTS.

SYNOVIAL JOINTS

The body's most numerous, versatile, and freely moving joints are known as synovial joints. They can work well for many decades if kept in good use, but not overused. Synovial joints are enclosed by a protective outer covering–the joint capsule. The capsule's inner lining, called the synovial membrane, produces a slippery, oil-like synovial fluid that keeps the joint well lubricated so that the joint surfaces slide against each other with minimal friction and wear. There are around 230 synovial joints in the body.

SEMIMOVABLE AND FIXED JOINTS

Not all joints have a wide range of movement. Some allow for growth or for greater stability. The bones in these joints are usually linked by cartilage or tough fibers made of substances such as the protein collagen. In the fixed joints of the skull, once growth is complete, the separate bone plates are securely connected by interlocking fibrous tissue, forming suture joints.

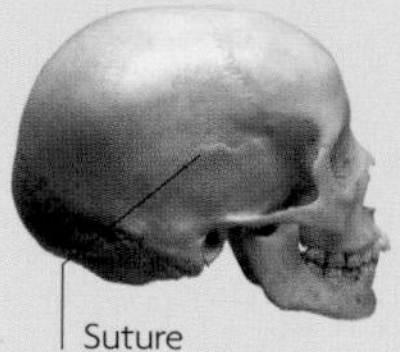
Suture

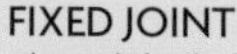

FIXED JOINT

The adult skull's suture joints show up as wiggling lines. In infancy, these joints are loosely attached to allow for expansion of the rapidly growing brain.

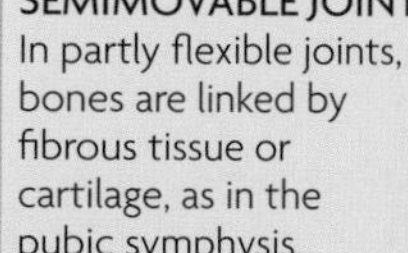

SEMIMOVABLE JOINT

In partly flexible joints, bones are linked by fibrous tissue or cartilage, as in the pubic symphysis.

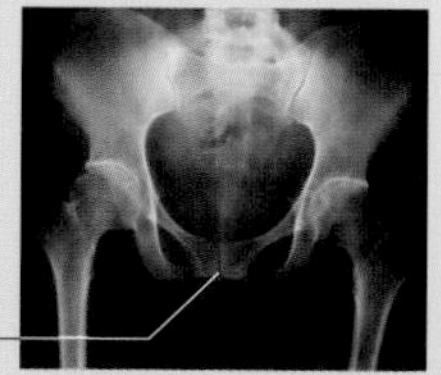
Pubic symphysis

TYPES OF SYNOVIAL JOINT

A synovial joint's range of movement is determined by the shape of its articular cartilage surfaces (see p.46) and how they fit together.

Pivot joint

A peglike projection from one bone turns in a ring-shaped socket of another bone; or, conversely, the ring turns around the peg. The pivot joint between the top two neck (cervical) vertebrae enables the skull to rotate on the spinal axis, and to move from side to side.

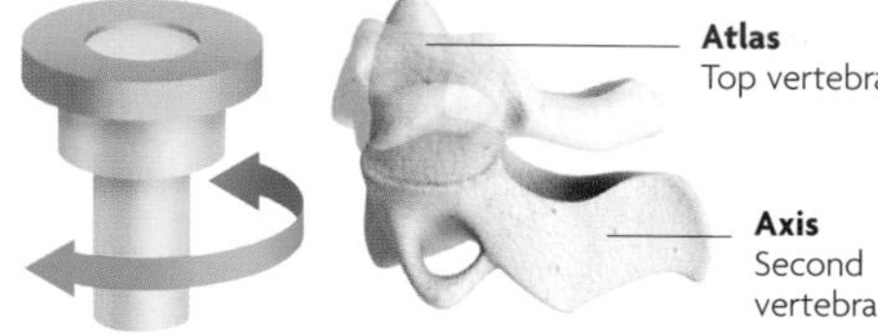
Atlas
Top vertebra
Axis
Second vertebra

Hinge joint

The convex surface of one bone fits into the concave surface of another bone to allow backward and forward movement, mainly in one plane. The elbow is a modified hinge joint that permits limited rotation of the arm bones.

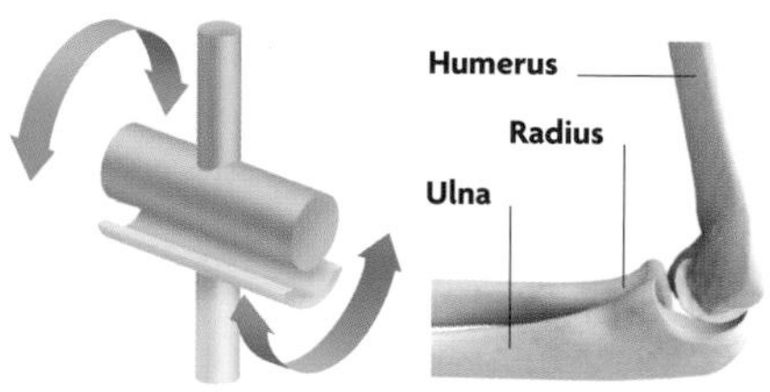
Humerus
Radius
Ulna

Gliding joint

The bone surfaces that meet in a gliding joint are almost flat and slide over one another. Movement is limited by ligaments. Some joints between the tarsals of the ankle and between the carpals in the wrist move in this way.

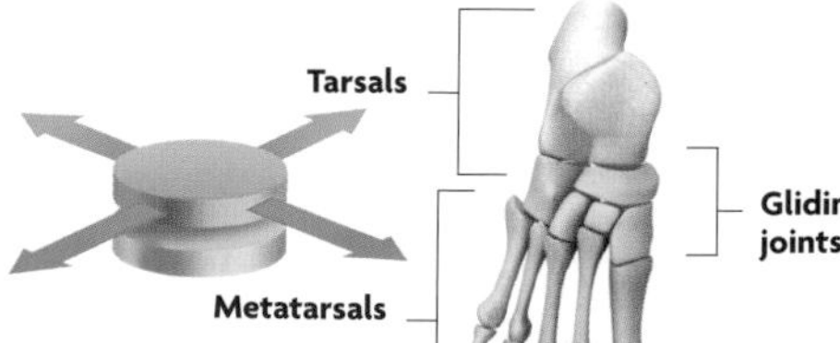
Tarsals
Metatarsals
Gliding joints

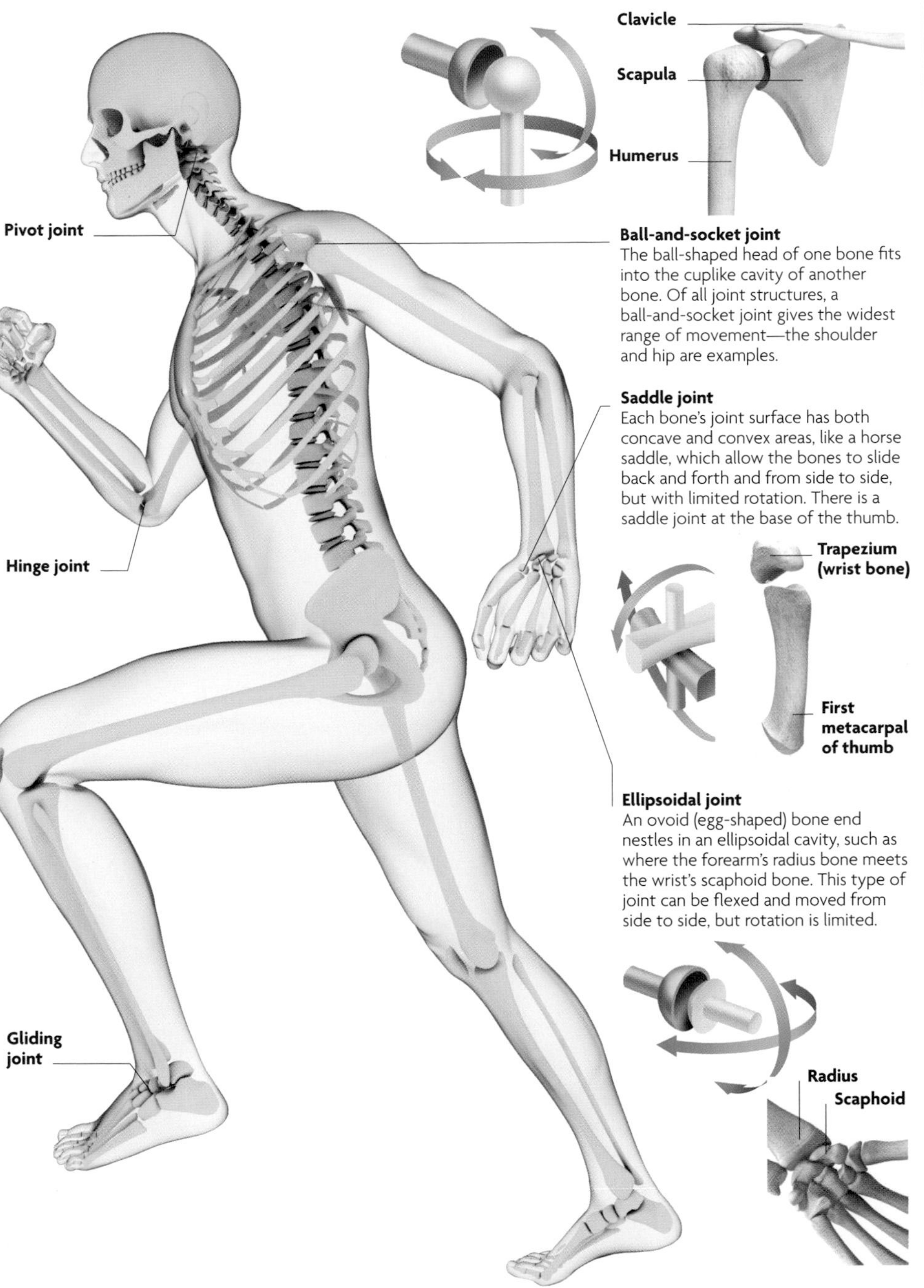
Clavicle
Scapula
Humerus
Pivot joint
Hinge joint
Gliding joint
Trapezium (wrist bone)
First metacarpal of thumb
Radius
Scaphoid

Ball-and-socket joint
The ball-shaped head of one bone fits into the cuplike cavity of another bone. Of all joint structures, a ball-and-socket joint gives the widest range of movement—the shoulder and hip are examples.

Saddle joint
Each bone's joint surface has both concave and convex areas, like a horse saddle, which allow the bones to slide back and forth and from side to side, but with limited rotation. There is a saddle joint at the base of the thumb.

Ellipsoidal joint
An ovoid (egg-shaped) bone end nestles in an ellipsoidal cavity, such as where the forearm's radius bone meets the wrist's scaphoid bone. This type of joint can be flexed and moved from side to side, but rotation is limited.

INSIDE A JOINT

The bone ends in a synovial joint are covered by a smooth, slightly compressible tissue called articular cartilage. Surrounding the joint is the joint capsule, which is made of strong connective tissue and is attached to the bone ends. Its delicate inner lining, the synovial membrane, secretes viscous synovial fluid into the synovial cavity to keep the joint well oiled. The fluid also nourishes the cartilage with fats and proteins, and is constantly reabsorbed. Fibrous thickenings of the capsule, called ligaments, are anchored to the bones at each end and prevent unnatural movement of the joint. Muscles around the joint, which are connected to the bones by tendons, provide stability and produce movement.

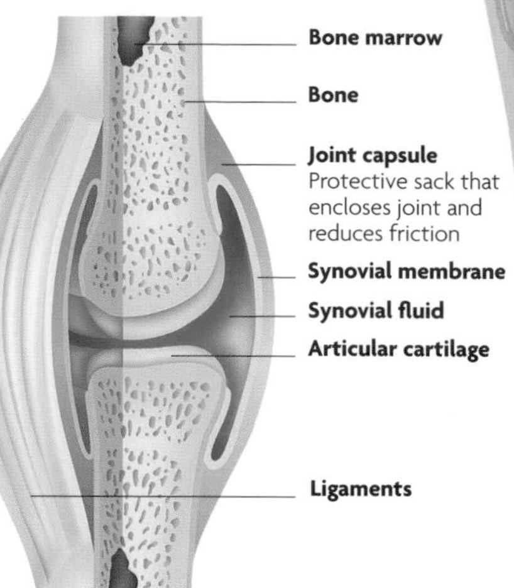
Bone marrow
Bone
Joint capsule
Protective sack that encloses joint and reduces friction
Synovial membrane
Synovial fluid
Articular cartilage
Ligaments

INSIDE A SYNOVIAL JOINT
A mere film of synovial fluid separates bone ends. There are just 3/100–7/100fl oz (1–2ml) of this liquid in the knee joint.

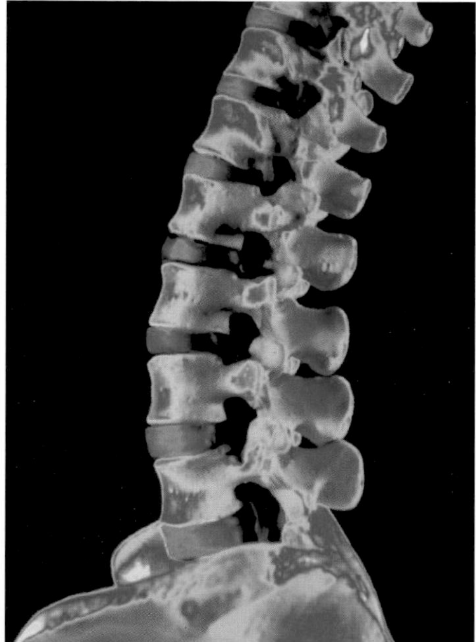

CARTILAGE AS A SHOCK ABSORBER

The articular cartilage that coats the bone ends in a synovial joint is also known as hyaline cartilage (see p.43). If sudden knocks or vibrations jolt the joint, this cartilage works as a shock absorber to dissipate the force of the impact and prevent jarring damage to the much more rigid bones. In certain joints, the cartilage has especially tough fibers. Examples include the fibrocartilaginous pads, called intervertebral disks, that act as cushions between the vertebrae of the backbone. Fibrocartilage also occurs in the jaw and wrist joints and the menisci in the knee.

SPINAL CARTILAGE
The fibrocartilage disks (blue) between the vertebrae help stabilize and cushion the spinal column.

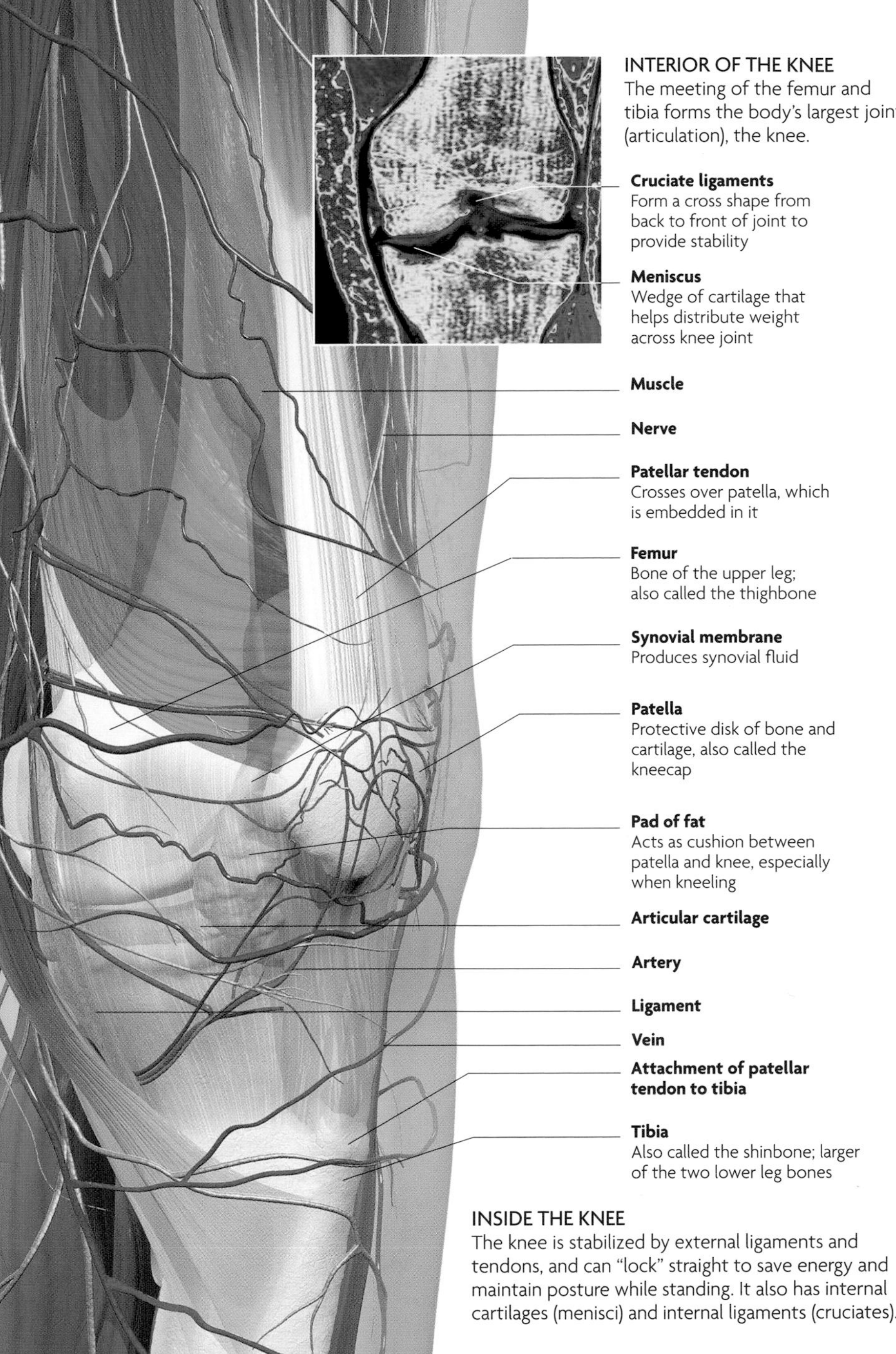
Cruciate ligaments
Form a cross shape from back to front of joint to provide stability
Meniscus
Wedge of cartilage that helps distribute weight across knee joint
Muscle
Nerve
Patellar tendon
Crosses over patella, which is embedded in it
Femur
Bone of the upper leg; also called the thighbone
Synovial membrane
Produces synovial fluid
Patella
Protective disk of bone and cartilage, also called the kneecap
Pad of fat
Acts as cushion between patella and knee, especially when kneeling
Articular cartilage
Artery
Ligament
Vein
Attachment of patellar tendon to tibia
Tibia
Also called the shinbone; larger of the two lower leg bones

INTERIOR OF THE KNEE

The meeting of the femur and tibia forms the body's largest joint (articulation), the knee.

INSIDE THE KNEE

The knee is stabilized by external ligaments and tendons, and can "lock" straight to save energy and maintain posture while standing. It also has internal cartilages (menisci) and internal ligaments (cruciates).

SKULL

THERE ARE 29 BONES IN THE SKULL—22 BONES FORM THE SKULL ITSELF, WITH 21 OF THEM, EXCLUDING THE LOWER JAW, FUSED TOGETHER. THE OTHER BONES ARE THE HYOID BONE IN THE NECK AND THREE PAIRS OF TINY EAR BONES, CALLED OSSICLES.

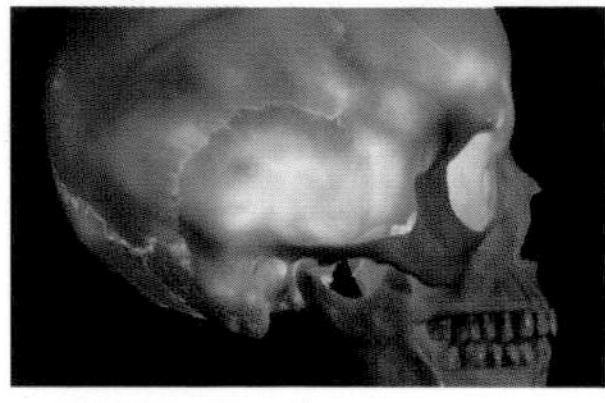

SKULL SUTURES
Lines on the skull's surface, highlighted here, are the fused margins of the skull bones.

SKULL

Two groups of bones make up the skull. The upper set of eight bones forms the domelike cranium (cranial skull or cranial vault), which encloses and protects the brain. The other 14 bones make the skeleton of the face. During growth in childhood, 21 of the 22 bones become strongly fused at faint joint lines, known as sutures. The lower jaw, or mandible, remains unfixed and is linked to the rest of the skull at the two jaw, or temporomandibular, joints.

SINUSES

The four pairs of sinuses, known as paranasal sinuses, are air-filled cavities within the skull bones. They are named after the bones in which they are located: maxillary, frontal, sphenoidal, and ethmoidal sinuses. The first three pairs have fairly well-defined shapes. The ethmoidal sinuses are more honeycomb-like and variable.

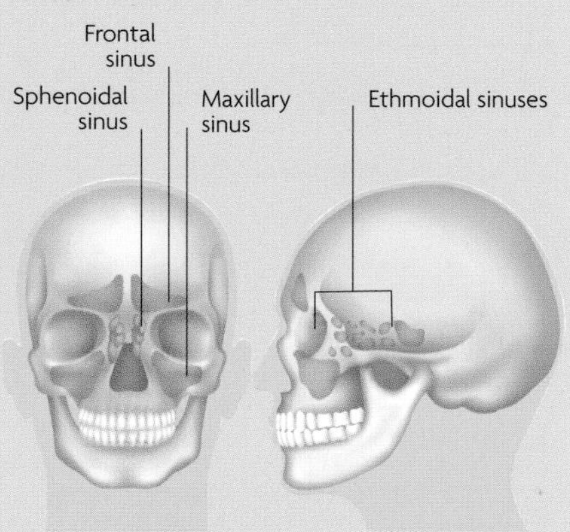

Frontal sinus
Sphenoidal sinus
Maxillary sinus
Ethmoidal sinuses

RESONANT WEIGHT-SAVERS
The sinuses help lighten the skull's overall weight, and also act as resonating chambers to give each person's voice an individual character.

SKULL AND HEAD REGIONS
Two sets of bones form the structure of the skull. The eight bones that enclose the brain are called the cranial vault.

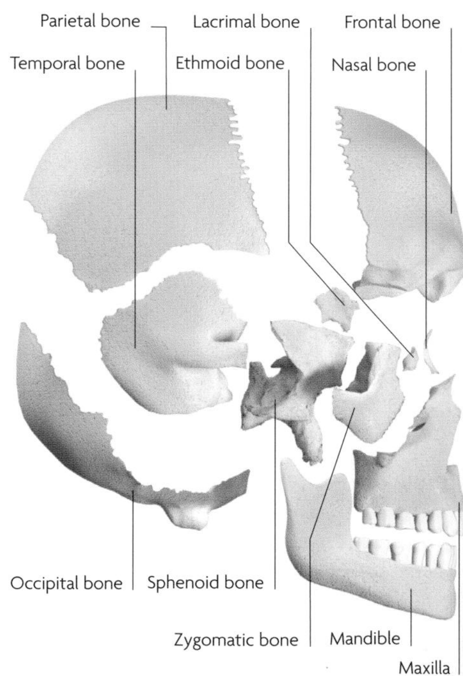

Parietal bone
Lacrimal bone
Frontal bone
Temporal bone
Ethmoid bone
Nasal bone
Occipital bone
Sphenoid bone
Zygomatic bone
Mandible
Maxilla

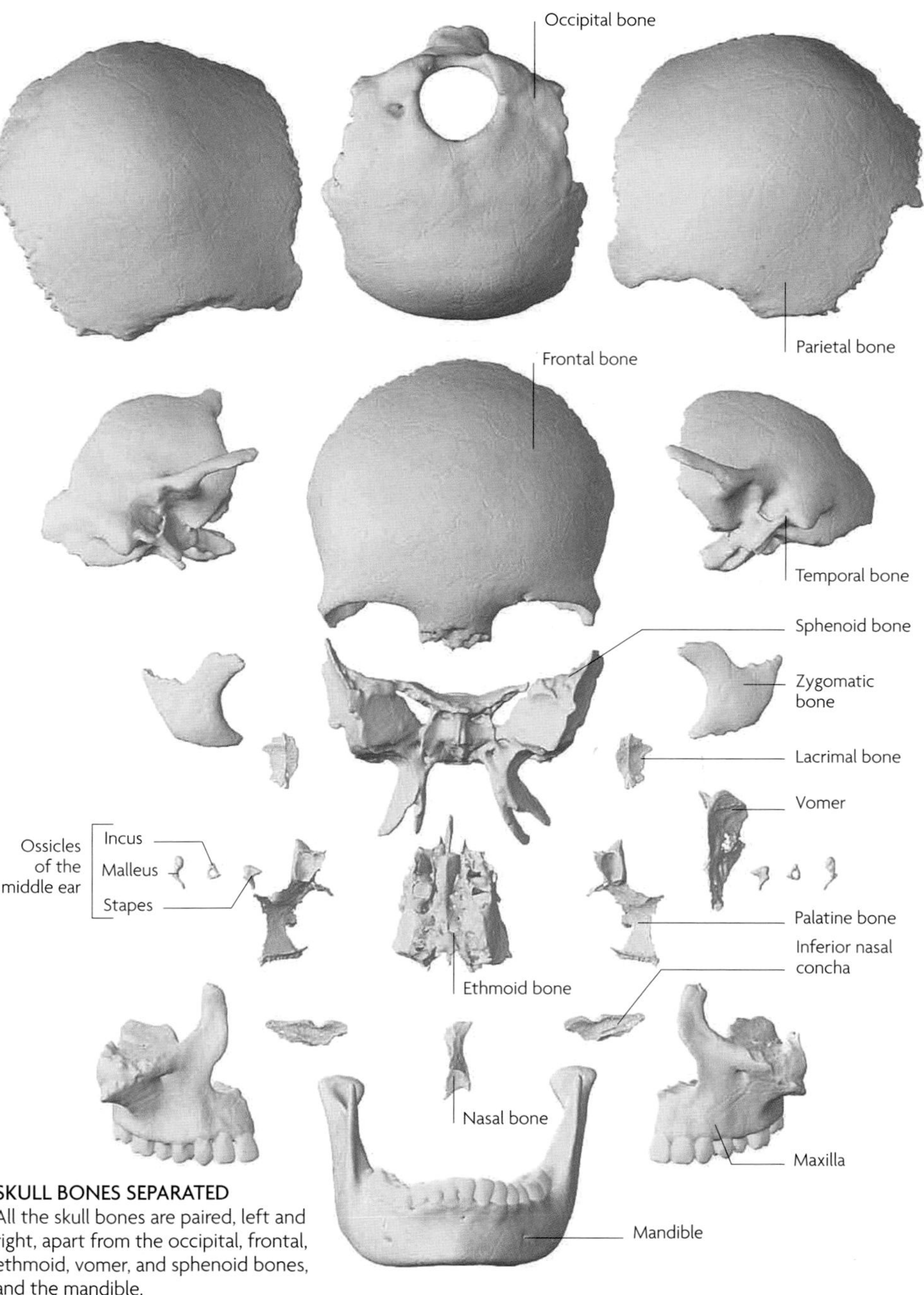
Occipital bone
Parietal bone
Frontal bone
Temporal bone
Sphenoid bone
Zygomatic bone
Lacrimal bone
Vomer
Ossicles of the middle ear
Incus
Malleus
Stapes
Palatine bone
Inferior nasal concha
Ethmoid bone
Nasal bone
Maxilla
Mandible

SKULL BONES SEPARATED
All the skull bones are paired, left and right, apart from the occipital, frontal, ethmoid, vomer, and sphenoid bones, and the mandible.

SPINE

THE SPINE IS ALSO KNOWN AS THE SPINAL OR VERTEBRAL COLUMN, OR SIMPLY "THE BACKBONE." THIS STRONG BUT FLEXIBLE CENTRAL SUPPORT HOLDS THE HEAD AND TORSO UPRIGHT, YET IT ALLOWS THE NECK AND BACK TO BEND AND TWIST.

SPINE FUNCTION

The spine consists of 33 ringlike bones called vertebrae. The bottom nine vertebrae are fused into two larger bones termed the sacrum and the coccyx. The 26 movable components of the spine are linked by a series of mobile joints. Between the bones of each joint is an intervertebral disk–a springy pad of tough, fibrous cartilage that squashes slightly under pressure to absorb shocks. Ligaments and muscles around the spine provide stability and help control movement. The spinal column also protects the spinal cord (see p.98).

FLEXIBLE COLUMN
Owing to the shape of the vertebrae, the spine can bend farther forward than back, and twist on its axis.

HYOID BONE

The single U-shaped hyoid bone is located at the root of the tongue, just above the larynx. It is one of the few bones in the body that does not join directly to another bone. It is held in position by muscles and by the strong stylohyoid ligament on each side of the bone, which links to the styloid process of the skull's temporal bone. The hyoid stabilizes several sets of muscles used in swallowing and speech.

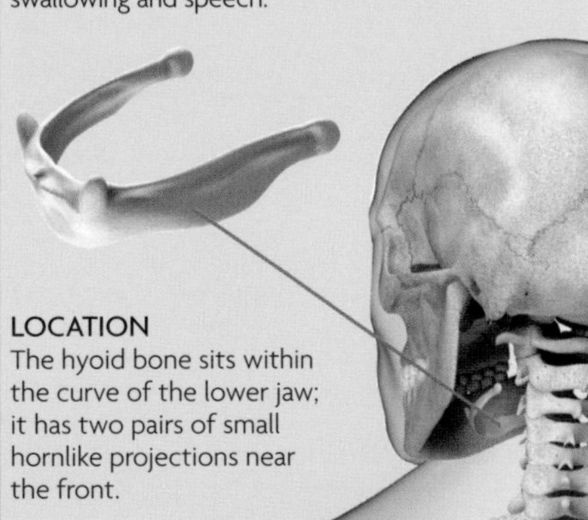

LOCATION
The hyoid bone sits within the curve of the lower jaw; it has two pairs of small hornlike projections near the front.

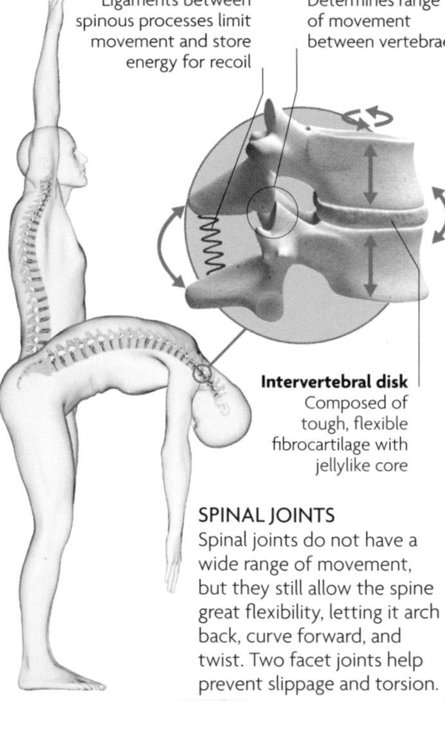
Springy ligament
Ligaments between spinous processes limit movement and store energy for recoil
Facet joint
Determines range of movement between vertebrae
Intervertebral disk
Composed of tough, flexible fibrocartilage with jellylike core

SPINAL JOINTS
Spinal joints do not have a wide range of movement, but they still allow the spine great flexibility, letting it arch back, curve forward, and twist. Two facet joints help prevent slippage and torsion.

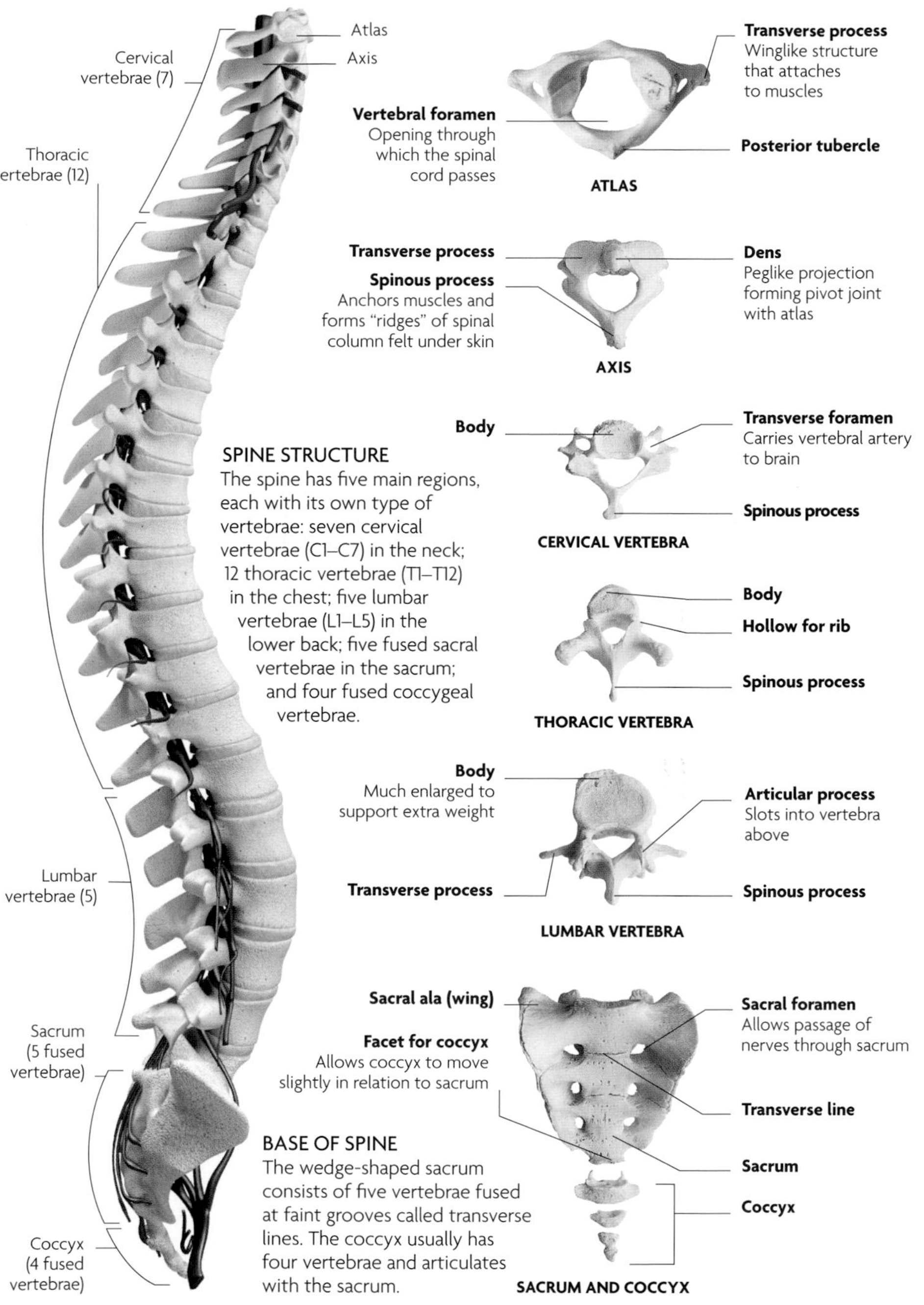
Cervical vertebrae (7)
Atlas
Axis
Thoracic vertebrae (12)
Lumbar vertebrae (5)
Sacrum (5 fused vertebrae)
Coccyx (4 fused vertebrae)
Transverse process
Winglike structure that attaches to muscles
Vertebral foramen
Opening through which the spinal cord passes
Posterior tubercle
ATLAS
Transverse process
Spinous process
Anchors muscles and forms "ridges" of spinal column felt under skin
Dens
Peglike projection forming pivot joint with atlas
AXIS
Body
Transverse foramen
Carries vertebral artery to brain
Spinous process
CERVICAL VERTEBRA
Body
Hollow for rib
Spinous process
THORACIC VERTEBRA
Body
Much enlarged to support extra weight
Articular process
Slots into vertebra above
Transverse process
Spinous process
LUMBAR VERTEBRA
Sacral ala (wing)
Facet for coccyx
Allows coccyx to move slightly in relation to sacrum
Sacral foramen
Allows passage of nerves through sacrum
Transverse line
Sacrum
Coccyx
SACRUM AND COCCYX

SPINE STRUCTURE

The spine has five main regions, each with its own type of vertebrae: seven cervical vertebrae (C1–C7) in the neck; 12 thoracic vertebrae (T1–T12) in the chest; five lumbar vertebrae (L1–L5) in the lower back; five fused sacral vertebrae in the sacrum; and four fused coccygeal vertebrae.

BASE OF SPINE

The wedge-shaped sacrum consists of five vertebrae fused at faint grooves called transverse lines. The coccyx usually has four vertebrae and articulates with the sacrum.

RIBS AND PELVIS

THE RIBS AND HIP BONE (PELVIS) GUARD VITAL CHEST AND ABDOMINAL ORGANS, AND DEMONSTRATE THE SKELETON'S FUNCTIONS OF SUPPORT AND PROTECTION. THE PELVIS PROVIDES SURFACES FOR ANCHORING THE HIP AND THIGH MUSCLES.

RIB CAGE

Most people have 12 pairs of ribs, but about 1 in 20 is born with one or more extra pairs. All ribs attach to the spinal column at the rear. The upper seven pairs of "true ribs" link directly to the breastbone (sternum) by their cartilage extensions (costal cartilages). The next two or three pairs of "false ribs" connect to the cartilages of the ribs above. The remaining "floating ribs" do not link to the sternum.

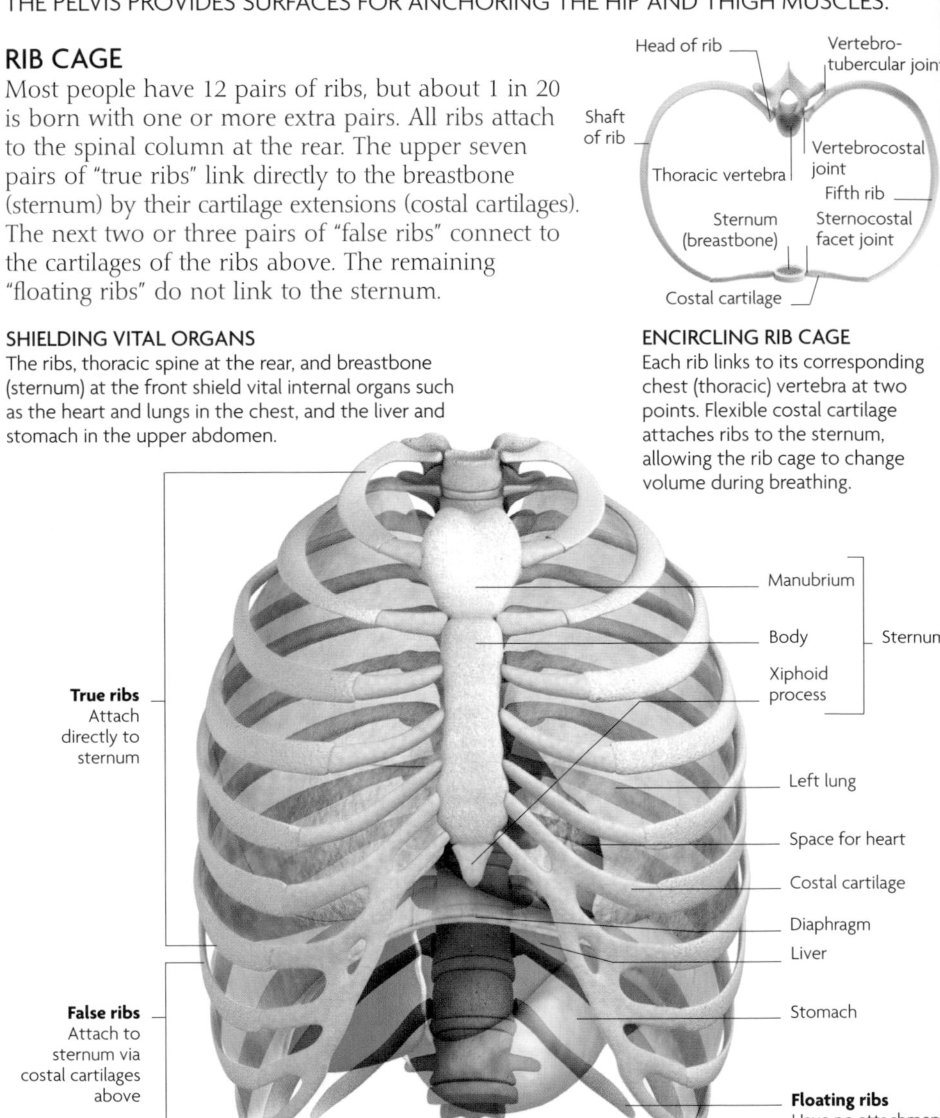
Head of rib
Vertebro-tubercular joint
Shaft of rib
Vertebrocostal joint
Thoracic vertebra
Fifth rib
Sternum (breastbone)
Sternocostal facet joint
Costal cartilage
Manubrium
Body
Sternum
Xiphoid process
True ribs
Attach directly to sternum
Left lung
Space for heart
Costal cartilage
Diaphragm
Liver
Stomach
False ribs
Attach to sternum via costal cartilages above
Floating ribs
Have no attachment at front

ENCIRCLING RIB CAGE
Each rib links to its corresponding chest (thoracic) vertebra at two points. Flexible costal cartilage attaches ribs to the sternum, allowing the rib cage to change volume during breathing.

SHIELDING VITAL ORGANS
The ribs, thoracic spine at the rear, and breastbone (sternum) at the front shield vital internal organs such as the heart and lungs in the chest, and the liver and stomach in the upper abdomen.

PELVIS

Often referred to as the hip bone, the pelvis is a bowl-like structure consisting of the left and right innominate bones or ossa coxae, and the wedge-shaped sacrum and coccyx, which make up the "tailbone" at the rear. Each innominate bone has three fused bony elements: the large, flaring ilium at the rear, which forms the hip bone that you feel under the skin; the ischium at the lower front; and the pubis above it. There are paired sacroiliac joints at the rear and the pubic symphysis, a semimovable joint made of fibrocartilage, at the front. The shape of the pelvis is shallower and wider in females than in males, with a larger gap, or pelvic inlet, and a greater pelvic outlet, to allow a baby to pass through at birth.

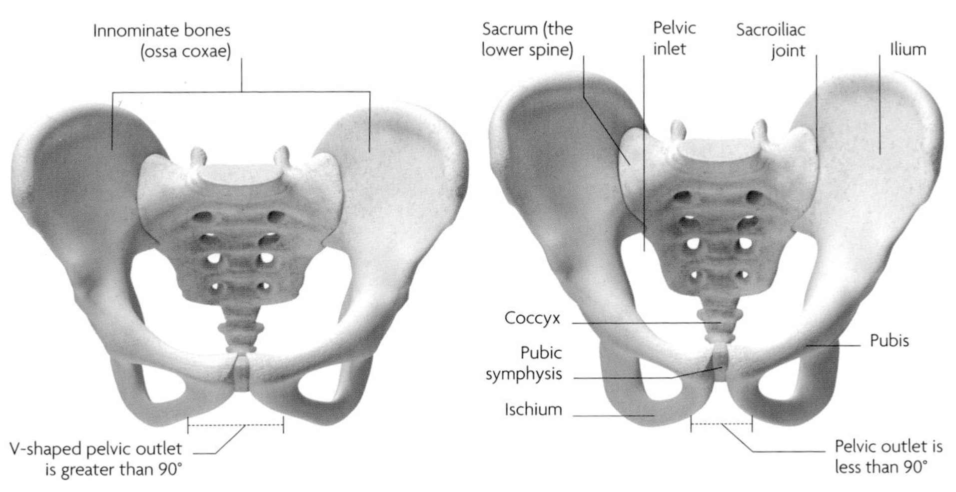
Innominate bones (ossa coxae)
V-shaped pelvic outlet is greater than 90°
Sacrum (the lower spine)
Pelvic inlet
Sacroiliac joint
Ilium
Coccyx
Pubic symphysis
Ischium
Pubis
Pelvic outlet is less than 90°

FEMALE PELVIS FRONT

MALE PELVIS FRONT

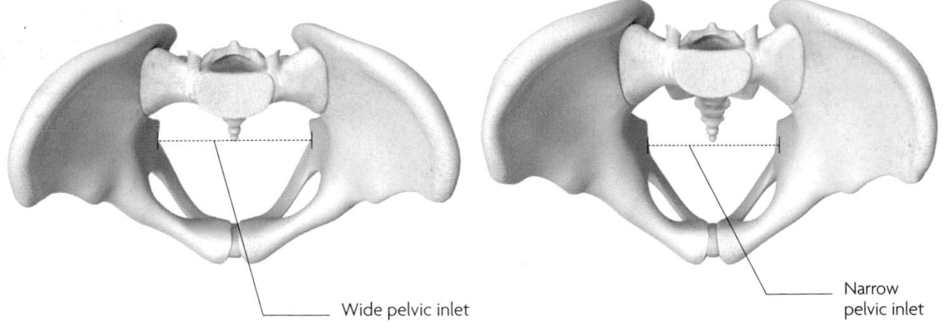
Wide pelvic inlet
Narrow pelvic inlet

FEMALE PELVIS TOP

MALE PELVIS TOP

HANDS AND FEET

THE WRISTS, HANDS, ANKLES, AND FEET COLLECTIVELY CONTAIN 106 BONES, WHICH IS MORE THAN HALF OF ALL THE 206 BONES IN THE BODY. TOGETHER WITH THEIR ASSOCIATED MUSCLES, THEY ARE VITAL FOR COORDINATED MOVEMENT.

WRIST AND HAND

The wrist is made up of the eight carpal bones, arranged roughly in two rows of four. They are linked to each other chiefly by plane or gliding joints (see p.44), and to the forearm bones by the radiocarpal joint. The palm of the hand contains five metacarpal bones. Each of these joins at its outer end to a finger bone (phalanx), of which there are two in the thumb (first digit, or pollex) and three each in the other four digits. The entire structure is moved by more than 50 muscles, including several in the forearm, to provide great flexibility and delicate manipulation.

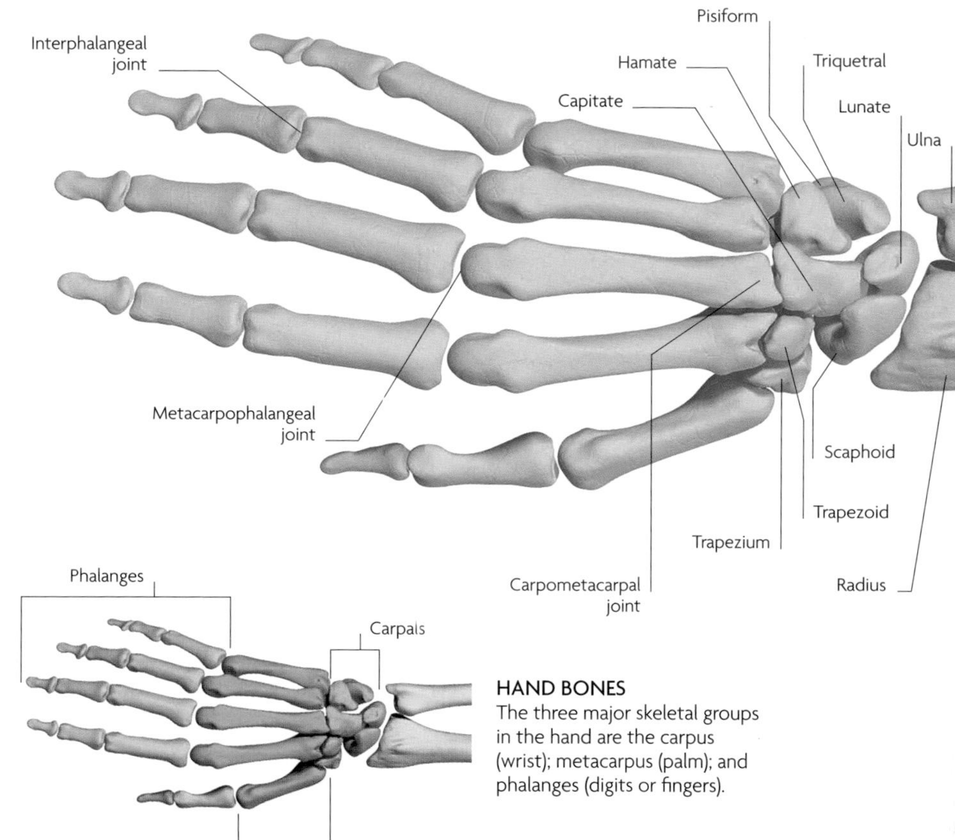
Interphalangeal joint
Pisiform
Hamate
Triquetral
Capitate
Lunate
Ulna
Metacarpophalangeal joint
Scaphoid
Trapezoid
Trapezium
Carpometacarpal joint
Radius
Phalanges
Carpals
Metacarpals

HAND BONES

The three major skeletal groups in the hand are the carpus (wrist); metacarpus (palm); and phalanges (digits or fingers).

ANKLE AND FOOT

The ankle and foot have a similar bone arrangement to the wrist and hand, except that there are only seven tarsal (ankle) bones. The build of the ankle and foot bones is heavier, for strength and weight-bearing stability. The sole is supported by the five metatarsal bones. The hallux (first digit or big toe) has two phalanges (toe bones), and the others have three each. The bony prominence commonly called the "heel bone" is formed by the calcaneus.

Calcaneus
Talus
Cuboid
Navicular
Lateral cuneiform
Intermediate cuneiform
Medial cuneiform

Tarsals
Metatarsals
Phalanges

FOOT BONES
The three main bone groups in the foot are the tarsus (ankle); metatarsus (sole); and phalanges (digits or toes).

LIGAMENTS

Ligaments are strong bands or straps of fibrous tissue that provide support to the bones and link bone ends together in and around joints. Ligaments are made of collagen–a tough, elastic protein. A large number of ligaments bind together the complex wrist and ankle joints. Each ligament is named after the bones it links; for example, the calcaneofibular ligament links the calcaneus ("heel bone") and the fibula.

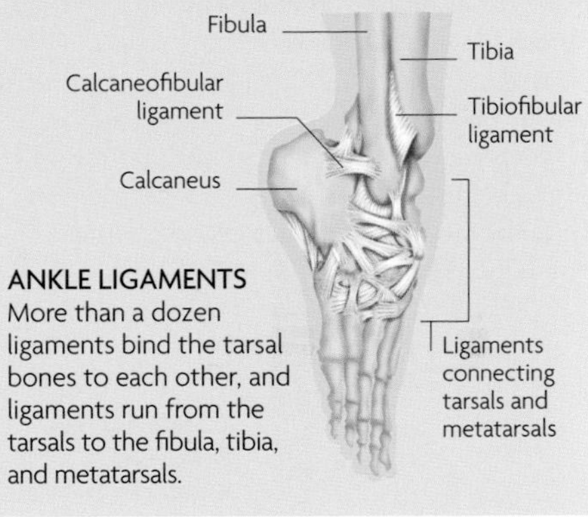
Fibula
Tibia
Calcaneofibular ligament
Tibiofibular ligament
Calcaneus
Ligaments connecting tarsals and metatarsals

ANKLE LIGAMENTS
More than a dozen ligaments bind the tarsal bones to each other, and ligaments run from the tarsals to the fibula, tibia, and metatarsals.

WALKING PRESSURE

With each step, the weight of the body moves from the rear to the front of the foot. The heel bears the initial pressure as the foot is put down. The force passes along the arch, which transfers energy and pressure to the ball of the foot, and finally to the big toe for the push-off.

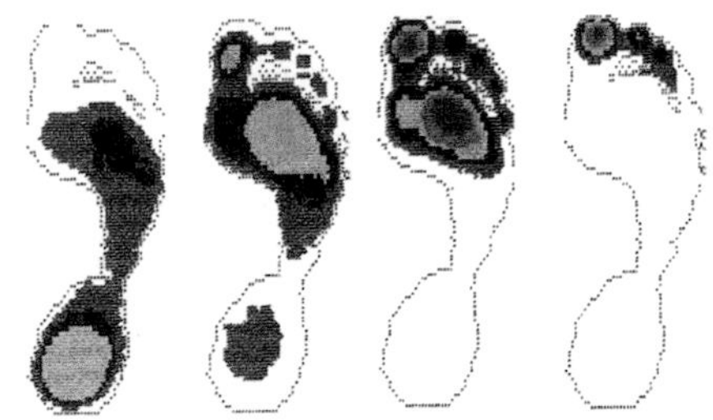

LOAD AREAS ON THE FOOT
These footprint impressions show (from left to right) how the body's weight transfers from the heel to the ball to the big toe when walking.

BONE AND JOINT DISORDERS

BONES AND JOINTS ARE VULNERABLE TO INJURIES SUCH AS FRACTURES AND, DUE TO CONSTANT WEAR, TO DISORDERS SUCH AS OSTEOARTHRITIS. BONES MAY BE WEAKENED BY OSTEOPOROSIS, AND JOINTS MAY BE AFFECTED BY INFLAMMATORY CONDITIONS SUCH AS RHEUMATOID ARTHRITIS.

FRACTURE

Fractures may be caused by a sudden impact, by compression, or by repeated stress. A displaced fracture occurs when the broken surfaces of bone are forced from their normal positions. There are various types of displaced fracture, depending on the angle and strength of the blow. A compression fracture occurs when spongy bone, such as in the vertebrae, is crushed. Stress fractures are caused by prolonged or repeated force straining the bone; they occur in long-distance runners and in the elderly, in whom minor stress, such as coughing, may cause a fracture. Nutritional deficiencies or certain chronic diseases such as osteoporosis, which can weaken bone, may increase the likelihood of fractures. If a broken bone remains beneath the skin, the fracture is described as closed or simple, and there is a low risk of infection. If the ends of the fractured bone project out through the skin, the injury is described as open or compound, and there is a danger of dirt entering the bone tissue and causing microbial contamination.

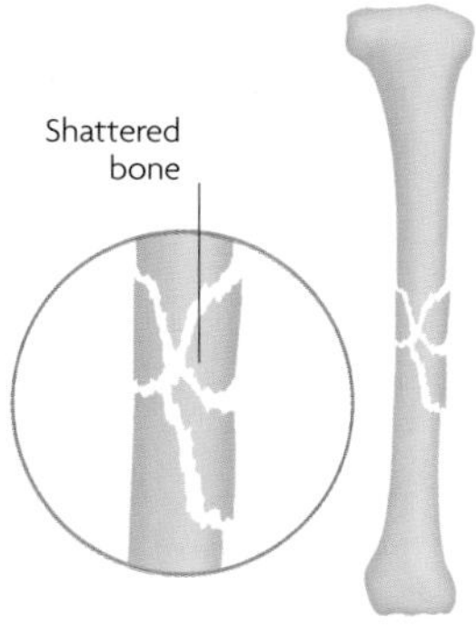
Shattered bone

COMMINUTED FRACTURE
A direct impact can shatter a bone into several fragments or pieces. This type of fracture is likely to occur during a traffic accident.

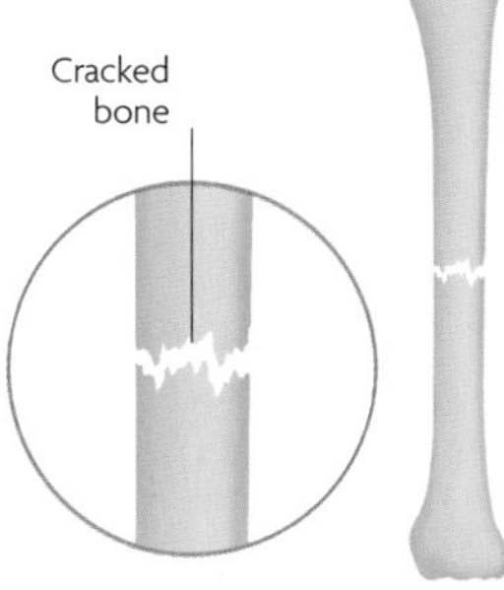
Cracked bone

TRANSVERSE FRACTURE
A powerful force may cause a break across the bone width. The injury is usually stable; the broken surfaces are unlikely to move.

GREENSTICK FRACTURE
If a long bone bends under force, a crack may occur on one side. This type of fracture is common in children, whose bones are flexible.

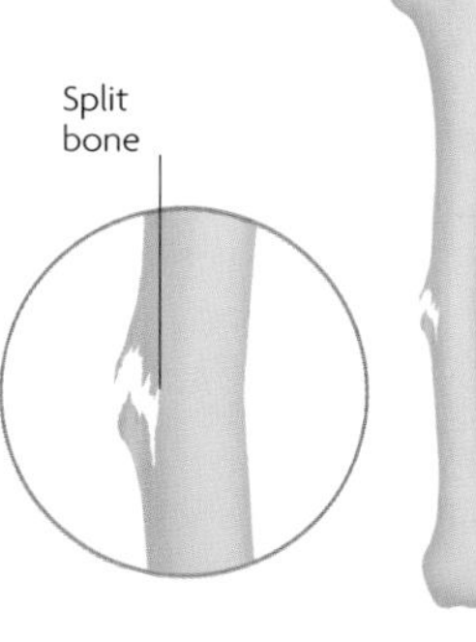
Split bone

SPIRAL FRACTURE
A sharp, twisting force may break a bone diagonally across the shaft. The jagged ends may be difficult to reposition.

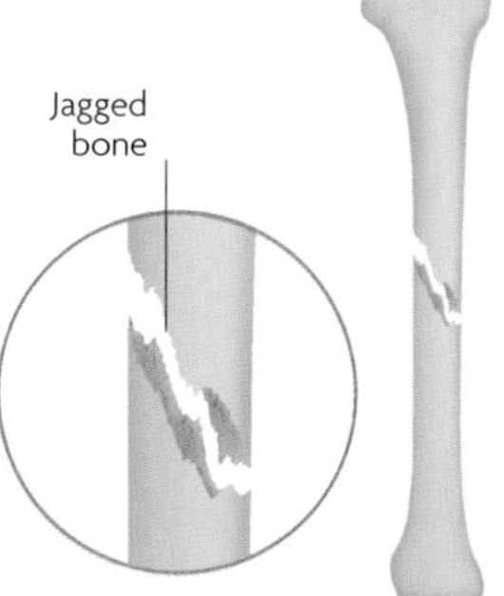
Jagged bone

BONE REPAIR

Despite its image as dry, brittle, and even lifeless, bone is an active tissue with an extensive blood supply and its own restorative processes. After a fracture, blood clots as it does elsewhere in the body. Fibrous tissue, and then new bone growth, bridge the break and eventually restore strength. However, medical treatment is often required to ensure that the repair process is effective and the result is not misshapen. If the bones are displaced, manipulation to restore their normal position—known as reduction—may be performed under anesthesia. The bone will also be immobilized to allow the ends to heal correctly.

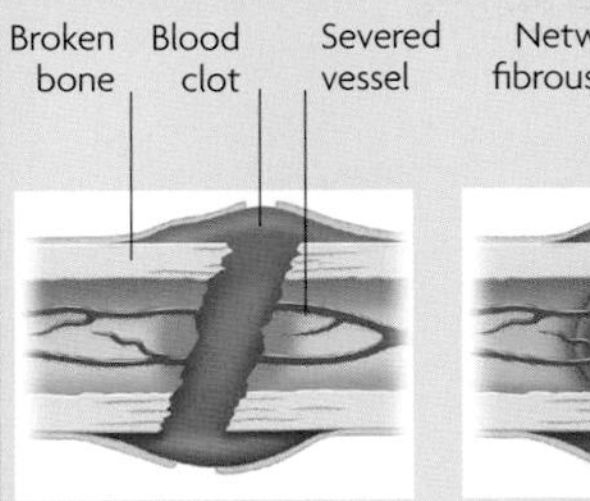
Broken bone
Blood clot
Severed vessel
Network of fibrous tissue

IMMEDIATE RESPONSE
Blood leaks from the blood vessels and clots. White blood cells gather at the area to scavenge damaged cells and debris.

AFTER SEVERAL DAYS
Fibroblast cells construct new fibrous tissue across the break. The limb is immobilized, usually in a plaster cast or splint.

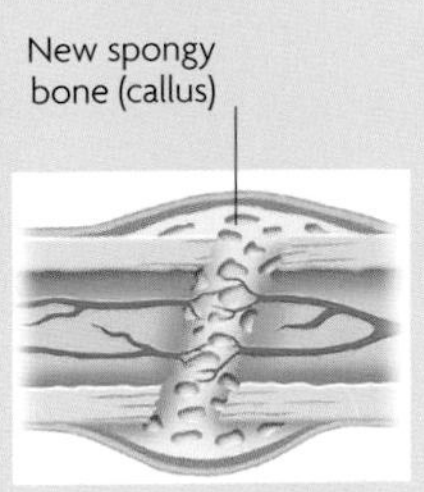
New spongy bone (callus)

AFTER 1–2 WEEKS
Bone-building cells (osteoblasts) multiply and form new bone tissue. Initially spongy, the new tissue infiltrates the site of the fracture as a callus.

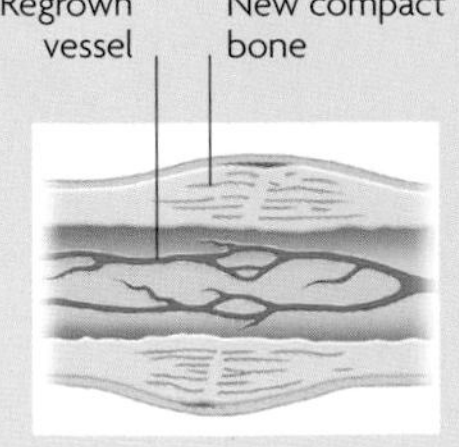
Regrown vessel
New compact bone

AFTER 2–3 MONTHS
Blood vessels reconnect across the fracture. The callus reshapes while the new bone tissue is "remodeled" into dense, compact bone.

DISK PROLAPSE

The cushionlike cartilage disks that separate adjacent vertebrae have a hard outer covering and a jellylike center. An accident, wear and tear, or excessive pressure when lifting awkwardly, may rupture the outer layer. This forces some of the core material to bulge out, or prolapse. The prolapsed (or herniated) portion may cause pressure on the nearby spinal nerve root. Symptoms of disk prolapse include dull pain, muscle spasm and stiffness in the affected area of the back, and pain, tingling, numbness, or weakness in the body part supplied by the nerve.

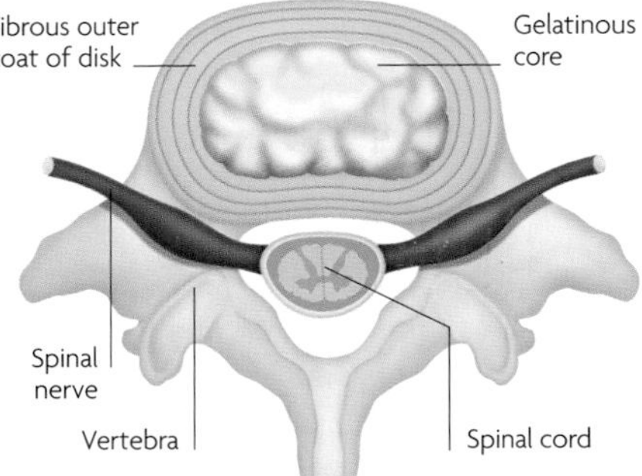
Fibrous outer coat of disk
Gelatinous core
Spinal nerve
Vertebra
Spinal cord

NORMAL DISK
The outer casing (capsule) of the intervertebral disk is intact and encloses its gelatinous core. The disk sits between the bodies (centra) of adjacent vertebrae.

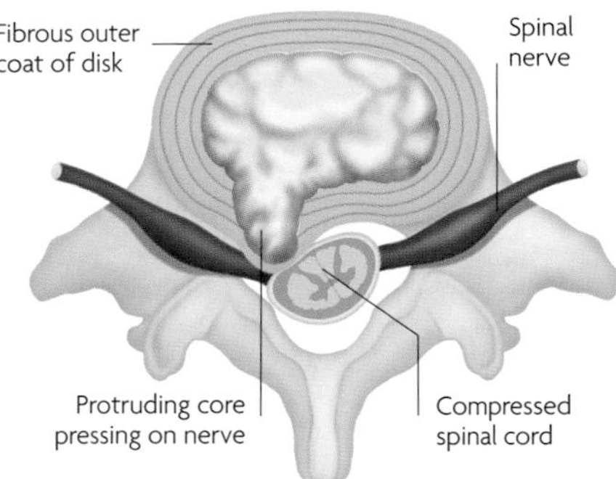
Fibrous outer coat of disk
Spinal nerve
Protruding core pressing on nerve
Compressed spinal cord

PROLAPSED DISK
A weak site in the outer casing allows the gelatinous core to bulge through as the disk is compressed. The resulting pressure on the spinal nerve causes pain.

OSTEOPOROSIS

For bones to stay healthy, bone tissue is continually being broken down and replaced. Sex hormones are essential for this process. With the decline in production of sex hormones in both sexes after middle age, bones become thinner and more porous. Estrogen levels fall rapidly in women after menopause, which can lead to severe thinning, or osteoporosis. The decline in testosterone in men is gradual and, in general, males are less prone to osteoporosis. Exercise is essential for maintaining bone health, and a lack of activity is a predisposing factor to developing osteoporosis. Other factors influencing the development of osteoporosis include smoking, corticosteroid treatment, rheumatoid arthritis, an overactive thyroid, and long-term kidney failure. The decreased density of osteoporotic bones makes them more likely to fracture.

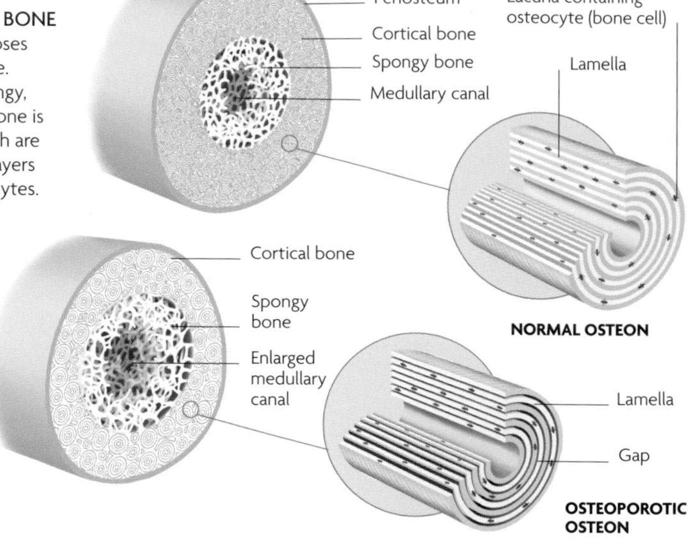
Periosteum
Lacuna containing osteocyte (bone cell)
Cortical bone
Spongy bone
Lamella
Medullary canal
Cortical bone
Spongy bone
Enlarged medullary canal
Lamella
Gap

STRUCTURE OF NORMAL BONE
The outer periosteum encloses a band of hard cortical bone. Within this is a layer of spongy, or cancellous, bone. Hard bone is composed of osteons, which are tightly packed, concentric layers (lamellae) formed by osteocytes.

NORMAL OSTEON

STRUCTURE OF OSTEOPOROTIC BONE
The mineral density (mainly calcium and phosphorus) is reduced from two-thirds to one-third. The medullary canal through the bone's center is enlarged, while gaps between the lamellae contribute to the fragility of the bone

OSTEOPOROTIC OSTEON

WHY OSTEOPOROSIS OCCURS

Bone tissue is built up by the deposition of minerals (mainly calcium salts) on a framework of collagen fibers. It is continually broken down and rebuilt in order to allow growth and repair. Osteoporosis develops when the rate at which fibers, minerals, and cells are broken down becomes much greater than the formation of new tissue.

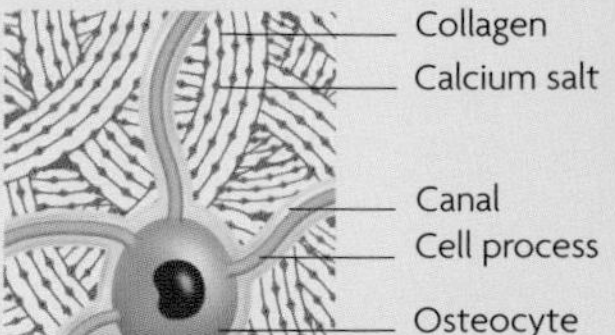
Collagen
Calcium salt
Canal
Cell process
Osteocyte

NORMAL BONE
Osteocytes form collagen fibers and aid calcium deposition. Calcium moves in canals between bone and blood in response to hormones.

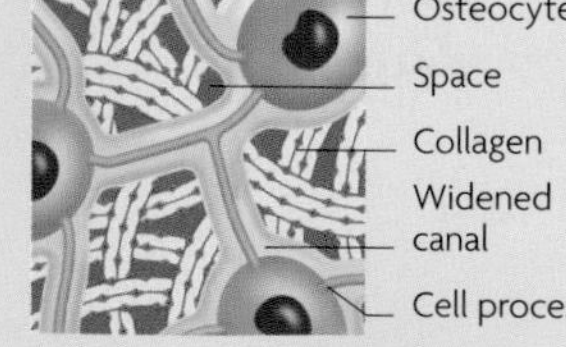
Osteocyte
Space
Collagen
Widened canal
Cell process

OSTEOPOROTIC BONE
In osteoporosis, collagen and minerals are broken down faster than they form. Canals widen, new spaces appear, and bone weakens.

LIGAMENT INJURIES

Ligaments are strong, flexible bands of fibrous tissue that link bone ends together around a joint. If the bones in a joint are pulled too far apart, often as a result of a sudden or forceful movement, the ligament fibers may overstretch or tear. This commonly results in swelling, pain, and muscle spasm. A joint "sprain" is usually due to partial tearing of a ligament. Rest, ice, compression, and elevation of the joint is the usual treatment if a sprain is not serious. A severe injury may result in joint instability or dislocation.

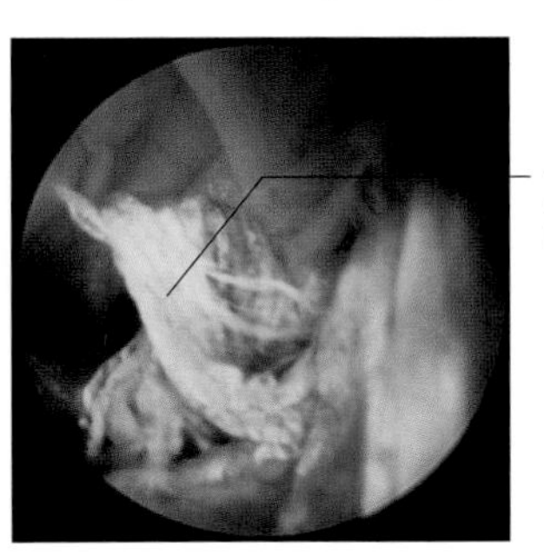
Anterior cruciate ligament

LIGAMENT FIBERS
This view through an arthroscope (a telescope-like tube for seeing into joints) reveals torn fibers of a knee's anterior cruciate ligament.

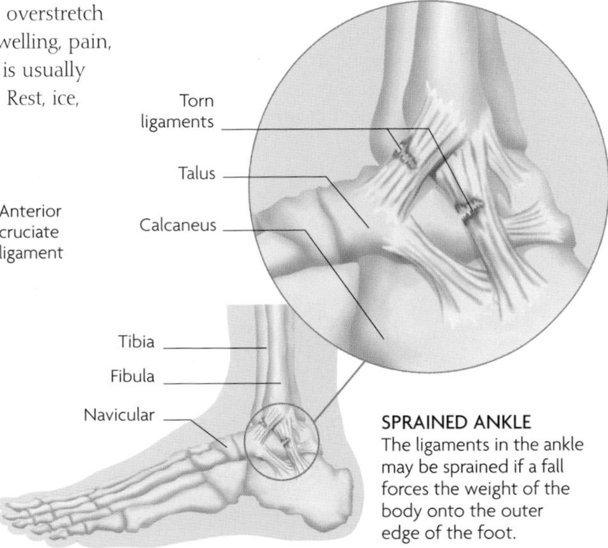
Torn ligaments
Talus
Calcaneus
Tibia
Fibula
Navicular

SPRAINED ANKLE
The ligaments in the ankle may be sprained if a fall forces the weight of the body onto the outer edge of the foot.

TORN CARTILAGE

The knee joint contains padlike, curved "disks" of cartilage called menisci. These are almost C-shaped and made of tough fibrous cartilage. The disks are situated between the lower end of the femur and upper end of the tibia, with the medial meniscus on the knee's inner side and the lateral meniscus on the outside. These disks stabilize the joint, helping it "lock" straight while standing, and cushion the bones. A meniscus may be crushed or torn by rapid twisting of the knee, often while playing a sport. If such an injury is painful, surgery can remove the damaged piece of cartilage.

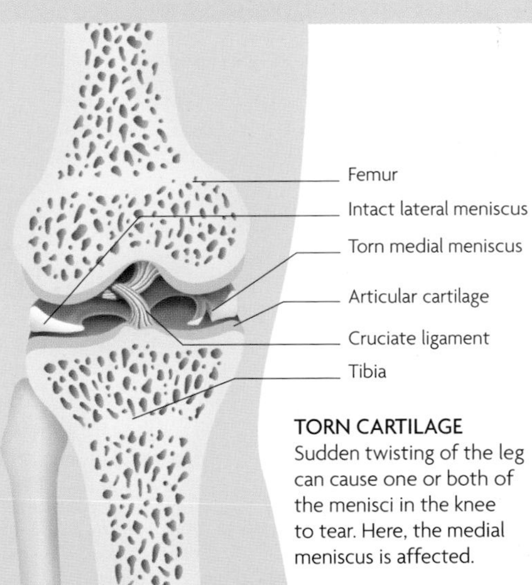
Femur
Intact lateral meniscus
Torn medial meniscus
Articular cartilage
Cruciate ligament
Tibia

TORN CARTILAGE
Sudden twisting of the leg can cause one or both of the menisci in the knee to tear. Here, the medial meniscus is affected.

OSTEOARTHRITIS

Osteoarthritis is often confused with rheumatoid arthritis (see opposite), but the two disorders have different causes and progressions. Osteoarthritis may affect only a single joint and can be triggered by localized wear and tear, resulting in painful inflammation from time to time. Joint degeneration may be hastened by a congenital defect, injury, infection, or obesity. Because cartilage normally wears away as the body ages, a mild form of osteoarthritis affects many people after about the age of 60 years. Typical symptoms of osteoarthritis include pain and swelling in the affected joint that worsen with activity and fade with rest; joint stiffness for a short time after rest; restricted movement of the joint; crepitus (crackling noises) when moving the joint; and referred pain (pain in areas remote from the site of damage but on the same nerve pathway as the affected joint).

OSTEOARTHRITIS OF THE HIP
The right hip, on the left of this X-ray, is badly eroded by osteoarthritis. The head of the femur, which is normally round, is flattened.

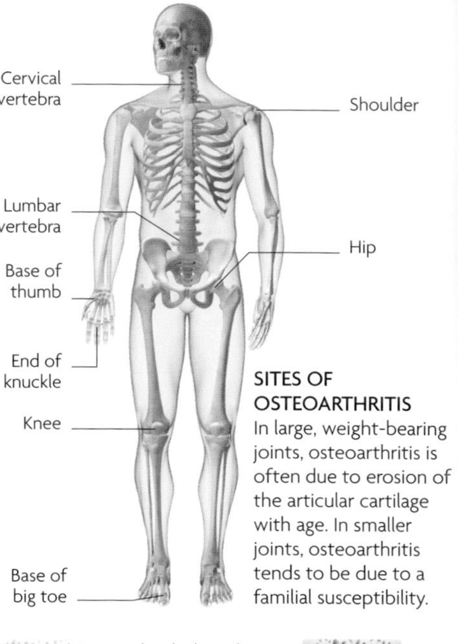
Cervical vertebra
Shoulder
Lumbar vertebra
Hip
Base of thumb
End of knuckle
Knee
Base of big toe

SITES OF OSTEOARTHRITIS
In large, weight-bearing joints, osteoarthritis is often due to erosion of the articular cartilage with age. In smaller joints, osteoarthritis tends to be due to a familial susceptibility.

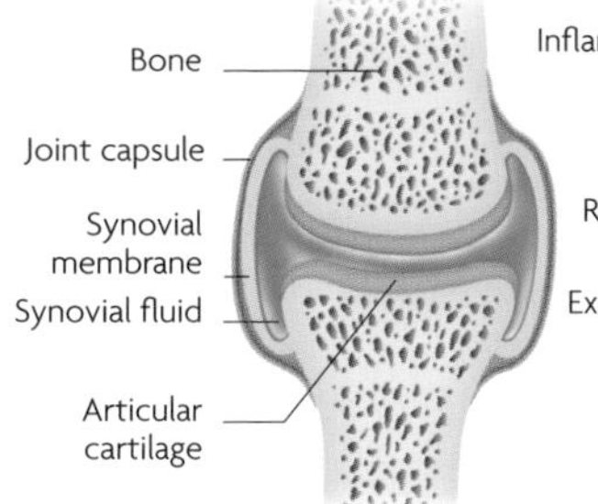
Bone
Joint capsule
Synovial membrane
Synovial fluid
Articular cartilage

HEALTHY JOINT
The articular cartilages coating the ends of the bones are smooth and compressible. They are lubricated by synovial fluid and slip past each other with minimal friction.

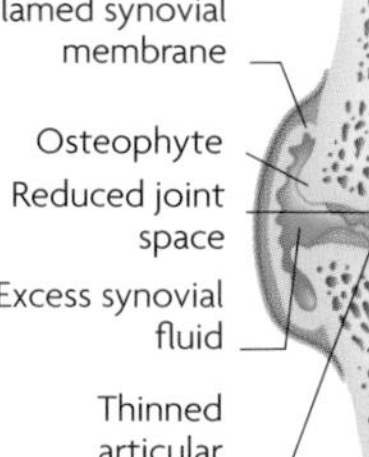
Inflamed synovial membrane
Osteophyte
Reduced joint space
Excess synovial fluid
Thinned articular cartilage

EARLY OSTEOARTHRITIS
The articular cartilage becomes thin and rough, with fissures in its surface. Bony outgrowths (osteophytes) form, and the synovial lining is inflamed, producing excess fluid.

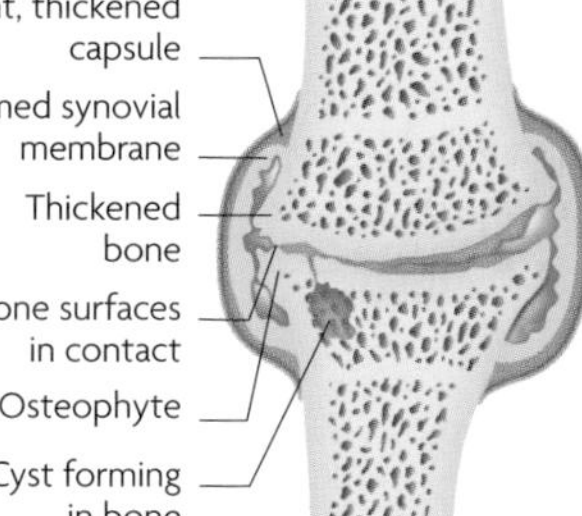
Tight, thickened capsule
Inflamed synovial membrane
Thickened bone
Bone surfaces in contact
Osteophyte
Cyst forming in bone

LATE OSTEOARTHRITIS
The articular cartilage and underlying bone crack and erode. The bones rub together, thicken, and overgrow, causing extreme discomfort. The joint capsule thickens.

RHEUMATOID ARTHRITIS

Rheumatoid arthritis develops when the immune system produces antibodies that attack its own body tissues–especially the synovial membranes inside joints. The joints become swollen and deformed, with painful and restricted movement. Early general symptoms include fever, pale skin, and weakness. Characteristically, many of the small joints are affected in a symmetrical pattern; for example, the hands and feet may become inflamed to the same degree on both sides. Painless small lumps or nodules form in areas of pressure, and the skin over the joint is thin and fragile. Stiffness is often worse in the mornings but eases during the day. The condition may flare up, then fade for a time. The diagnosis is supported if a blood test detects an antibody, rheumatoid factor (RhF), associated with rheumatoid arthritis. The disease can also affect the eyes, skin, heart, nerves, and lungs.

JOINT INFLAMMATION
In this X-ray, the middle knuckles of the hands are severely damaged by rheumatoid arthritis (red). Inflammation of the joints causes abnormal bending of the fingers.

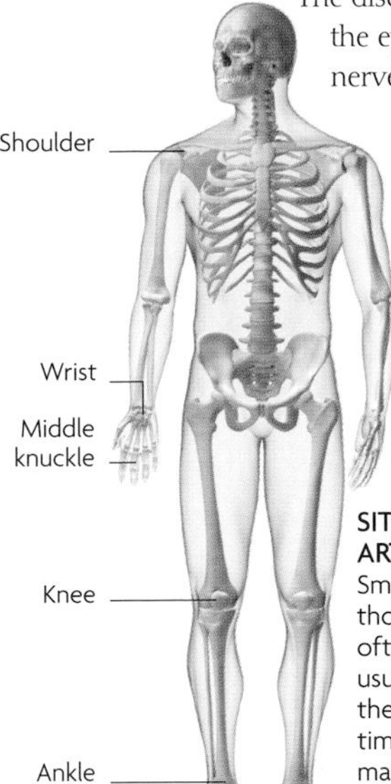
Shoulder
Wrist
Middle knuckle
Knee
Ankle
Toe

SITES OF RHEUMATOID ARTHRITIS
Smaller joints, especially those in the hands, are often affected first, usually on both sides of the body at the same time. The inflammation may then transfer or "flit" to other, larger joints, such as the wrist.

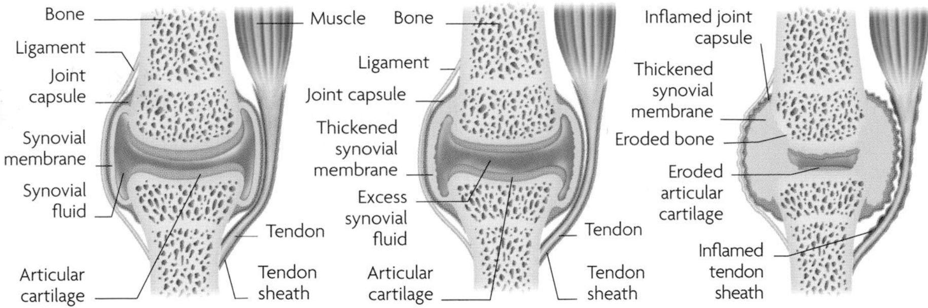
Bone
Ligament
Joint capsule
Synovial membrane
Synovial fluid
Articular cartilage
Muscle
Tendon
Tendon sheath
Bone
Ligament
Joint capsule
Thickened synovial membrane
Excess synovial fluid
Articular cartilage
Tendon
Tendon sheath
Inflamed joint capsule
Thickened synovial membrane
Eroded bone
Eroded articular cartilage
Inflamed tendon sheath

HEALTHY JOINT
Cartilage is smooth and intact in a healthy joint. Ligaments aid stability, and tendons slide in sheaths as muscles pull on them.

EARLY RHEUMATOID ARTHRITIS
The synovial membrane becomes inflamed and thickens, spreading across the joint. Excess synovial fluid accumulates in the joint.

LATE RHEUMATOID ARTHRITIS
As the synovial membrane thickens, the cartilage and bone ends are eroded. The joint capsule and tendon sheath become inflamed.

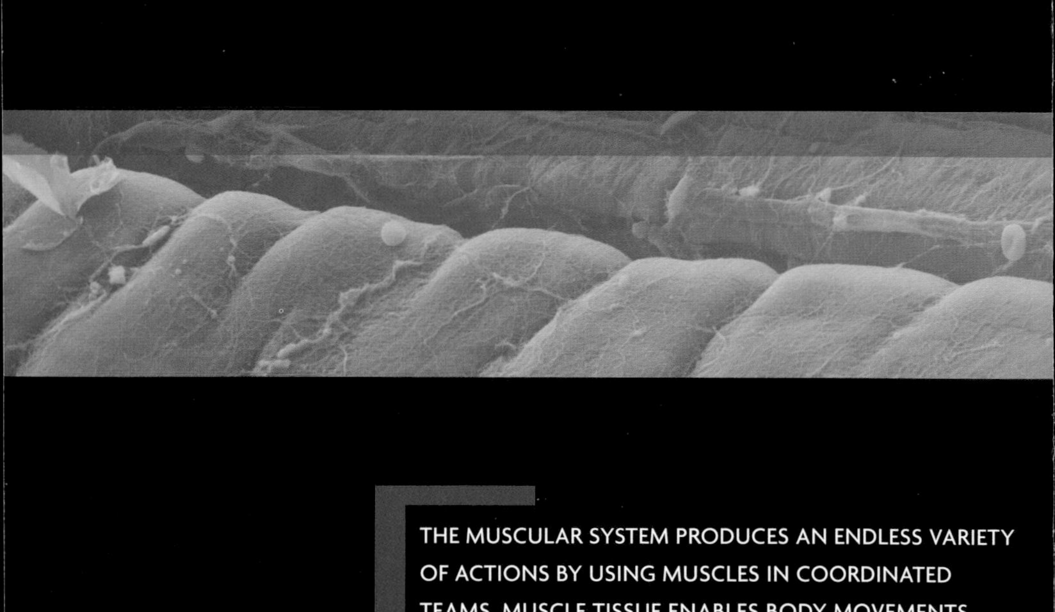

THE MUSCULAR SYSTEM PRODUCES AN ENDLESS VARIETY OF ACTIONS BY USING MUSCLES IN COORDINATED TEAMS. MUSCLE TISSUE ENABLES BODY MOVEMENTS AND ALSO POWERS INTERNAL PROCESSES, FROM THE

MUSCULAR SYSTEM

MUSCLES OF THE BODY

THE MUSCLES ARE RESPONSIBLE FOR MOVEMENT. THEY ACHIEVE THIS BY CONTRACTING AND PULLING ON THE BONES TO WHICH THEY ARE ATTACHED.

The typical male body contains approximately 640 muscles, which compose about two-fifths of its total weight. The same number in a female body make up a slightly smaller proportion. A typical muscle spans a joint and tapers at each end into a fibrous tendon anchored to a bone. Some muscles divide to attach to different bones. Superficial muscles, those located just beneath the skin, are pictured here on the left side of a male body. On the right side of this body are the deeper layers–the intermediate and deep muscles.

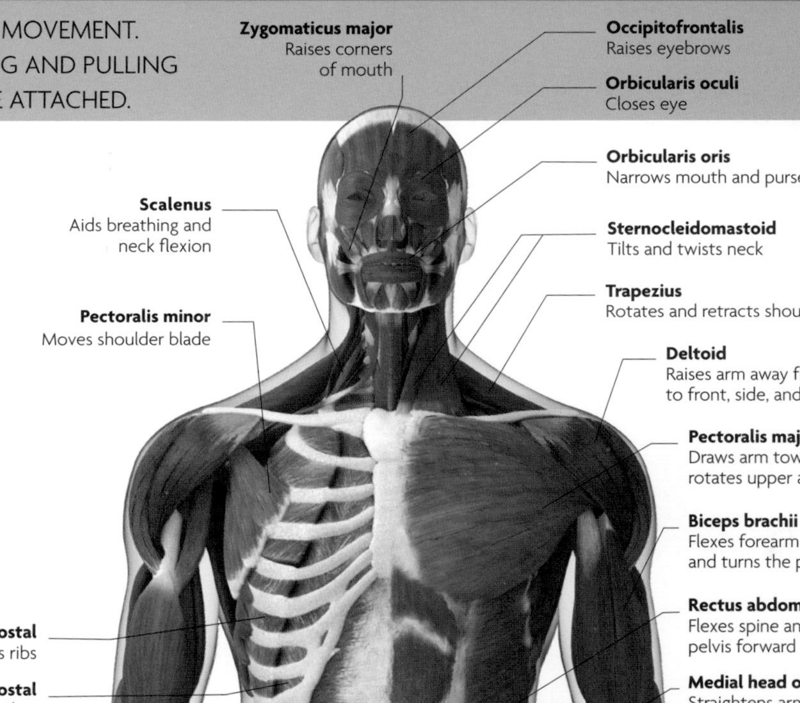
Zygomaticus major
Raises corners of mouth
Scalenus
Aids breathing and neck flexion
Pectoralis minor
Moves shoulder blade
External intercostal
Elevates ribs
Internal intercostal
Pulls adjacent ribs together
Internal oblique abdominal
Flexes and rotates trunk
Linea alba
Tendinous structure dividing left and right abdominal muscles
Inguinal ligament
Occipitofrontalis
Raises eyebrows
Orbicularis oculi
Closes eye
Orbicularis oris
Narrows mouth and purses lips
Sternocleidomastoid
Tilts and twists neck
Trapezius
Rotates and retracts shoulder blade
Deltoid
Raises arm away from body to front, side, and rear
Pectoralis major
Draws arm toward body; rotates upper arm inward
Biceps brachii
Flexes forearm at elbow and turns the palm upward
Rectus abdominis
Flexes spine and draws pelvis forward
Medial head of triceps
Straightens arm at elbow
External oblique abdominal
Flexes and rotates trunk
Brachioradialis
Flexes arm at elbow

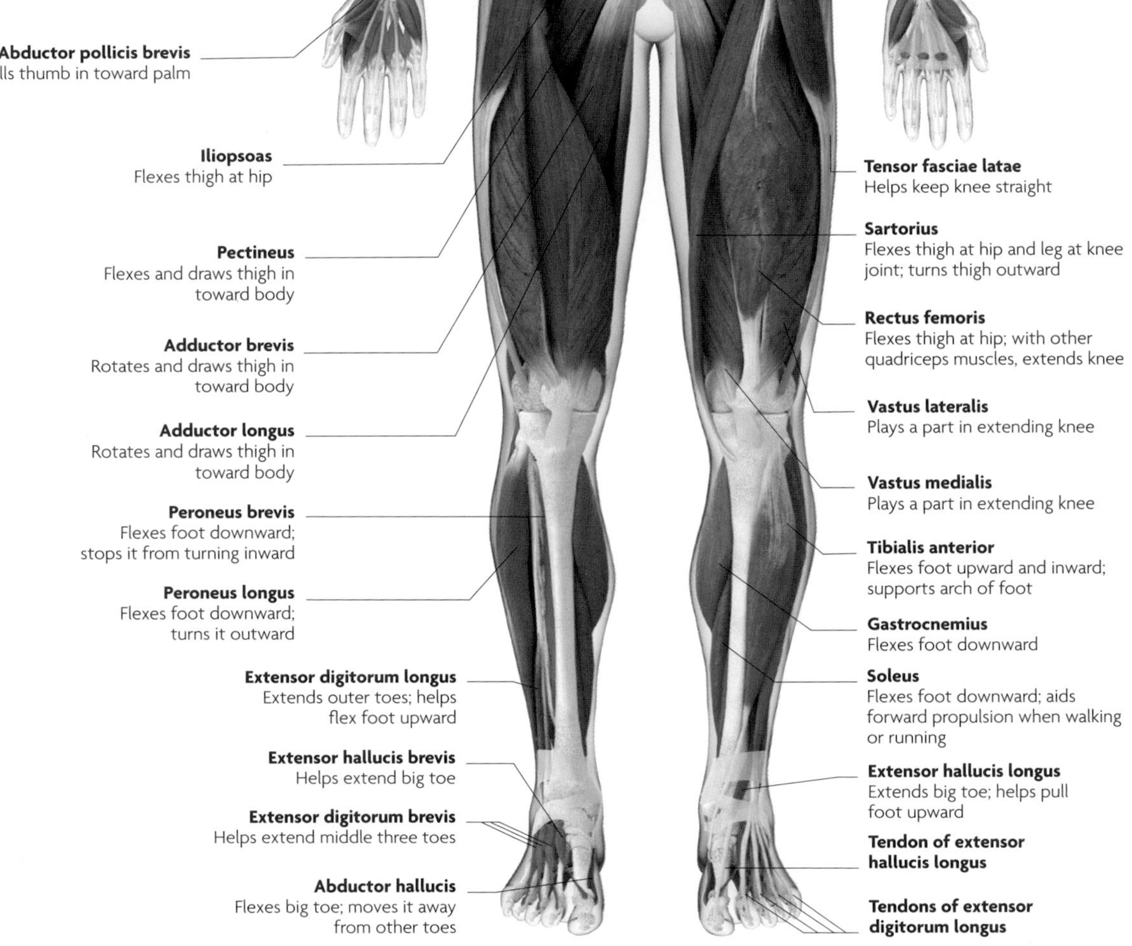

Abductor pollicis brevis
Pulls thumb in toward palm
Iliopsoas
Flexes thigh at hip
Pectineus
Flexes and draws thigh in toward body
Adductor brevis
Rotates and draws thigh in toward body
Adductor longus
Rotates and draws thigh in toward body
Peroneus brevis
Flexes foot downward; stops it from turning inward
Peroneus longus
Flexes foot downward; turns it outward
Extensor digitorum longus
Extends outer toes; helps flex foot upward
Extensor hallucis brevis
Helps extend big toe
Extensor digitorum brevis
Helps extend middle three toes
Abductor hallucis
Flexes big toe; moves it away from other toes
Tensor fasciae latae
Helps keep knee straight
Sartorius
Flexes thigh at hip and leg at knee joint; turns thigh outward
Rectus femoris
Flexes thigh at hip; with other quadriceps muscles, extends knee
Vastus lateralis
Plays a part in extending knee
Vastus medialis
Plays a part in extending knee
Tibialis anterior
Flexes foot upward and inward; supports arch of foot
Gastrocnemius
Flexes foot downward
Soleus
Flexes foot downward; aids forward propulsion when walking or running
Extensor hallucis longus
Extends big toe; helps pull foot upward
Tendon of extensor hallucis longus
Tendons of extensor digitorum longus

Some muscles are named for their shapes (e.g. the rhomboideus major), whereas others are named according to their bone attachments (e.g. the intercostals between the ribs, or costae) or the movement they produce. Flexor muscles bend a limb at the joint they span, while their opposing equivalents, the extensors, straighten the joint. Abductor muscles cause movement away from the midline of the body, as when holding the arm out to the side. Their adductor partners are responsible for the reverse movement back toward the midline. In this illustration the superficial muscles are on the right side, with deeper ones to the left.

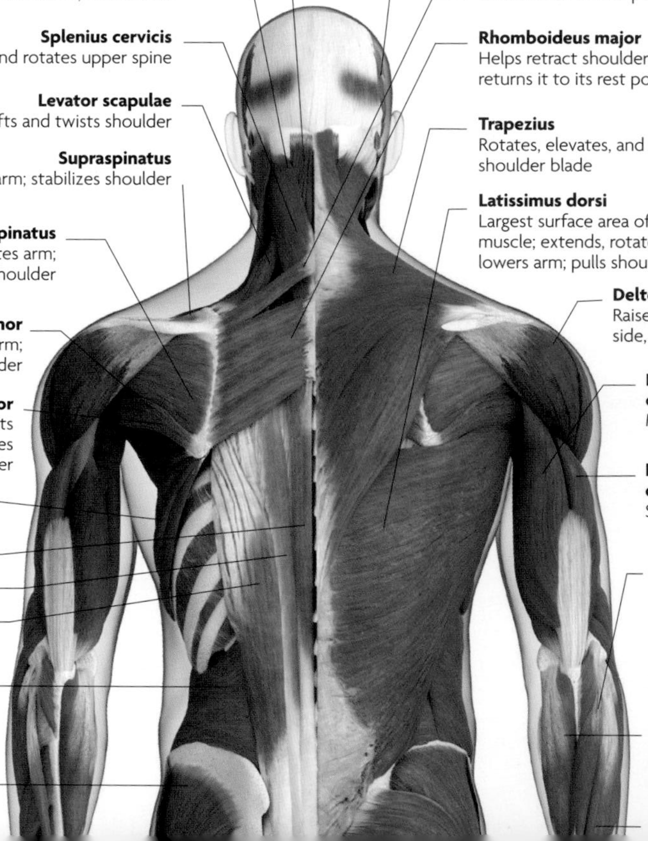

Semispinalis capitis
Extends head and neck; flexes them from side to side

Splenius capitis
Moves head; twists neck

Splenius cervicis
Flexes and rotates upper spine

Levator scapulae
Lifts and twists shoulder

Supraspinatus
Raises arm; stabilizes shoulder

Infraspinatus
Rotates arm; stabilizes shoulder

Teres minor
Lifts and twists arm; stabilizes shoulder

Teres major
Lifts and twists arm, stabilizes shoulder

Serratus anterior
Rotates and extends shoulder blade

Erector spinae
Elevates and straightens spine

Spinalis

Longissimus

Iliocostalis

Internal oblique abdominal
Supports abdominal wall; assists forced breathing; aids raising intra-abdominal pressure; helps flex and rotate trunk

Gluteus minimus
Raises thigh away from body at hip, rotates thigh; when walking, tilts pelvis

Temporoparietalis (auricularis)
Wiggles ears

Rhomboideus minor
Helps retract shoulder blade and returns it to its rest position

Rhomboideus major
Helps retract shoulder blade and returns it to its rest position

Trapezius
Rotates, elevates, and retracts shoulder blade

Latissimus dorsi
Largest surface area of any body muscle; extends, rotates, and lowers arm; pulls shoulder back

Deltoid
Raises arm to front, side, and rear

Long head of triceps
Moves arm down

Lateral head of triceps
Straightens elbow

Extensor carpi ulnaris
Pulls wrist toward body

Flexor carpi ulnaris
Pulls wrist toward body

Extensor digitorum
Extends all finger joints

MUSCLE TISSUE

There are three main types of muscle tissue. What we usually think of as “muscles” are skeletal muscles. Also called voluntary or striated muscles, most are joined to bones and produce bodily movements under conscious control. Smooth muscles, also called involuntary muscles because they are not under conscious control, occur in the walls of body parts such as the airways and blood vessels. Cardiac muscle forms the walls of the heart.

SKELETAL
A microscope view shows pronounced stripes, bands, or striations, created by the alignment of muscle fibrils.

SMOOTH
The light microscope reveals few features: principally tapered muscle cells with dark nuclei.

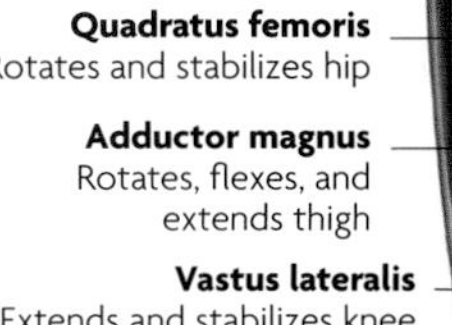

CARDIAC
Fibers in heart (cardiac) muscle are short and branching, often Y- or V-shaped, with faint banding or striations.

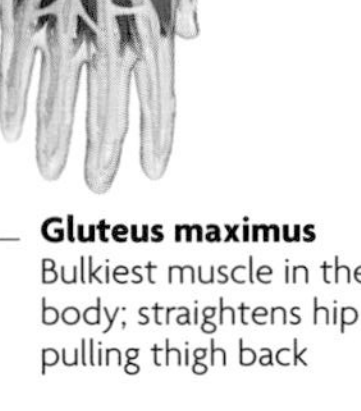

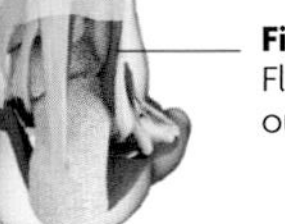

Quadratus femoris
Rotates and stabilizes hip
Adductor magnus
Rotates, flexes, and extends thigh
Vastus lateralis
Extends and stabilizes knee
Gracilis
Moves thigh towards body; flexes and rotates thigh
Plantaris
Assists in knee flexion
Popliteus
Flexes and turns leg to unlock extended knee
Tibialis posterior
Main muscle in turning foot inward
Flexor digitorum longus
Flexes and turns in foot; flexes toes
Flexor hallucis longus
The “push-off” muscle in walking
Fibularis longus
Flexes and turns foot outward
Abductor digiti minimi
Moves little toe outward
Gluteus maximus
Bulkiest muscle in the body; straightens hip by pulling thigh back
Biceps femoris
Extends thigh at hip; flexes knee; rotates leg
Semitendinosus
Extends thigh at hip; flexes knee; rotates leg
Semimembranosus
Extends thigh; flexes knee; rotates leg
Hamstrings
Gastrocnemius
Main calf muscle; flexes ankle and pulls up heel; flexes knee
Soleus
Flexes foot; important during running and walking
Achilles (calcaneal) tendon
Fibularis brevis
Flexes and turns foot outward

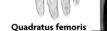

MUSCLES OF THE FACE, HEAD, AND NECK

THE MUSCLES OF THE FACE, HEAD, AND NECK INTERACT TO STEADY AND MOVE THE HEAD AND TO MOVE THE FACIAL FEATURES. THE MUSCULATURE INVOLVED IS HIGHLY COMPLEX, MAKING POSSIBLE A HUGE RANGE OF FACIAL EXPRESSIONS.

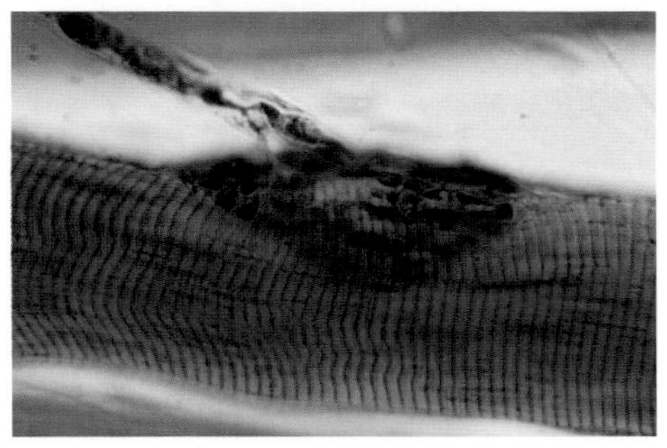

NERVE–MUSCLE JUNCTION
In this microscope image, a nerve cell (top left) joins a facial muscle fiber. At the point of contact between the two is the motor end plate (center), an area of highly excitable muscle fiber.

FACIAL MUSCLES

Some facial muscles are anchored to bones. Others are joined to tendons or to dense, sheetlike clusters of fibrous connective tissue called aponeuroses. This means that some facial muscles are joined to each other. Many of these muscles have their other end inserted into deeper layers of the skin. The advantage of this complex system is that even a slight degree of muscle contraction produces movement of the facial skin, which reveals itself as a show of expression or emotion. Almost all facial muscles are controlled by the facial nerve called cranial VII (see p.102).

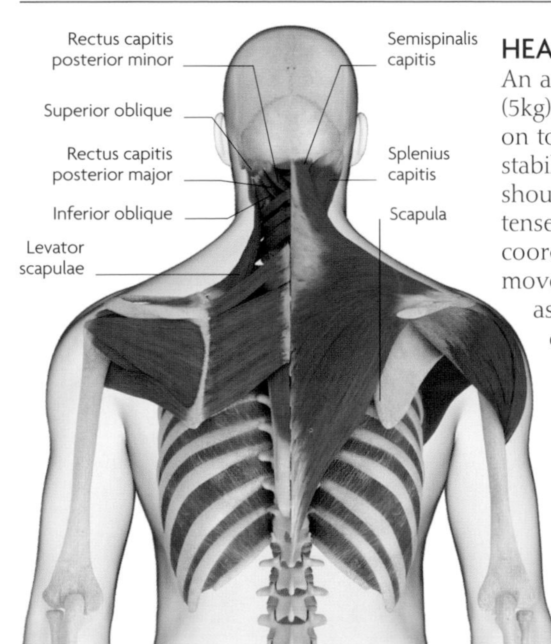
Rectus capitis posterior minor
Superior oblique
Rectus capitis posterior major
Inferior oblique
Levator scapulae
Semispinalis capitis
Splenius capitis
Scapula

HEAD AND NECK MUSCLES

An adult's head weighs more than 11lb (5kg) and is, to some extent, "balanced" on top of the vertebral column. Strong, stabilizing muscles in the neck, inner shoulders, and upper back constantly tense to steady the head and contract in coordinated teams to produce complex movements of the neck. These muscles assist facial expressions and nonverbal communication, such as emphasizing doubt by cocking the head slightly to one side, or moving the head to indicate "yes" or "no."

BACK MUSCLES
The neck and shoulder muscles support and steady the head. Upper-back muscles that attach to the shoulder blade (scapula) help stabilize the shoulders.

FACE AND NECK MUSCLES

Intermeshing muscles around the lips are involved in speech, nonverbal expression, eating, and drinking. Some facial muscles act as sphincters to open and close orifices, such as the eyelids, nostrils, and lips.

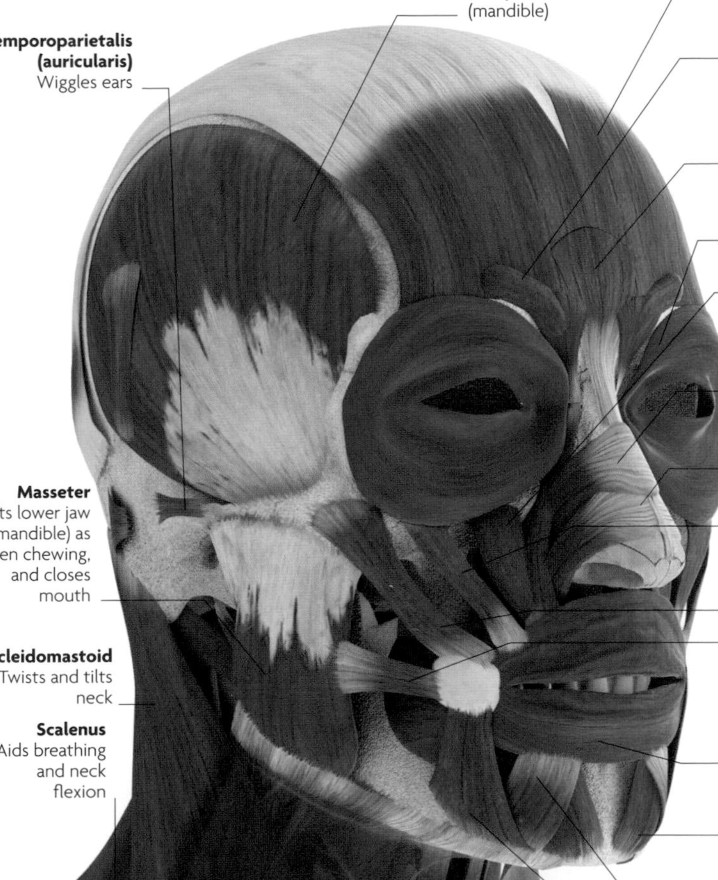
Temporalis
Lifts jawbone (mandible)
Temporoparietalis (auricularis)
Wiggles ears
Masseter
Lifts lower jaw (mandible) as when chewing, and closes mouth
rnocleidomastoid
Twists and tilts neck
Scalenus
Aids breathing and neck flexion
Occipitofrontalis
Raises eyebrows
Corrugator supercilii
Pulls eyebrows together and wrinkles lower forehead
Procerus
Pulls eyebrows down and together
Orbicularis oculi
Closes eyelid
Levator labii superioris
Raises and pushes out upper lip
Compressor naris
Closes nostrils
Dilator naris
Opens and flares nostrils
Zygomaticus minor
Raises upper lip
Zygomaticus major
Pulls corner of mouth up and out
Risorius
Pulls corner of mouth outward
Orbicularis oris
Narrows mouth and purses lips
Mentalis
Raises lower lip and wrinkles chin
Depressor labii inferioris
Pulls down lower lip
Depressor anguli oris
Lowers corner of mouth
Sternohyoid
Depresses larynx

MUSCLES AND TENDONS

MUSCLES CAN ONLY CONTRACT AND SHORTEN. TO RETURN TO THEIR ORIGINAL SHAPE, THEY RELAX AND LENGTHEN PASSIVELY AS OTHER MUSCLES CONTRACT. THE CONTRACTION OF SKELETAL MUSCLES AND TENDONS PRODUCES BODY MOVEMENTS.

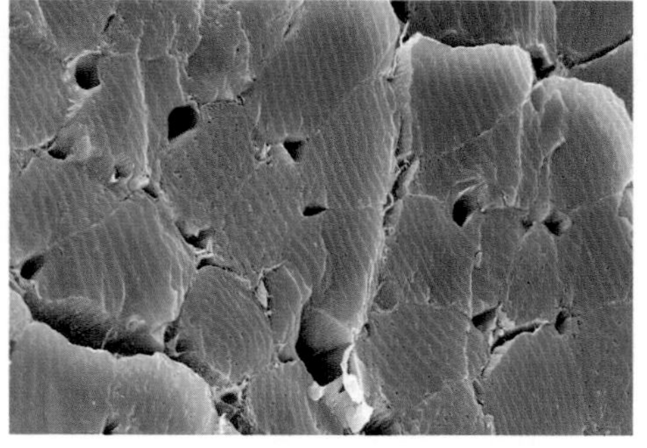

STRIATED MUSCLE
This electron micrograph shows a crosssection through skeletal muscle. The bundles of myofibers are interspersed with capillaries (dark areas).

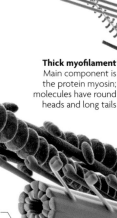

MUSCLE STRUCTURE

Skeletal (striated or voluntary) muscle consists of densely packed groups of hugely elongated cells called myofibers. These are grouped into bundles (fascicles). A typical myofiber is $\frac{3}{4}$–$1\frac{1}{5}$in (2–3cm) long and $\frac{1}{500}$in (0.05mm) in diameter and is composed of narrower structures called myofibrils. These contain thick and thin myofilaments made up mainly of the proteins actin and myosin. Numerous capillaries keep the muscle supplied with the oxygen and glucose needed to fuel contraction.

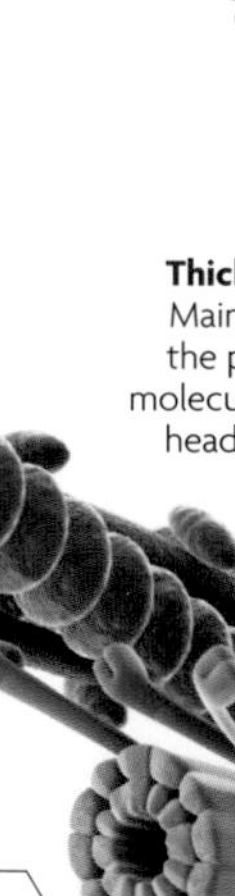
Z band
Where the contractile units (sarcomeres) join end to end
Capillary
M band
Connects neighbouring strands of myosin
Thick myofilament
Main component is the protein myosin; molecules have round heads and long tails
Actin
Thin myofilament
Consists of twisted strands of actin and tropomyosin (protein that inhibits contraction), plus occasional troponin complexes
Tropomyosin
Tail of myosin molecule
Head of myosin molecule

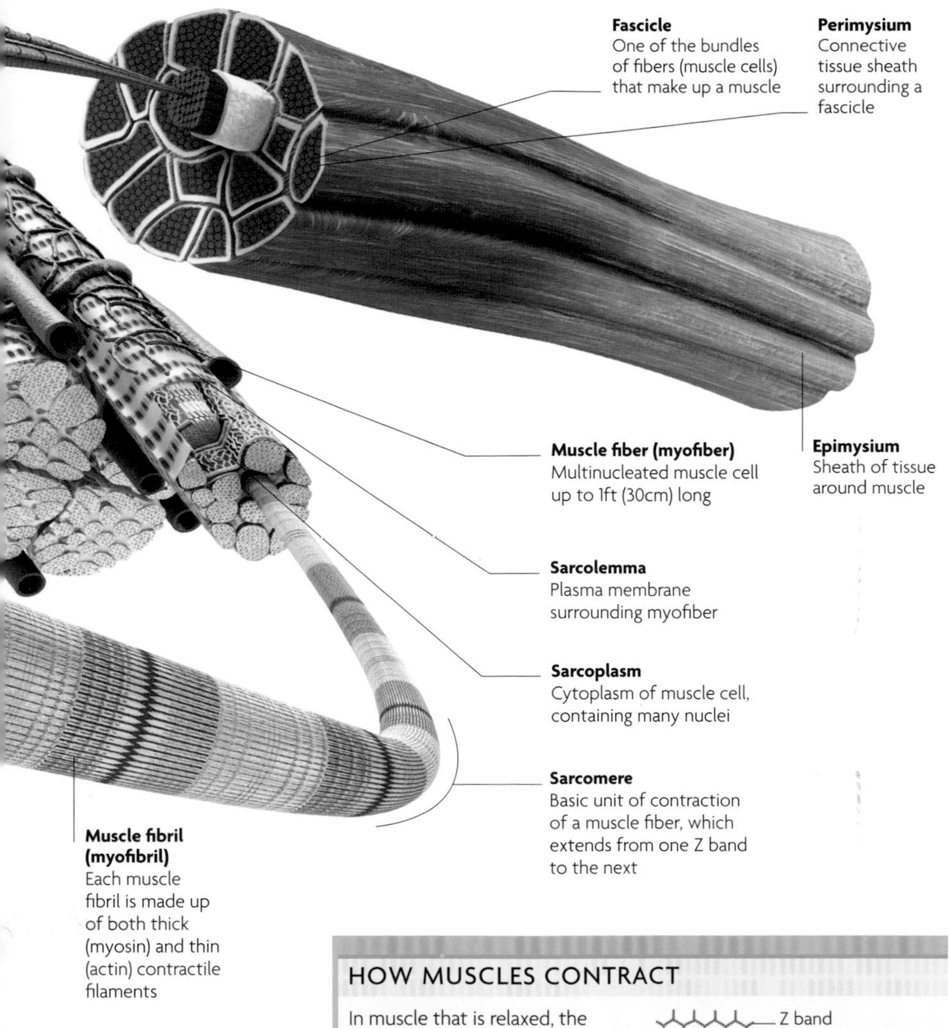

Fascicle
One of the bundles of fibers (muscle cells) that make up a muscle

Perimysium
Connective tissue sheath surrounding a fascicle

Muscle fiber (myofiber)
Multinucleated muscle cell up to 1ft (30cm) long

Epimysium
Sheath of tissue around muscle

Sarcolemma
Plasma membrane surrounding myofiber

Sarcoplasm
Cytoplasm of muscle cell, containing many nuclei

Sarcomere
Basic unit of contraction of a muscle fiber, which extends from one Z band to the next

Muscle fibril (myofibril)
Each muscle fibril is made up of both thick (myosin) and thin (actin) contractile filaments

HOW MUSCLES CONTRACT

In muscle that is relaxed, the myofilaments only partly overlap. When a muscle contracts, the myosin filaments slide between the actin filaments, shortening the myofibrils and the entire muscle fiber. The more shortened muscle fibers there are, the greater the contraction in the muscle as a whole.

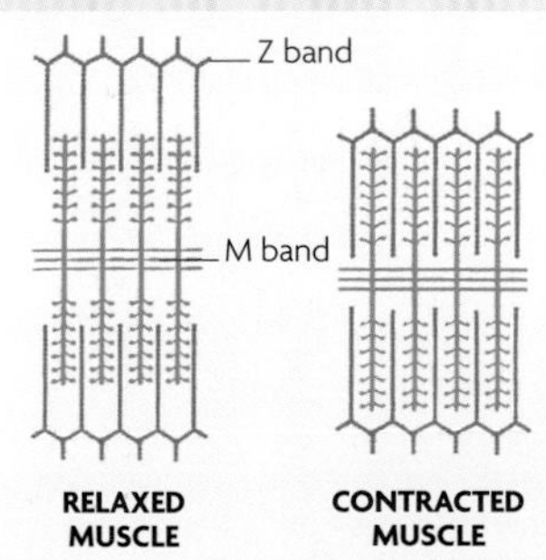
Z band
M band
RELAXED MUSCLE
CONTRACTED MUSCLE

BODY PARTS AS LEVERS

Body movements employ the mechanical principles of applying a force to one part of a rigid lever, which tilts at a pivot point (fulcrum) to move a weight (load) that is elsewhere on the lever. The muscles apply force, bones serve as levers, and joints function as fulcrums. The various lever systems in the body allow a wide range of movement as well as making it possible to lift and carry things.

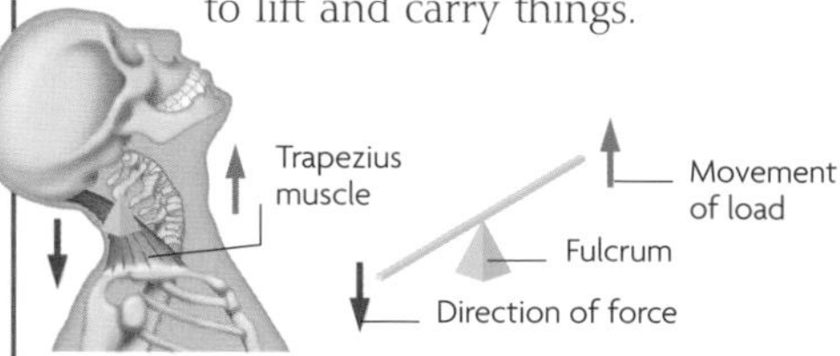
Trapezius muscle
Movement of load
Fulcrum
Direction of force

FIRST-CLASS LEVER

The fulcrum is positioned between the force and the load, like a see-saw. An example of this type of lever in the body is seen in the posterior neck muscles that tilt back the head on the cervical vertebrae.

SECOND-CLASS LEVER

The load lies between the force and the fulcrum. Standing on tip-toe, the calf muscles provide the force, the heel and foot form the lever, and the toes provide the fulcrum.

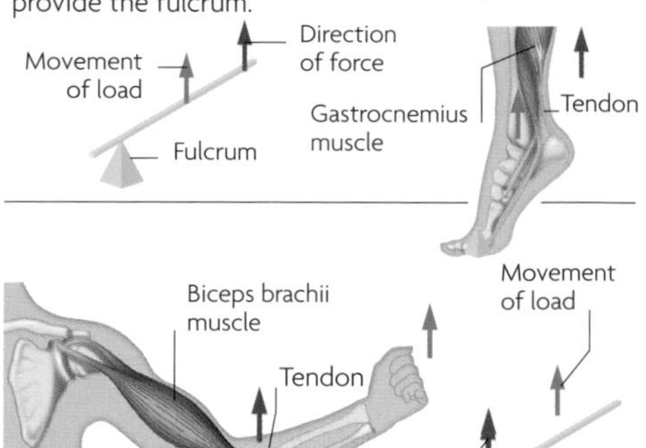
Direction of force
Movement of load
Gastrocnemius muscle
Tendon
Fulcrum
Biceps brachii muscle
Movement of load
Tendon
Direction of force
Fulcrum

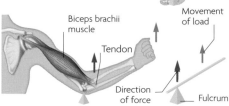

THIRD-CLASS LEVER

The most common type of lever in the body; the force is applied between the load and the fulcrum. An example is flexing the elbow joint (the fulcrum) by contracting the biceps brachii muscle.

TENDONS

Tendons are tough, fibrous cords of connective tissue that link skeletal muscles to bones. Within them, Sharpey's fibers pass through the bone covering (periosteum) to embed in the bone. Tendons in the hands and feet are enclosed in self-lubricating sheaths to protect them from rubbing against the bones. From the hand bones, tendons extend up to muscles near the elbow.

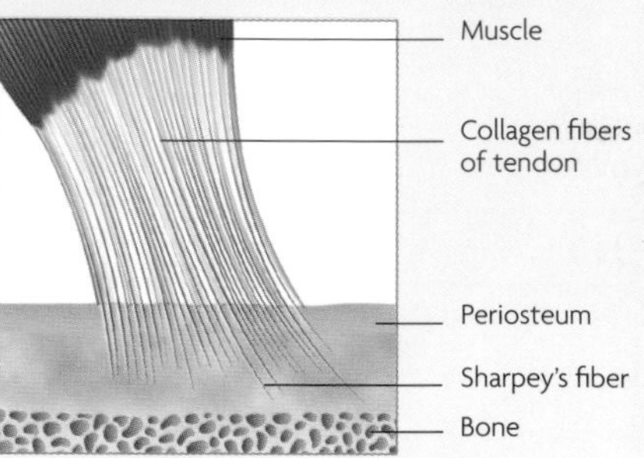
Muscle
Collagen fibers of tendon
Periosteum
Sharpey's fiber
Bone

BONE–TENDON ATTACHMENT

Sharpey's fibers, which are also known as perforating fibers, are extensions of the tendon's proteinaceous collagen fibers.

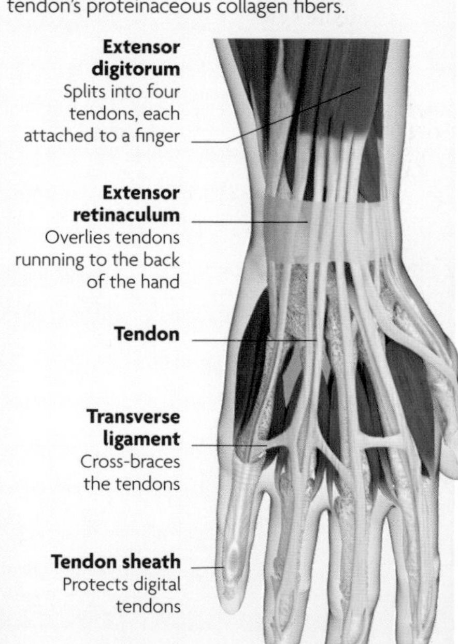
Extensor digitorum
Splits into four tendons, each attached to a finger
Extensor retinaculum
Overlies tendons runnning to the back of the hand
Tendon
Transverse ligament
Cross-braces the tendons
Tendon sheath
Protects digital tendons

HOW MUSCLES WORK TOGETHER

Muscles can only pull, not push, and so are arranged in pairs that act in opposition to one other. The movement produced by one muscle can be reversed by its opposing partner. When a muscle contracts to produce movement, it is called the agonist, while its opposite partner, called the antagonist, relaxes and is passively stretched. In reality, few movements are achieved by a single muscle contraction. Usually, whole teams of muscles act as agonists to give the precisely required degree and direction of motion, while the antagonists tense to prevent the movement overextending.

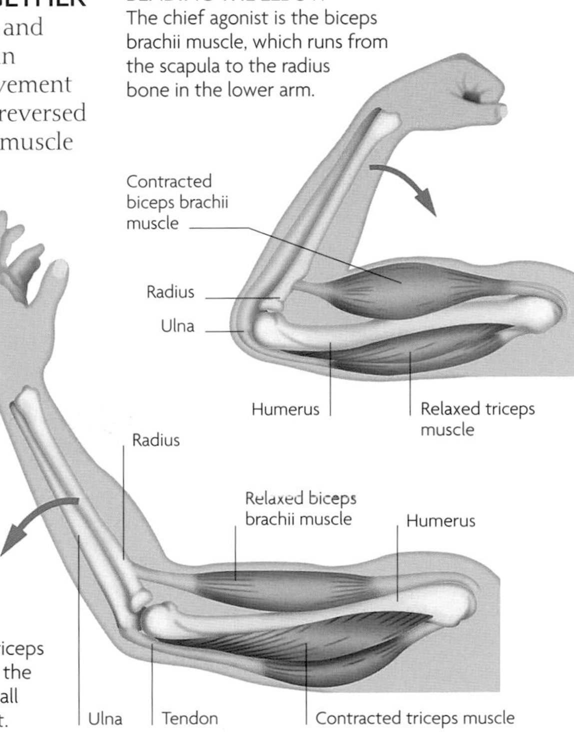

Contracted biceps brachii muscle
Radius
Ulna
Humerus
Relaxed triceps muscle
Radius
Relaxed biceps brachii muscle
Humerus
Ulna
Tendon
Contracted triceps muscle

BENDING THE ELBOW
The chief agonist is the biceps brachii muscle, which runs from the scapula to the radius bone in the lower arm.

STRAIGHTENING THE ELBOW
The biceps brachii relaxes and the triceps brachii, attached at its lower end to the ulna, contracts. It is aided by the small anconeus muscle on the elbow joint.

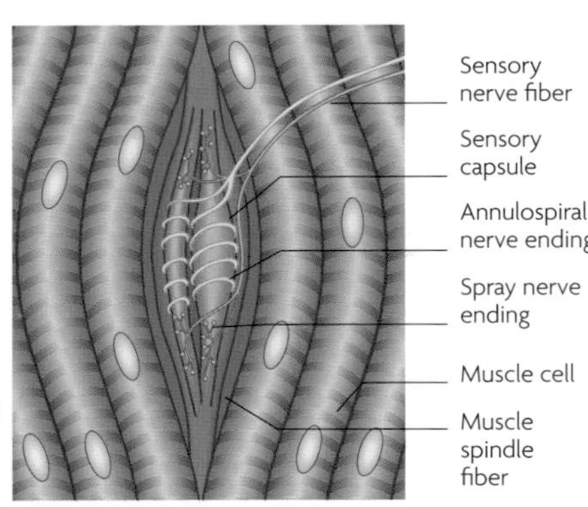

Sensory nerve fiber
Sensory capsule
Annulospiral nerve ending
Spray nerve ending
Muscle cell
Muscle spindle fiber

NEUROMUSCULAR SPINDLE
The effect of motor signals sent to the spindle's muscle fibers is fed back to the brain by the sensory nerve fibers, thereby allowing the brain to gauge the muscle's tension and elongation.

POSITIONAL SENSE

Muscles contain many tiny sensors, known as neuromuscular spindles. These are modified muscle fibers with a spindle-shaped sheath or capsule and several types of nerve supply. The sensory or afferent nerve fibers, which are wrapped around the modified muscle fibers, relay information to the brain about muscle length and tension as the muscle stretches. The motor neurons stimulate the opposite reaction, causing the muscle to contract and shorten, and restoring muscle tension to normal. Similar receptors are found in ligaments and tendons. Together they provide the body's innate sense of its own position and posture, called proprioception.

MUSCLE AND TENDON DISORDERS

MUSCLES AND TENDONS MAY BE DAMAGED AS A RESULT OF PHYSICAL EXERTION DURING DAILY ACTIVITIES, FROM SUDDEN PULLING OR TWISTING MOVEMENTS SUCH AS THOSE OCCURRING IN SPORTS OR AN ACCIDENT, OR FROM REPETITIVE ACTIONS, FOR EXAMPLE, DUE TO EMPLOYMENT.

MUSCLE STRAINS AND TEARS

Muscle strain is the term used for a moderate amount of soft-tissue damage to muscle fibers, which is usually caused by sudden, strenuous movements. Limited bleeding inside the muscle causes tenderness and swelling, which may be accompanied by painful spasms or contractions. Visible bruising may follow. More serious damage, involving a larger number of torn or ruptured fibers, is called a muscle tear. A torn muscle produces severe pain and swelling. Following a medical check to gauge the severity of the injury, the usual treatment is rest, anti inflammatory medication, and sometimes physical therapy. Rarely, surgery may be needed if a muscle has been badly torn.

TORN HAMSTRING

The hamstring muscles (rear of the thigh) may be torn by vigorous movements, such as the rapid acceleration common in athletics.

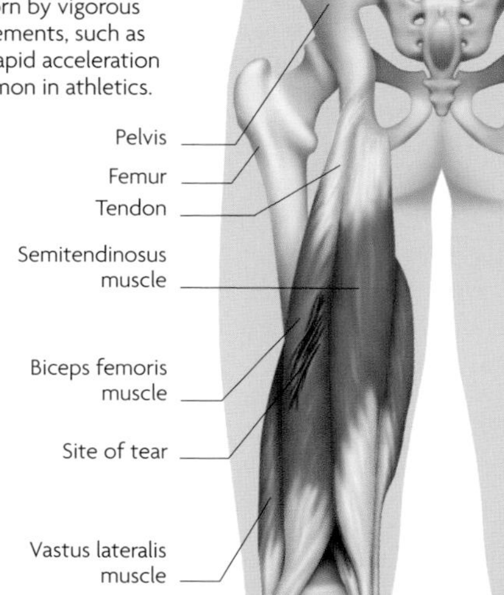
Pelvis
Femur
Tendon
Semitendinosus muscle
Biceps femoris muscle
Site of tear
Vastus lateralis muscle

TENDINITIS AND TENOSYNOVITIS

Tendinitis may occur when strong or repeated movement creates excessive friction between the tendon's outer surface and an adjacent bone. Tenosynovitis may be the result of overstretching or repeated movement causing inflammation of the lubricating sheaths that enclose some tendons. Both of these problems can occur together and may be part of the group of disorders known collectively as repetitive strain injuries (RSIs).

TENDINITIS

Repeated arm-lifting, such as in racquet sports, may force the supraspinous tendon to rub against the shoulder blade's acromion process, causing tendinitis.

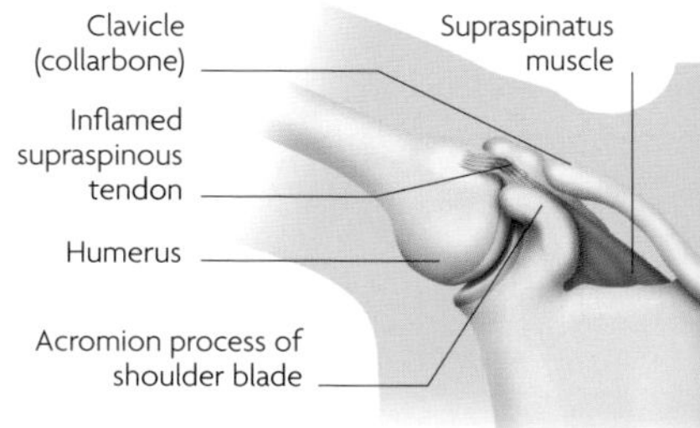
Clavicle (collarbone)
Supraspinatus muscle
Inflamed supraspinous tendon
Humerus
Acromion process of shoulder blade

TENOSYNOVITIS

The complex, weight-bearing nature of the foot makes it susceptible to tendon damage. Activities such as running, kicking, or dancing may cause inflammation.

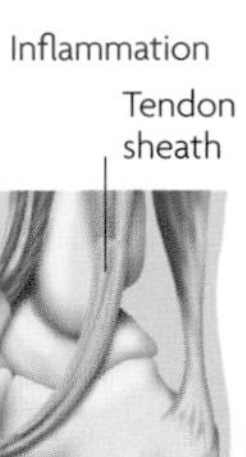
Inflammation
Tendon sheath

Tendon sheaths
Tendons

RUPTURED TENDON

Playing sports and lifting heavy weights may result in torn, or ruptured, tendons. Examples are tearing of the tendons attached to the biceps brachii muscle in the upper arm, or of the quadriceps tendon at the front of the thigh that stretches over the knee. A sudden impact that bends a fingertip toward the palm may snap the extensor tendon on the back of the finger. In severe cases, the tendon may be torn away from the bone. Symptoms include a snapping sensation, pain, swelling, and impaired movement.

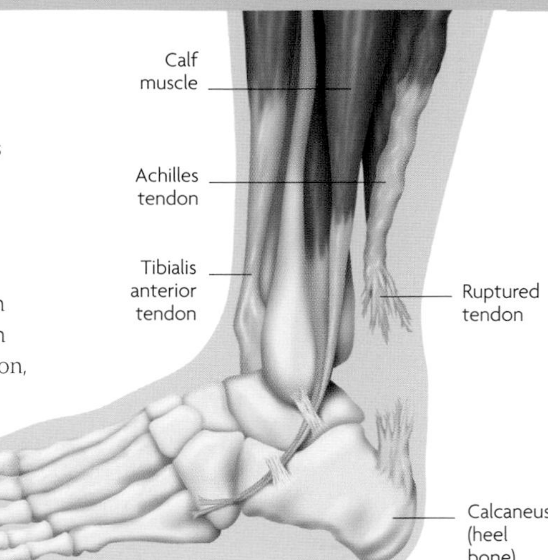
Calf muscle
Achilles tendon
Tibialis anterior tendon
Ruptured tendon
Calcaneus (heel bone)

TORN ACHILLES TENDON
The Achilles tendon attaches the calf muscle to the heel bone. It can snap after sudden exertion and may need to be treated by surgery and immobilization in a cast.

CARPAL TUNNEL SYNDROME

The carpal tunnel is a narrow passage formed by the carpal ligament (flexor retinaculum), on the inside of the wrist, and the underlying wrist bones (carpals). Tendons run through the tunnel from the forearm muscles to the hand. The median nerve also passes through the carpal tunnel, to control hand muscles and convey sensations from the fingers. In carpal tunnel syndrome (CTS), the median nerve is compressed by swelling of the tissues around it in the tunnel. Causes include diabetes mellitus, pregnancy, a wrist injury, rheumatoid arthritis, and repetitive movements.

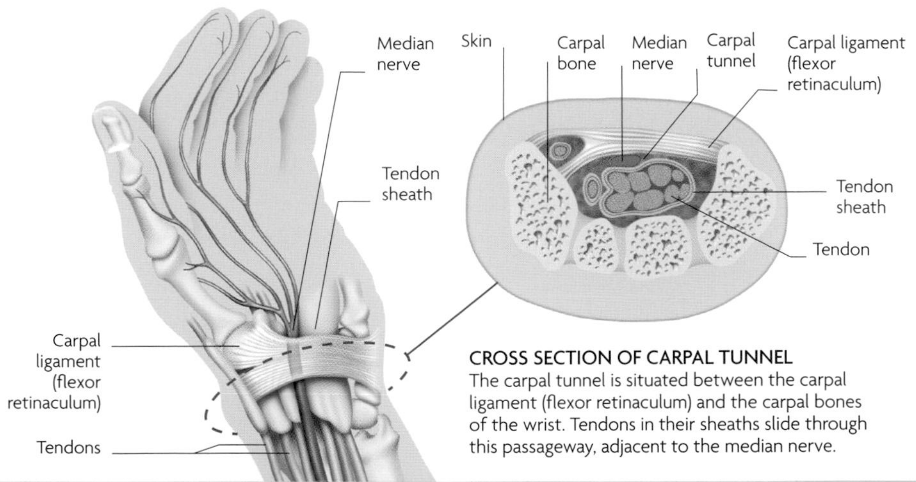
Median nerve
Tendon sheath
Carpal ligament (flexor retinaculum)
Tendons
Skin
Carpal bone
Median nerve
Carpal tunnel
Carpal ligament (flexor retinaculum)
Tendon sheath
Tendon

CROSS SECTION OF CARPAL TUNNEL
The carpal tunnel is situated between the carpal ligament (flexor retinaculum) and the carpal bones of the wrist. Tendons in their sheaths slide through this passageway, adjacent to the median nerve.

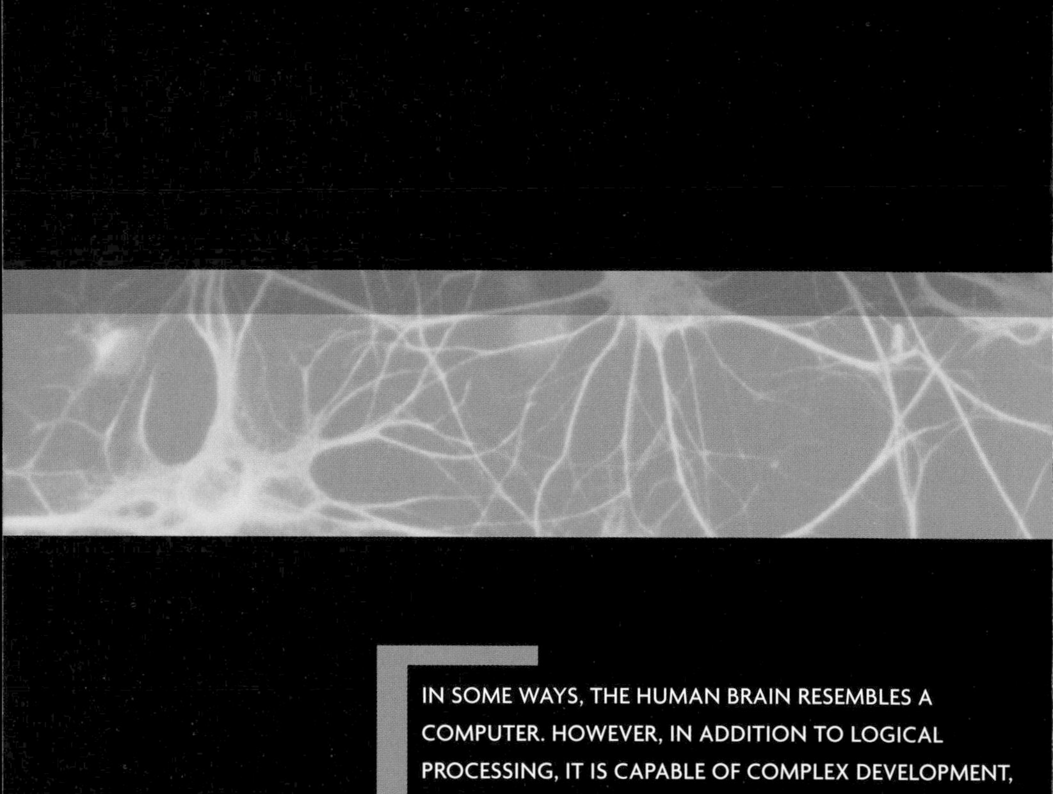

IN SOME WAYS, THE HUMAN BRAIN RESEMBLES A COMPUTER. HOWEVER, IN ADDITION TO LOGICAL PROCESSING, IT IS CAPABLE OF COMPLEX DEVELOPMENT, LEARNING, SELF-AWARENESS, EMOTION, AND CREATIVITY.

NERVOUS SYSTEM

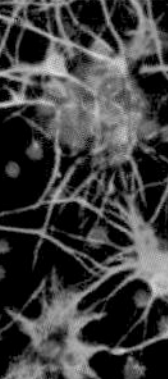

NERVOUS SYSTEM

CONSTANTLY ALIVE WITH ELECTRICITY, THE EXTENSIVE AND COMPLEX NERVOUS SYSTEM IS THE BODY'S PRIME COMMUNICATION AND COORDINATION NETWORK.

The body's nervous system has three parts. The central nervous system, or CNS, is made up of the brain and spinal cord. The peripheral nervous system, or PNS, comprises 12 pairs of nerves from the brain and 31 pairs from the cord. These branching nerves go to each part of the body, relaying information to and from the CNS, which has the roles of coordination and decision-making. The autonomic nervous system, or ANS, has nerve pathways in the CNS and PNS. Its work is primarily "automatic": it deals with activities such as heart rate adjustment and blood pressure control, of which we are rarely aware.

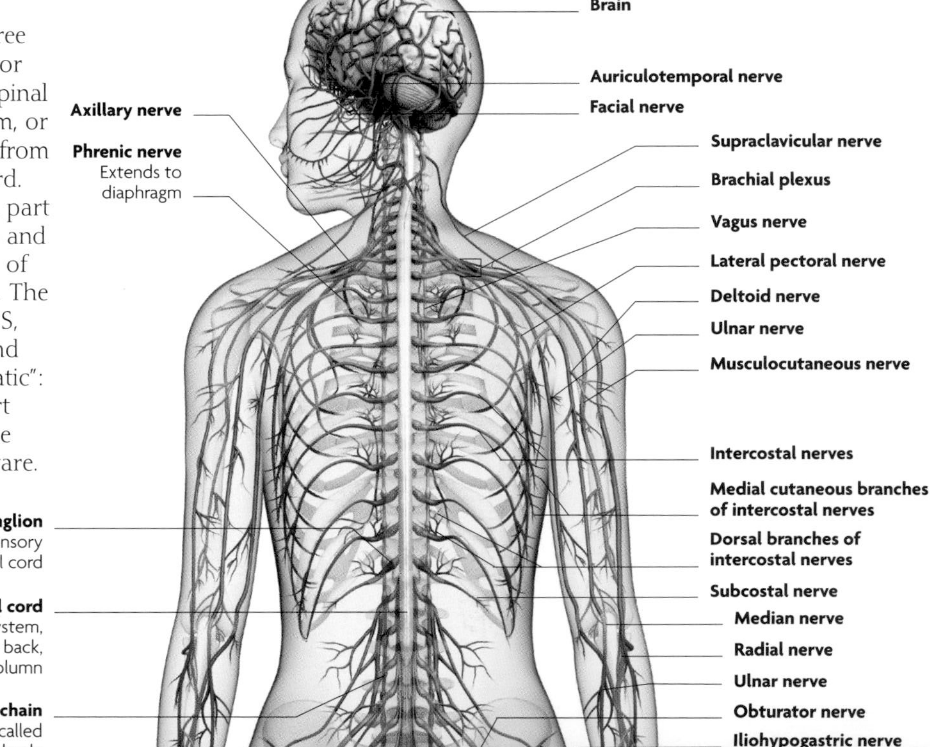
Brain
Auriculotemporal nerve
Facial nerve
Axillary nerve
Phrenic nerve
Extends to diaphragm
Supraclavicular nerve
Brachial plexus
Vagus nerve
Lateral pectoral nerve
Deltoid nerve
Ulnar nerve
Musculocutaneous nerve
Intercostal nerves
Medial cutaneous branches of intercostal nerves
Dorsal branches of intercostal nerves
Subcostal nerve
Median nerve
Radial nerve
Ulnar nerve
Obturator nerve
Iliohypogastric nerve
Spinal ganglion
One of many nodules that send sensory information to brain via spinal cord
Spinal cord
Part of central nervous system, extends from brain down the back, protected by vertebral column
Sympathetic ganglia chain
Part of sympathetic nervous system, also called paravertebral ganglia; conveys stress signals to body

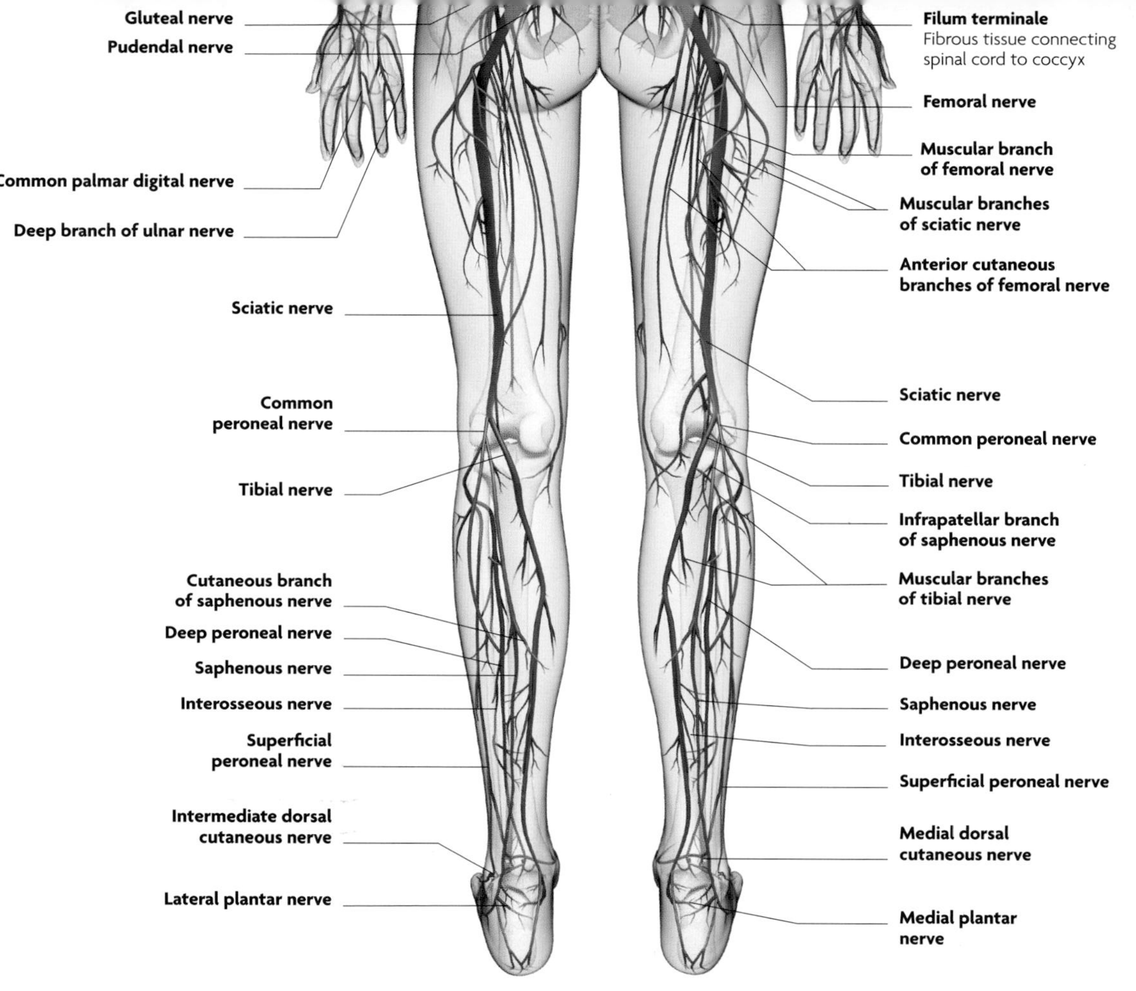
Gluteal nerve
Pudendal nerve
Common palmar digital nerve
Deep branch of ulnar nerve
Sciatic nerve
Common peroneal nerve
Tibial nerve
Cutaneous branch of saphenous nerve
Deep peroneal nerve
Saphenous nerve
Interosseous nerve
Superficial peroneal nerve
Intermediate dorsal cutaneous nerve
Lateral plantar nerve
Filum terminale
Fibrous tissue connecting spinal cord to coccyx
Femoral nerve
Muscular branch of femoral nerve
Muscular branches of sciatic nerve
Anterior cutaneous branches of femoral nerve
Sciatic nerve
Common peroneal nerve
Tibial nerve
Infrapatellar branch of saphenous nerve
Muscular branches of tibial nerve
Deep peroneal nerve
Saphenous nerve
Interosseous nerve
Superficial peroneal nerve
Medial dorsal cutaneous nerve
Medial plantar nerve

NERVES AND NEURONS

NERVE CELLS, OR NEURONS, ARE HIGHLY SPECIALIZED IN THEIR STRUCTURE, FUNCTION, AND COMMUNICATION LINKS.

NEURON STRUCTURE

Like all other cells, a typical neuron has a main cell body with a nucleus. But a neuron also has long, wirelike processes that reach out to transmit messages to other neurons at junctions called synapses. These processes are of two main kinds. Dendrites receive messages from other neurons, or from nervelike cells in sense organs, and conduct them toward the cell body of the neuron. Axons convey messages away from the cell body, to other neurons or to muscle or gland cells. Dendrites tend to be short and have many branches, while axons are usually longer and branch less along their length. Neurons in the brain and spinal cord are protected and nurtured by supporting nerve cells known as glial cells.

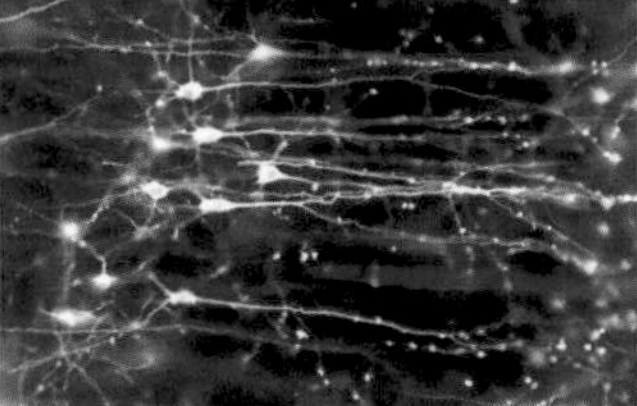

MICROSCOPE VIEW
Nerve cells under the microscope display their cell bodies with nuclei (left) and processes (right).

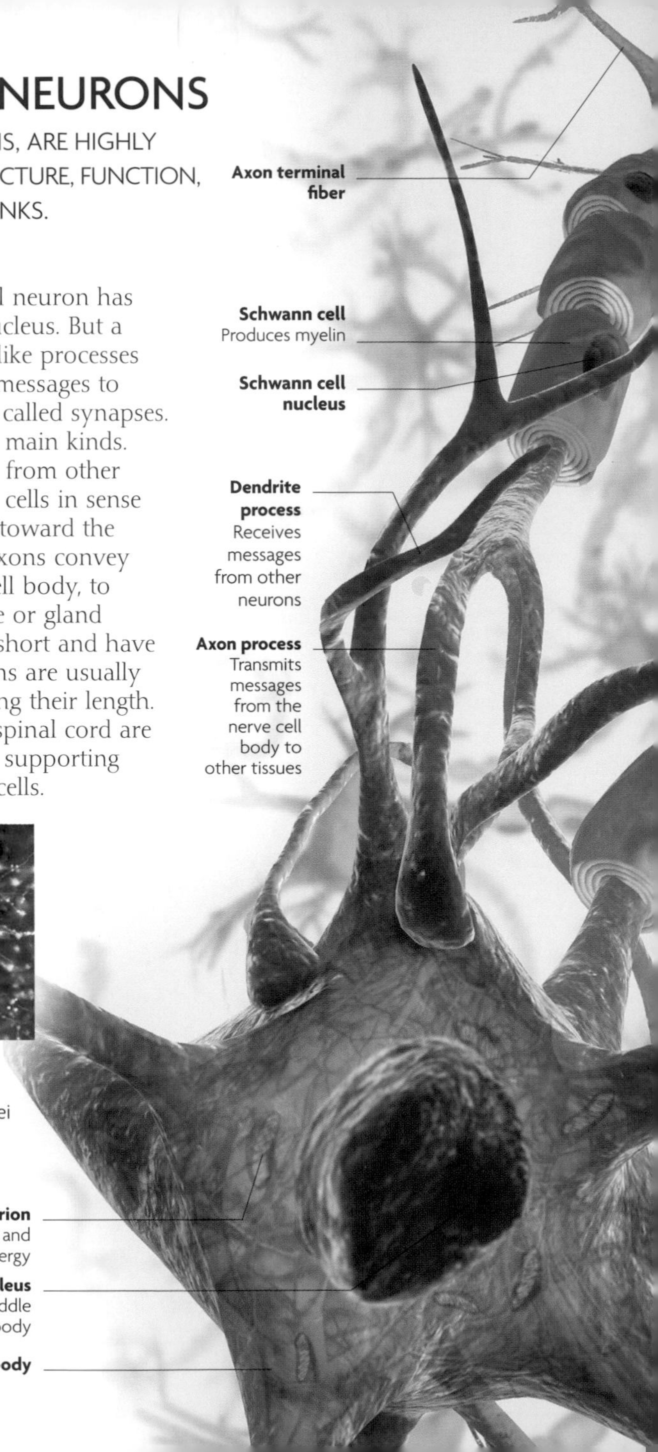

Axon terminal fiber

Schwann cell Produces myelin

Schwann cell nucleus

Dendrite process Receives messages from other neurons

Axon process Transmits messages from the nerve cell body to other tissues

Mitochondrion Involved in cell respiration and production of energy

Nucleus Located toward the middle of the cell body

Cell body

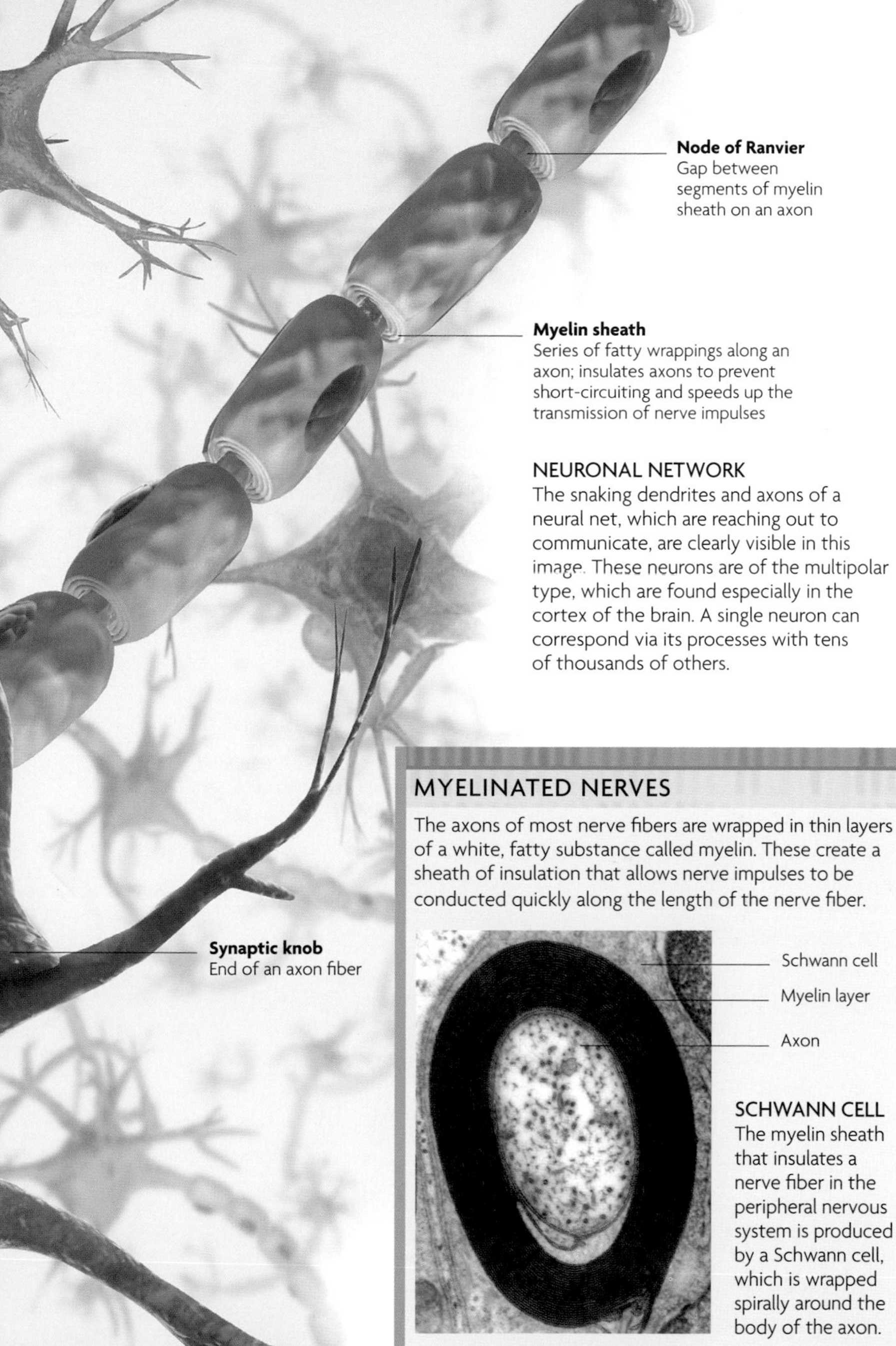
Node of Ranvier
Gap between segments of myelin sheath on an axon
Myelin sheath
Series of fatty wrappings along an axon; insulates axons to prevent short-circuiting and speeds up the transmission of nerve impulses
Synaptic knob
End of an axon fiber

NEURONAL NETWORK
The snaking dendrites and axons of a neural net, which are reaching out to communicate, are clearly visible in this image. These neurons are of the multipolar type, which are found especially in the cortex of the brain. A single neuron can correspond via its processes with tens of thousands of others.

MYELINATED NERVES

The axons of most nerve fibers are wrapped in thin layers of a white, fatty substance called myelin. These create a sheath of insulation that allows nerve impulses to be conducted quickly along the length of the nerve fiber.

Schwann cell
Myelin layer
Axon

SCHWANN CELL
The myelin sheath that insulates a nerve fiber in the peripheral nervous system is produced by a Schwann cell, which is wrapped spirally around the body of the axon.

TYPES OF NEURON

Neuron cell bodies vary greatly in size and shape, as do the type, number, and length of their projections. Neurons can be unipolar, bipolar, or multipolar. Unipolar neurons are found mainly in the sensory nerves of the PNS. Bipolar neurons exist mostly in the embryo, but adults have some in the retina of the eye and the olfactory nerve in the nose. Most neurons in the brain and spinal cord are multipolar.

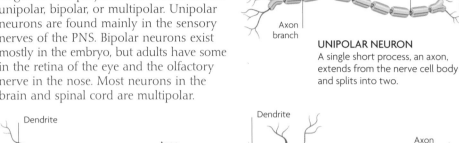

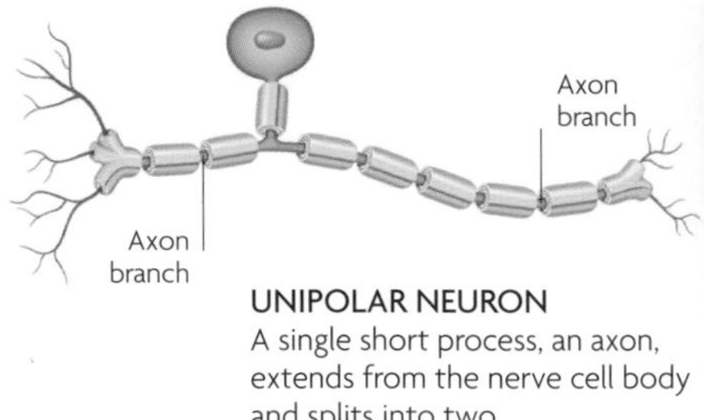
Axon branch
Axon branch

UNIPOLAR NEURON
A single short process, an axon, extends from the nerve cell body and splits into two.

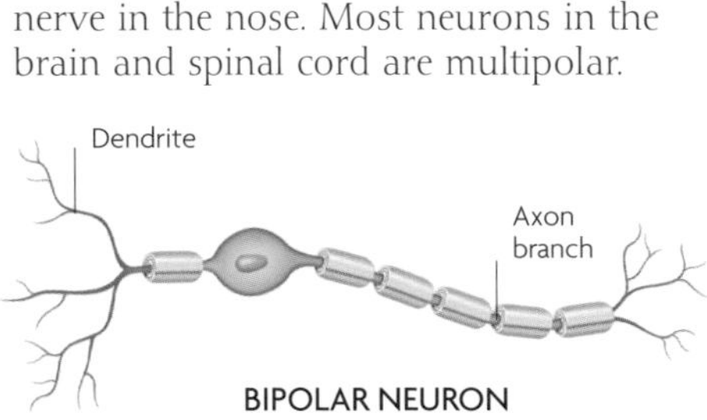
Dendrite
Axon branch

BIPOLAR NEURON
The nerve cell body is located between two processes—an axon and a dendrite.

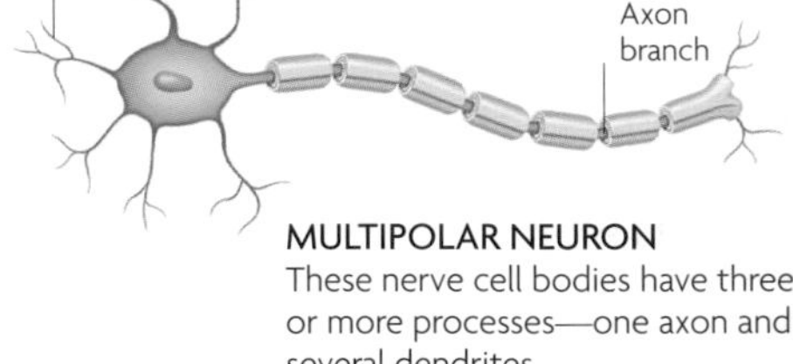
Dendrite
Axon branch

MULTIPOLAR NEURON
These nerve cell bodies have three or more processes—one axon and several dendrites.

NERVES

Nerves, which resemble ropelike cords, pass between and branch into the body's organs and tissues. They are composed of bundles of communication strands–the elongated axons or nerve fibers of neurons. Each bundle is known as a fascicle. Most nerves carry two types of fiber. Sensory, or afferent, fibers bring messages from receptors in the sense organs and other structures to the spinal cord and brain. Motor, or efferent, fibers convey signals from the brain or spinal cord to a muscle or gland. Some nerves contain just sensory fibers, such as the optic nerve, while others have solely motor fibers.

INSIDE A NERVE
Bundles of nerve fibers are embedded within tough connective tissue to protect them from damage.

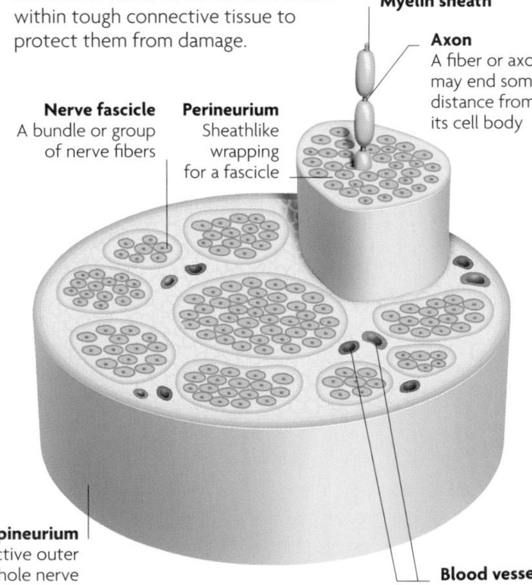
Myelin sheath
Axon
A fiber or axon may end some distance from its cell body
Nerve fascicle
A bundle or group of nerve fibers
Perineurium
Sheathlike wrapping for a fascicle
Epineurium
Strong, protective outer covering for the whole nerve
Blood vessels

NERVE REGENERATION

Peripheral nerve fibers that have been crushed or partly cut may slowly regenerate if the cell body is undamaged. The damaged section of fiber degenerates, leaving the myelin sheath hollow. The healthy remaining fiber begins to grow along the empty sheath at a rate of $^1/_{25}$–$^2/_{25}$ in (1–2 mm) daily. Natural regeneration is much less likely in the nerve fibers of the brain and spinal cord, where the neurons are so specialized that generally they cannot replicate themselves or recreate their highly developed connections.

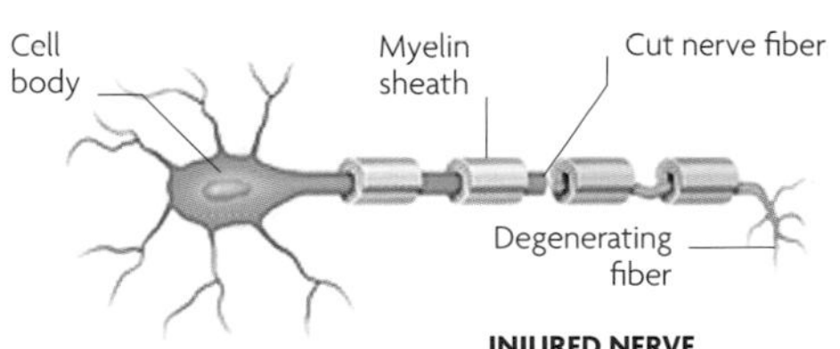
Cell body
Myelin sheath
Cut nerve fiber
Degenerating fiber

INJURED NERVE

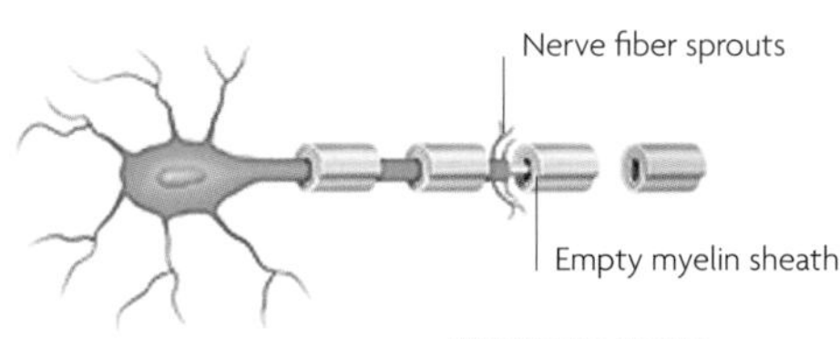
Nerve fiber sprouts
Empty myelin sheath

ATTEMPTED REPAIR

REGROWTH

The stump end of a damaged nerve fiber sends out several sproutlike growths. One of these finds the empty but intact myelin sheath and grows inside it.

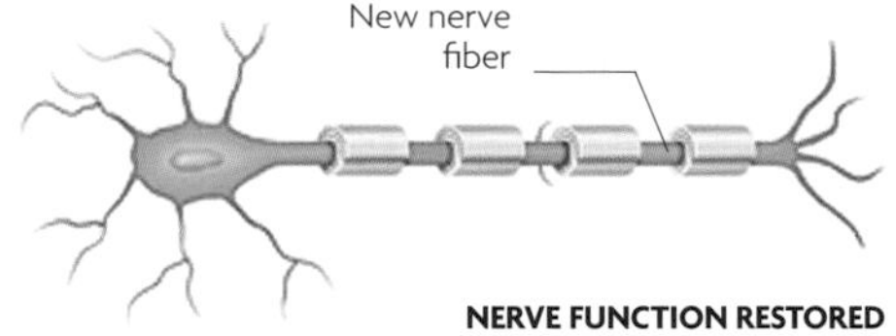
New nerve fiber

NERVE FUNCTION RESTORED

SUPPORT CELLS

Supporting nerve cells, known as glial cells or neuroglia, protect and nourish the neurons. There are several types of glial cell. The smallest are microglia, which destroy microorganisms, foreign particles, and cell debris from disintegrating neurons. Ependymal cells line cavities that are filled with cerebrospinal fluid, which surrounds both the brain and spinal cord (see p.89). Other glial cells insulate the axons and dendrites or regulate the flow of cerebrospinal fluid.

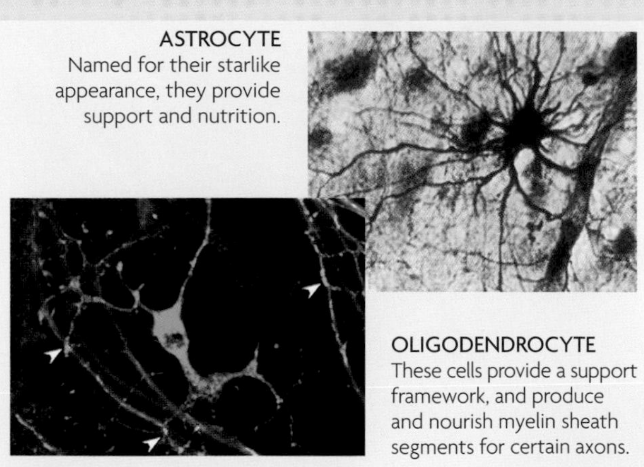

ASTROCYTE

Named for their starlike appearance, they provide support and nutrition.

OLIGODENDROCYTE

These cells provide a support framework, and produce and nourish myelin sheath segments for certain axons.

NERVE IMPULSE

WHEN NERVE CELLS, OR NEURONS, ARE STIMULATED THEY UNDERGO CHEMICAL CHANGES THAT PRODUCE TINY WAVES OF ELECTRICITY—NERVE IMPULSES.

Information is conveyed throughout the nervous system as nerve impulses, or action potentials. Impulses are about 0.1 volts (100 millivolts) in strength and last just 1/1000 s (1 millisecond). The information carried depends on the location of the impulses in the nervous system, and on their frequency–from one impulse every few seconds to several hundreds per second.

When impulses reach a junction known as a synapse, they trigger the release of chemicals called neurotransmitters. Molecules of the neurotransmitter cross the synapse and stimulate the receiving neuron to fire an impulse of its own, as wavelike movements of ions (electrically charged particles). Neurotransmitters may also actively inhibit a receiving neuron from firing.

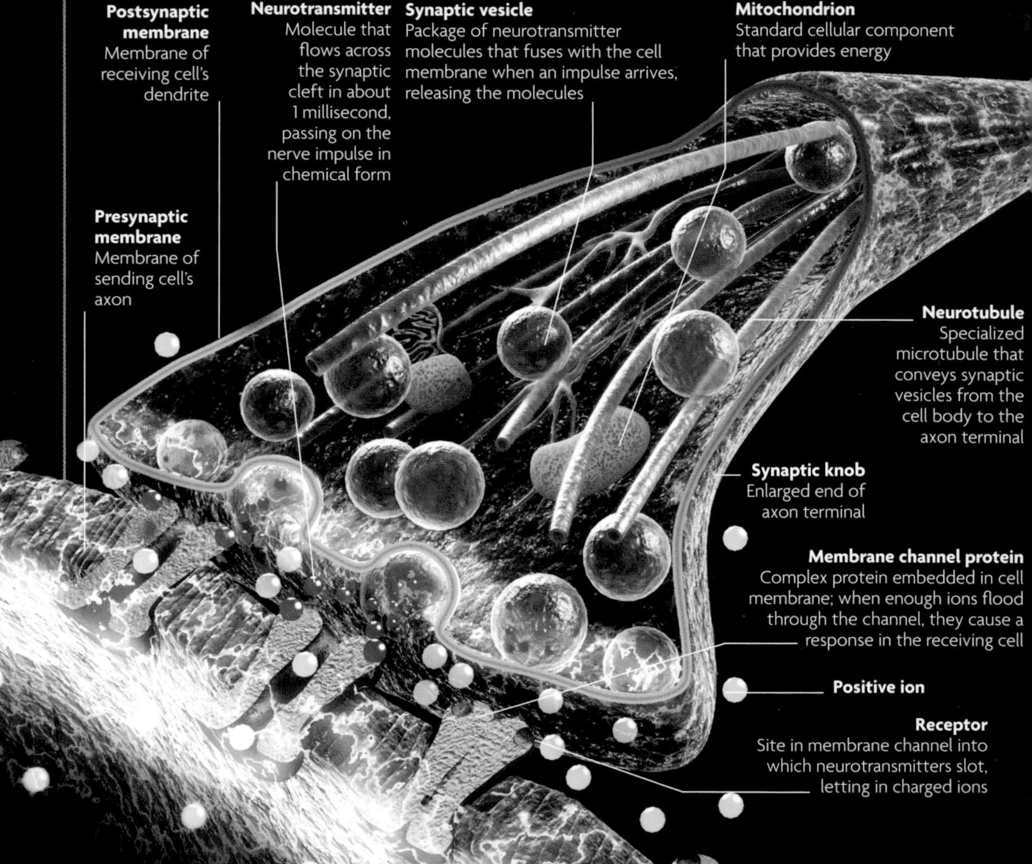
Postsynaptic membrane
Membrane of receiving cell's dendrite
Neurotransmitter
Molecule that flows across the synaptic cleft in about 1 millisecond, passing on the nerve impulse in chemical form
Synaptic vesicle
Package of neurotransmitter molecules that fuses with the cell membrane when an impulse arrives, releasing the molecules
Mitochondrion
Standard cellular component that provides energy
Presynaptic membrane
Membrane of sending cell's axon
Neurotubule
Specialized microtubule that conveys synaptic vesicles from the cell body to the axon terminal
Synaptic knob
Enlarged end of axon terminal
Membrane channel protein
Complex protein embedded in cell membrane; when enough ions flood through the channel, they cause a response in the receiving cell
Positive ion
Receptor
Site in membrane channel into which neurotransmitters slot, letting in charged ions

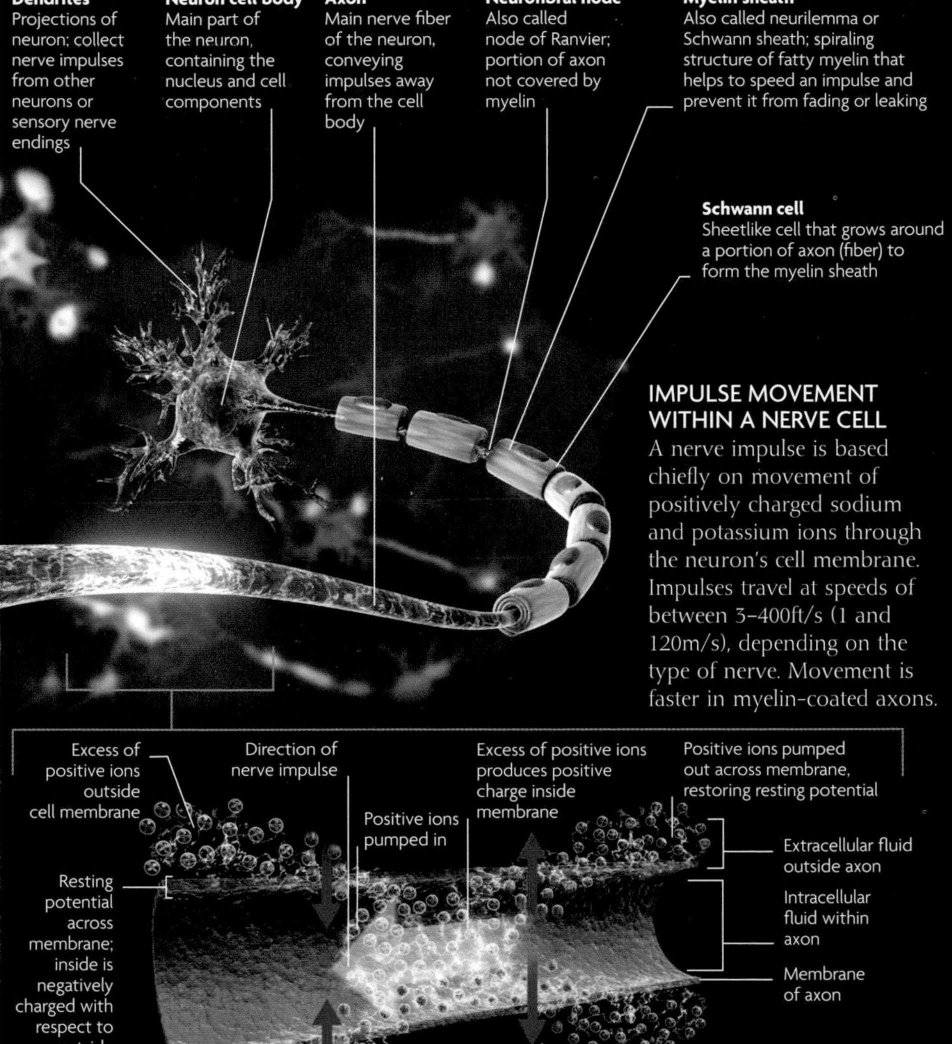
Dendrites
Projections of neuron; collect nerve impulses from other neurons or sensory nerve endings
Neuron cell body
Main part of the neuron, containing the nucleus and cell components
Axon
Main nerve fiber of the neuron, conveying impulses away from the cell body
Neurofibral node
Also called node of Ranvier; portion of axon not covered by myelin
Myelin sheath
Also called neurilemma or Schwann sheath; spiraling structure of fatty myelin that helps to speed an impulse and prevent it from fading or leaking
Schwann cell
Sheetlike cell that grows around a portion of axon (fiber) to form the myelin sheath
Excess of positive ions outside cell membrane
Direction of nerve impulse
Positive ions pumped in
Excess of positive ions produces positive charge inside membrane
Positive ions pumped out across membrane, restoring resting potential
Extracellular fluid outside axon
Intracellular fluid within axon
Membrane of axon
Resting potential across membrane; inside is negatively charged with respect to outside

IMPULSE MOVEMENT WITHIN A NERVE CELL

A nerve impulse is based chiefly on movement of positively charged sodium and potassium ions through the neuron's cell membrane. Impulses travel at speeds of between 3–400ft/s (1 and 120m/s), depending on the type of nerve. Movement is faster in myelin-coated axons.

RESTING POTENTIAL

There are more positive ions outside the cell and more negative ions inside, producing a "resting potential" of -70 millivolts.

DEPOLARIZATION

Positive ions rush in through ion channels. The membrane is first depolarized, then its polarity is reversed, resulting in an "action potential" of +30 millivolts inside.

REPOLARIZATION

Positively charged potassium ions flow in the opposite direction, restoring the charge balance. This stimulates an adjacent area of membrane, and the next, and the impulse moves along like a wave.

BRAIN

THE BRAIN, TOGETHER WITH THE SPINAL CORD, REGULATES NONCONSCIOUS PROCESSES AND COORDINATES MOST VOLUNTARY MOVEMENT. IT IS THE SITE OF CONSCIOUSNESS, ALLOWING HUMANS TO THINK AND LEARN.

BRAIN STRUCTURE

The largest part of the brain is the cerebrum, which has a heavily folded surface–the pattern of which is unique in each person. The grooves are called sulci when shallow and fissures when deep. Fissures and some of the large sulci outline four functional areas, called lobes: frontal, parietal, occipital, and temporal (see p.90). A ridge on the surface of the brain is called a gyrus. The center of the brain contains the thalamus, which acts as the brain's information relay station. Surrounding this is a group of structures known as the limbic system (see p.94), which is involved in survival instincts, behavior, and emotions. Closely linked with the limbic system is the hypothalamus (see p.95, which receives sensory information.

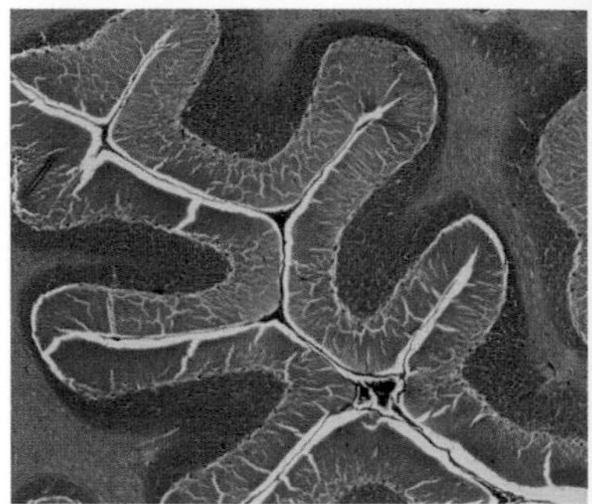

CEREBELLUM
The cerebellum (section shown above) contains billions of neurons that link up with other regions of the brain and spinal cord to facilitate precise movement.

BLOOD SUPPLY TO THE BRAIN

The brain forms 2 percent of the body's weight but needs 20 percent of its blood. Without oxygen and glucose, brain function quickly deteriorates, leading to dizziness and loss of consciousness. Within only four to eight minutes of oxygen deprivation, brain damage or death results. The brain has an abundant supply of blood from a vast network of blood vessels that stem from the carotid arteries, which run up each side of the neck, and from two vertebral arteries that run alongside the spinal cord.

CIRCLE OF WILLIS
A ring of arteries, the Circle of Willis, encircles the base of the brain and provides multiple pathways to supply oxygenated blood to all parts of the brain. If a pathway becomes blocked, blood can be redirected from another pathway.

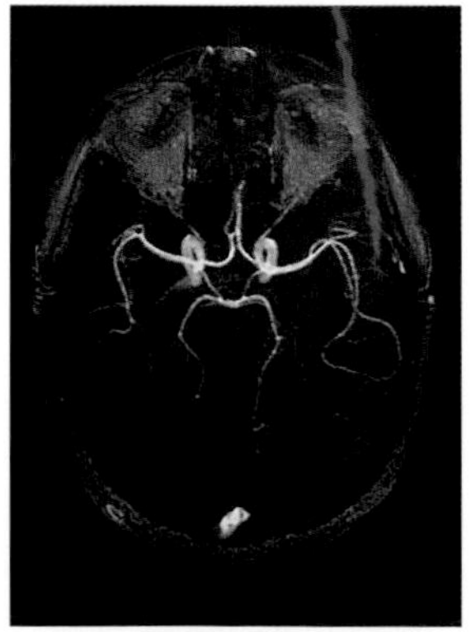

BLOOD SUPPLY
The brain has an extensive blood supply from two front and two rear arteries, as illustrated in this color, three-dimensional magnetic resonance angiography (MRA) scan. The blood vessels are colored red; here, they are seen supplying oxygenated blood to various parts of the brain, which is shown as the blue area.

INNER STRUCTURES

A section down the middle of the brain reveals its inner structures. Although these structures look very different in the diagram below, they are all made up of brain tissue, which is composed of billions of neurons. There are two types of brain tissue—gray matter and white matter.

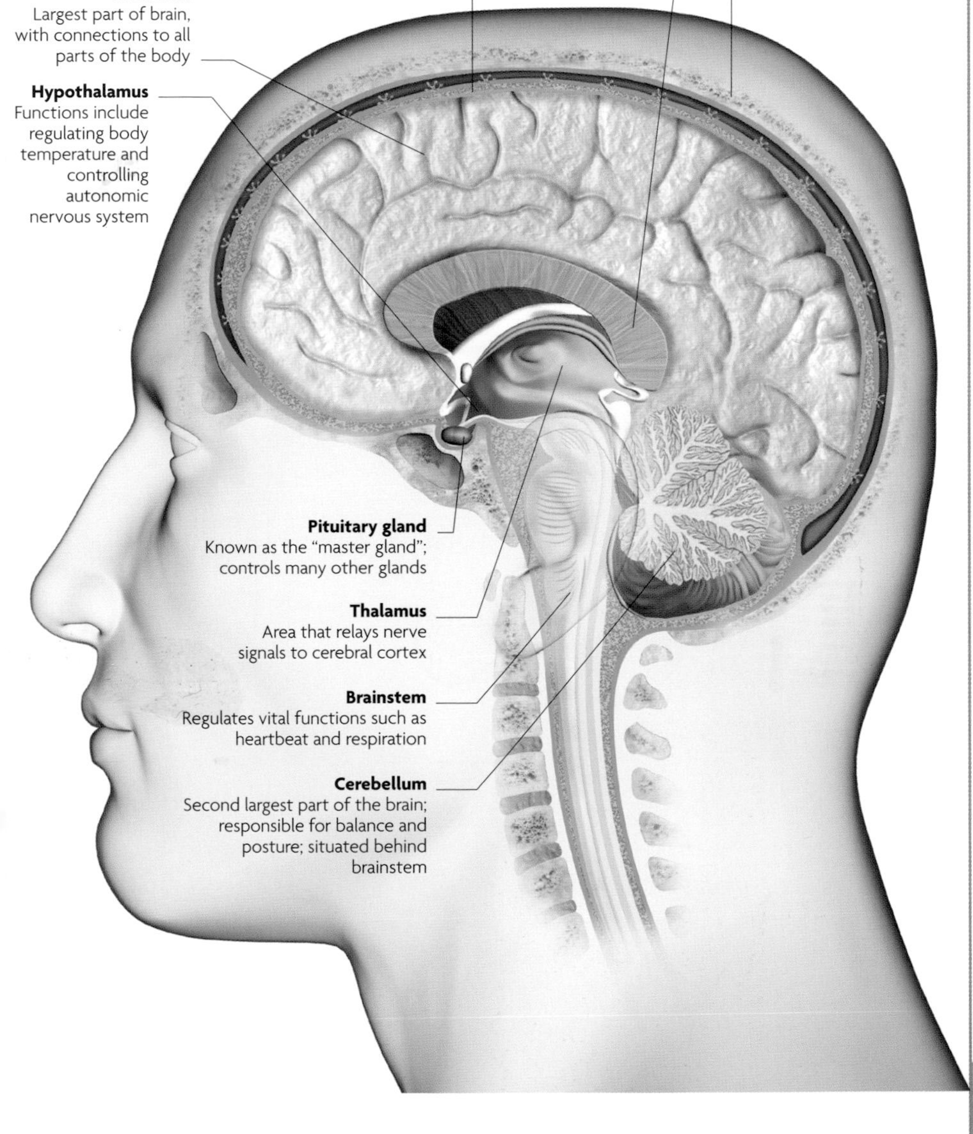

Meninges
Three membranes that surround and protect the brain and spinal cord; made of connective tissue

Corpus callosum
Largest of several bundles of nerve fibers that connect the two brain hemispheres

Skull

Cerebrum
Largest part of brain, with connections to all parts of the body

Hypothalamus
Functions include regulating body temperature and controlling autonomic nervous system

Pituitary gland
Known as the "master gland"; controls many other glands

Thalamus
Area that relays nerve signals to cerebral cortex

Brainstem
Regulates vital functions such as heartbeat and respiration

Cerebellum
Second largest part of the brain; responsible for balance and posture; situated behind brainstem

PROTECTION

The brain has several forms of protection. First and foremost is the skull (see p.48). Between the skull and the gray matter of the cerebrum lie three protective membranes that also protect the spinal cord (see p.98). The dura mater lines the inside of the skull, where it is attached to the bones; it is a thick, inelastic layer that provides support and protection. The arachnoid membrane lies beneath the dura, and was named for its resemblance to a spider's web. The pia mater adheres to the convolutions of the cerebrum, and supplies brain tissue with blood vessels. Protection also comes from the cerebrospinal fluid (see opposite) in the subarachnoid space, between the arachnoid membrane and the pia mater. It absorbs and disperses excessive mechanical forces that might otherwise cause serious injury. Analyses of its chemical constituents and flow pressure have offered vital clues for diagnosing diseases and disorders of the brain and spinal cord, such as meningitis.

Skull

Brain

MENINGES OF THE BRAIN

A section through the skull and brain reveals the three meninges. The dura mater lines the skull, and sends four projections inward to give support. The arachnoid membrane cushions the brain, and the pia mater closely envelops the cerebrum.

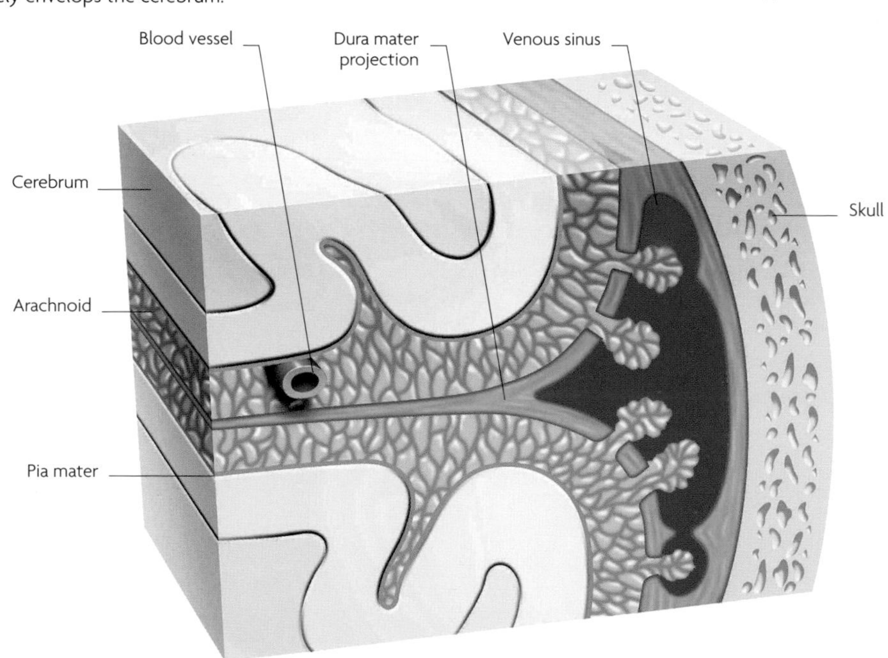
Blood vessel
Dura mater projection
Venous sinus
Cerebrum
Skull
Arachnoid
Pia mater

CEREBROSPINAL FLUID FLOW

The tissue of the brain floats in cerebrospinal fluid (CSF) within the skull. CSF is a clear liquid, which is renewed four to five times a day. CSF protects and nourishes the brain and spinal cord as it flows around them. It contains proteins and glucose that provide energy for brain cell function, as well as lymphocytes that guard against infection. CSF is produced by the choroid plexuses in the lateral ventricles, and drains into the third ventricle. It then flows into the fourth ventricle, which is located in front of the cerebellum. Circulation of the fluid is aided by pulsations of the cerebral arteries.

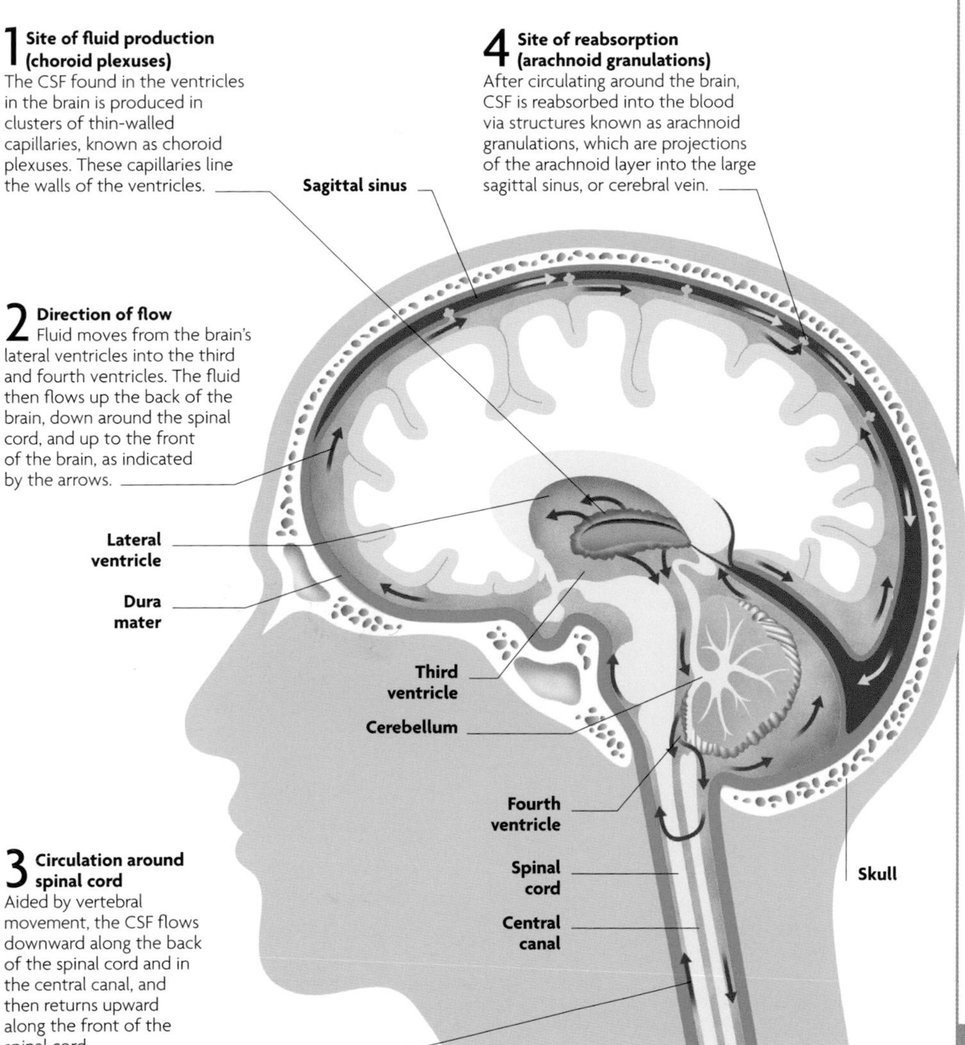
1 Site of fluid production (choroid plexuses)
The CSF found in the ventricles in the brain is produced in clusters of thin-walled capillaries, known as choroid plexuses. These capillaries line the walls of the ventricles.
Sagittal sinus
4 Site of reabsorption (arachnoid granulations)
After circulating around the brain, CSF is reabsorbed into the blood via structures known as arachnoid granulations, which are projections of the arachnoid layer into the large sagittal sinus, or cerebral vein.
2 Direction of flow
Fluid moves from the brain's lateral ventricles into the third and fourth ventricles. The fluid then flows up the back of the brain, down around the spinal cord, and up to the front of the brain, as indicated by the arrows.
Lateral ventricle
Dura mater
Third ventricle
Cerebellum
Fourth ventricle
Spinal cord
Central canal
Skull
3 Circulation around spinal cord
Aided by vertebral movement, the CSF flows downward along the back of the spinal cord and in the central canal, and then returns upward along the front of the spinal cord.

BRAIN STRUCTURES

THE BRAIN HAS FOUR MAIN STRUCTURES: THE LARGE, DOMED CEREBRUM; THE INNER, DEEPER DIENCEPHALON (CONSISTING OF THE THALAMUS AND NEARBY STRUCTURES); THE CEREBELLUM, TO THE LOWER REAR; AND THE BRAINSTEM, AT THE BASE.

EXTERNAL BRAIN FEATURES

The cerebrum makes up more than four-fifths of the brain's tissue. Its heavily folded surface forms the lobes of the cerebral cortex. The cerebrum partly envelops the thalamus and nearby structures, and also the brainstem. The smaller cerebellum forms about one-tenth of the brain's volume; it is mainly concerned with organizing motor information sent to muscles so that body movements are smooth and coordinated.

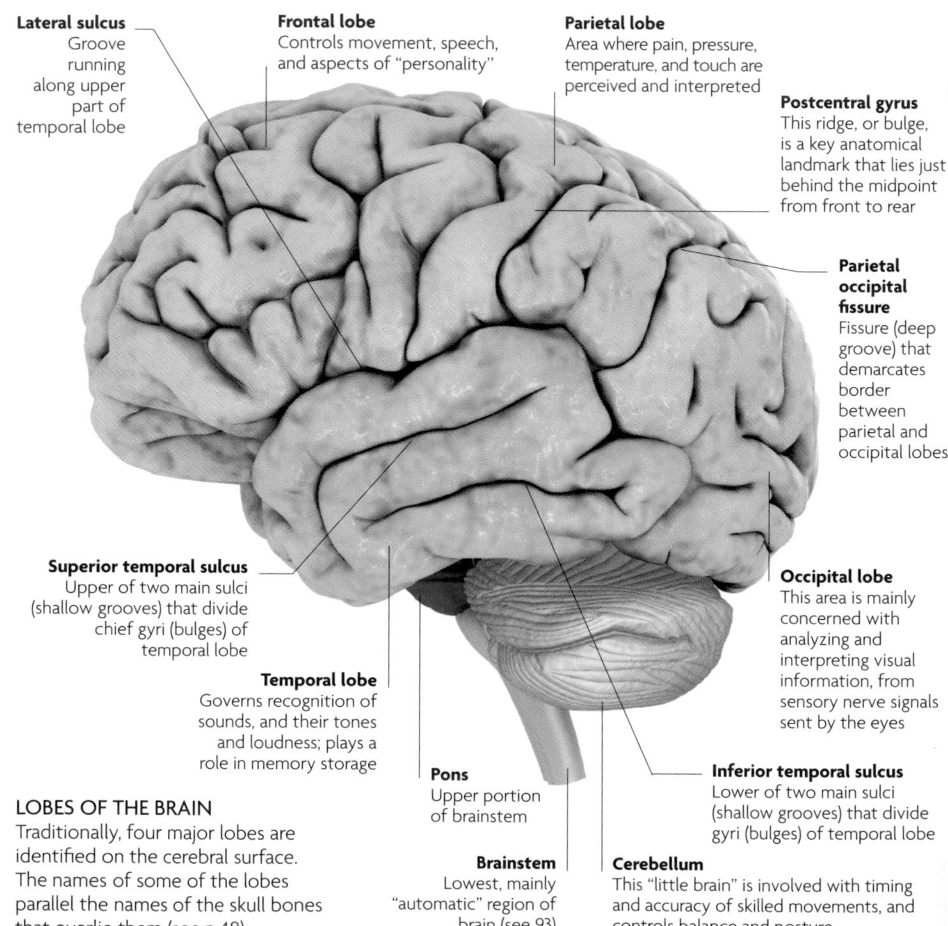
Lateral sulcus
Groove running along upper part of temporal lobe
Frontal lobe
Controls movement, speech, and aspects of "personality"
Parietal lobe
Area where pain, pressure, temperature, and touch are perceived and interpreted
Postcentral gyrus
This ridge, or bulge, is a key anatomical landmark that lies just behind the midpoint from front to rear
Parietal occipital fissure
Fissure (deep groove) that demarcates border between parietal and occipital lobes
Superior temporal sulcus
Upper of two main sulci (shallow grooves) that divide chief gyri (bulges) of temporal lobe
Occipital lobe
This area is mainly concerned with analyzing and interpreting visual information, from sensory nerve signals sent by the eyes
Temporal lobe
Governs recognition of sounds, and their tones and loudness; plays a role in memory storage
Pons
Upper portion of brainstem
Inferior temporal sulcus
Lower of two main sulci (shallow grooves) that divide gyri (bulges) of temporal lobe
Brainstem
Lowest, mainly "automatic" region of brain (see 93)
Cerebellum
This "little brain" is involved with timing and accuracy of skilled movements, and controls balance and posture

LOBES OF THE BRAIN

Traditionally, four major lobes are identified on the cerebral surface. The names of some of the lobes parallel the names of the skull bones that overlie them (see p.48).

OUTER BRAIN STRUCTURES

The cerebrum is partly separated into two halves (cerebral hemispheres) by the deep longitudinal fissure. The cerebellum is the smaller bulbous structure beneath, responsible for muscle control. Below the cerebellum is the brainstem, which controls basic life processes.

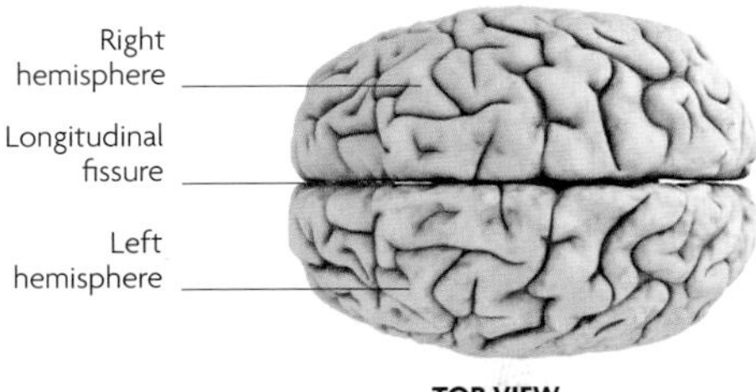
Right hemisphere
Longitudinal fissure
Left hemisphere

TOP VIEW

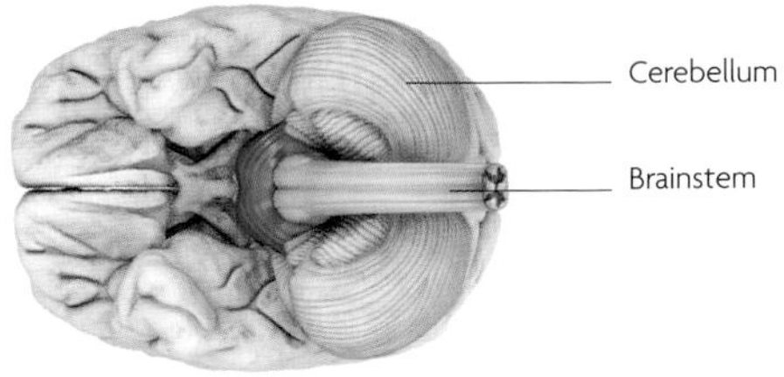
Cerebellum
Brainstem

BOTTOM VIEW

THE HOLLOW BRAIN

The brain is, in a sense, hollow: it contains four chambers known as ventricles, which are filled with cerebrospinal fluid, or CSF (see p.89). There are two lateral ventricles, one in each hemisphere, and the CSF fluid is produced here. It then drains via the interventricular foramen into the third ventricle, which is situated close to the thalamus and occupies a more central position. From here it flows through the cerebral aqueduct and into the fourth ventricle, which extends down between the pons and cerebellum into the medulla. The total volume of CSF in the ventricles is about 9/10 fl oz (25 ml). Circulation is aided by head movements and pulsations of the cerebral arteries.

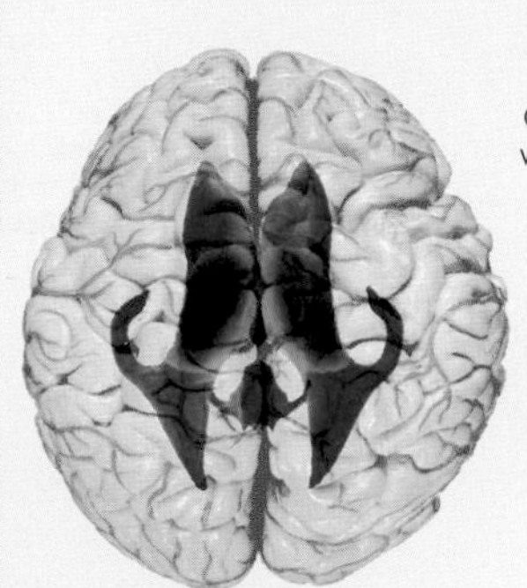

VIEW FROM ABOVE
The lateral ventricles have frontward-, backward-, and side-facing horns, or cornua. The central third ventricle lies between them in this view.

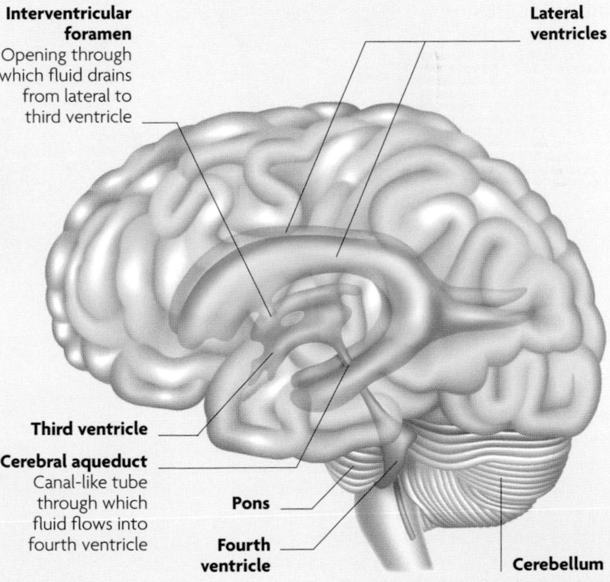
Interventricular foramen
Opening through which fluid drains from lateral to third ventricle
Lateral ventricles
Third ventricle
Cerebral aqueduct
Canal-like tube through which fluid flows into fourth ventricle
Pons
Fourth ventricle
Cerebellum

GRAY AND WHITE MATTER

Most of the cerebrum has two main layers. The outer layer, often known as "gray matter," is the cerebral cortex. It follows the folds and bulges of the cerebrum to cover its entire surface. Its average thickness is $\frac{1}{10}$–$\frac{2}{10}$ in (3–5 mm), and, spread out flat, it would cover about the same area as a pillowcase. Deeper within the cerebrum are small islands of gray matter. These, and the cerebral cortex, are composed chiefly of the cell bodies and projections (dendrites) of neurons. The paler "white matter" forms the bulk of the cerebrum's interior. It is composed mainly of nerve fibers.

VERTICAL SECTION

A vertical "slice" through the middle of the brain reveals the paired structures, outer gray layer, and inner white matter. The corpus callosum contains more than 100 million nerve fibers, and is the main "bridge" between the two hemispheres.

Gray matter
Outermost layer of cerebral cortex; contains an estimated 50 billion neurons and perhaps 10 times as many supporting cells

White matter interior
Here, axons, or fibers, of neurons run up from lower areas and project down from neuron cell bodies of cortex

Corpus callosum
Largest of several bundles of nerve fibers, called commissures, which connect specific areas of the two halves, or cerebral hemispheres, of the upper brain

Basal ganglia
"Islands" of gray matter deep in cerebrum

Motor nerve tracts
Large bundles of nerve fibers carry instructions for movements down to the spinal cord; they cross over in lower brainstem

Brainstem

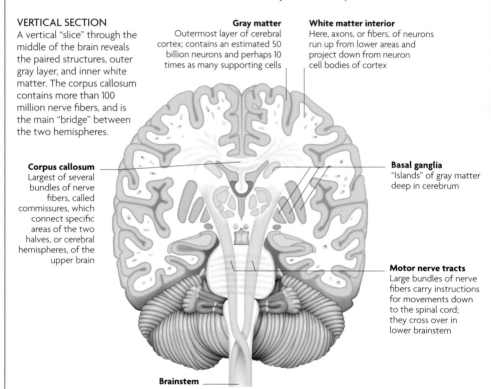

BASAL GANGLIA

These structures include the lentiform nucleus (putamen and globus pallidus), caudate nucleus, subthalamic nucleus, and substantia nigra (the latter two not seen in this view). They are a complex interface between sensory inputs and motor skills, especially for semiautomatic movements, such as walking.

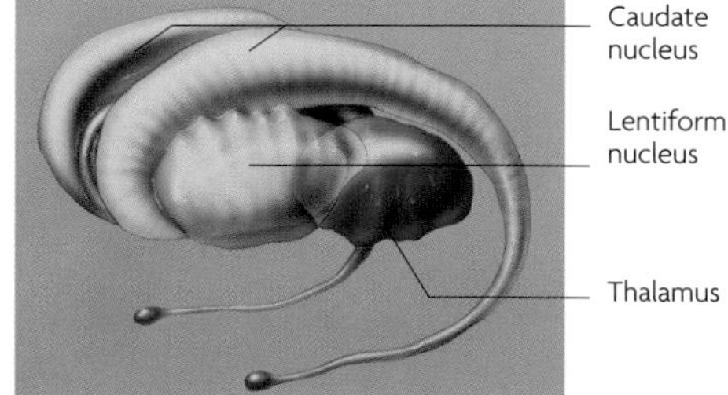
Caudate nucleus
Lentiform nucleus
Thalamus

VERTICAL LINKS

Sheathed (myelinated) nerve fibers are organized into bundles called projection tracts. These nerve fibers transmit impulses between the spinal cord and lower brain areas and the cerebral cortex above. The nerve tracts pass through a communication link called the internal capsule, and also intersect the corpus callosum. In addition, similar bundles pass through the upper, outer zones of the white matter, from one area of the cerebral cortex to another. These bundles, called association tracts, convey nerve signals directly between different regions or centers of the cortex.

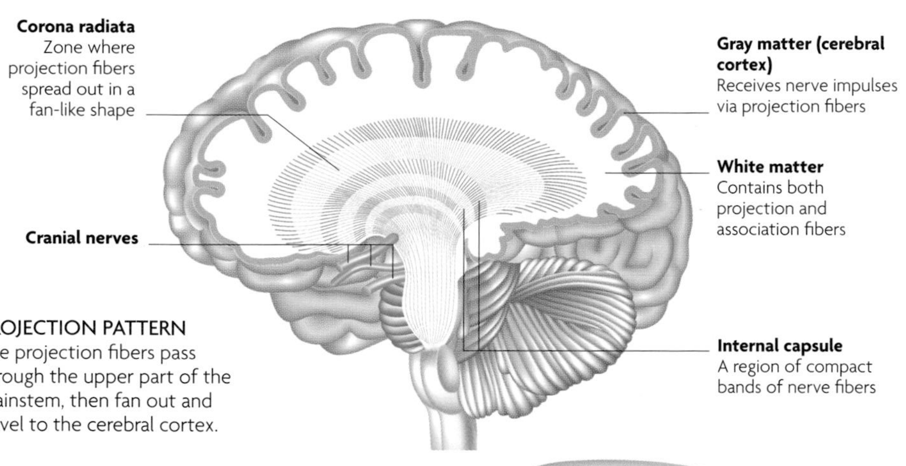
Corona radiata
Zone where projection fibers spread out in a fan-like shape
Cranial nerves
Gray matter (cerebral cortex)
Receives nerve impulses via projection fibers
White matter
Contains both projection and association fibers
Internal capsule
A region of compact bands of nerve fibers

PROJECTION PATTERN
The projection fibers pass through the upper part of the brainstem, then fan out and travel to the cerebral cortex.

THE THALAMUS AND BRAINSTEM

The thalamus sits on top of the brainstem, and is shaped like two eggs placed side by side. It lies almost at the "heart" of the brain, and acts as a major relay station that monitors and processes incoming information before this is sent to the upper regions of the brain. The brainstem contains centers that regulate several functions vital for survival: these functions include the heartbeat, respiration, blood pressure, and some reflex actions, such as swallowing and vomiting.

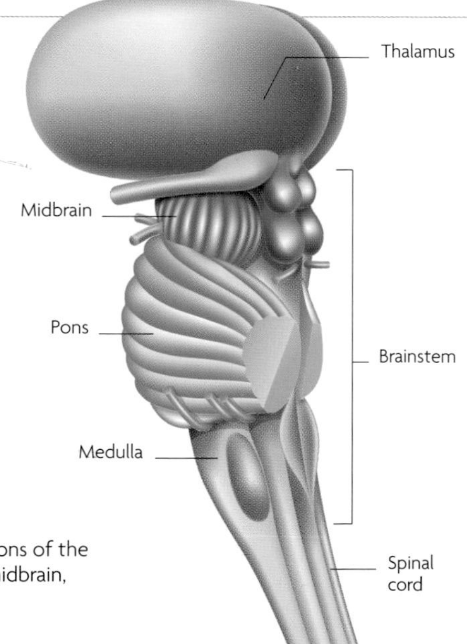
Thalamus
Midbrain
Pons
Brainstem
Medulla
Spinal cord

BRAINSTEM
The three main regions of the brainstem are the midbrain, pons, and medulla.

THE PRIMITIVE BRAIN

IN TIMES OF STRESS OR CRISIS, DEEP-SEATED INSTINCTS WELL UP FROM WITHIN US AND TAKE OVER OUR AWARENESS. SUCH EVENTS INVOLVE THE "PRIMITIVE BRAIN," WHICH IS BASED MAINLY IN A SERIES OF PARTS KNOWN AS THE LIMBIC SYSTEM.

THE LIMBIC SYSTEM

The limbic system influences subconscious, instinctive behavior, similar to animal responses that relate to reproduction and survival. In humans, many of these innate, "primitive" behaviors are modified by conscious, thoughtful considerations based in upper regions of the brain. However, when primal urges prevail, the limbic system and its associated structures take over. At other times they play lesser, but still complex and important, roles in the expression of instincts, drives, and emotions.

LIMBIC STRUCTURES

The ring-shaped limbic system is located in the lower center of the brain, between the cortical "thinking" areas and the lowest parts, which control automatic functions.

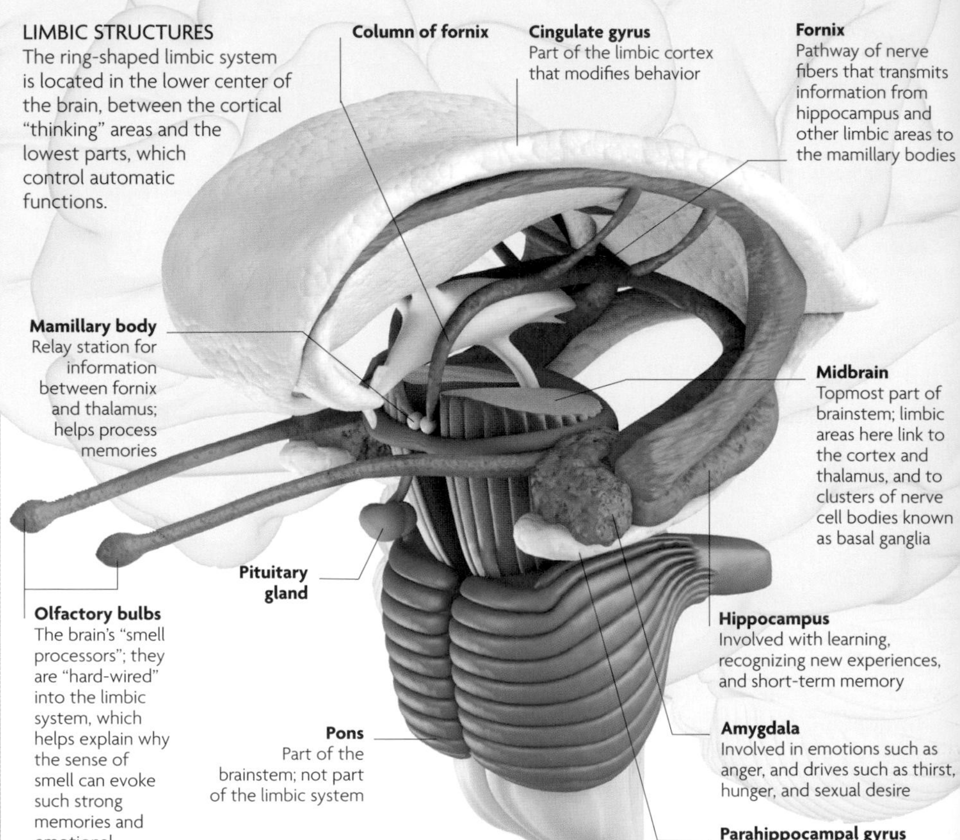

Column of fornix
Cingulate gyrus Part of the limbic cortex that modifies behavior
Fornix Pathway of nerve fibers that transmits information from hippocampus and other limbic areas to the mamillary bodies
Mamillary body Relay station for information between fornix and thalamus; helps process memories
Midbrain Topmost part of brainstem; limbic areas here link to the cortex and thalamus, and to clusters of nerve cell bodies known as basal ganglia
Pituitary gland
Olfactory bulbs The brain's "smell processors"; they are "hard-wired" into the limbic system, which helps explain why the sense of smell can evoke such strong memories and emotional responses
Hippocampus Involved with learning, recognizing new experiences, and short-term memory
Pons Part of the brainstem; not part of the limbic system
Amygdala Involved in emotions such as anger, and drives such as thirst, hunger, and sexual desire
Parahippocampal gyrus Modifies expression of forceful emotions; forms and recalls memories of scenes and views

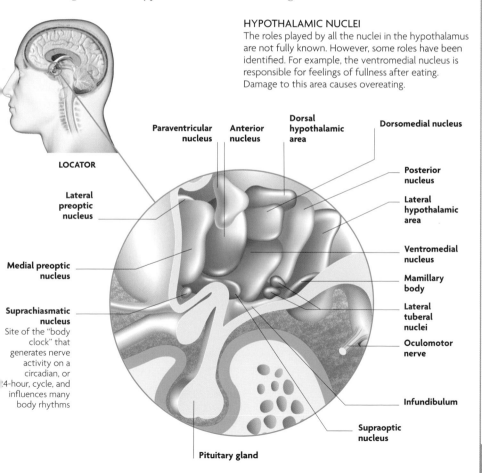

THE HYPOTHALAMUS

The hypothalamus, which literally means "below the thalamus," is about the size of a sugar cube and contains numerous tiny clusters of neurons called nuclei. It forms an important part of the relationship between the brain and the body, and is usually regarded as the vital integrating center of the limbic system. A stalk below links it to the pituitary gland (see p.134), which helps regulate the activity of the endocrine system, including the thyroid and adrenal glands. The hypothalamus also has complex associations with the rest of the limbic system around it, and with the autonomic parts of the general nervous system. Functions of the hypothalamus include monitoring and regulating vital internal conditions such as nutrient levels, body temperature, water–salt balance, blood flow, the sleep–wake cycle, and the levels of hormones such as sex hormones. The hypothalamus also initiates feelings, actions, and emotions such as hunger, thirst, rage, and terror.

HYPOTHALAMIC NUCLEI

The roles played by all the nuclei in the hypothalamus are not fully known. However, some roles have been identified. For example, the ventromedial nucleus is responsible for feelings of fullness after eating. Damage to this area causes overeating.

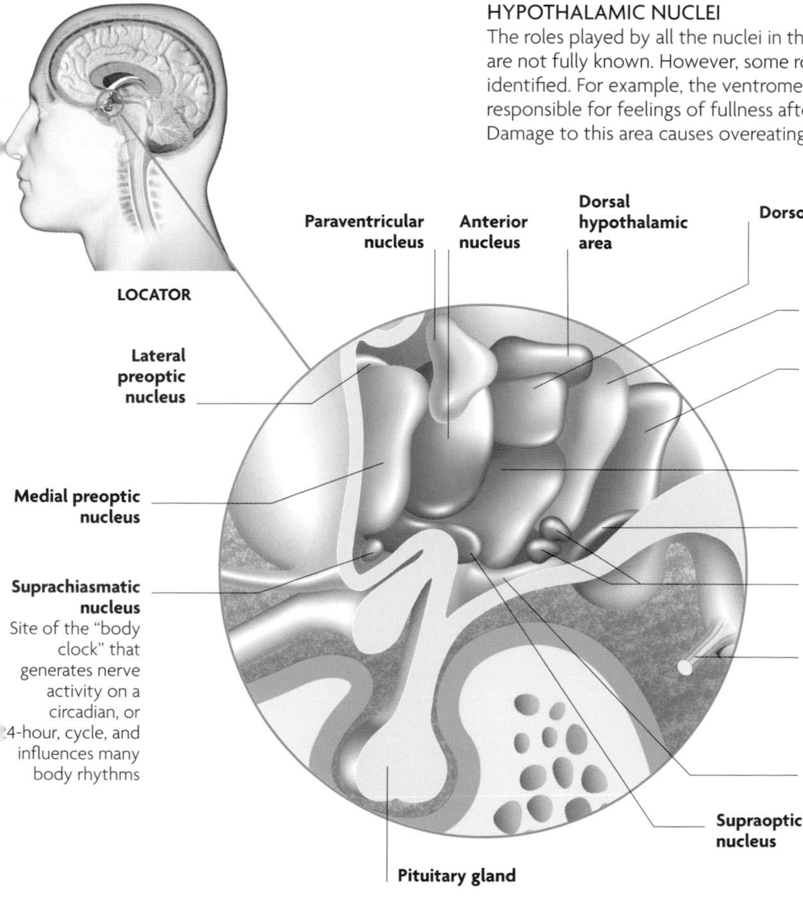
LOCATOR
Lateral preoptic nucleus
Medial preoptic nucleus
Suprachiasmatic nucleus
Site of the "body clock" that generates nerve activity on a circadian, or 24-hour, cycle, and influences many body rhythms
Paraventricular nucleus
Anterior nucleus
Dorsal hypothalamic area
Dorsomedial nucleus
Posterior nucleus
Lateral hypothalamic area
Ventromedial nucleus
Mamillary body
Lateral tuberal nuclei
Oculomotor nerve
Infundibulum
Supraoptic nucleus
Pituitary gland

THE RETICULAR FORMATION

The reticular formation is a series of long, slim nerve tracts that are located in much of the length of the brainstem (see p.93). Its fibers extend to the cerebellum behind, the diencephalon above, and the spinal cord below. The reticular formation comprises several distinct neural systems, each with its own neurotransmitter (the chemical that passes on nerve signals at the tiny junctions, or synapses, between neurons). One of the reticular formation's functions is to operate an arousal system, known as the reticular activating system (RAS), that keeps the brain awake and alert. The reticular formation also includes the cardioregulatory and respiratory centers that control heart rate and breathing, and other essential centers.

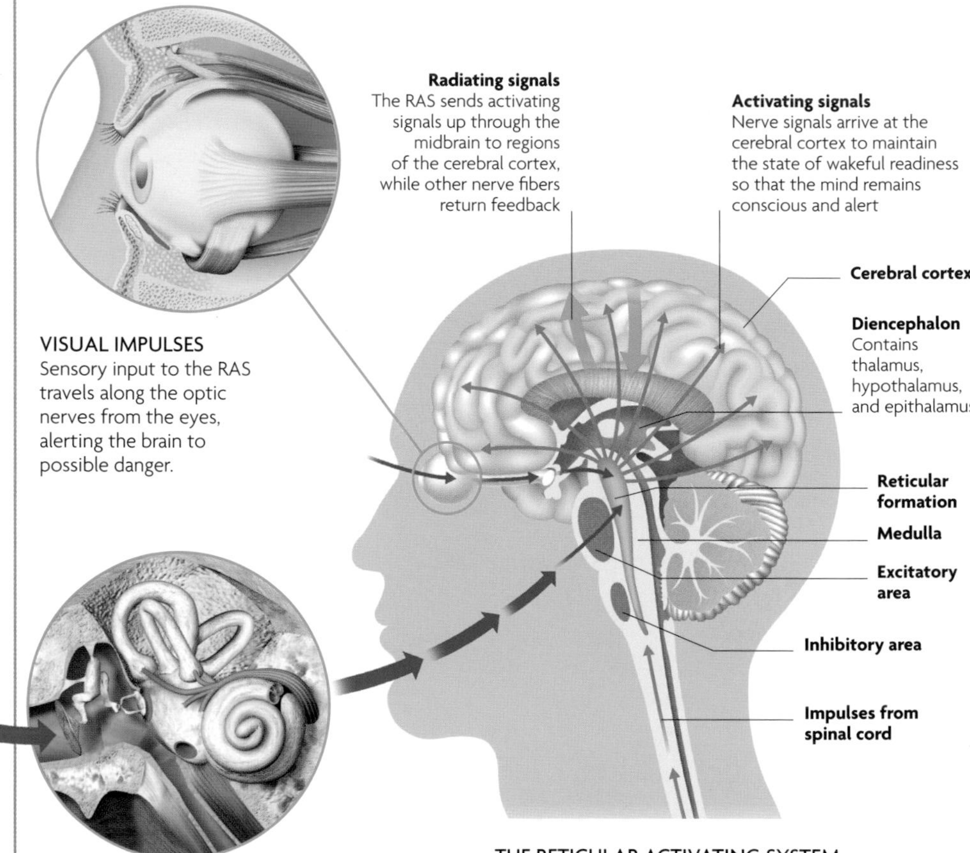
Radiating signals
The RAS sends activating signals up through the midbrain to regions of the cerebral cortex, while other nerve fibers return feedback
Activating signals
Nerve signals arrive at the cerebral cortex to maintain the state of wakeful readiness so that the mind remains conscious and alert
Cerebral cortex
Diencephalon
Contains thalamus, hypothalamus, and epithalamus
Reticular formation
Medulla
Excitatory area
Inhibitory area
Impulses from spinal cord

VISUAL IMPULSES
Sensory input to the RAS travels along the optic nerves from the eyes, alerting the brain to possible danger.

AUDITORY IMPULSES
The RAS filters out insignificant sensory information, such as background noise, and reacts if there is a change in input.

THE RETICULAR ACTIVATING SYSTEM
Long, slender pathways of nerve fibers in the reticular formation, which is located within the brainstem, detect incoming sensory information from many sources. They send activating signals to the higher centers of the brain.

SLEEP CYCLES

During sleep, much of the body rests, but not the brain. Its billions of neurons continue to send signals, as shown by EEG traces. Sleep occurs in cycles, which are made up of lengthening phases of REM (rapid eye movement) sleep, when dreaming occurs, and four stages of NREM (nonrapid eye movement sleep), which is dreamless. In stage 1, sleep is light: people wake relatively easily, and brain waves are active. In stage 2, brain waves begin to slow down. In stage 3, fast and slow waves are interspersed. In stage 4, the deepest stage, there are slow waves only.

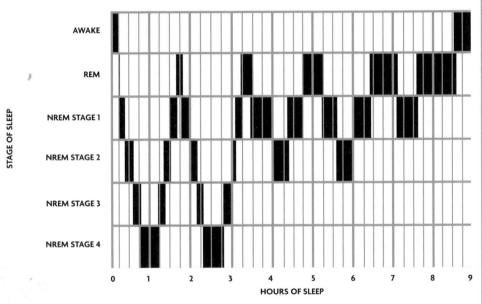

STAGE OF SLEEP

AWAKE
REM
NREM STAGE 1
NREM STAGE 2
NREM STAGE 3
NREM STAGE 4

0 1 2 3 4 5 6 7 8 9

HOURS OF SLEEP

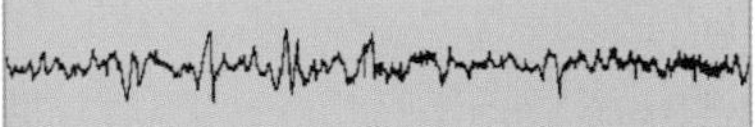

NREM SLEEP: STAGE 1

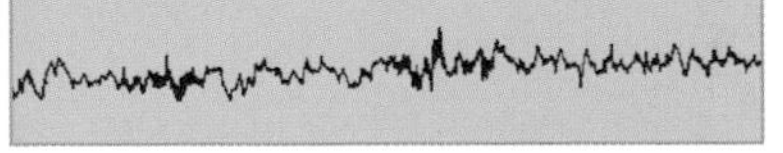

NREM SLEEP: STAGE 2

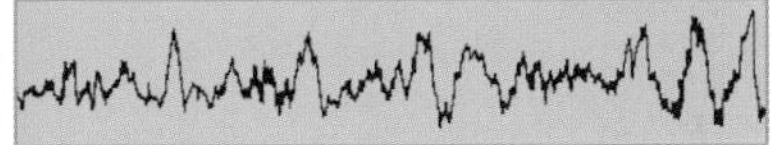

NREM SLEEP: STAGE 3

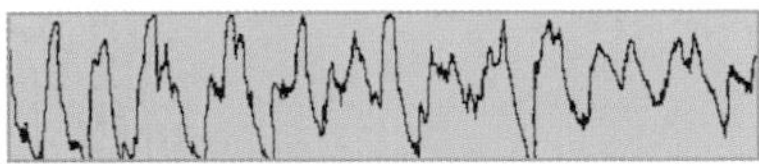

NREM SLEEP: STAGE 4

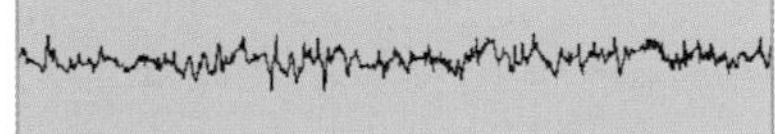

REM SLEEP

SLEEP STAGES

EEG traces show different waveforms of brain activity for each sleep stage. As the body reaches the later stages, body temperature, heart rate, breathing rate, and blood pressure all reduce. During REM sleep, these functions increase slightly and dreams usually occur.

SPINAL CORD

THE NERVE FIBERS OF THE SPINAL CORD LINK THE BRAIN WITH THE TORSO, ARMS, AND LEGS. THE CORD IS MORE THAN A PASSIVE CONDUIT FOR NERVE SIGNALS. WHEN NECESSARY, IT CAN BYPASS THE BRAIN: FOR EXAMPLE, IN REFLEX ACTIONS.

SPINAL CORD ANATOMY

The spinal cord is a bundle of nerve fibers (axons) about 16–18 in (40–45 cm) long. It is only slightly wider than a pencil for most of its length, tapering to a thread-like tail at the lower (lumbosacral) part of the spine. Branching out from the cord are 31 pairs of spinal nerves, which carry sensory information to the cord about conditions within the body and transmit the sense of touch from the skin. They also convey mot information to muscles throughout the boc and to glands within the chest and abdome

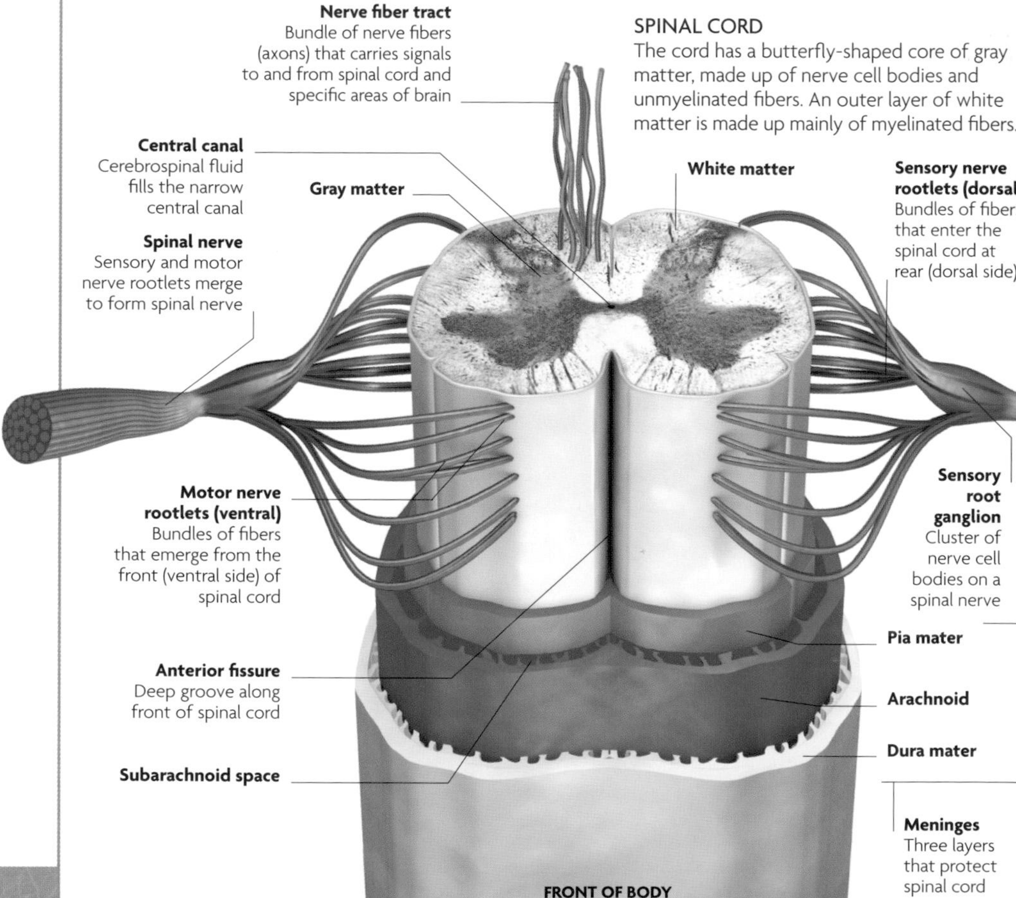

Nerve fiber tract
Bundle of nerve fibers (axons) that carries signals to and from spinal cord and specific areas of brain
SPINAL CORD
The cord has a butterfly-shaped core of gray matter, made up of nerve cell bodies and unmyelinated fibers. An outer layer of white matter is made up mainly of myelinated fibers.
Central canal
Cerebrospinal fluid fills the narrow central canal
Gray matter
White matter
Sensory nerve rootlets (dorsal
Bundles of fiber that enter the spinal cord at rear (dorsal side)
Spinal nerve
Sensory and motor nerve rootlets merge to form spinal nerve
Motor nerve rootlets (ventral)
Bundles of fibers that emerge from the front (ventral side) of spinal cord
Sensory root ganglion
Cluster of nerve cell bodies on a spinal nerve
Pia mater
Anterior fissure
Deep groove along front of spinal cord
Arachnoid
Dura mater
Subarachnoid space
Meninges
Three layers that protect spinal cord
FRONT OF BODY

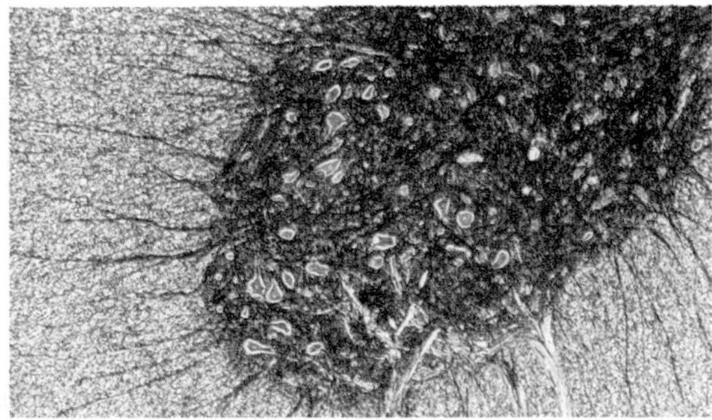

SPINAL GRAY MATTER
This microscopic view of a cross section through the spinal cord shows a brown-stained "wing" of the butterfly-shaped gray matter, which lies at the cord's center.

NERVE CROSSOVER

Bundles of nerve fibers (axons) in the left and right sides of the spinal cord do not all pass straight up into the left and right sides of the brain. In the uppermost portion of the spinal cord and the lower brainstem (the part called the medulla; see p.93), many of the fibers cross over, or decussate, to the other side—left to right, and right to left. This means that nerve signals about, for example, touch sensations on the left side of the body reach the touch center (somatosensory cortex) on the right side of the brain. Likewise, motor signals from the right motor cortex in the brain and the right side of the cerebellum travel to the muscles on the left side of the body. Different major bundles, or tracts, of fibers decussate at slightly different levels. About one-tenth of those that cross over do so in the upper spinal cord, and the remainder cross over in the medulla.

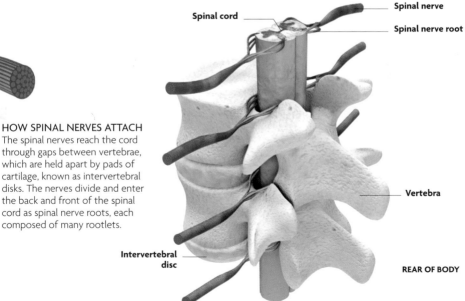

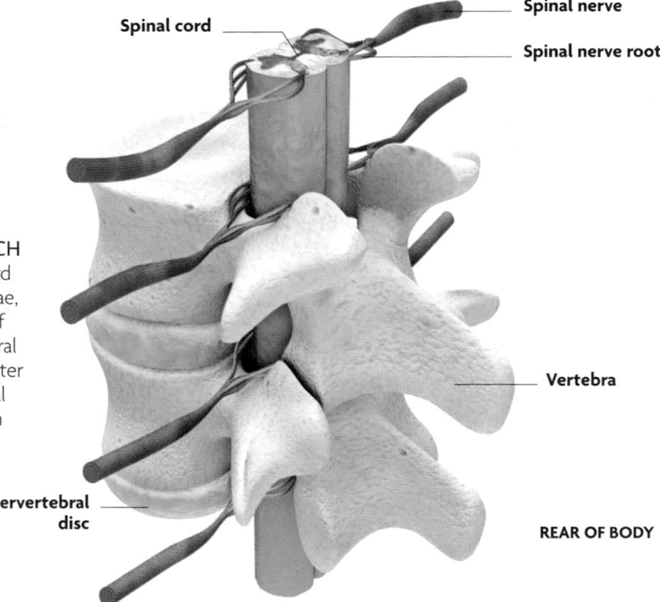
Spinal cord
Spinal nerve
Spinal nerve root
Vertebra
Intervertebral disc
REAR OF BODY

HOW SPINAL NERVES ATTACH
The spinal nerves reach the cord through gaps between vertebrae, which are held apart by pads of cartilage, known as intervertebral disks. The nerves divide and enter the back and front of the spinal cord as spinal nerve roots, each composed of many rootlets.

PROTECTION OF THE SPINAL CORD

The spinal cord is located inside the spinal canal, which is a long tunnel within the aligned column of backbones (vertebrae). This vertebral column, along with its strengthening ligaments and muscles, bends and flexes the cord, but also guards it from direct knocks and blows. Within the spinal canal, the circulating cerebrospinal fluid (see p.89) acts as a shock-absorber and the epidural space provides a cushioning layer of fat and connective tissue. The epidural tissues lie between the periosteum (the membrane that lines the bone of the spinal canal) and the dura mater, the outer layer of the meninges.

INSIDE THE SPINAL CANAL

A cross section of the vertebral column in the neck (cervical) region shows how the spinal cord nestles in the well-padded bony cavity. Although the vertebrae shift position as the trunk of the body moves, the spinal cord remains well supported and protected.

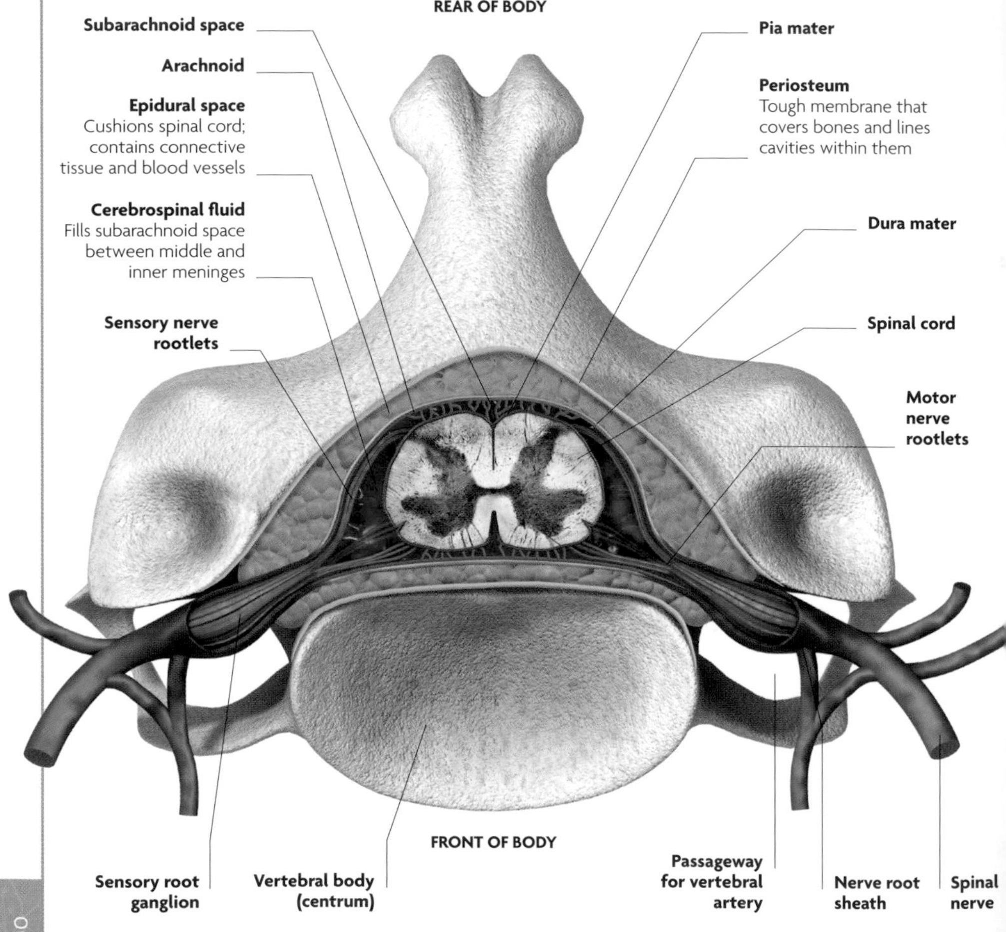
REAR OF BODY
Subarachnoid space
Arachnoid
Epidural space Cushions spinal cord; contains connective tissue and blood vessels
Cerebrospinal fluid Fills subarachnoid space between middle and inner meninges
Sensory nerve rootlets
Pia mater
Periosteum Tough membrane that covers bones and lines cavities within them
Dura mater
Spinal cord
Motor nerve rootlets
FRONT OF BODY
Sensory root ganglion
Vertebral body (centrum)
Passageway for vertebral artery
Nerve root sheath
Spinal nerve

EXTENT OF THE SPINAL CORD

While the body is growing, the spinal cord does not continue to lengthen the way that the spinal bones do. By adulthood, it extends from the brain down to the first lumbar vertebra (L1) in the lower back. Here, the cord forms a cone-shaped ending that tapers to a slender, tail-like filament, known as the filum terminale. This extends down through the lumbar and sacral vertebrae to the coccyx.

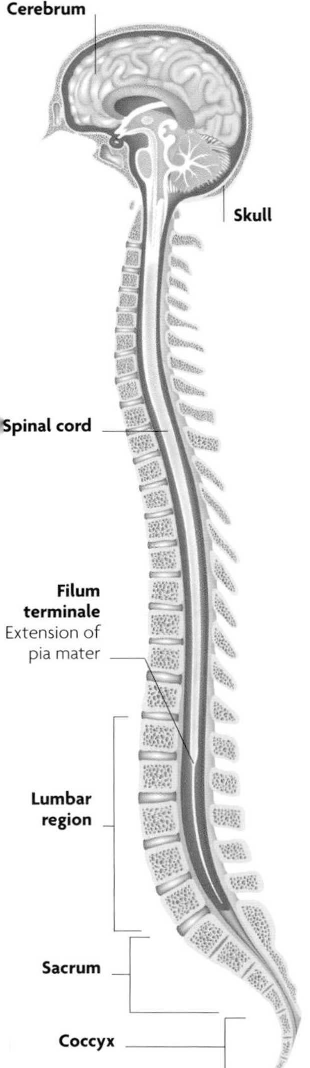

Cerebrum
Skull
Spinal cord
Filum terminale
Extension of pia mater
Lumbar region
Sacrum
Coccyx

NERVE TRACTS OF SPINAL CORD

In the white matter of the spinal cord, nerve fibers are grouped into main bundles, or tracts, according to the direction of the nerve signals that they carry and the type of signals they transmit and respond to, such as pain sensations or temperature. Some tracts connect and relay impulses between a few local pairs of spinal nerves, without sending fibers up to the brain. The central grey matter of the cord is organized into horns, or columns.

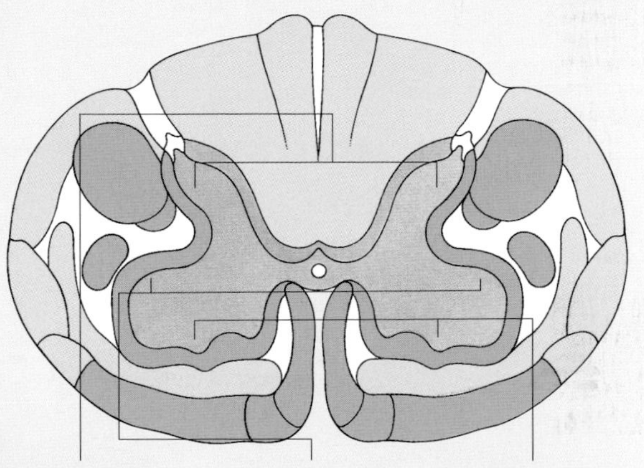

Dorsal (back) horns
Neurons here receive sensory information about touch, balance, muscle activity, and temperature

Lateral (side) horns
Neurons here monitor and regulate internal organs, such as the heart, lungs, stomach, and intestines

Ventral (front) horns
Neurons here send signals along motor fibers to skeletal muscles, causing them to contract and move

ASCENDING TRACTS
These ascending tracts are bundles of nerve fibers that relay impulses about body sensations, and inner sensors such as pain, up the spinal cord to the brain.

DESCENDING TRACTS
These descending tracts convey motor signals from the brain to the skeletal muscles of the torso and limbs in order to bring about voluntary movements.

PERIPHERAL NERVES

THE BODY'S NETWORK OF PERIPHERAL NERVES COMMUNICATES WITH THE BRAIN AND SPINAL CORD. SENSORY FIBERS CARRY MESSAGES FROM SENSE AND INTERNAL ORGANS, WHILE MOTOR FIBERS CONTROL MUSCLE AND GLAND ACTIVITY.

CRANIAL NERVES

The 12 pairs of cranial nerves connect to the brain directly, not via the spinal cord. Some perform sensory functions for organs and tissues in the head and neck, while others have motor functions. The nerves with predominantly motor fibers also contain some sensory fibers that convey information to the brain about the amount of stretch and tension in the muscles they serve, as part of the proprioceptive sense (see p.73). Most of the cranial nerves are named according to the body parts they serve, such as the optic nerves (eyes). By convention, the nerves are also identified by Roman numerals, so the trigeminal nerve, for example, is cranial V (five).

Olfactory nerve (I, sensory)
Relays information about smells from the olfactory epithelium inside the nose, via the olfactory bulbs and tracts, to the brain's limbic centers.

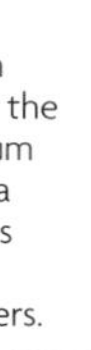

Trigeminal nerve (V, two sensory and one mixed branch)
Ophthalmic and maxillary branches send signals from the eye, face, and teeth; mandibular fibers control chewing and send sensory signals from the lower jaw.

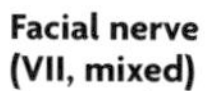

Facial nerve (VII, mixed)
Sensory branches come from the taste buds of the front two-thirds of the tongue; motor fibers run to the muscles of facial expression and to the salivary and lacrimal glands.

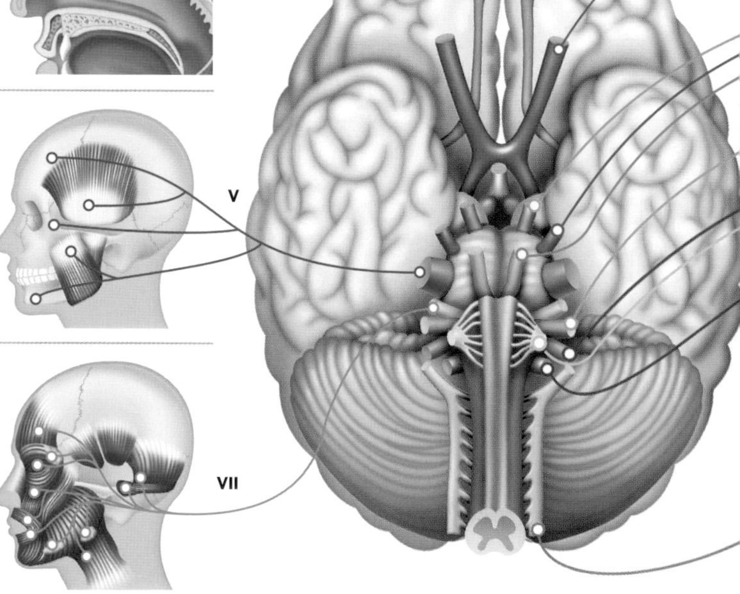
I
V
VII

VIEW FROM BELOW

In this view of the underside of the brain, the pairs of cranial nerves are revealed as joining mainly to the lower regions of the brain. Some of these cranial nerves have a sensory function, taking impulses to the brain. Others have a motor function, carrying nerve signals from the brain to various muscles and glands. Some cranial nerves are mixed, with both sensory and motor nerve fibers.

Optic nerve (II, sensory)
The optic nerve sends visual messages from the rod and cone cells in the retina of the eye to the visual cortex in the brain; parts of the two nerves cross at the optic chiasm (see p.122), where they form bands of nerve fibers, called optic tracts. Each nerve consists of a bundle of about one million sensory fibers – it carries the most information of any cranial nerve.

II

Oculomotor, trochlear, and abducens nerves (III, IV, VI, mainly motor)
These three nerves regulate the voluntary movements of the eye muscles, to move the eyeball and eyelids; the oculomotor also controls pupil constriction by the iris muscles and focusing changes in the lens by the ciliary muscles.

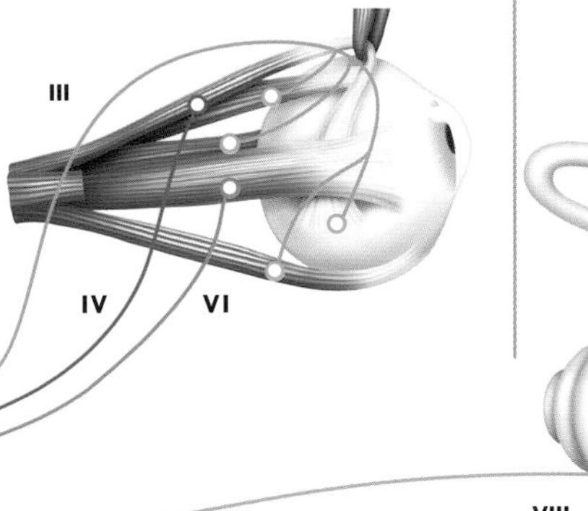
III
IV
VI

Vestibulocochlear nerve (VIII, sensory)
The vestibular branch sends nerve signals from the inner ear about head orientation and balance; the cochlear branch brings signals from the ear concerning sound and hearing.

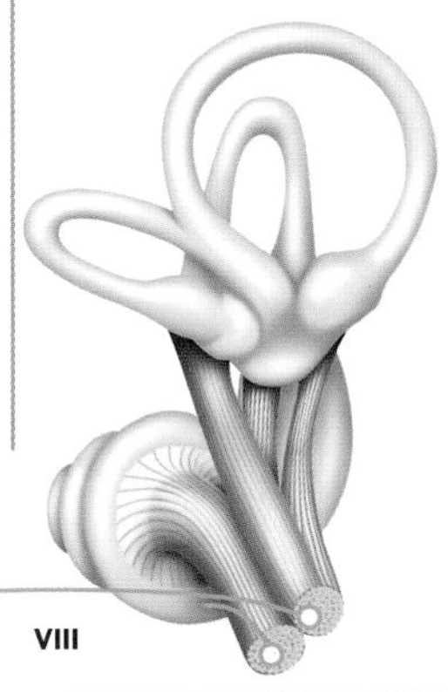
VIII

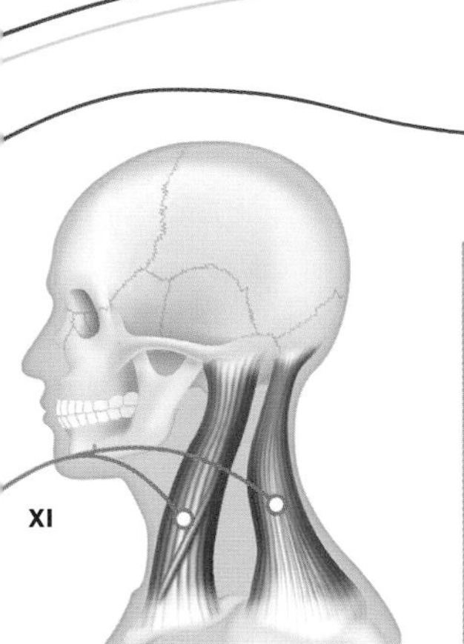
XI

Spinal accessory nerve (XI, mainly motor)
This nerve controls muscles and movements in the head, neck, and shoulders. It also stimulates the muscles of the pharynx and larynx, which are involved in swallowing.

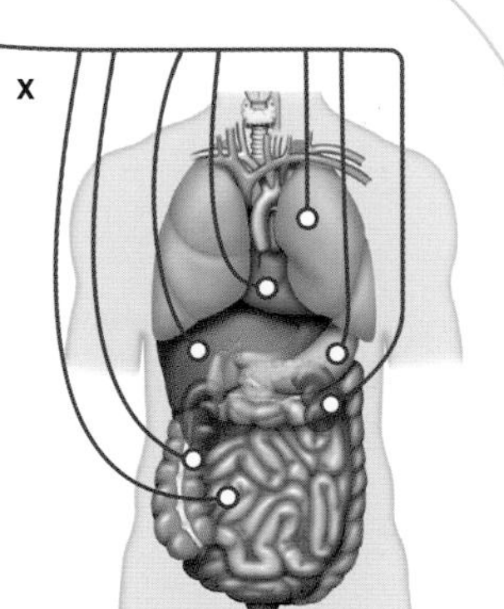
X

Vagus nerve (X, mixed)
The longest and most branched cranial nerve, the vagus has sensory, motor, and autonomic fibers that pass to the lower head, throat, neck, chest, and abdomen; they are involved in many vital body functions, including swallowing, breathing, and heartbeat.

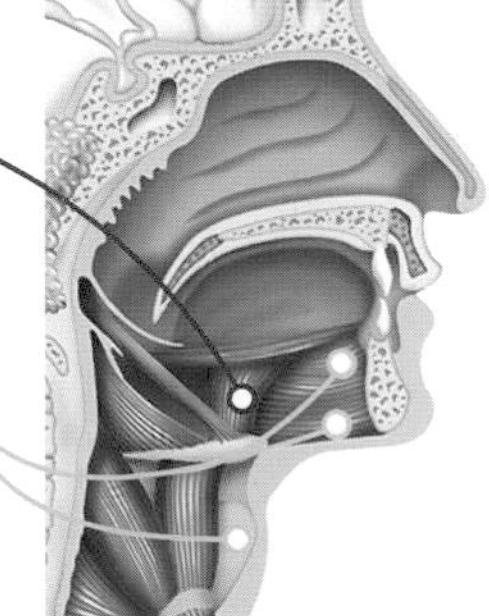
IX
XII

Glossopharyngeal and hypoglossal nerves (IX, XII, both mixed)
Motor fibers of these nerves are involved in tongue movement and swallowing, while sensory fibers relay information about taste, touch, and temperature from the tongue and pharynx.

SPINAL NERVES

The 31 pairs of peripheral spinal nerves emerge from the spinal cord through spaces between the vertebrae (see p.99). Each nerve divides and subdivides into a number of branches; the dorsal branches serve the rear portion of the body, while the ventral branches serve the front and sides. The branches of one spinal nerve may join with other nerves to form meshes called plexuses, where information is shared. The plexuses send signals along secondary nerve branches to areas of complex function or movement.

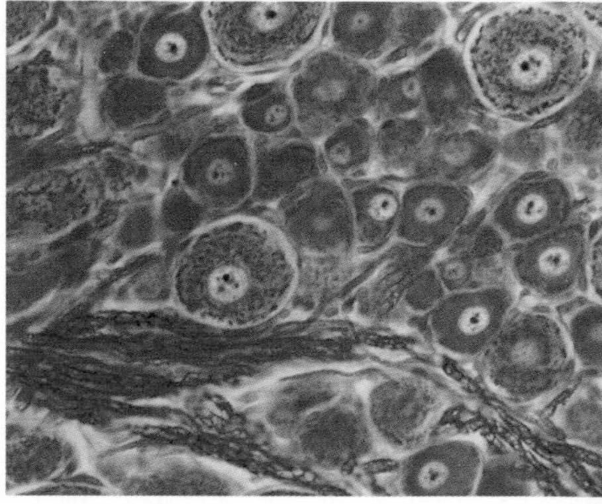

SPINAL NERVE GANGLION

This microscope image shows a section through a cluster of spinal nerve cells (ganglion), where nerve impulses are coordinated. Each neuron (purple) is surrounded by support cells (light blue).

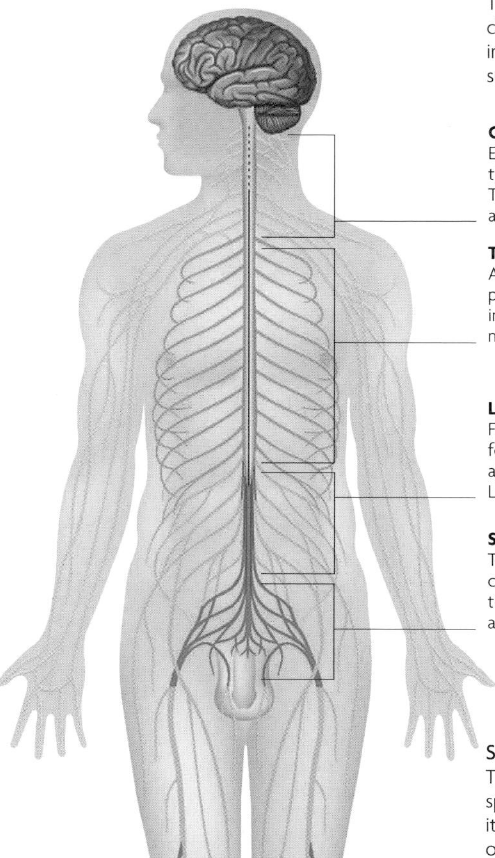

Cervical region (C1–C8)
Eight pairs of cervical spinal nerves form two networks, the cervical (C1–C4) and brachial plexuses (C5–C8/T1). These run to the chest, head, neck, shoulders, arms, and hands, and to the diaphragm.

Thoracic region (T1–T12)
Apart from T1, which is considered part of the brachial plexus, thoracic spinal nerves are connected to the intercostal muscles between the ribs, the deep back muscles, and the abdominal muscles.

Lumbar region (L1–L5)
Four of the five pairs of lumbar spinal nerves (L1–L4) form the lumbar plexus, which supplies the lower abdominal wall and parts of the thighs and legs. L4 and L5 interconnect with the first four sacral nerves (S1–S4).

Sacral region (S1–S5)
Two nerve networks, the sacral plexus (L5–S3) and the coccygeal plexus (S4/S5/Co 1), send branches to the thighs, buttocks, muscles and skin of the legs and feet, and anal and genital areas.

SPINAL REGIONS

The organization and naming of the four main spinal nerve regions reflect the regions of the spine itself—cervical or neck, thoracic or chest, lumbar or lower back, and sacral or base of spine.

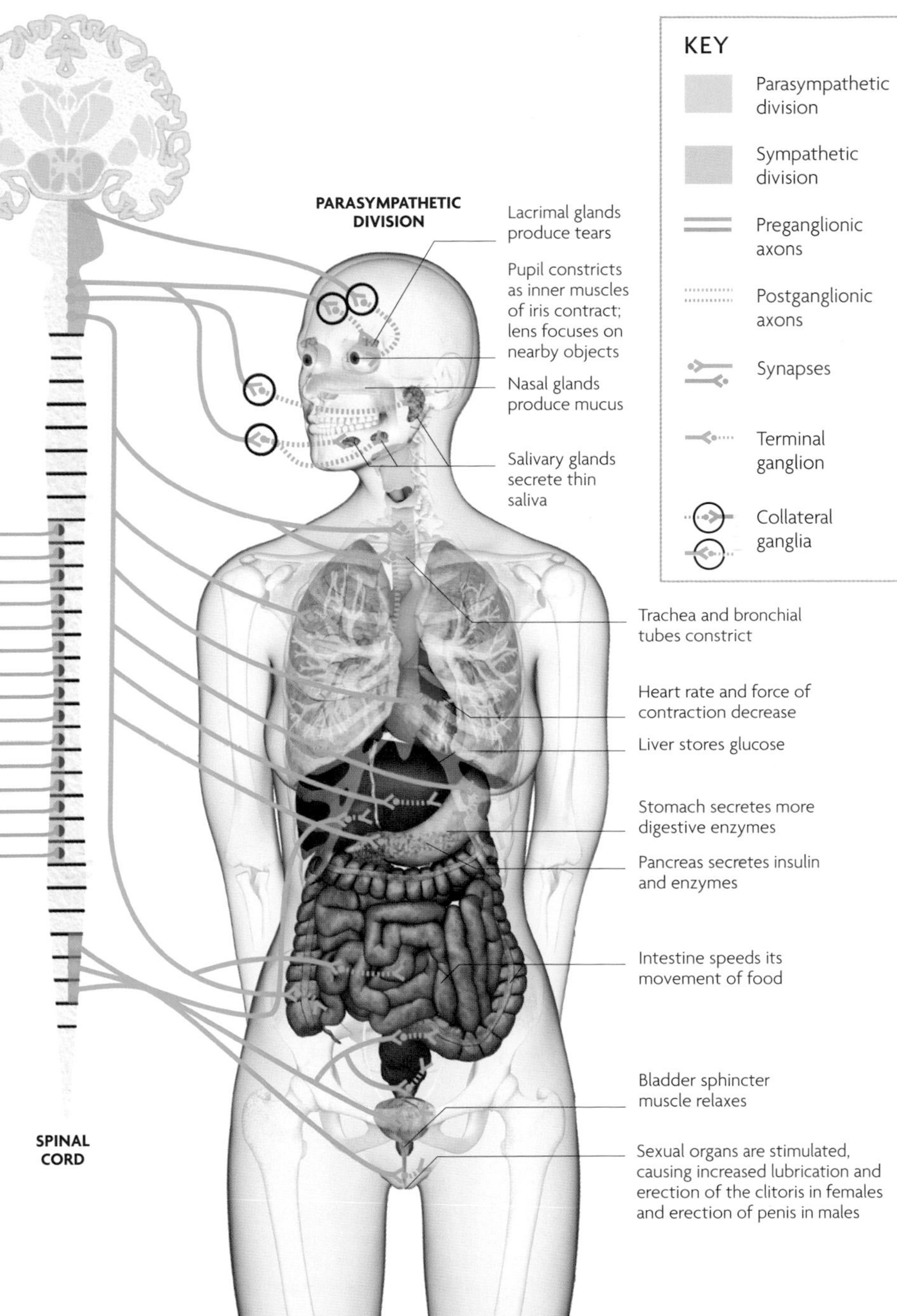
PARASYMPATHETIC DIVISION
Lacrimal glands produce tears
Pupil constricts as inner muscles of iris contract; lens focuses on nearby objects
Nasal glands produce mucus
Salivary glands secrete thin saliva
KEY
Parasympathetic division
Sympathetic division
Preganglionic axons
Postganglionic axons
Synapses
Terminal ganglion
Collateral ganglia
Trachea and bronchial tubes constrict
Heart rate and force of contraction decrease
Liver stores glucose
Stomach secretes more digestive enzymes
Pancreas secretes insulin and enzymes
Intestine speeds its movement of food
Bladder sphincter muscle relaxes
Sexual organs are stimulated, causing increased lubrication and erection of the clitoris in females and erection of penis in males
SPINAL CORD

INVOLUNTARY RESPONSES

There are two main types of involuntary, or automatic, responses, which do not usually involve conscious awareness. One involves reflex actions (see p.105). Reflexes mainly affect muscles normally under voluntary control. The other type of response includes autonomic motor actions. The initial nerve pathways for these responses run along spinal nerves into the spinal cord, then up ascending nerve tracts to the lower autonomic regions of the brain, particularly parts of the limbic system and the hypothalamus. These regions analyze and process the information received, and then use the autonomic pathways to send out motor impulses, as instructions for the involuntary muscles and the glands. Response signals for the parasympathetic and sympathetic systems have separate pathways.

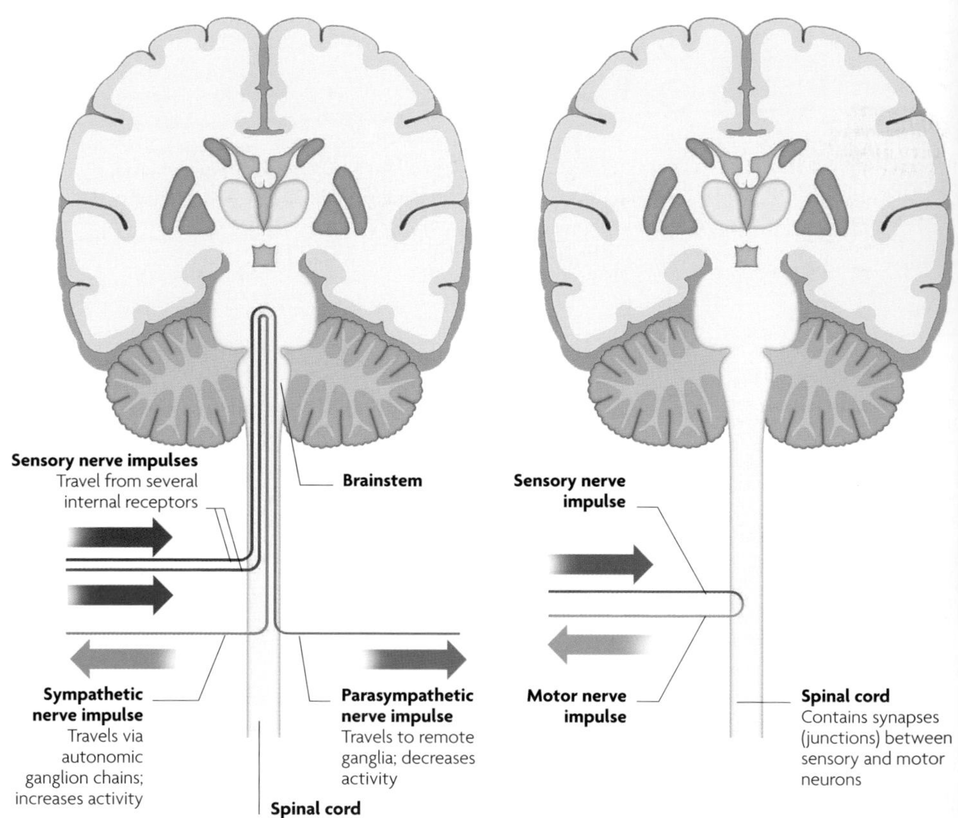
Sensory nerve impulses
Travel from several internal receptors
Brainstem
Sympathetic nerve impulse
Travels via autonomic ganglion chains; increases activity
Parasympathetic nerve impulse
Travels to remote ganglia; decreases activity
Spinal cord
Sensory nerve impulse
Motor nerve impulse
Spinal cord
Contains synapses (junctions) between sensory and motor neurons

AUTONOMIC RESPONSES

Nerve signals pass along spinal nerves and up the spinal cord to the lower autonomic regions of the brain, which output motor impulses in response.

REFLEXES

Sensory signals arrive, and motor signals depart, entirely within the spinal cord, and without the involvement of the brain—although the brain becomes aware soon afterward.

RESPONSES UNDER VOLUNTARY CONTROL

Nervous responses under voluntary control are the opposite of reactions controlled by the ANS. Stimulated by incoming sensory nerve messages, or by conscious thought and intention, the brain's cerebral cortex formulates a central motor plan for a particular movement, and sends out instructions as motor nerve signals to voluntary muscles. As the movement progresses, it is monitored by sensory endings in the muscles, tendons, and joints. The sensory endings update the cerebellum, so that the cerebral cortex can send corrective nerve signals back to the muscles in order to keep the movement coordinated and on course.

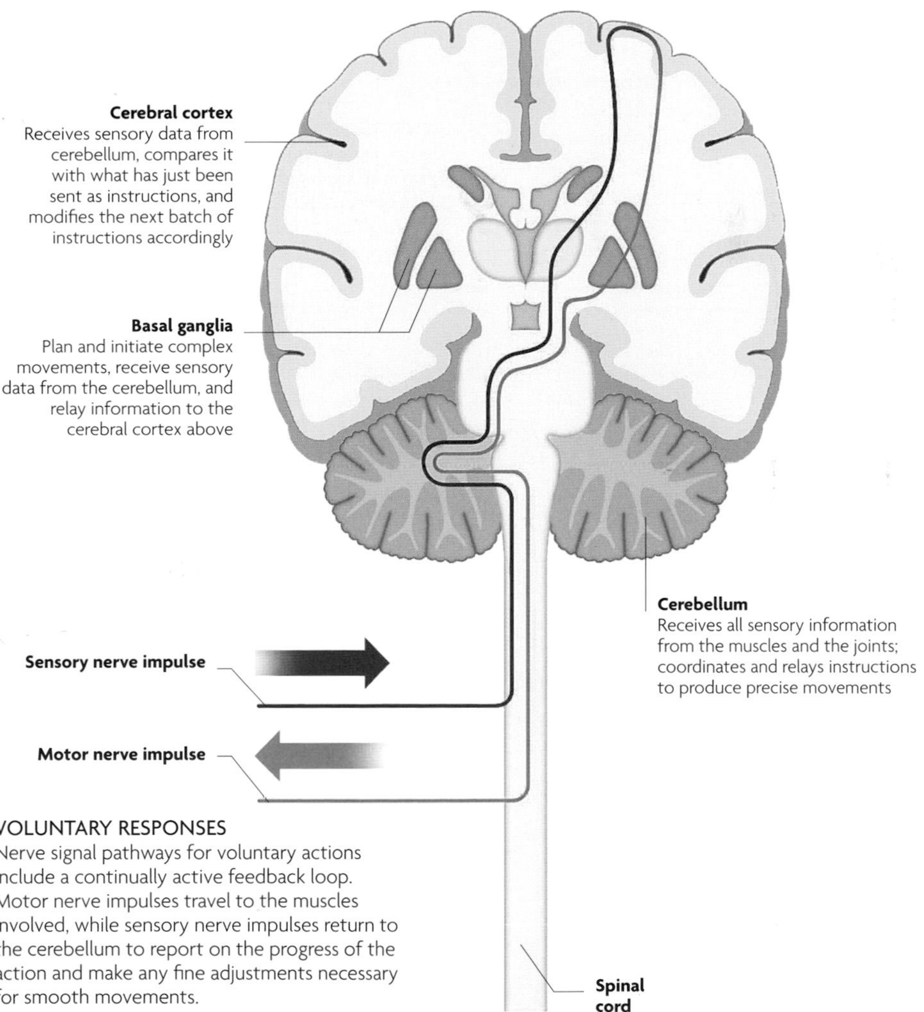
Cerebral cortex
Receives sensory data from cerebellum, compares it with what has just been sent as instructions, and modifies the next batch of instructions accordingly
Basal ganglia
Plan and initiate complex movements, receive sensory data from the cerebellum, and relay information to the cerebral cortex above
Cerebellum
Receives all sensory information from the muscles and the joints; coordinates and relays instructions to produce precise movements
Sensory nerve impulse
Motor nerve impulse
Spinal cord

VOLUNTARY RESPONSES
Nerve signal pathways for voluntary actions include a continually active feedback loop. Motor nerve impulses travel to the muscles involved, while sensory nerve impulses return to the cerebellum to report on the progress of the action and make any fine adjustments necessary for smooth movements.

MEMORIES, THOUGHTS, AND EMOTIONS

MANY MENTAL FACULTIES ARE NOT CONTROLLED BY JUST ONE AREA OF THE BRAIN. FOR EXAMPLE, THERE IS NO SINGLE "MEMORY CENTER". THOUGHTS, FEELINGS, EMOTIONS, AWARENESS, AND MEMORY INVOLVE MANY PARTS OF THE BRAIN.

MAP OF THE CORTEX

Certain regions of the brain's cortex are called primary sensory areas. Each of these receives sensory information from a specific sense. The primary visual cortex, for instance, analyzes data from the eyes. Around each region are association areas, where data from the specific sense is integrated with data from other senses, compared with memories and knowledge, and associated with feelings and emotions. In this way, seeing a particular scene allows us to recognize, identify, and name the objects in it; remember where we saw them previously, recall related sensory data, such as a certain smell, and experience associated emotions again.

CENTERS OF ACTIVITY

Certain areas of the cortex carry out specific brain functions, while others are more generalized. No areas have been identified as exact sites of consciousness or learning.

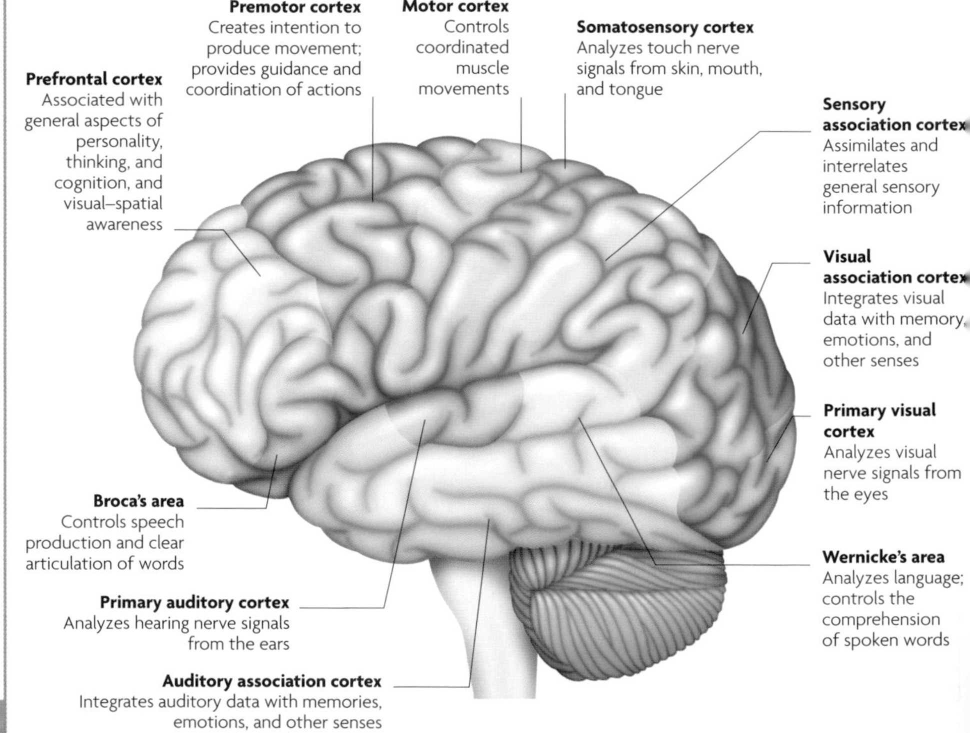
Premotor cortex
Creates intention to produce movement; provides guidance and coordination of actions
Motor cortex
Controls coordinated muscle movements
Somatosensory cortex
Analyzes touch nerve signals from skin, mouth, and tongue
Prefrontal cortex
Associated with general aspects of personality, thinking, and cognition, and visual–spatial awareness
Sensory association cortex
Assimilates and interrelates general sensory information
Visual association cortex
Integrates visual data with memory, emotions, and other senses
Primary visual cortex
Analyzes visual nerve signals from the eyes
Broca's area
Controls speech production and clear articulation of words
Wernicke's area
Analyzes language; controls the comprehension of spoken words
Primary auditory cortex
Analyzes hearing nerve signals from the ears
Auditory association cortex
Integrates auditory data with memories, emotions, and other senses

MEMORY AND RECALL

Memories are the brain's information store. No single region of the brain processes them as they are being established, nor acts as a storage site for all memories. These processes depend on the significance and time span of the memory, its depth of emotional impact, and its association with specific senses such as eyesight.

Putamen
Stores "subconscious" memories, such as motor skills gained by repetition

Cortex
Stores parts of a memory associated with specific senses and motor actions in their relevant areas

Prefrontal cortex
Controls grasp of passing situations, such as visual–spatial awareness of current surroundings

Amygdala
Recalls powerful emotions associated with memorable events, such as fear

Hippocampus
Establishes long-term memories and knowledge linked with spatial awareness

Temporal lobe
Stores language, words, vocabulary, and speech

AREAS INVOLVED IN MEMORY STORAGE

Various areas of the brain are involved in memory. The hippocampus, for example, helps transfer immediate thoughts and sensory data into short- and long-term stores. If it is damaged, a person can recall events from long ago, before the damage, but not what happened a few hours previously.

THOUGHT IN ACTION

The real-time scanning method fMRI (functional magnetic resonance imaging) reveals tiny localized increases in blood flow. As a result, the scans can pinpoint which areas of the brain are busy during well-defined mental activities, such as studying the visual details of an image, or listening to and understanding speech.

PLANNING A MOVEMENT

The subject of this fMRI scan was asked to think about performing a task during the scan. The image shows activity in both the left and right prefrontal areas and also in both the left and right auditory cortex.

RIGHT SIDE OF BRAIN **LEFT SIDE OF BRAIN**

MAKING THAT MOVEMENT

When actually performing the task, large parts of the premotor and motor cortex show up on the brain's left side. The cerebellum (at the base of the brain) helps control precise muscle coordination.

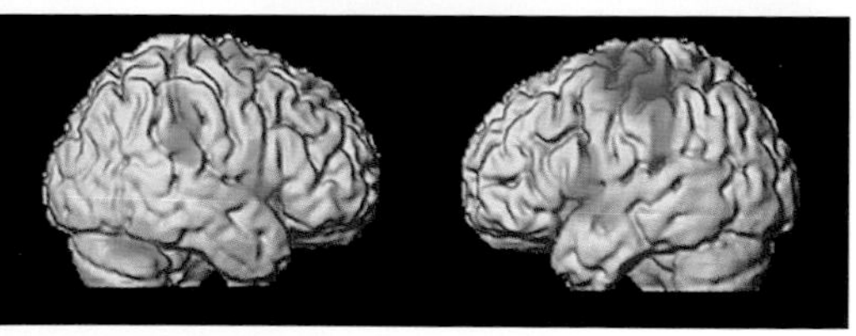

RIGHT SIDE OF BRAIN **LEFT SIDE OF BRAIN**

SMELL, TASTE, AND TOUCH

RECEPTORS THAT SENSE PRESSURE, PAIN, AND TEMPERATURE ARE WIDESPREAD IN THE BODY. TASTE AND SMELL, IN CONTRAST, ARE "SPECIAL SENSES" BECAUSE THEIR RECEPTORS ARE COMPLEX AND LOCALIZED, AND DETECT SPECIFIC STIMULI.

SMELL

Smell is a sense that can detect chemical molecules known as odorants floating in the air. Specialized epithelial tissue provides a smelling zone, known as the olfactory epithelium, on the roof of the nasal cavity. In humans, smell is more sensitive than taste and can distinguish more than 10,000 odors. Smell is important for warning of dangers, such as smoke and poisonous gas, and for appreciating food and drink. The sense of smell tends to deteriorate with age, so young people are able to distinguish a wider range of odors and experience them more vividly than older people.

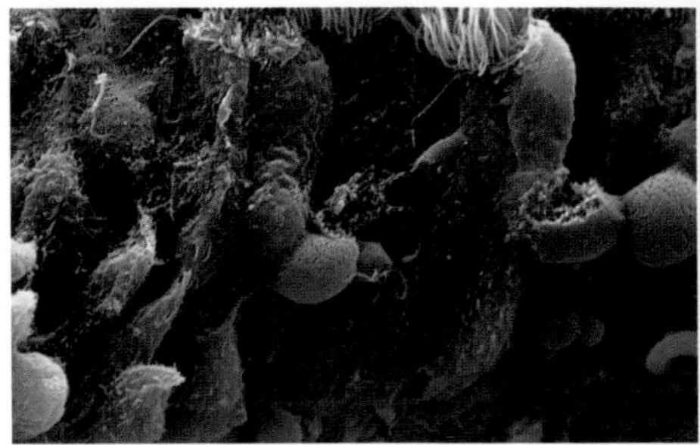

NASAL LINING
Epithelial cells in the lining of the nasal chamber have tufts of hairlike cilia, which wave germ- and odorant-trapping mucus toward the back of the chamber to be swallowed.

HOW WE SMELL

Odor molecules dissolve in the mucus lining the nasal chamber. In the roof of the chamber, they touch the cilia (microscopic, hairlike endings of olfactory receptor cells). If the correct molecule slots into the same-shaped receptor on the cilial membrane, like a key in a lock, a nerve impulse is generated. The impulses are partly processed by intermediate neurons called glomeruli in the olfactory bulb.

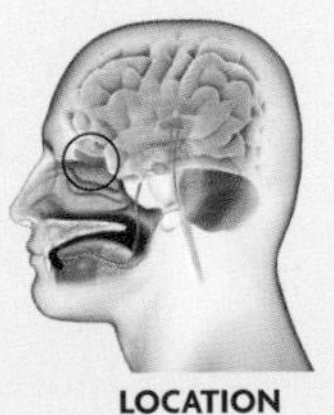

LOCATION

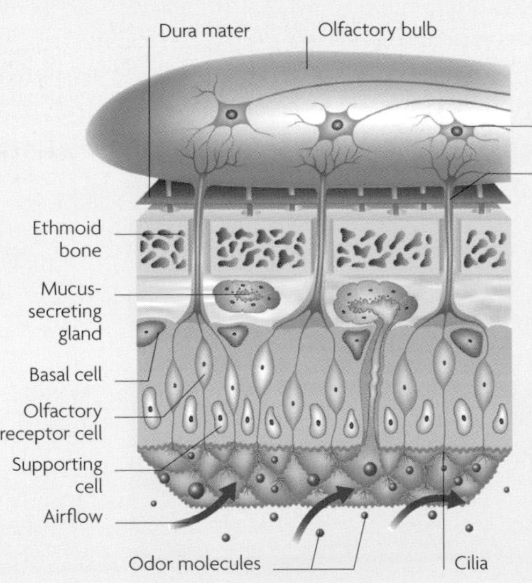
Dura mater
Olfactory bulb
Glomerulus
Nerve fiber
Ethmoid bone
Mucus-secreting gland
Basal cell
Olfactory receptor cell
Supporting cell
Airflow
Odor molecules
Cilia

OLFACTORY EPITHELIUM
When odor molecules make contact with cilia on olfactory receptor cells in the olfactory epithelium, nerve impulses travel along nerve fibers to glomeruli in the olfactory bulb.

NERVE PATHWAYS FOR SMELL AND TASTE

Both smell (olfactory) and taste (gustatory) sensations pass along cranial nerves directly to the brain. Smell signals travel from the olfactory bulbs along the olfactory nerve, which is made up of groups of nerve fibers, to a patch of the cortex located in the temporal lobe. Taste sensations travel along branches of the glossopharyngeal and facial nerves to the gustatory center in the cortex.

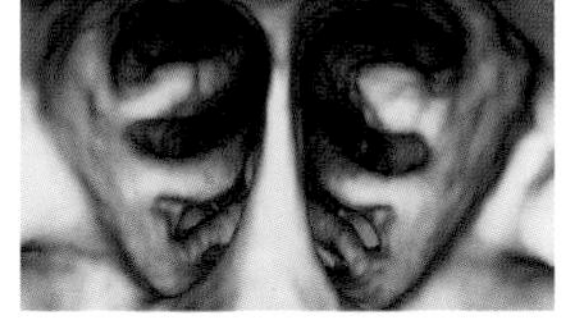

NASAL CHAMBER
A 3D CT scan shows the three shelves of bone known as the conchae on both sides of the nasal chamber.

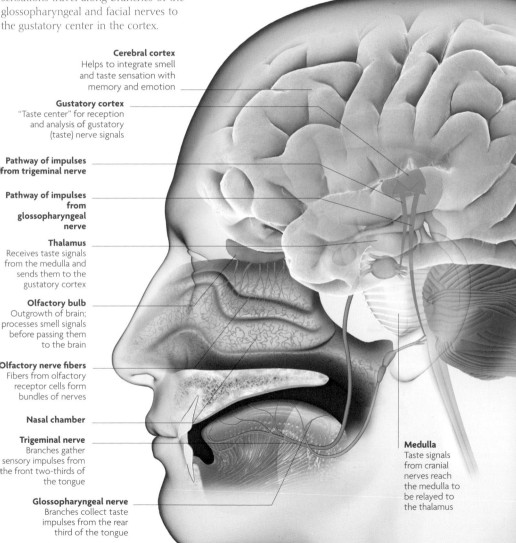

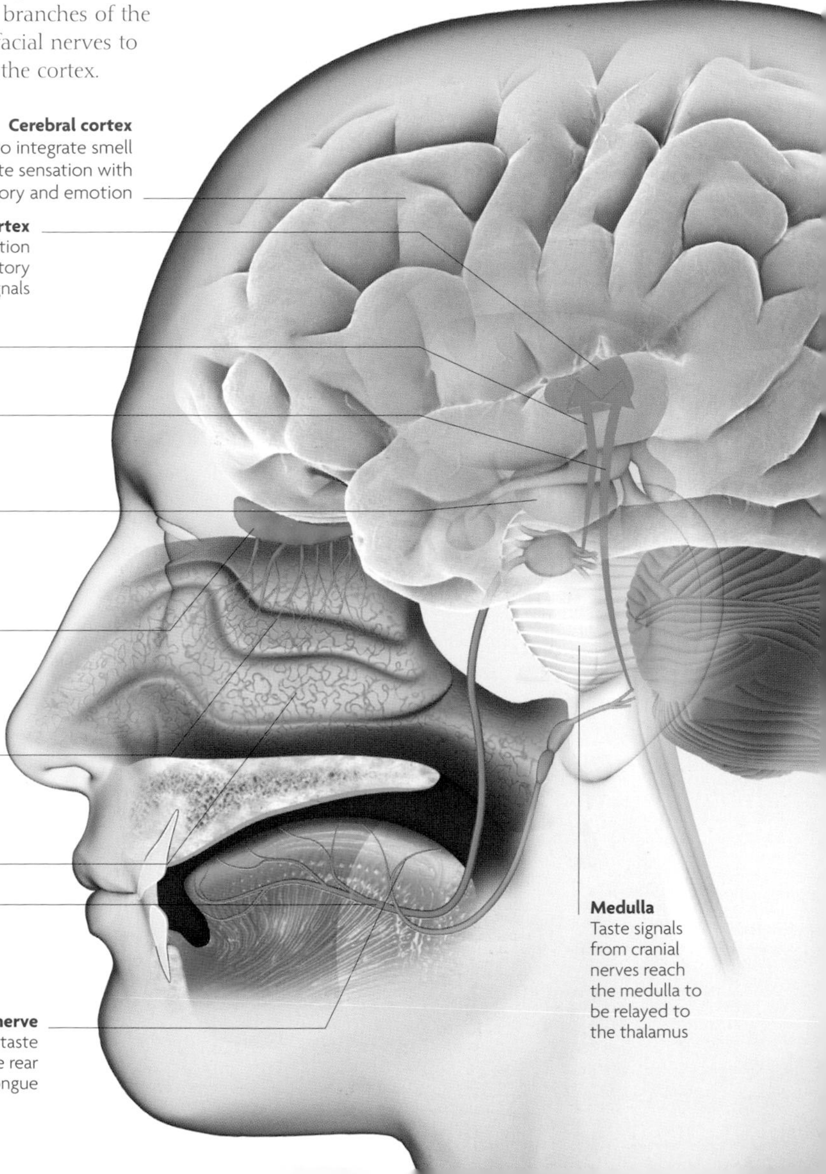

Cerebral cortex
Helps to integrate smell and taste sensation with memory and emotion

Gustatory cortex
"Taste center" for reception and analysis of gustatory (taste) nerve signals

Pathway of impulses from trigeminal nerve

Pathway of impulses from glossopharyngeal nerve

Thalamus
Receives taste signals from the medulla and sends them to the gustatory cortex

Olfactory bulb
Outgrowth of brain; processes smell signals before passing them to the brain

Olfactory nerve fibers
Fibers from olfactory receptor cells form bundles of nerves

Nasal chamber

Trigeminal nerve
Branches gather sensory impulses from the front two-thirds of the tongue

Glossopharyngeal nerve
Branches collect taste impulses from the rear third of the tongue

Medulla
Taste signals from cranial nerves reach the medulla to be relayed to the thalamus

TASTE

Taste works in a similar way to smell. Its gustatory cell (taste) receptors detect specific chemicals dissolved in saliva by a "lock-and-key" method (see p.112). Groups of receptor cells are known as taste buds. A child has about 10,000 taste buds, but with age, their numbers may fall to under 5,000. They are located mainly on and between the pimplelike papillae that dot regions of the tongue's upper surface. There are also some taste buds on the palate (roof of the mouth), throat, and epiglottis.

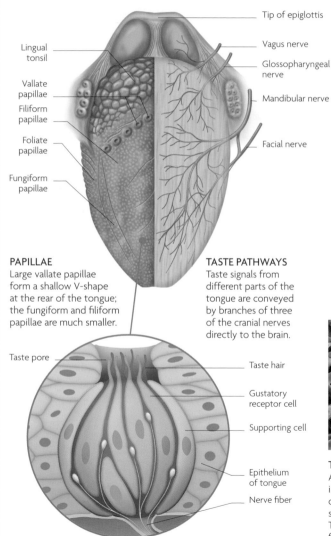

Lingual tonsil
Vallate papillae
Filiform papillae
Foliate papillae
Fungiform papillae
Tip of epiglottis
Vagus nerve
Glossopharyngeal nerve
Mandibular nerve
Facial nerve

TASTE BUDS

Each taste bud is structured much like an orange whose "segments" consist of roughly 25 "gustatory" receptor cells and numerous supporting cells. The receptor cells have hairlike tips that project into a hole (the taste pore) in the tongue's surface. Their nerve fibers gather at the bud base.

PAPILLAE

Large vallate papillae form a shallow V-shape at the rear of the tongue; the fungiform and filiform papillae are much smaller.

TASTE PATHWAYS

Taste signals from different parts of the tongue are conveyed by branches of three of the cranial nerves directly to the brain.

Taste pore
Taste hair
Gustatory receptor cell
Supporting cell
Epithelium of tongue
Nerve fiber

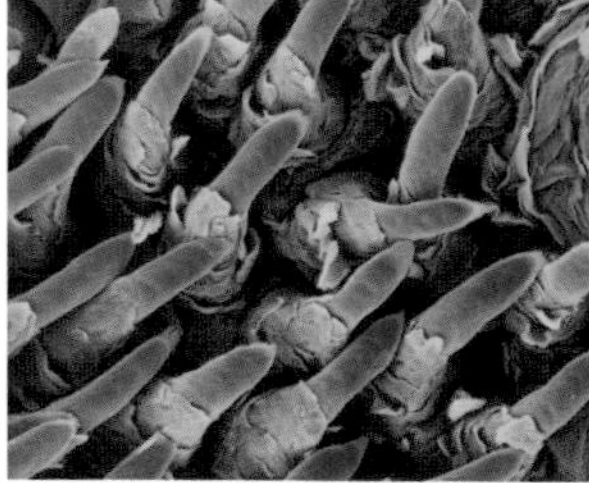

TASTE RECEPTORS

A scanning electron microscope image shows two different types of papillae. The purple conical structures are filiform papillae. The circular pink structure is a fungiform papilla.

TOUCH

The sense of touch comes from microscopic sensory receptors (specialized endings of nerve cells) in the skin or in deeper tissues (see p.182). Some receptors are enclosed in capsules of connective tissue, while others are uncovered. Different shapes and sizes of receptor detect a range of stimuli, such as light touch, heat, cold, pressure, and pain. The receptors relay their signals via the spinal cord and lower brain to a strip curving around the cerebral cortex, known as the somatosensory cortex, or "touch center."

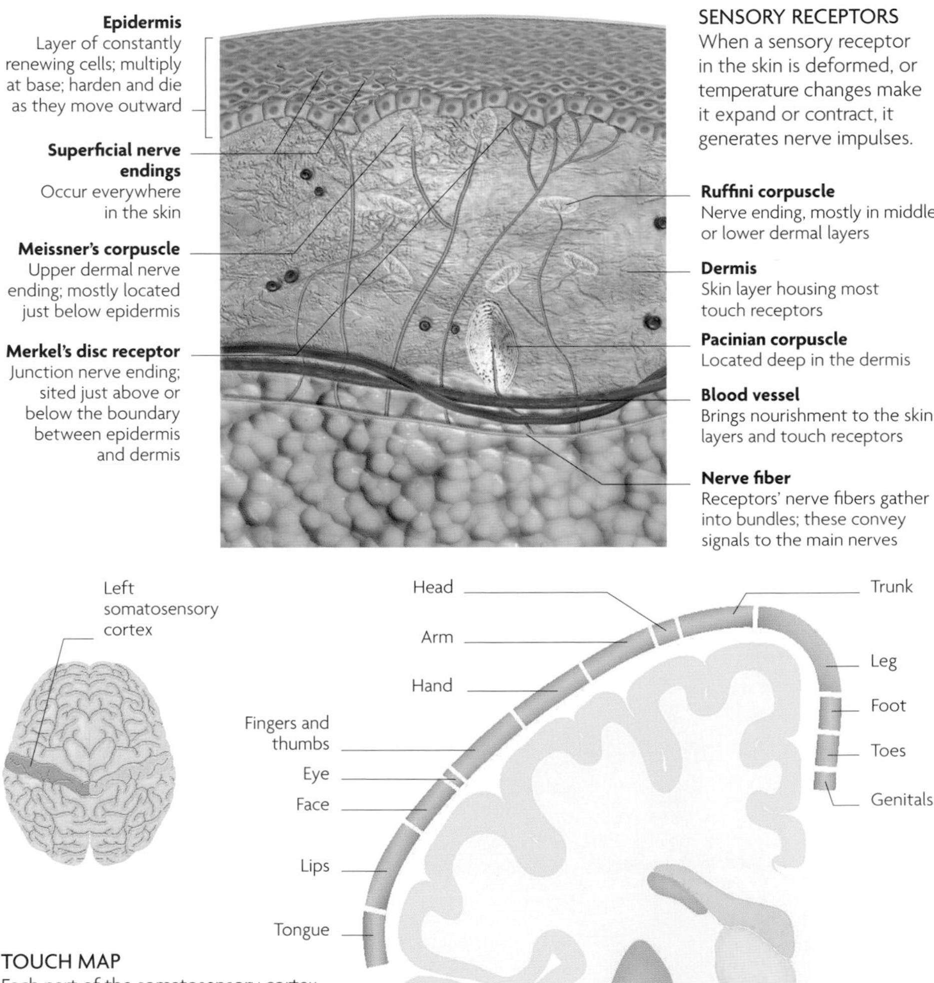
Epidermis
Layer of constantly renewing cells; multiply at base; harden and die as they move outward
Superficial nerve endings
Occur everywhere in the skin
Meissner's corpuscle
Upper dermal nerve ending; mostly located just below epidermis
Merkel's disc receptor
Junction nerve ending; sited just above or below the boundary between epidermis and dermis
Ruffini corpuscle
Nerve ending, mostly in middle or lower dermal layers
Dermis
Skin layer housing most touch receptors
Pacinian corpuscle
Located deep in the dermis
Blood vessel
Brings nourishment to the skin layers and touch receptors
Nerve fiber
Receptors' nerve fibers gather into bundles; these convey signals to the main nerves
Left somatosensory cortex
Head
Arm
Hand
Fingers and thumbs
Eye
Face
Lips
Tongue
Trunk
Leg
Foot
Toes
Genitals

SENSORY RECEPTORS

When a sensory receptor in the skin is deformed, or temperature changes make it expand or contract, it generates nerve impulses.

TOUCH MAP

Each part of the somatosensory cortex (above) receives touch messages from skin around the body, as mapped on this vertical cross section of the brain (right).

EARS, HEARING, AND BALANCE

THE EARS PROVIDE THE SENSE OF HEARING. THEY ALSO DETECT HEAD POSITION AND MOTION, SO THEY ARE ESSENTIAL TO BALANCE. THE FUNCTION OF BOTH HEARING AND BALANCE IS BASED ON THE ACTIVITY OF "HAIR CELL" RECEPTORS.

INSIDE THE EAR

The ear is divided into three parts. The outer ear comprises the ear flap and the slightly S-shaped outer ear canal, which guides sound waves to the second region, the middle ear. The elements of the middle ear amplify the sound waves and transfer them from the air into the fluid of the inner ear. They include the eardrum and the three smallest bones in the body: the auditory ossicles, which span the air-filled cavity of the middle ear. The fluid-filled inner ear changes sound waves to nerve signals inside the snail-shaped cochlea. The middle-ear cavity connects to the throat via the Eustachian tube, and so to the air outside. This connection allows atmospheric pressure to transfer to the cavity, equalizing the air pressure on either side of the eardrum and preventing it from bulging as the outside pressure changes.

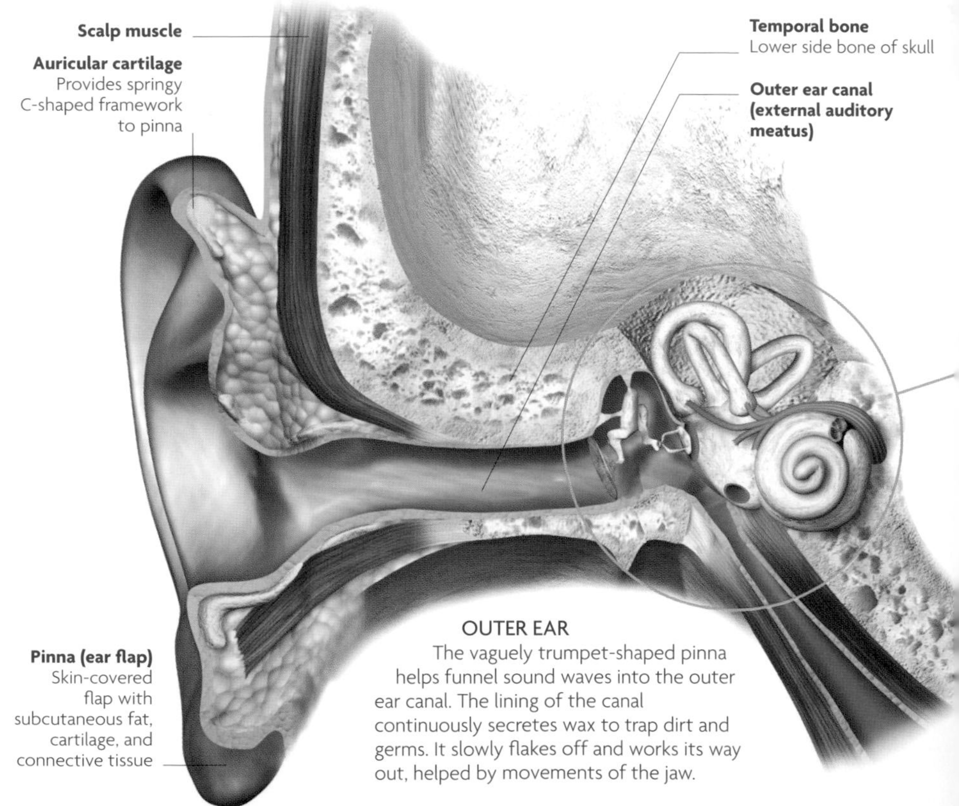
Scalp muscle
Auricular cartilage
Provides springy C-shaped framework to pinna
Temporal bone
Lower side bone of skull
Outer ear canal (external auditory meatus)
Pinna (ear flap)
Skin-covered flap with subcutaneous fat, cartilage, and connective tissue

OUTER EAR

The vaguely trumpet-shaped pinna helps funnel sound waves into the outer ear canal. The lining of the canal continuously secretes wax to trap dirt and germs. It slowly flakes off and works its way out, helped by movements of the jaw.

MIDDLE AND INNER EAR

The ossicles of the middle ear are positioned and connected by miniature ligaments, tendons, and joints, just like larger bones. The cochlea, semicircular canals, and vestibule of the inner ear are linked. They are all filled with fluid, and are encased and protected within the thickness of the skull's temporal bone. They occupy a complex series of tunnels and chambers known as the osseous labyrinth.

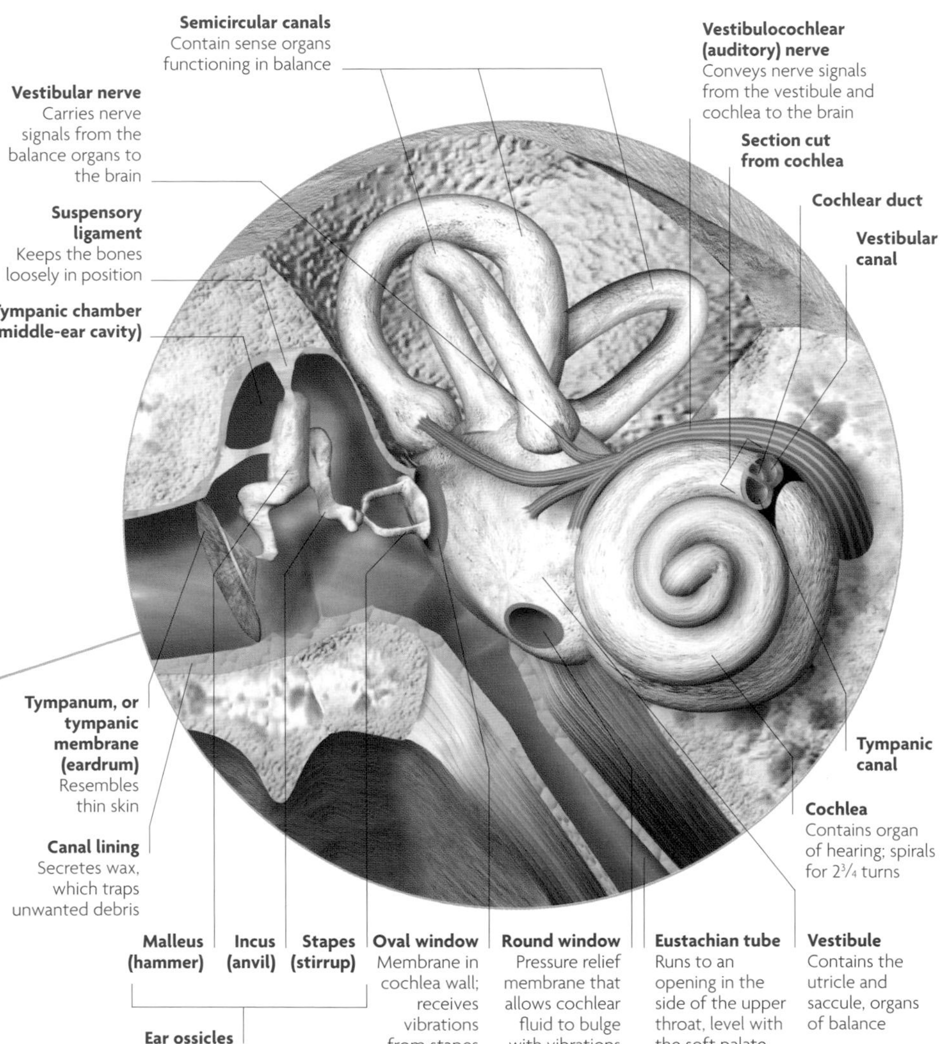

Semicircular canals Contain sense organs functioning in balance
Vestibulocochlear (auditory) nerve Conveys nerve signals from the vestibule and cochlea to the brain
Vestibular nerve Carries nerve signals from the balance organs to the brain
Section cut from cochlea
Suspensory ligament Keeps the bones loosely in position
Cochlear duct
Vestibular canal
Tympanic chamber (middle-ear cavity)
Tympanum, or tympanic membrane (eardrum) Resembles thin skin
Tympanic canal
Cochlea Contains organ of hearing; spirals for 2¾ turns
Canal lining Secretes wax, which traps unwanted debris
Malleus (hammer)
Incus (anvil)
Stapes (stirrup)
Ear ossicles
Oval window Membrane in cochlea wall; receives vibrations from stapes
Round window Pressure relief membrane that allows cochlear fluid to bulge with vibrations
Eustachian tube Runs to an opening in the side of the upper throat, level with the soft palate
Vestibule Contains the utricle and saccule, organs of balance

HOW WE HEAR

Ears act as energy converters, changing pressure differences in air, known as sound waves, into electrochemical nerve impulses. Sound waves usually occur as a complex pattern of frequencies, and they vibrate the eardrum in that same pattern. The vibrations are conducted along the ossicle chain, which rocks like a bent lever and forces the footplate of the stapes to act like a piston, pushing and pulling at the flexible oval window of the cochlea. The motions set off waves through the perilymph fluid inside the cochlea. These, in turn, transfer their vibrational energy to the organ of Corti, which coils within the cochlea.

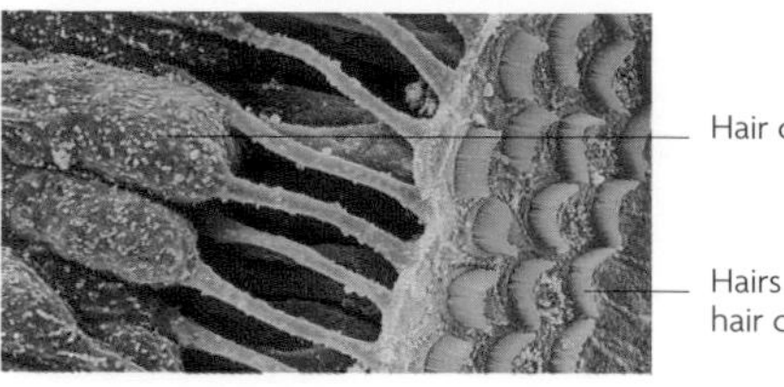
Hair cell
Hairs of hair cell

HAIR CELLS

Within the organ of Corti, with the tectorial membrane removed on the right, each hair cell is seen to have 40–100 hairs arranged in a curve. Nerve fibers run from the cell bases.

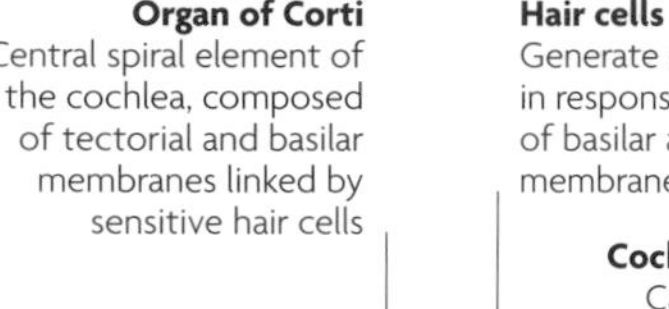

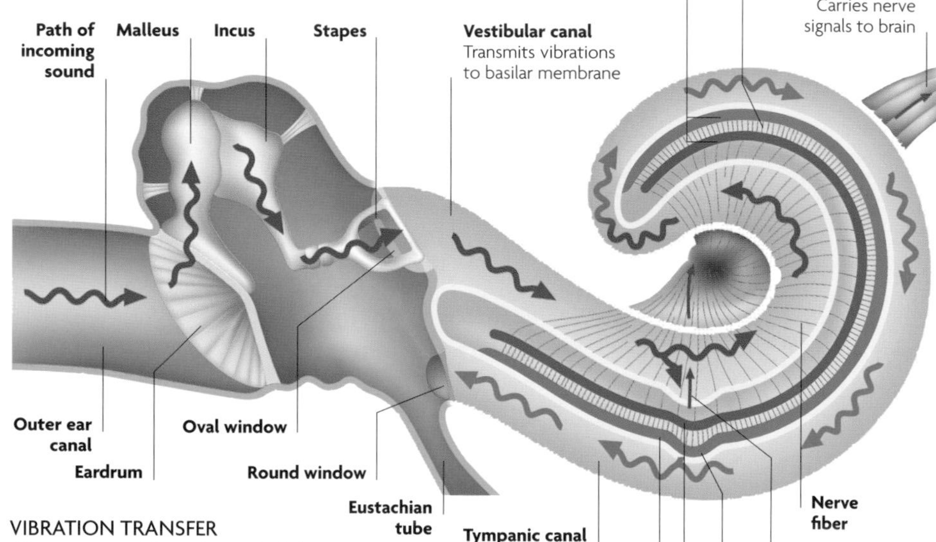
Organ of Corti
Central spiral element of the cochlea, composed of tectorial and basilar membranes linked by sensitive hair cells
Hair cells
Generate nerve signals in response to motion of basilar and tectorial membranes
Cochlear nerve
Carries nerve signals to brain
Path of incoming sound
Malleus
Incus
Stapes
Vestibular canal
Transmits vibrations to basilar membrane
Outer ear canal
Eardrum
Oval window
Round window
Eustachian tube
Tympanic canal
Conveys residual vibrations returning to round window
Tectorial membrane
Tips of hairs from hair cells embed in this membrane
Basilar membrane
Supports the bases of hair cells and their nerve fibers
Nerve fiber
Nerve signal
Frequency response
Organ of Corti "shakes" at a particular point along its length, according to frequency of vibration

VIBRATION TRANSFER

Vibrations travel from the oval window through the fluid in the vestibular canal to the organ of Corti. Here, hair cells on the basilar membrane vibrate, pulling the hairs and stimulating them to produce nerve impulses. These travel via the cochlear nerve to the auditory cortex in the brain for interpretation. Residual vibrations from the vestibular canal pass along the tympanic canal to the round window.

THE PROCESS OF BALANCE

Balance involves analyzing sensory inputs from the eyes, skin, and muscles, and then adjusting the body's position through motor outputs. The vestibule and semicircular canals of the inner ear play a key role, too. The vestibule responds mainly to the position of the head relative to gravity (static equilibrium), while the canals react chiefly to the speed and direction of head movements (dynamic equilibrium).

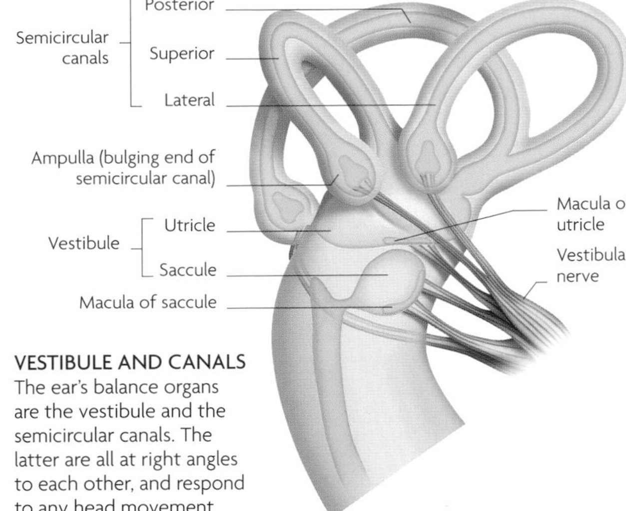
Semicircular canals
Posterior
Superior
Lateral
Ampulla (bulging end of semicircular canal)
Vestibule
Utricle
Saccule
Macula of saccule
Macula of utricle
Vestibular nerve

VESTIBULE AND CANALS

The ear's balance organs are the vestibule and the semicircular canals. The latter are all at right angles to each other, and respond to any head movement.

VESTIBULE

The vestibule's two parts, the utricle and saccule, each have a patch, the macula, containing hair cells. The tips of the cells extend into a membrane covered in heavy mineral crystals (otoliths). With the head level, the saccule's macula is vertical and the utricle's horizontal. As the head bends forward, the hair cells monitor the head's position in relation to the ground.

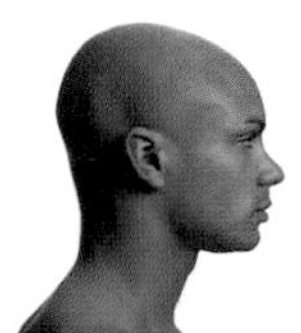
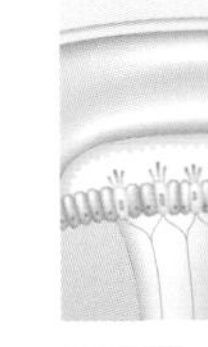
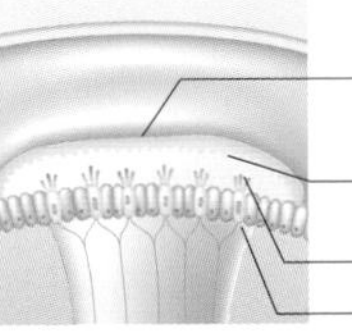
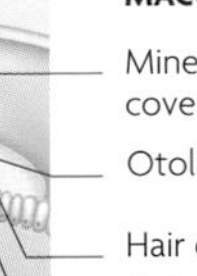
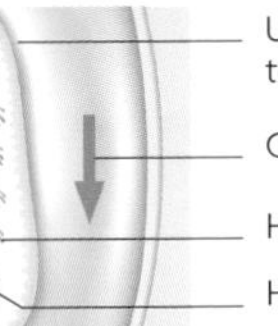

MACULA ACTION
Mineral crystals (otoliths) cover the membrane
Otolithic membrane
Hair of hair cell
Hair cell
Utricular macula rotated to vertical
Gravity pulls membrane
Hairs deflected
Hair cell stimulated

SEMICIRCULAR CANALS

Each semicircular canal has a bulge near one end, called the ampulla. This houses a low mound of hair cells, their hair ends set into a taller, jellylike mound, the cupula. As the head moves, fluid in the canal lags behind, swirls past the cupula, and bends it. This pulls the hairs and triggers their cells to fire nerve signals.

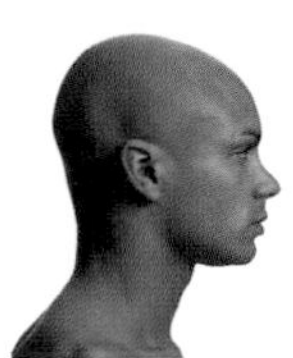

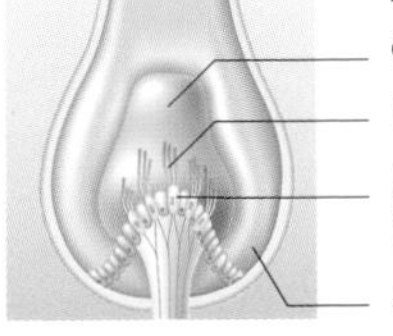
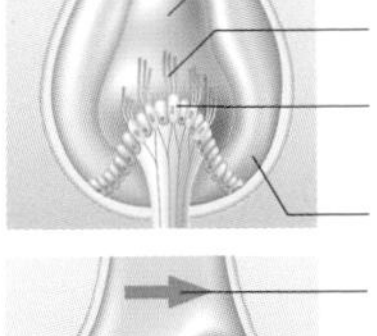
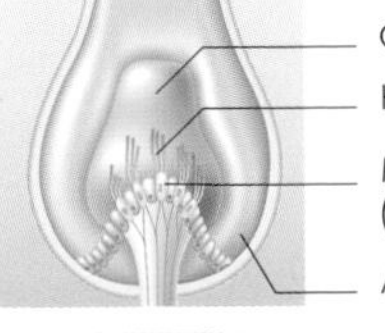
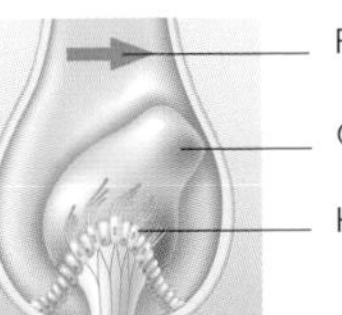
AMPULLA ACTION
Cupula
Hairs of hair cells
Mound of hair cells (crista ampullaris)
Ampulla
Fluid swirls due to head motion
Cupula bends
Hair cells stimulated

EYES AND VISION

EYESIGHT PROVIDES THE BRAIN WITH MORE INPUT THAN ALL OTHER SENSES COMBINED—MORE THAN HALF THE INFORMATION IN THE CONSCIOUS MIND IS ESTIMATED TO ENTER THROUGH THE EYES.

THE SEQUENCE OF VISION

Light rays enter the eye through the clear, domed front of the eyeball, the cornea, where they are partly bent (refracted). They then pass through the transparent lens, which changes shape to focus them (see Accommodation, right) as an upside-down image onto the retina. The retina contains many millions of light-sensitive cells called rods and cones, which convert light energy into nerve signals. Rods are scattered through the retina and detect low levels of light. Cones are concentrated in the fovea and distinguish colors and fine details. The signals of the image are sent along the optic nerves of each eye to the visual cortex in the brain.

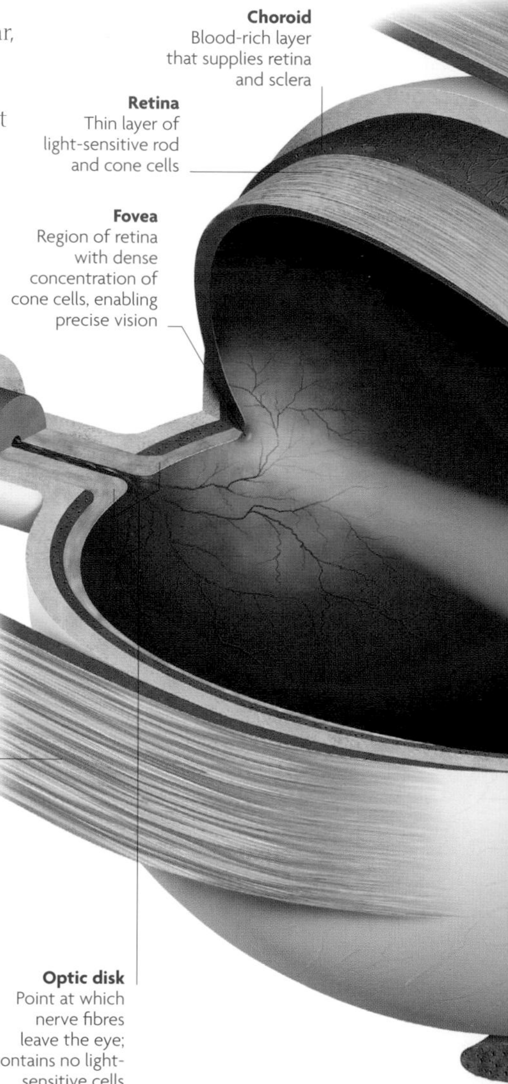
Sclera
Tough, white, protective outer sheath of eyeball
Choroid
Blood-rich layer that supplies retina and sclera
Retina
Thin layer of light-sensitive rod and cone cells
Fovea
Region of retina with dense concentration of cone cells, enabling precise vision
Optic nerve
Conveys nerve signals to brain
Lateral rectus
Small muscle that swivels the eye to look out to the side
Optic disk
Point at which nerve fibres leave the eye; contains no light-sensitive cells

INSIDE THE EYE

An average eyeball is 1 in (25 mm) in diameter, and has three main outer layers: the sclera, choroid, and retina. Near the front, the sclera can be seen as the white of the eye, and at the front it becomes the clear cornea. The main bulk of the eye, between the lens and the retina, is filled with a clear, jellylike fluid known as vitreous humor. This maintains the eyeball's spherical shape.

ACCOMMODATION

Most of the eye's focusing power comes from the cornea, but the lens alters in shape to fine-focus light rays, a process known as accommodation. To focus on nearby objects, the ring-shaped ciliary muscle around the lens contracts, making the lens thicker. To focus on more distant objects, the muscle relaxes, making the lens flatter and thinner.

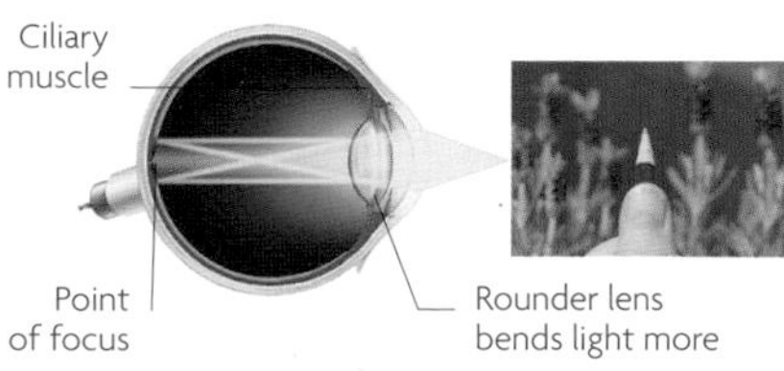
Ciliary muscle
Point of focus
Rounder lens bends light more

NEAR VISION

Light rays from close objects diverge more, so they need the extra focusing power of a fatter lens to bend the light rays so that they converge.

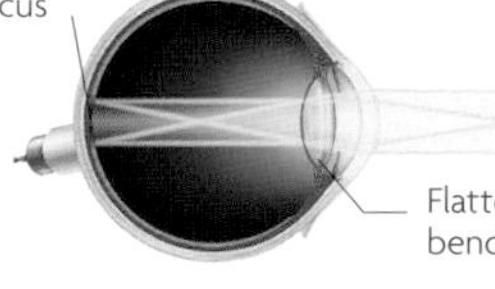
Point of focus
Flatter lens bends light less

DISTANT VISION

Light rays from distant objects are almost parallel and require less refracting power to focus, so the ciliary muscle relaxes to make the lens bulge less.

Superior rectus
Small muscle that swivels eye to look up

Suspensory ligaments
Hold lens within the ring of ciliary muscle

Posterior chamber
Fluid-filled cavity behind the iris

Iris
Ring of muscle that changes size of pupil to regulate amount of light entering the eye

Anterior chamber
Between cornea and iris, filled with aqueous humor (fluid)

Pupil
Hole in iris that becomes wider in dim light

Cornea
Domed, transparent "window" at front of eye

Conjunctiva
Delicate, sensitive covering of cornea and eyelid lining

Lens
Transparent disk of tissue that changes shape for near or far vision

Ciliary muscle
Ring of muscle that alters lens shape

VISUAL PATHWAYS

Nerve signals conducted along the left and right optic nerves converge at a crossover junction, called the optic chiasm, at the base of the brain. Here, fibers carrying signals from the left side of each retina join and proceed as the left optic tract to the left visual cortex at the back of the brain. Likewise fibers from the right side of each retina form the right optic tract and go to the right visual cortex. Because the eyes are set apart, each sees a slightly different view of an object. The combination of the views of both eyes into a single image is called binocular vision.

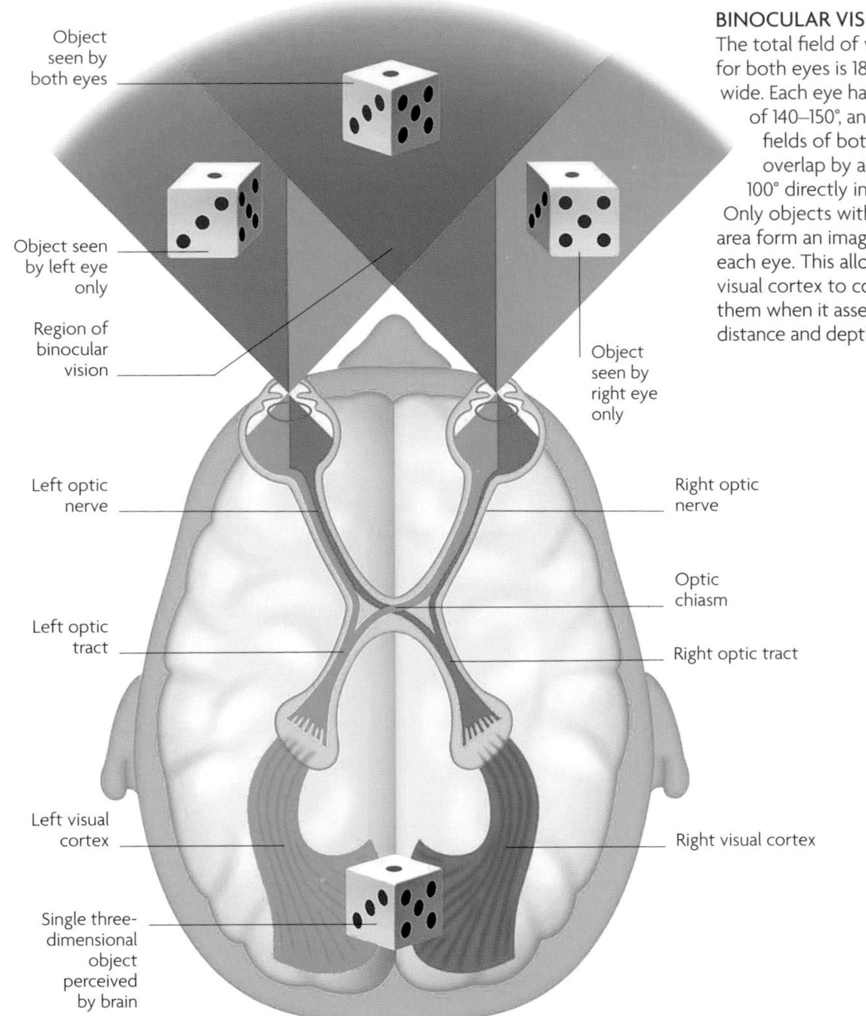
Object seen by both eyes
Object seen by left eye only
Region of binocular vision
Object seen by right eye only
Left optic nerve
Right optic nerve
Optic chiasm
Left optic tract
Right optic tract
Left visual cortex
Right visual cortex
Single three-dimensional object perceived by brain

BINOCULAR VISION
The total field of vision for both eyes is 180–200° wide. Each eye has a field of 140–150°, and the fields of both eyes overlap by about 100° directly in front. Only objects within this area form an image in each eye. This allows the visual cortex to compare them when it assesses distance and depth.

PUPILS

The size of the pupils constantly changes in response to changing levels of light. This is a function of the autonomic nervous system (see pp.106–07). Smooth muscle fibers in the iris are arranged as an inner, circular band and an outer, radial band. Sensory receptors in the eyes respond to light and send nerve signals to the brain, which sends messages to one or the other muscle band to adjust pupil size.

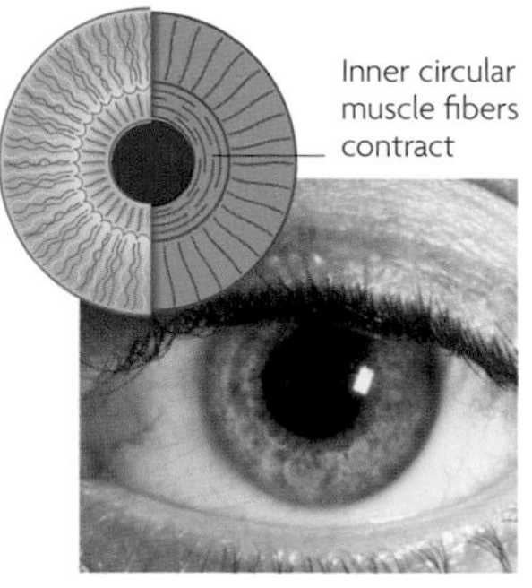
Inner circular
muscle fibers
contract

CONSTRICTED PUPIL
In bright light or to view nearby objects, the pupil constricts as the parasympathetic nervous system stimulates the inner circular muscle fibers to contract.

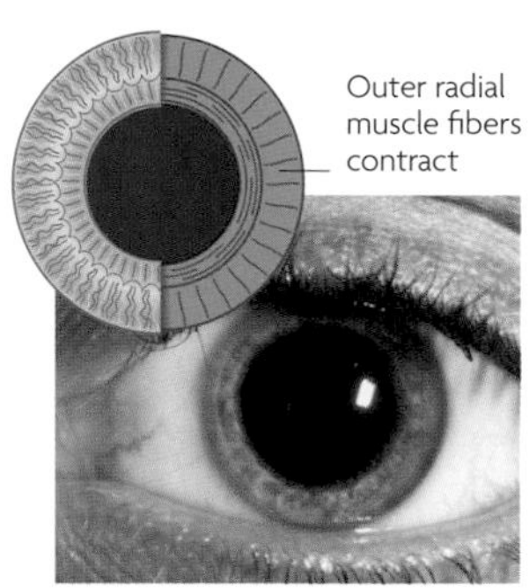
Outer radial
muscle fibers
contract

DILATED PUPIL
When light is dim and the eye needs more light to see, the pupil widens as the sympathetic nervous system causes the outer radial muscle fibers to shorten.

AROUND THE EYE

When it closes, the eyelid physically protects the eye and smears lacrimal fluid, or tears, over the conjunctiva. Tears wash away dirt and dust and protect against microbes. Around the eyeball, there are six small, straplike muscles that attach it to the socket (orbit) in the skull bone. These extraocular or extrinsic muscles are very fast-acting and swivel, or roll, the eyeball so the eye can look up or down, inward or out.

Lacrimal ducts
5–10 ducts convey fluid to the surface of the eye

Lacrimal gland
Secretes tears to keep eye clean and moist

Lacrimal canals
Collect tears draining through small holes in the corner of the eye

Lacrimal sac
Channels tears toward nose

Nasolacrimal duct
Opens into nasal cavity

TEAR APPARATUS
The tear (lacrimal) gland is under the soft tissues of the outer part of the upper eyelid. It produces $\frac{1}{3}$–$\frac{2}{3}$fl oz (1–2 ml) of fluid daily.

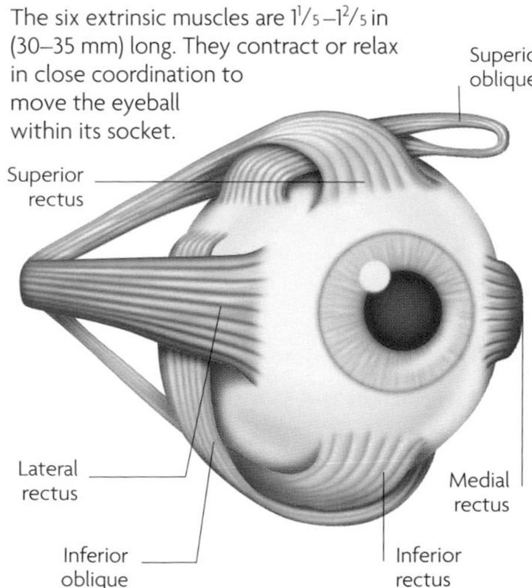
Superior
oblique
Superior
rectus
Lateral
rectus
Inferior
oblique
Medial
rectus
Inferior
rectus

EYE MUSCLES OF RIGHT EYE
The six extrinsic muscles are $1\frac{1}{5}$–$1\frac{2}{5}$ in (30–35 mm) long. They contract or relax in close coordination to move the eyeball within its socket.

NERVOUS SYSTEM DISORDERS

DISORDERS OF THE NERVOUS SYSTEM CAN DEVELOP FOR VARIOUS REASONS, SUCH AS PROBLEMS WITH THE BRAIN'S BLOOD SUPPLY, THE DETERIORATION OF BRAIN CELLS, ABNORMAL DEVELOPMENT IN THE WOMB, TISSUE INFECTION, CELL DAMAGE, AND THE AGING PROCESS.

STROKE

Disruption of blood supply to nerve cells in the brain results in a temporary or permanent loss of function of the body parts they serve. In most people, symptoms of stroke develop rapidly, and may include weakness or numbness on one side of the body, visual disturbances, slurred speech, and difficulty maintaining balance. Immediate admission to a hospital is essential to try to prevent brain damage. Long-term treatment to reduce the risk of further strokes usually consists of drugs. Rehabilitation, such as physical therapy and speech therapy, are often needed. The after effects of a stroke range from mild, temporary symptoms to lifelong disability.

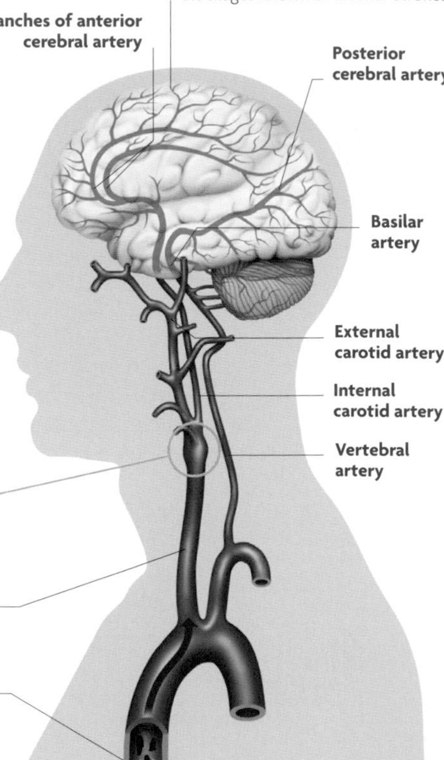
Blockage of tiny vessels
Prolonged high blood pressure or diabetes may damage tiny blood vessels, which may lead to localized blockages known as lacunar strokes
Branches of anterior cerebral artery
Posterior cerebral artery
Basilar artery
External carotid artery
Internal carotid artery
Vertebral artery
Thrombus
Fat deposits in artery walls reduce flow and may encourage a blood clot, or thrombus, to form; if this blocks an artery to the brain, a stroke follows
Common carotid artery
Embolus
A piece of material called an embolus may travel in the blood and block a cerebral artery, causing a stroke

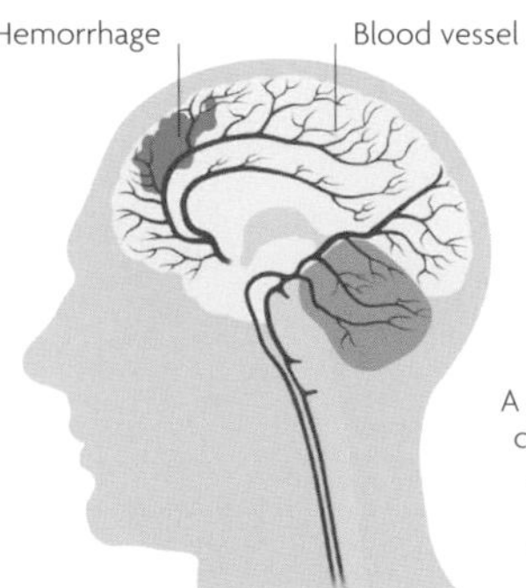
Hemorrhage
Blood vessel

BLEEDING WITHIN THE BRAIN

An intracerebral hemorrhage, bleeding within brain tissue, is a main cause of stroke in older people who have hypertension. High blood presssure may put extra strain on small arteries in the brain, which causes them to rupture.

BLOCKED BLOOD VESSELS

Blocked arteries that cause a stroke can occur for several reasons, ranging from localized blockages in tiny blood vessels deep within the brain to a blockage caused by a fragment of material that has traveled to the brain from elsewhere.

DEMENTIA

Dementia combines memory loss, confusion, and general intellectual decline. The disorder mainly occurs in people over the age of 65, but young people are sometimes affected. In the early stages of dementia, a person is prone to becoming anxious or depressed due to awareness of the memory loss. As the dementia worsens, the person may eventually need full-time care in a nursing home Caregivers may also need support.

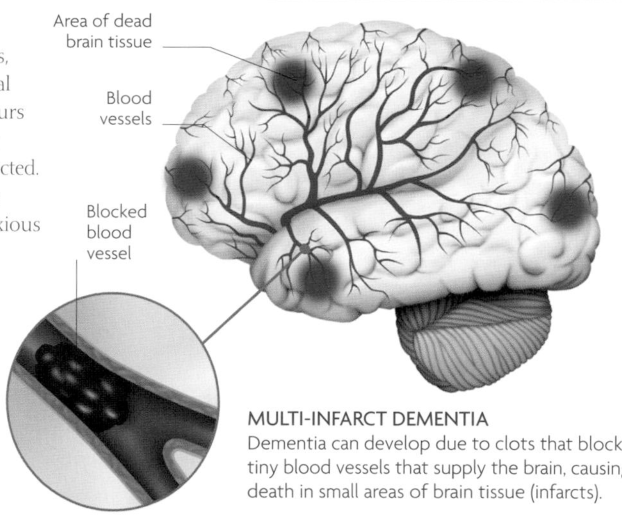
Area of dead brain tissue
Blood vessels
Blocked blood vessel

MULTI-INFARCT DEMENTIA
Dementia can develop due to clots that block tiny blood vessels that supply the brain, causing death in small areas of brain tissue (infarcts).

ALZHEIMER'S DISEASE

The most common form of dementia is Alzheimer's disease. Brain damage occurs due to the abnormal production of a protein called amyloid, which builds up in the brain. No cure has been found, but drugs can slow the progress of the disease in some people.

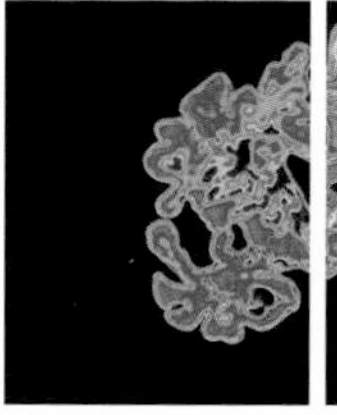

ALZHEIMER'S DISEASE

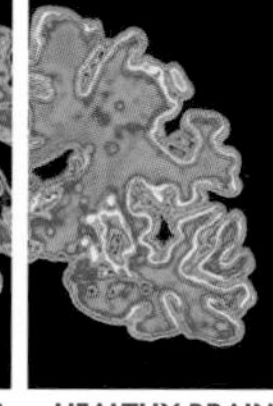

HEALTHY BRAIN

BRAIN IN ALZHEIMER'S DISEASE
This computer graphic shows a slice through the brain of a person with Alzheimer's disease compared to a slice of healthy brain. The diseased brain is considerably shrunken due to the degeneration and death of nerve cells. The surface of a brain affected by Alzheimer's disease may be more deeply folded than normal.

MIGRAINE

About 1 in 10 people are migraine sufferers, with episodes of severe headache often associated with visual disturbances, nausea, and vomiting. The underlying cause of a migraine is unknown, but changes in the diameter of the blood vessels in the scalp and brain are known to occur, and the brain chemical serotonin seems to play a role. Triggers for a migraine attack include stress, missed meals, lack of sleep, and certain foods, such as cheese or chocolate. In many women, migraines are associated with menstruation.

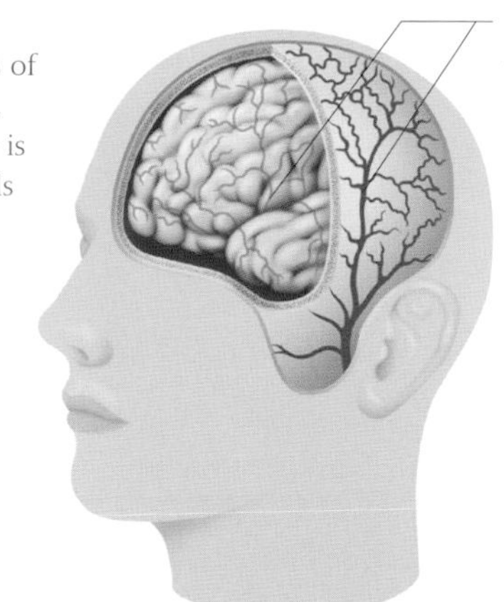
Dilated blood vessels

HEADACHE PHASE
During migraine, severe, throbbing pain may affect half or all of the head as blood vessels in the scalp and brain widen (dilate). The neurotransmitter serotonin controls the diameter of blood vessels.

BRAIN INFECTIONS

Infection of brain tissue or its protective layers can be caused by a variety of viruses, bacteria, and tropical parasites. Infection of the brain, or encephalitis, is a rare complication of a viral infection, such as mumps or measles. It can be fatal, with babies and elderly people most at risk.

SITES OF INFECTION

Infectious organisms can affect the brain itself, the three membranes (meninges) that surround the brain, or both. Infections can reach the brain through the blood, but can also spread from a nearby infection (such as an ear infection) or through a skull wound.

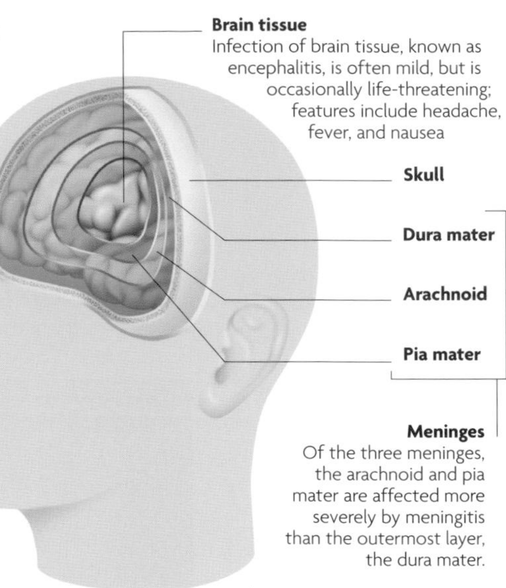
Brain tissue
Infection of brain tissue, known as encephalitis, is often mild, but is occasionally life-threatening; features include headache, fever, and nausea
Skull
Dura mater
Arachnoid
Pia mater
Meninges
Of the three meninges, the arachnoid and pia mater are affected more severely by meningitis than the outermost layer, the dura mater.

MENINGITIS

Inflammation of the meninges is usually caused by a virus or bacterium. Initially, meningitis may cause vague flulike symptoms. More pronounced symptoms may also develop, such as headache, fever, nausea, vomiting, stiff neck, and a dislike of bright light. In meningitis due to *Meningococcus* bacteria, there is a distinctive reddish purple rash. If meningitis is suspected, immediate admission to a hospital is necessary for tests. If bacterial meningitis is confirmed, treatment in intensive care is often required, and a complete recovery may take weeks or months. It can be fatal despite treatment. Recovery from viral meningitis usually takes up to two weeks. No specific treatment is needed.

BRAIN ABSCESS

An abscess is a collection of pus. Brain abscesses are rare, and are usually caused by bacteria that have spread to the brain from an infection in nearby tissues in the skull. Treatment consists of high doses of antibiotics and possibly corticosteroids to control swelling of the brain. Surgery may be needed to drain pus through a hole drilled in the skull. If given early treatment, many people with a brain abscess recover. However, some have persistent problems, such as seizures, slurred speech, or weakness of a limb.

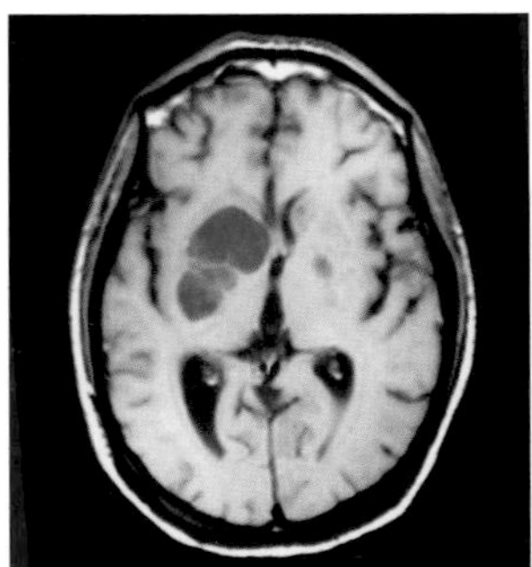

BRAIN ABSCESS
This MRI scan of the brain shows an abscess (blue area) due to a fungal infection in a person who has AIDS. People with AIDS are at increased risk of developing a brain abscess.

TESTING A MENINGITIS RASH
In meningococcal meningitis, bacteria in the blood may cause dark-red or purple spots that turn into blotches. The rash does not fade when pressed with a glass.

SPINA BIFIDA

There are three main forms of spina bifida: spina bifida occulta, meningocele, and myelomeningocele. Spina bifida occulta may require surgery to avoid serious neurological complications later in life. Meningoceles usually have a good prognosis after surgery. Myelomeningocele has effects that may include paralysis or weakness in the legs, and lack of bladder and bowel control. Children with this form are permanently disabled and require lifelong care. Folic acid helps prevent spina bifida, and women are advised to take supplements when planning to conceive and during the first 12 weeks of pregnancy.

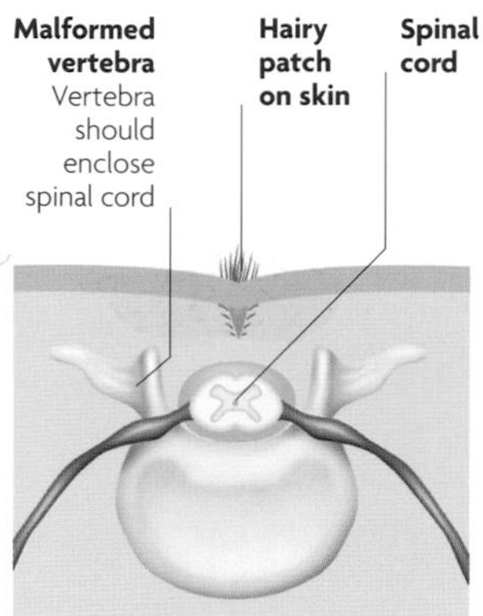
Malformed vertebra
Vertebra should enclose spinal cord
Hairy patch on skin
Spinal cord

SPINA BIFIDA OCCULTA
One or more vertebrae in the spine are malformed, but the cord is not damaged. On the skin, there may be dimpling, a tuft of hair, a fatty lump (lipoma), or a birthmark.

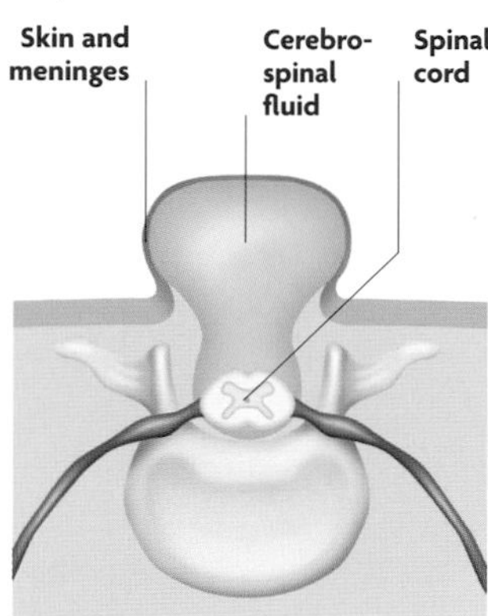
Skin and meninges
Cerebro-spinal fluid
Spinal cord

MENINGOCELE
The meninges protrude through a malformed vertebra as a visible fluid-filled sac called a meningocele. The spinal cord remains intact, and the defect can be repaired.

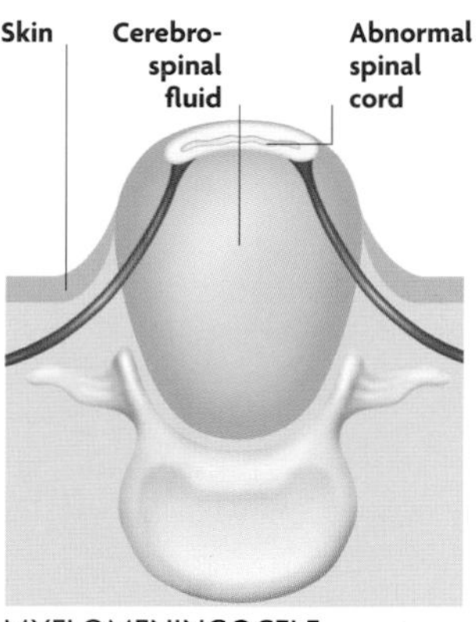
Skin
Cerebro-spinal fluid
Abnormal spinal cord

MYELOMENINGOCELE
A part of the spinal cord, contained within a sac of fluid, protrudes through the skin. This is the most severe form of spina bifida, and affected children need lifelong care.

MULTIPLE SCLEROSIS

Multiple sclerosis (MS) is due to immune system damage to the sheaths that insulate nerve fibers. It affects sensation, movement, body functions, and balance. In some people, symptoms may last for days or weeks, then clear up for months or years. In others, symptoms gradually get worse. Drugs may help to lengthen remission periods and shorten attacks.

EARLY STAGE
At first, there are only small patches of damage. Macrophages, a type of scavenging cell, remove damaged areas of the myelin sheaths, exposing the fibers and impairing nerve conduction.

Macrophage
Myelin sheath
Nerve fiber

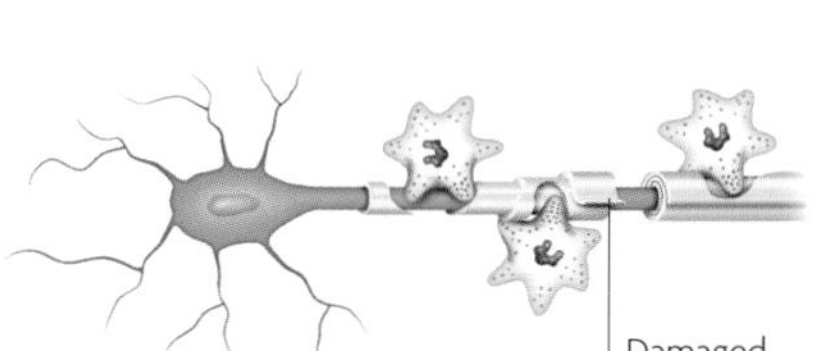
Damaged myelin sheath

LATE STAGE
As MS progresses, the amount of damage to the sheaths increases, and affects conduction in more fibers. As the damage spreads, the symptoms become progressively worse.

DEAFNESS

There are two types of hearing loss: conductive and sensorineural. Conductive hearing loss results from impaired transmission of sound waves to the inner ear, and is often temporary. In children, the most common cause is glue ear (see below). In adults, it is most commonly due to blockage by earwax. Other causes include damage to the eardrum or, rarely, stiffening of a bone in the middle ear so that it cannot transmit sound. Sensorineural hearing loss is most commonly due to deterioration of the cochlea with age. It may also result from damage to the cochlea by excessive noise or by Ménière's disease. Rarely, hearing loss is caused by an acoustic neuroma or by certain drugs. Simple measures can be effective for treating conductive deafness, such as syringing the ear to remove earwax. Surgery may be required for glue ear or otosclerosis. Sensorineural deafness usually cannot be cured, but hearing aids can help. A cochlear implant, in which electrodes are surgically implanted in the cochlea, may help in profound deafness.

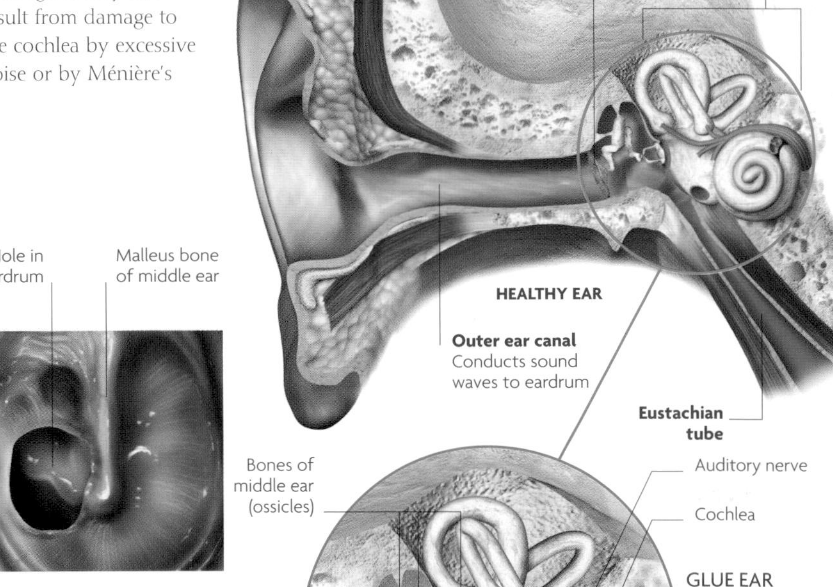
Eardrum
Vibrates in response to sound waves
Semicircular canals
Play a role in balance
HEALTHY EAR
Outer ear canal
Conducts sound waves to eardrum
Eustachian tube
Hole in eardrum
Malleus bone of middle ear
Bones of middle ear (ossicles)
Auditory nerve
Cochlea
Glue-like fluid

PERFORATED EARDRUM

A tear or hole in the eardrum may occur due to pressure from a build-up of pus or fluid in the middle ear during an infection. It may also occur due to unequal pressures between the middle and outer ear, as may happen when flying. Healing usually takes about a month.

GLUE EAR

A persistent collection of fluid in the middle ear, which occurs more commonly in children, can cause difficulty in hearing. Fluid buildup is caused by a blockage of the Eustachian tube, which ventilates the middle ear, often as a result of infection.

FOCUSING PROBLEMS

Farsightedness (hypermetropia) and nearsightedness (myopia) result from the eyeball being either too short or too long (see right). In astigmatism, vision is blurred because the cornea is irregularly curved, and the lens cannot focus all light rays on the retina. Aging often affects near vision, because the lens loses its elasticity and cannot easily adjust its shape. Refractive errors can usually be corrected by glasses or contact lenses, or by surgical techniques such as laser-assisted in-situ keratomileusis (LASIK) and photorefractive keratectomy (PRK). In LASIK, the middle layers of the cornea are reshaped by a laser, while in PRK, areas of the cornea's surface are shaved away by a laser to alter its shape.

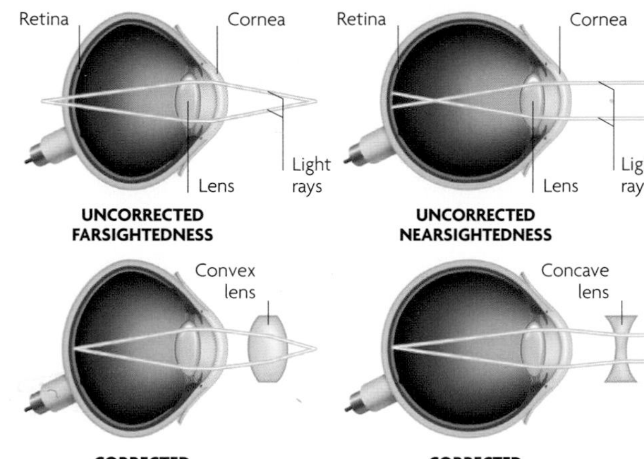
Retina
Cornea
Light rays
Lens
UNCORRECTED FARSIGHTEDNESS
Retina
Cornea
Light rays
Lens
UNCORRECTED NEARSIGHTEDNESS
Convex lens
CORRECTED FARSIGHTEDNESS
Concave lens
CORRECTED NEARSIGHTEDNESS

FARSIGHTEDNESS

In farsightedness, the eyeball is too short, so the cornea and lens focus light rays behind the retina, and the image is blurred. Convex lenses make the light rays converge so that they are focused on the retina, correcting vision.

NEARSIGHTEDNESS

In nearsightedness, the eyeball is too long, so the cornea and lens focus light rays in front of the retina, and the image is blurred. Concave lenses are required, which make the light rays diverge so that they are focused on the retina.

GLAUCOMA

Glaucoma occurs when there is an abnormally high pressure inside the eyeball that is caused by a buildup of fluid. The pressure may permanently damage nerve fibers in the retina or the optic nerve, affecting vision. In acute glaucoma, the condition develops suddenly and is accompanied by severe pain. Chronic glaucoma (see right) comes on slowly and painlessly over many years.

CHRONIC GLAUCOMA

Fluid continually moves into and out of the eye to nourish its tissues and maintain the shape of the eye. Normally, the fluid flows out through the pupil and drains out of the trabecular meshwork within the drainage angle. In chronic glaucoma, the meshwork is blocked, and pressure builds up.

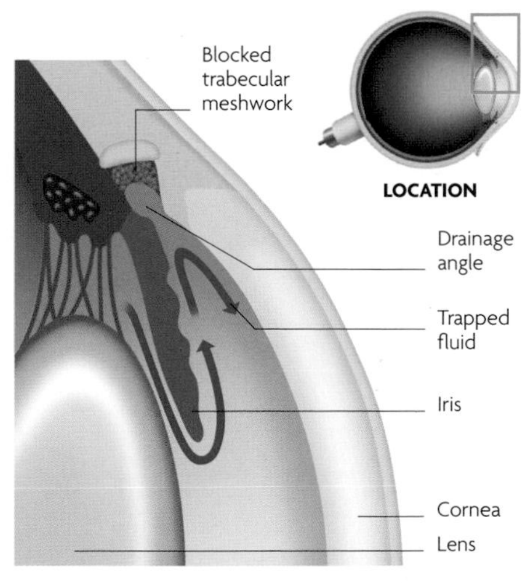
Blocked trabecular meshwork
LOCATION
Drainage angle
Trapped fluid
Iris
Cornea
Lens

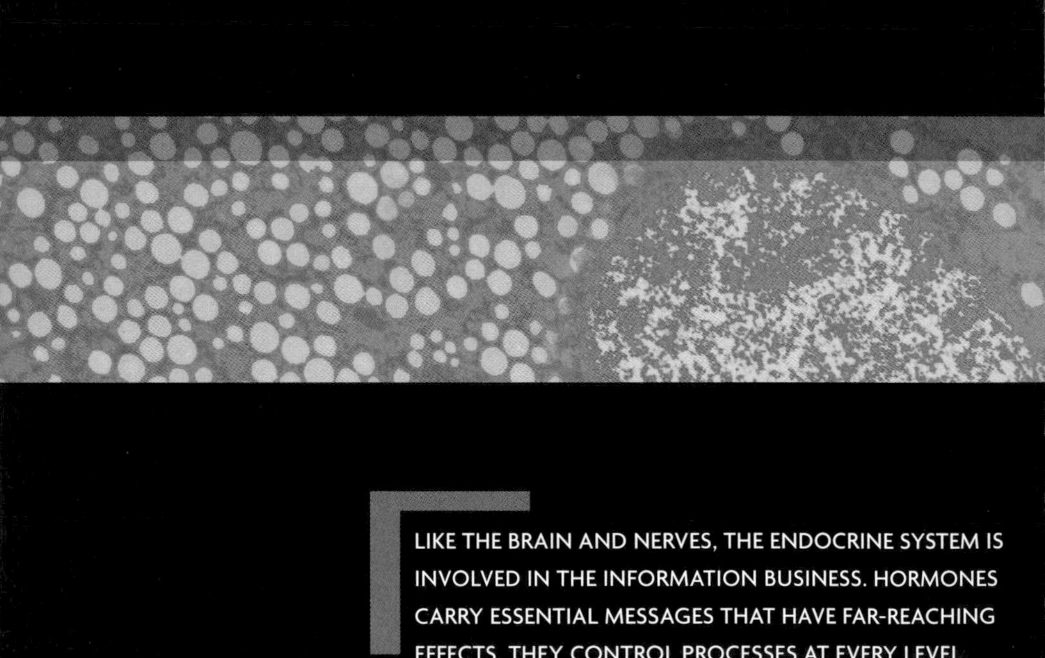

LIKE THE BRAIN AND NERVES, THE ENDOCRINE SYSTEM IS INVOLVED IN THE INFORMATION BUSINESS. HORMONES CARRY ESSENTIAL MESSAGES THAT HAVE FAR-REACHING EFFECTS. THEY CONTROL PROCESSES AT EVERY LEVEL,

ENDOCRINE SYSTEM

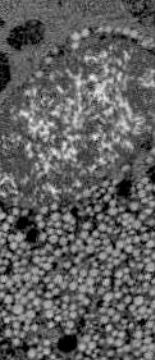

ENDOCRINE ANATOMY

THE BODY'S CHEMICAL MESSENGERS, HORMONES, ARE PRODUCED BY ENDOCRINE GLANDS. HORMONES TARGET CERTAIN TISSUES TO REGULATE THEIR ACTIVITIES.

The endocrine system is composed of bodies of glandular tissue, such as the thyroid, and also includes glands within certain organs, including the testes, ovaries, and heart. These glands and tissues secrete hormones that control and coordinate body functions such as the breakdown of chemical substances in metabolism, fluid balance and urine production, growth and development, and sexual reproduction. Hormones travel in the blood, so each hormone reaches every body part. However, the specific molecular shape of each hormone slots only into receptors on its target tissues or organs.

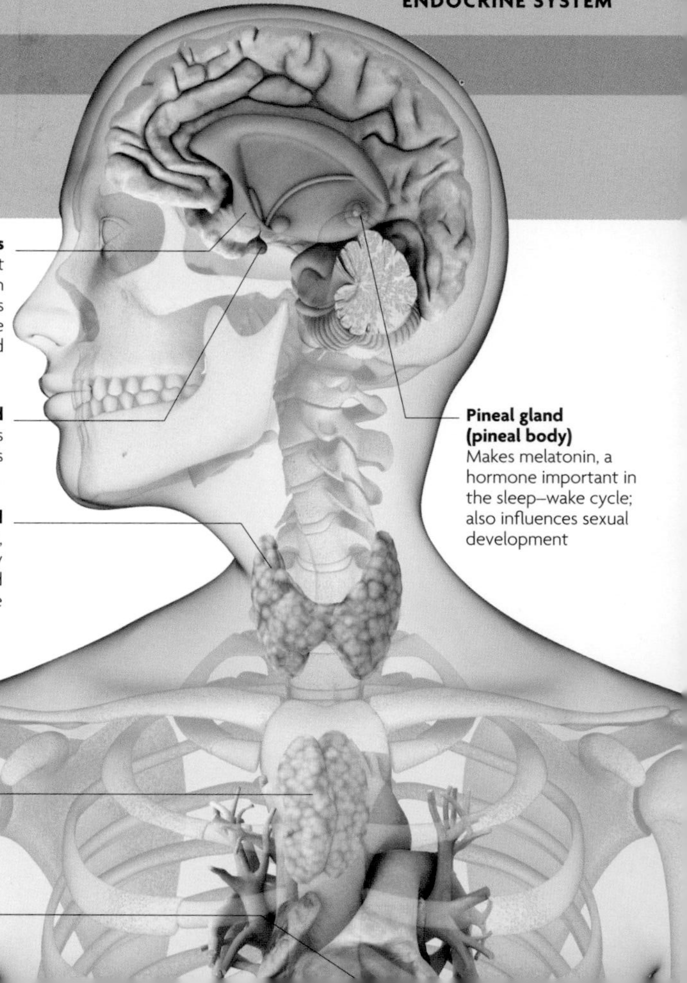

Hypothalamus
Cluster of nerve cells that serves as the main link between nerves and hormones; produces hormones that control the pituitary gland

Pituitary gland
Called the "master gland"; controls many other endocrine glands

Thyroid gland
Controls aspects of metabolism, including maintenance of body weight, rate of energy use, and heart rate

Thymus gland
Produces hormones involved in development of white blood cells, called T-cells, that are part of the immune system

Heart
Produces a hormone called atriopeptin that reduces blood volume and pressure and helps regulate fluid balance

Pineal gland (pineal body)
Makes melatonin, a hormone important in the sleep–wake cycle; also influences sexual development

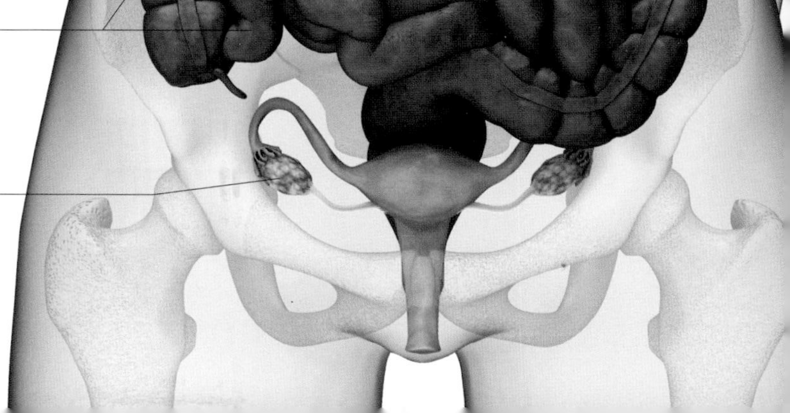
Adrenal gland
Outer layer manufactures steroid hormones that regulate metabolism and maintain fluid balance; inner layer produces epinephrine
Kidney
Secretes erythropoietin, which stimulates production of red blood cells in bone marrow
Stomach
Makes hormones that stimulate production or release of digestive enzymes
Pancreas
Contains clusters of cells that produce the hormones insulin and glucagon, which control blood glucose levels
Intestines
Like the stomach, make hormones that stimulate production or release of digestive enzymes
Ovary
Makes the female sex hormones estrogen and progesterone, which regulate the menstrual cycle

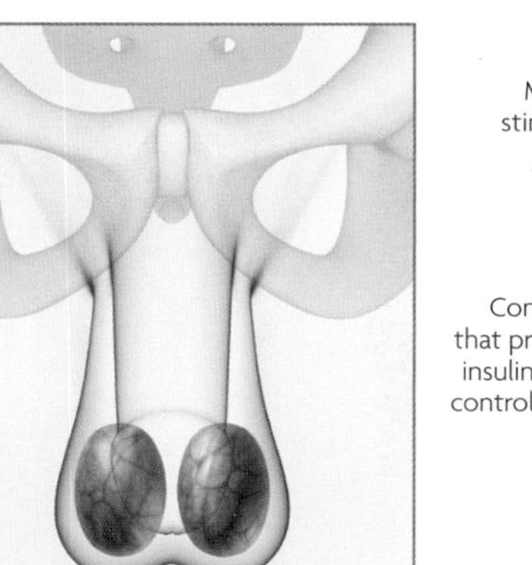

TESTES

In males, the two testes produce androgens: male sex hormones, including testosterone. Androgens stimulate the growth and development of the male sexual organs and the production of sperm, and influence secondary sexual characteristics, such as facial hair and deepening of the voice.

HORMONE PRODUCERS

HORMONES CARRY THE CHEMICAL DATA THAT CONTROL THE RATE AT WHICH GLANDS AND ORGANS WORK. HORMONE-PRODUCING CELLS ARE FOUND ALL AROUND THE BODY, MANY IN GLANDS THAT HAVE SPECIALIZED FUNCTIONS.

MASTER GLAND: THE PITUITARY

The pituitary, or hypophysis, is the most influential gland in the endocrine system. It is actually two distinct glands in one. The front (anterior) lobe, also known as the adenohypophysis, forms the larger part. Behind is the posterior lobe, or neurohypophysis. The anterior pituitary manufactures eight major hormones on site and releases them into the bloodstream. The posterior pituitary receives its two main hormones from the hypothalamus, which lies above it; there, they are made by neurosecretory cells. Other neurosecretory cells make regulatory hormones, which travel via capillaries to the anterior lobe and control the release of hormones there.

Hypophyseal portal system
System of blood vessels that carry regulatory hormones (releasing factors) from hypothalamus to anterior pituitary lobe

Anterior lobe of pituitary
Contains cells that manufacture about eight main hormones; secretion of these hormones is regulated by hypothalamus

Vein

SKIN

The action of melanocyte-stimulating hormone (MSH) – produced in a thin layer between the two pituitary lobes – causes cells called melanocytes in skin tissue to produce more melanin pigment, making the skin darken.

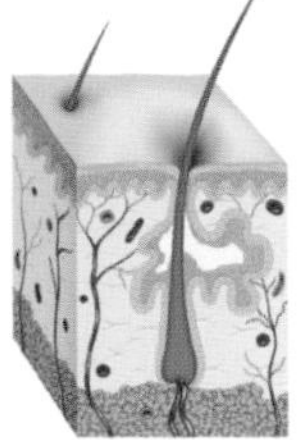

ADRENAL GLAND

Adrenocorticotropic hormone (ACTH) triggers the adrenals to produce steroid hormones that control stress response and the body's use of fats, carbohydrates, proteins, and minerals.

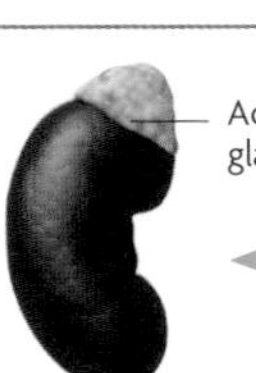
Adrenal gland

THYROID

Thyrotropin-releasing hormone from the hypothalamus controls the release of thyroid-stimulating hormone (TSH). This encourages the thyroid to become more active and affects metabolism.

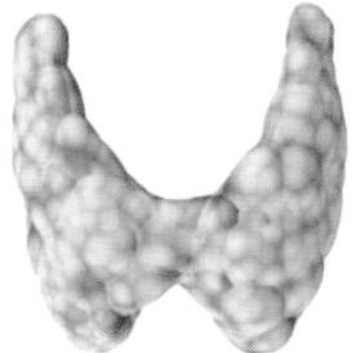

BONE AND GENERAL GROWTH

Growth hormone (GH) acts on the whole body to promote protein manufacture, bone growth, and building of new tissues throughout life, but is especially important for development in children.

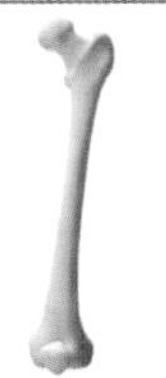

SEX GLANDS

Luteinizing hormone (LH) and follicle-stimulating hormone (FSH) trigger the sex glands to make their own hormones, and also to produce ripe egg cells in females and mature sperm cells in males.

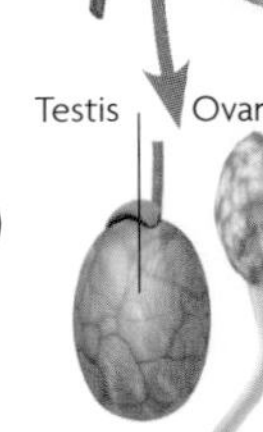
Testis
Ovary

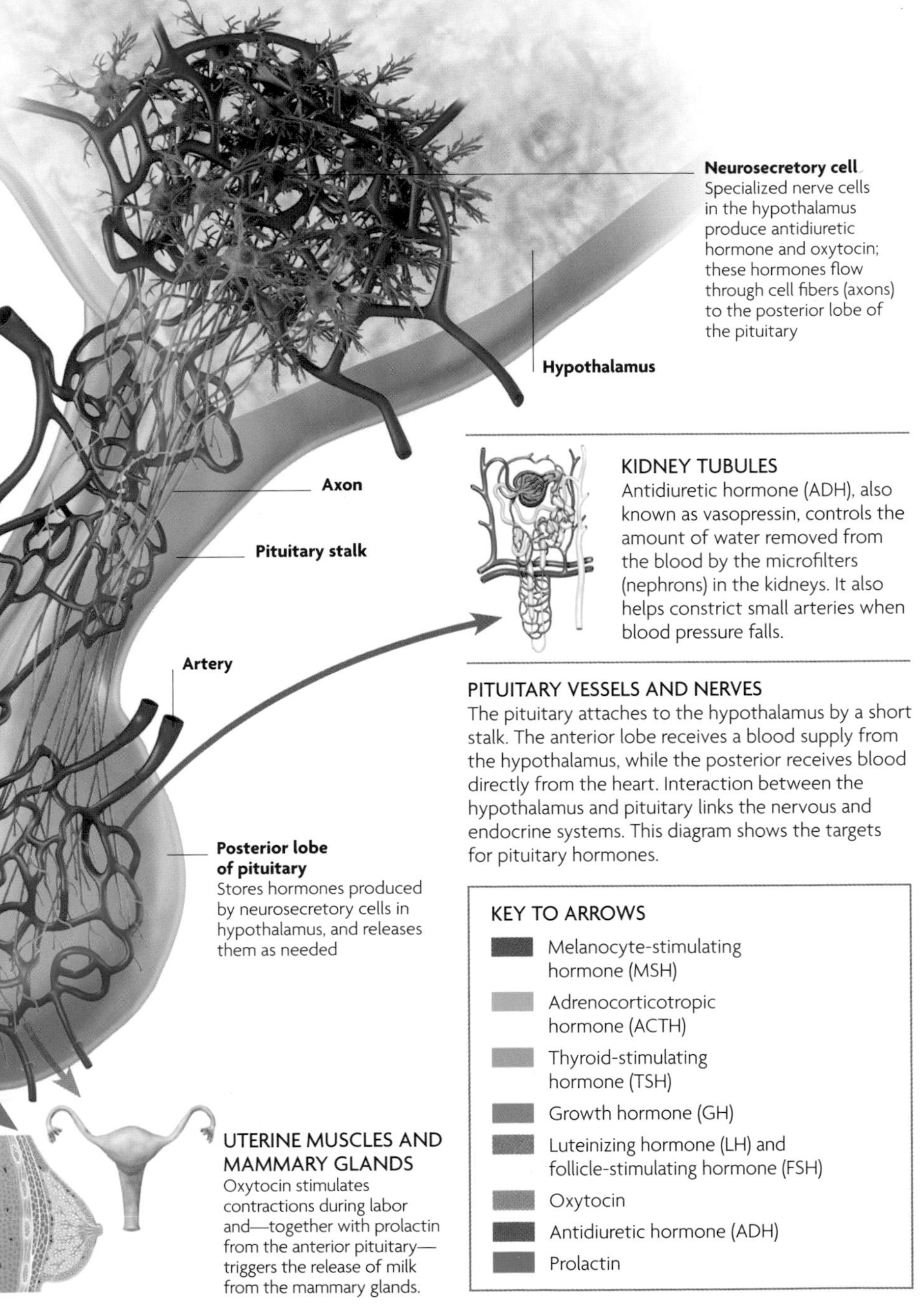
Neurosecretory cell
Specialized nerve cells in the hypothalamus produce antidiuretic hormone and oxytocin; these hormones flow through cell fibers (axons) to the posterior lobe of the pituitary
Hypothalamus
Axon
Pituitary stalk
Artery
Posterior lobe of pituitary
Stores hormones produced by neurosecretory cells in hypothalamus, and releases them as needed

KIDNEY TUBULES

Antidiuretic hormone (ADH), also known as vasopressin, controls the amount of water removed from the blood by the microfilters (nephrons) in the kidneys. It also helps constrict small arteries when blood pressure falls.

PITUITARY VESSELS AND NERVES

The pituitary attaches to the hypothalamus by a short stalk. The anterior lobe receives a blood supply from the hypothalamus, while the posterior receives blood directly from the heart. Interaction between the hypothalamus and pituitary links the nervous and endocrine systems. This diagram shows the targets for pituitary hormones.

KEY TO ARROWS

- Melanocyte-stimulating hormone (MSH)
- Adrenocorticotropic hormone (ACTH)
- Thyroid-stimulating hormone (TSH)
- Growth hormone (GH)
- Luteinizing hormone (LH) and follicle-stimulating hormone (FSH)
- Oxytocin
- Antidiuretic hormone (ADH)
- Prolactin

UTERINE MUSCLES AND MAMMARY GLANDS

Oxytocin stimulates contractions during labor and—together with prolactin from the anterior pituitary—triggers the release of milk from the mammary glands.

PANCREAS

The pancreas is a dual-purpose gland. It produces digestive enzymes in cells called acini, but also has an endocrine function. Within the acinar tissues are cell clusters known as islets of Langerhans, which produce hormones involved in controlling glucose (blood sugar), the body's main energy source. Beta cells make the hormone insulin, which promotes glucose uptake by body cells and speeds conversion of glucose into glycogen for storage in the liver. In this way, insulin lowers blood glucose levels. Another hormone, glucagon, is produced by alpha cells and has opposing actions, raising blood glucose levels. Delta cells make somatostatin, which regulates the alpha and beta cells.

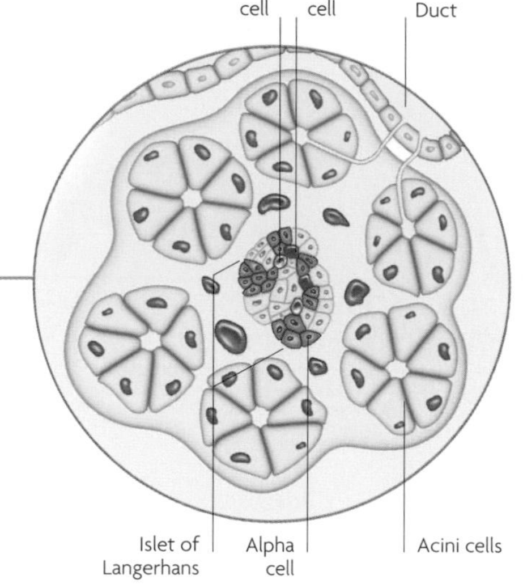
Delta cell
Beta cell
Duct
Islet of Langerhans
Alpha cell
Acini cells

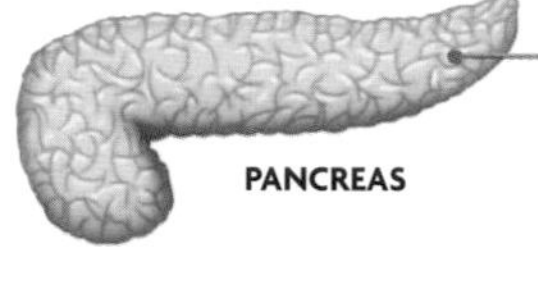
PANCREAS

PANCREATIC ISLETS

Surrounded by enzyme-producing acini cells, the tiny pancreatic islets contain three types of cells: alpha, beta, and delta. The secretions of the latter help regulate insulin and glucagon production.

ADRENAL GLANDS

The inner layer (medulla) and outer layer (cortex) of the adrenal gland secrete different hormones. The cortical hormones are steroids (see p.139) and include glucocorticoids, such as cortisol, which affect metabolism; mineralocorticoids, such as aldosterone, which influence salt and mineral balance; and gonadocorticoids, which act on the ovaries and testes. The medulla functions as a separate gland. Its nerve fibers link to the sympathetic nervous system, and it makes the fight-or-flight hormones, such as epinephrine.

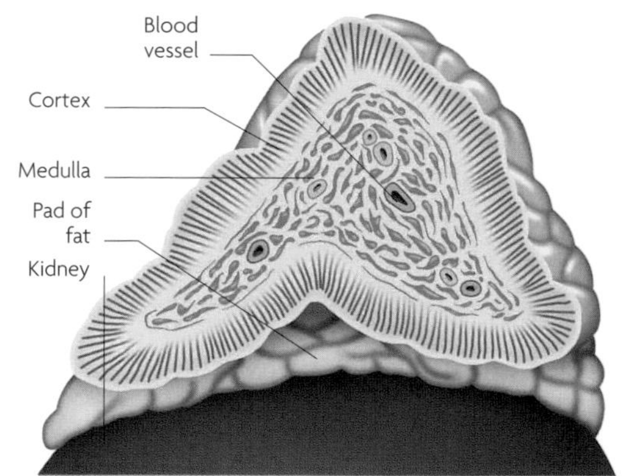
Blood vessel
Cortex
Medulla
Pad of fat
Kidney

ADRENAL ANATOMY

Each adrenal gland is shaped like a cone or pyramid and sits on top of the kidney, cushioned by a pad of fat. The glands consist of two parts: the cortex, with three layers, and the medulla, containing nerve fibers and blood vessels.

THYROID AND PARATHYROID GLANDS

The thyroid is located in the front of the neck, and has four tiny parathyroid glands embedded at the back. The hormones it produces have wide-ranging effects on body chemistry, including the maintenance of body weight, the rate of energy use from blood glucose, and heart rate. Unlike other glands, the thyroid can store its hormones. The parathyroids make parathormone (PTH), which increases the levels of calcium in the blood. PTH acts on bones to release their stored calcium, on the intestines to increase calcium absorption, and on the kidneys to prevent calcium loss.

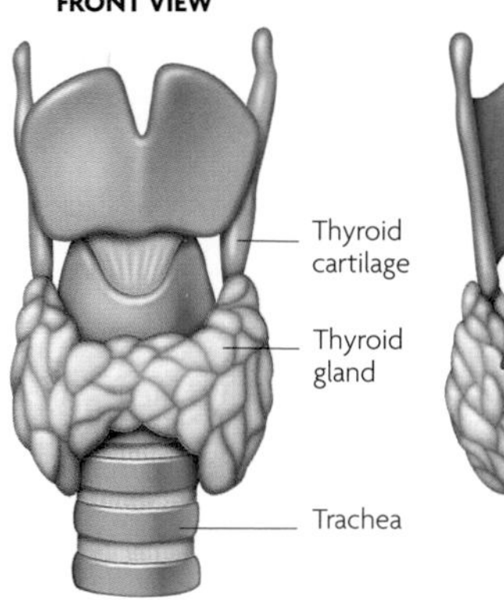
FRONT VIEW
Thyroid cartilage
Thyroid gland
Trachea

THYROID
The thyroid wraps around the upper windpipe (trachea). It produces two hormones that regulate the body's metabolism: thyroxine (T4) and triiodothyronine (T3).

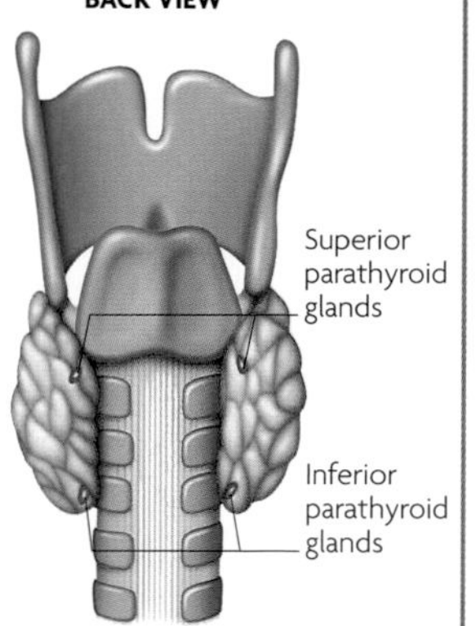
BACK VIEW
Superior parathyroid glands
Inferior parathyroid glands

PARATHYROIDS
The small parathyroid glands are set into the rear corners of the thyroid's lobes, at the back of the trachea. There are usually four, but their number and exact locations vary.

SEX GLANDS AND HORMONES

The main sex glands are the ovaries and testes. The hormones they produce stimulate the production of eggs and sperm respectively, and influence a developing embryo's sex. Until puberty, levels of sex hormones remain low. Then, in males, the testes increase their output of androgens (male sex hormones), such as testosterone. In females, the ovaries produce more estrogen and progesterone.

Interstitial cell
Blood vessel
Seminiferous tubule

TESTOSTERONE PRODUCERS
The cells shown in pink in this microscopic image of the testis secrete testosterone. They are found in the connective tissue between seminiferous tubules.

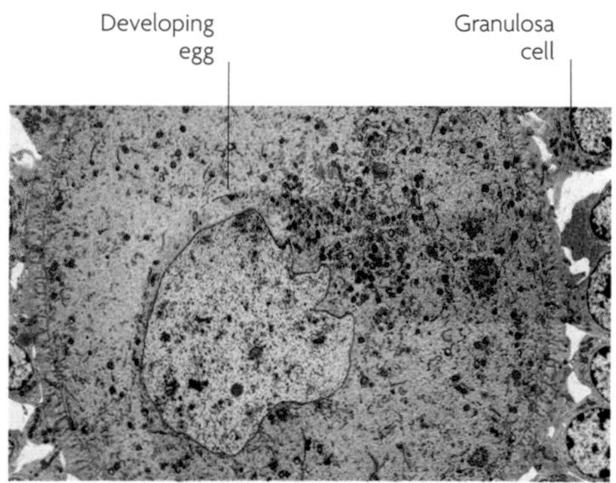
Developing egg
Granulosa cell

ESTROGEN PRODUCERS
This electron microscope shows a developing egg (pink) in an ovary surrounded by granulosa cells (blue and green). These cells secrete estrogen.

HORMONAL ACTION

HORMONES REGULATE THE FUNCTION OF THEIR TARGET CELLS BY ADJUSTING THE RATE AT WHICH A CELL'S BIOCHEMICAL REACTIONS OCCUR. DIFFERENT HORMONES ARE RELEASED ACCORDING TO DIFFERENT TRIGGER MECHANISMS.

HORMONAL TRIGGERS

Various stimuli cause an endocrine gland to release more of its hormone. In some cases, the gland responds to the level of a certain substance in the blood, using a feedback loop (see opposite). In other cases, there is an intermediate mechanism, such as the hypothalamus–pituitary complex. The adrenal gland is controlled both by adrenocorticotropic hormone (ACTH), released by the pituitary on cue from the hypothalamus, and by nerve impulses direct from the hypothalamus. The pea-sized pineal gland, near the center of the brain, is triggered by darkness to release the sleep hormone melatonin. Pineal activity is inhibited by light, which is detected by the eye and sent to the gland as a series of nerve impulses.

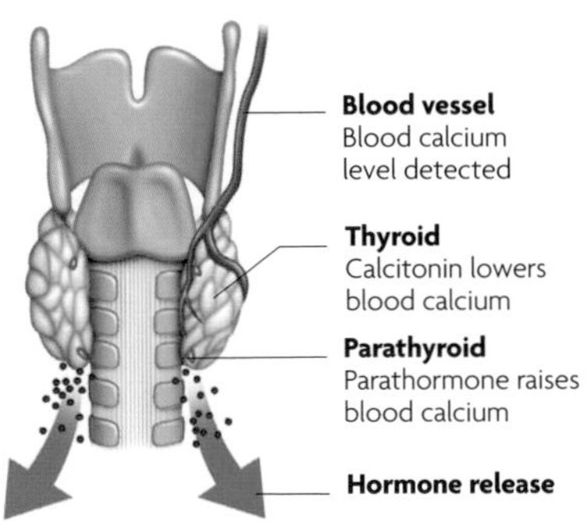
Blood vessel
Blood calcium level detected
Thyroid
Calcitonin lowers blood calcium
Parathyroid
Parathormone raises blood calcium
Hormone release

BLOOD LEVEL STIMULATION
Low blood calcium levels inhibit release of calcitonin from the thyroid and stimulate the parathyroids to release parathormone; calcium levels are raised.

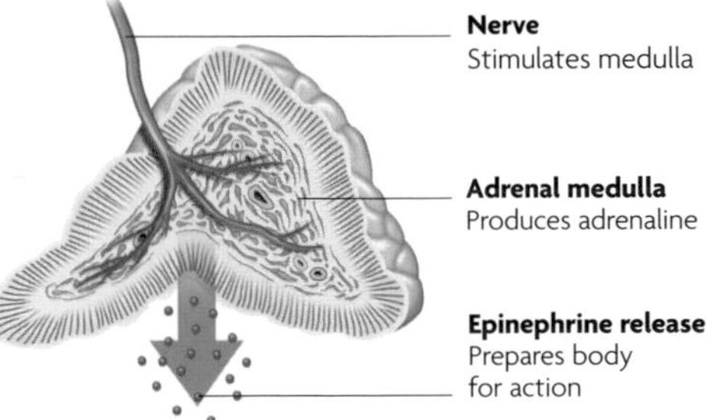
Nerve
Stimulates medulla
Adrenal medulla
Produces adrenaline
Epinephrine release
Prepares body for action

DIRECT INNERVATION
The adrenal medulla receives nerve fibers (is innervated) from the hypothalamus via the sympathetic nervous system.

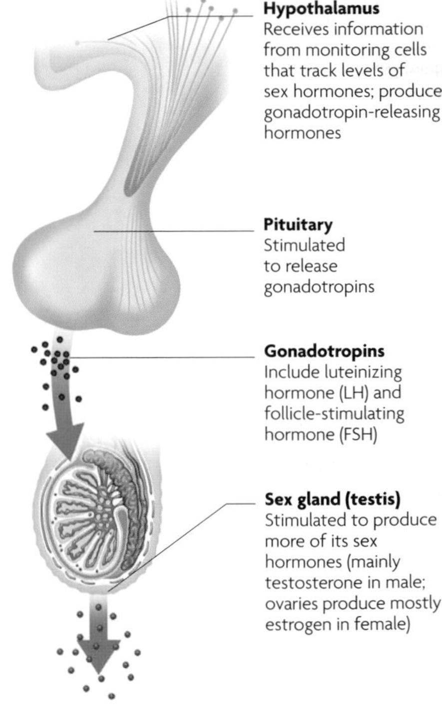
Hypothalamus
Receives information from monitoring cells that track levels of sex hormones; produces gonadotropin-releasing hormones
Pituitary
Stimulated to release gonadotropins
Gonadotropins
Include luteinizing hormone (LH) and follicle-stimulating hormone (FSH)
Sex gland (testis)
Stimulated to produce more of its sex hormones (mainly testosterone in male; ovaries produce mostly estrogen in female)

HYPOTHALAMIC–PITUITARY CONTROL
As sex hormone levels fall, gonadotropin-releasing hormones (GnRH) are sent from the hypothalamus to the pituitary, which releases more gonadotropins.

HORMONE CONTROL MECHANISMS

Chemically, there are two main types of hormones: those made of protein and amine molecules, and those made of steroids. The two groups work in a similar way, biochemically altering production rates of certain substances, but at a cellular level they have different mechanisms. Protein and amine hormones act on receptor sites at a cell's surface; steroid hormones act on receptors inside the cell.

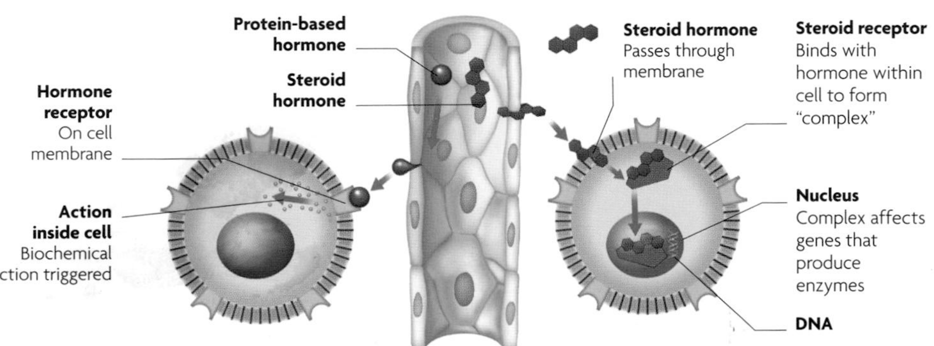
Protein-based hormone
Steroid hormone
Hormone receptor On cell membrane
Action inside cell Biochemical action triggered
Steroid hormone Passes through membrane
Steroid receptor Binds with hormone within cell to form "complex"
Nucleus Complex affects genes that produce enzymes
DNA

PROTEIN-BASED HORMONES

These hormones cannot pass through the cell membrane. They bind to receptors on the membrane, triggering biochemical action inside the cell.

STEROID-BASED HORMONES

Steroids pass into the cell, then bind to receptors and enter the cell nucleus. This triggers genes to produce enzymes that prompt biochemical action.

FEEDBACK MECHANISMS

Hormone levels are controlled by feedback mechanisms, or loops. The amount of a hormone in the blood is detected and passed on to a control unit, which in many cases is the hypothalamus–pituitary complex (as with the thyroid hormones, see below). If a hormone level is too high, the control unit reduces hormone production. If the level is too low, the control unit stimulates production.

Hypothalamus
Receives messages about blood levels of thyroid hormones, makes TRH

Pituitary gland
TRH causes increased release of TSH into blood

Thyroid gland
Triggered by TSH to produce more of its hormones

INCREASING LEVELS

Thyrotropin-releasing hormone (TRH) from the hypothalamus causes the pituitary to make thyroid-stimulating hormone (TSH); hormone levels rise.

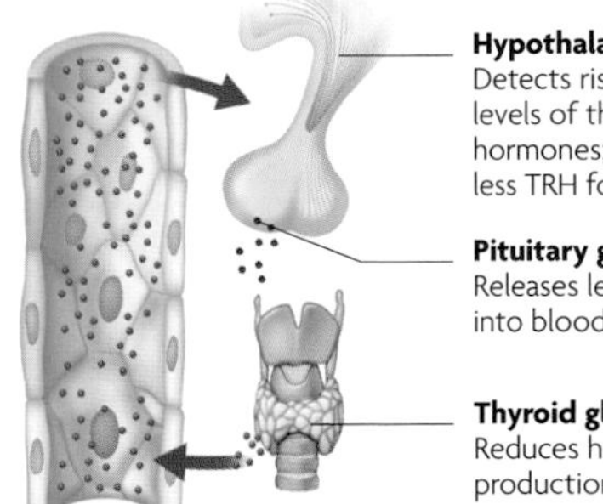
Hypothalamus
Detects rising blood levels of thyroid hormones; produces less TRH for pituitary
Pituitary gland
Releases less TSH into blood
Thyroid gland
Reduces hormone production

DECREASING LEVELS

High hormone levels prompt negative feedback, so the hypothalamus produces less TRH. This reduces TSH levels and the thyroid produces fewer hormones.

ENDOCRINE DISORDERS

SOME HORMONES HAVE WIDESPREAD EFFECTS, SO HORMONAL DISORDERS CAN CAUSE PROBLEMS AROUND THE BODY. THE PREFIX "HYPER-" IMPLIES AN EXCESS OF HORMONE, MAKING ITS TARGETS TOO ACTIVE; "HYPO-" IMPLIES TOO LITTLE HORMONE AND UNDERACTIVITY OF ITS TARGETS.

PITUITARY TUMORS

The central role of the pituitary in the endocrine system is reflected in the problems caused by a pituitary tumor, which may grow in any part of the gland; those in the anterior lobe are more likely to be benign (noncancerous). One result may be excess growth hormone, which causes enlargement of certain bones, such as those in the face, hands, and feet, and of some tissues, such as the tongue, as well as the development of coarse body hair and deepening of the voice. This condition is known as acromegaly. Some tumors cause excessive prolactin secretion or overstimulate the adrenal cortex.

PROLACTINOMAS

About 40 percent of pituitary tumors are prolactinomas—slow-growing, noncancerous tumors that cause the anterior lobe to secrete excessive prolactin. Normally this hormone promotes breast development and milk production in pregnancy. Symptoms of excess prolactin include irregular periods and lowered fertility in women; breast enlargement and impotence in men; and fluid leakage from the nipples, along with reduced sexual desire. In most cases, medication helps to shrink the tumor and reduce prolactin output; otherwise, surgery or radiation therapy may be necessary.

PITUITARY TUMOR
An enlarging tumor may press on the optic nerves that pass just above it, causing headaches and visual disturbances, such as losing part of the visual field.

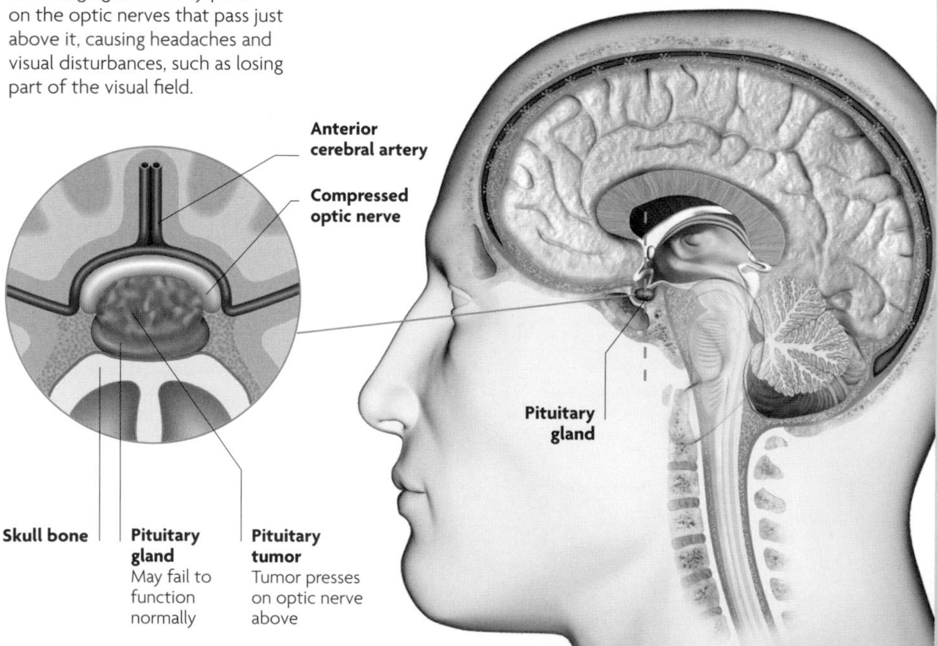
Anterior cerebral artery
Compressed optic nerve
Pituitary gland
Skull bone
Pituitary gland
May fail to function normally
Pituitary tumor
Tumor presses on optic nerve above

HYPERTHYROIDISM

Three-quarters of overstimulated thyroid cases are due to Graves' disease, an autoimmune disorder in which antibodies attack the thyroid, causing excessive hormone production. It is one of the most common hormonal disorders, especially in women aged 20-50. A less common cause is small lumps (nodules) in the gland. Raised hormone levels push up the metabolic rate, causing weight loss due to increased energy usage, rapid irregular heartbeat, trembling, sweating, anxiety, insomnia, weakness, and more frequent bowel movements. The enlarged thyroid may show as a swelling in the neck (goiter). Drug treatment can usually control the condition.

GRAVES' DISEASE
Hyperthyroidism due to Graves' disease can cause bulging eyes, giving a staring appearance and possibly blurred vision.

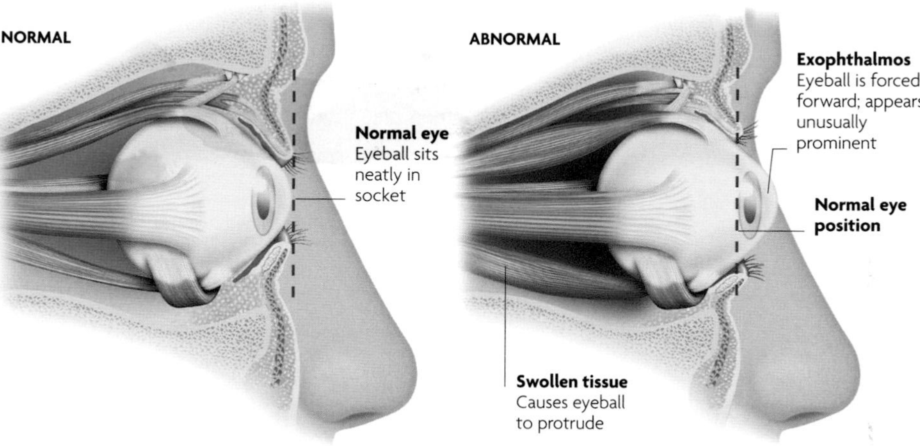
NORMAL
Normal eye
Eyeball sits neatly in socket
ABNORMAL
Exophthalmos
Eyeball is forced forward; appears unusually prominent
Normal eye position
Swollen tissue
Causes eyeball to protrude

HYPOTHYROIDISM

In hypothyroidism, the thyroid hormones, tri-iodothyronine and thyroxine, are underproduced. Because these hormones govern the speed of many metabolic processes, a lack of them leads to a slowing of bodily functions. Symptoms of hypothyroidism include fatigue, weight gain, slow bowel activity and constipation, swollen face, puffy eyes, thickened skin, thinned hair, hoarse voice, and inability to cope with cold. The most common cause of hypothyroidism is inflammation of the thyroid gland due to an autoimmune condition called Hashimoto's thyroiditis, in which antibodies mistakenly damage the gland. The thyroid gland may swell considerably as a lump, or goitre, in the neck. A less common cause of hypothyroidism is a lack of the mineral iodine–needed to make the thyroid hormones–in the diet. A rarer possibility is damage to the pituitary gland by a tumor.

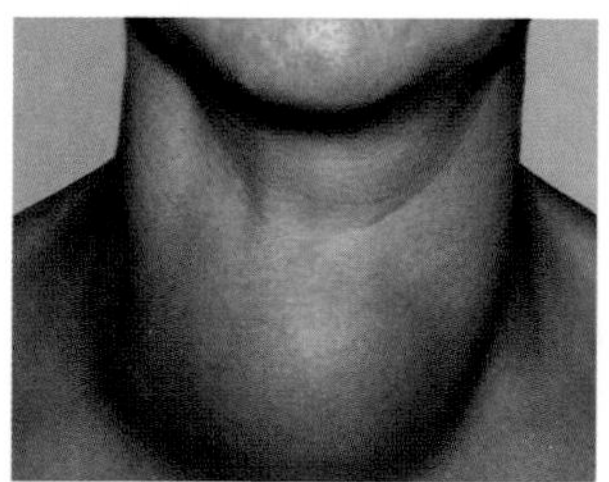

GOITER
A swollen thyroid (goiter) may be due to thyroiditis, hyperthyroidism, hypothyroidism, thyroid nodules, or cancer of the thyroid.

DIABETES MELLITUS

THE MAIN ENERGY SOURCE FOR CELLS IS GLUCOSE, WHICH THE CELLS ABSORB FROM THE BLOOD WITH THE HELP OF THE HORMONE INSULIN. IN DIABETES MELLITUS, THIS PROCESS DOES NOT WORK PROPERLY. THERE ARE TWO MAIN TYPES OF DIABETES MELLITUS: TYPE 1 AND TYPE 2.

TYPE 1 DIABETES

Type 1 diabetes mellitus is an autoimmune disorder. It occurs when the immune system misidentifies beta cells in the islets of Langerhans in the pancreas as foreign and destroys them. The cause is unknown, but the disease may be triggered by a viral infection or inflammation in the pancreas. It usually develops in childhood or adolescence. Symptoms include thirst, dry mouth, hunger, frequent urination, fatigue, blurred vision, and weight loss. If untreated, the disorder can cause ketoacidosis, in which toxic chemicals called ketones build up in the blood. Affected people need urgent medical attention; otherwise they can fall into a coma. There can also be long-term complications (see Type 2 diabetes, opposite). Treatment involves insulin injections.

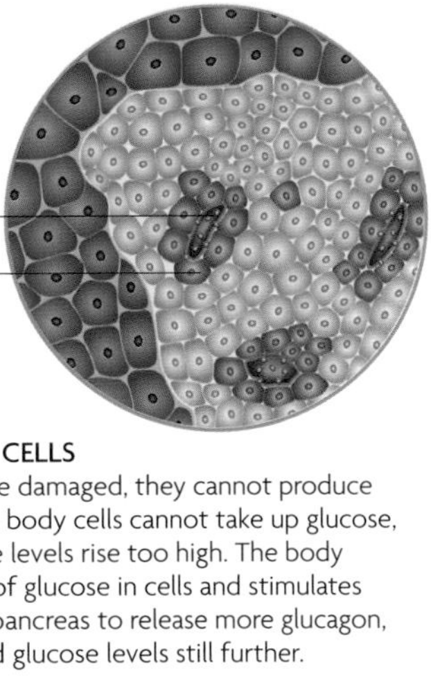

Insulin
Insulin secreted into capillaries

Beta cells
Insulin-producing cells

NORMAL BETA-CELL FUNCTION

As food and drink are digested, the presence of glucose, amino acids, and fatty acids in the intestine stimulates beta cells to release insulin into the bloodstream via tiny blood vessels called capillaries, which run through the islets of Langerhans.

Damaged beta cells
Insulin-producing cells destroyed

Capillary
No insulin is secreted into capillaries

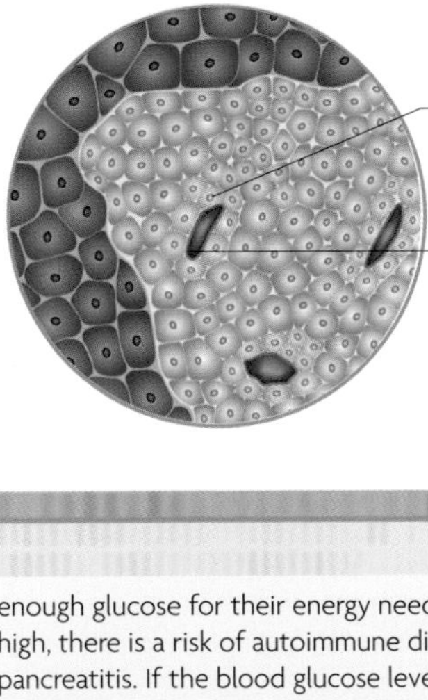

DAMAGED BETA CELLS

If the beta cells are damaged, they cannot produce insulin. As a result, body cells cannot take up glucose, and blood glucose levels rise too high. The body registers the lack of glucose in cells and stimulates alpha cells in the pancreas to release more glucagon, which raises blood glucose levels still further.

BLOOD SUGAR REGULATION

During digestion food is broken down to provide substances that cells can use to fuel and repair themselves. The main source of fuel is glucose, which is carried in the bloodstream to cells. Any excess is stored in the liver, muscle cells, and fat cells. The body needs to keep the blood glucose level steady. If it is too low, cells will not have enough glucose for their energy needs. If it is too high, there is a risk of autoimmune disease and pancreatitis. If the blood glucose level is too low, alpha cells in the islets of Langerhans secrete glucagon, which stimulates the release of stored glucose. If the level is too high, beta cells in the islets secrete insulin, which reduces the level.

TYPE 2 DIABETES

In type 2 diabetes, the pancreas secretes insulin, but the body cells are unable to respond to it. The cause is unknown, but some people may be predisposed to developing it. This form of diabetes is often associated with obesity and is a growing problem in affluent societies. The disorder develops slowly. There may be initial symptoms such as thirst, fatigue, and frequent urination, but in some cases the diabetes goes unnoticed for several years. As a result, complications may arise. Persistent high glucose levels can cause damage to small blood vessels around the body. People with type 2 diabetes are also more prone to high cholesterol levels, atherosclerosis (see p.157), and high blood pressure. The condition can be controlled with a healthy diet, regular exercise, and daily monitoring of blood glucose. However, in some cases, drugs are needed to boost insulin production or help the cells absorb glucose.

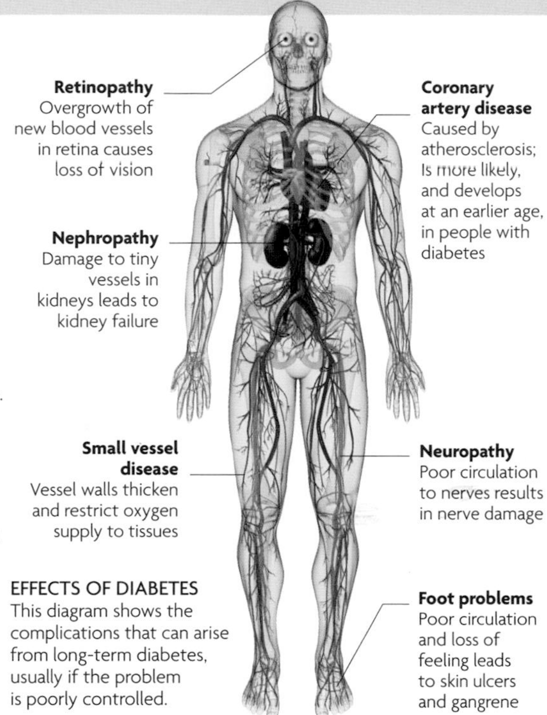
Retinopathy
Overgrowth of new blood vessels in retina causes loss of vision
Coronary artery disease
Caused by atherosclerosis; Is more likely, and develops at an earlier age, in people with diabetes
Nephropathy
Damage to tiny vessels in kidneys leads to kidney failure
Small vessel disease
Vessel walls thicken and restrict oxygen supply to tissues
Neuropathy
Poor circulation to nerves results in nerve damage
Foot problems
Poor circulation and loss of feeling leads to skin ulcers and gangrene

EFFECTS OF DIABETES
This diagram shows the complications that can arise from long-term diabetes, usually if the problem is poorly controlled.

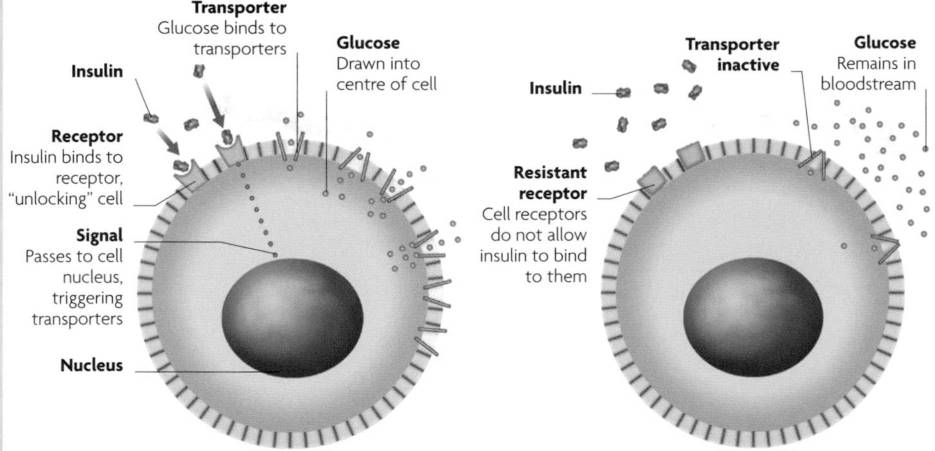
Transporter
Glucose binds to transporters
Insulin
Glucose
Drawn into centre of cell
Receptor
Insulin binds to receptor, "unlocking" cell
Signal
Passes to cell nucleus, triggering transporters
Nucleus
Transporter inactive
Glucose
Remains in bloodstream
Insulin
Resistant receptor
Cell receptors do not allow insulin to bind to them

NORMAL RECEPTORS
Insulin binds with receptors on a cell to allow glucose to enter the cell. This, in turn, triggers transporters in the cell to draw glucose inside.

MALFUNCTIONING RECEPTORS
People with type 2 diabetes produce enough insulin, but the receptors are resistant to it and glucose cannot be taken into cells.

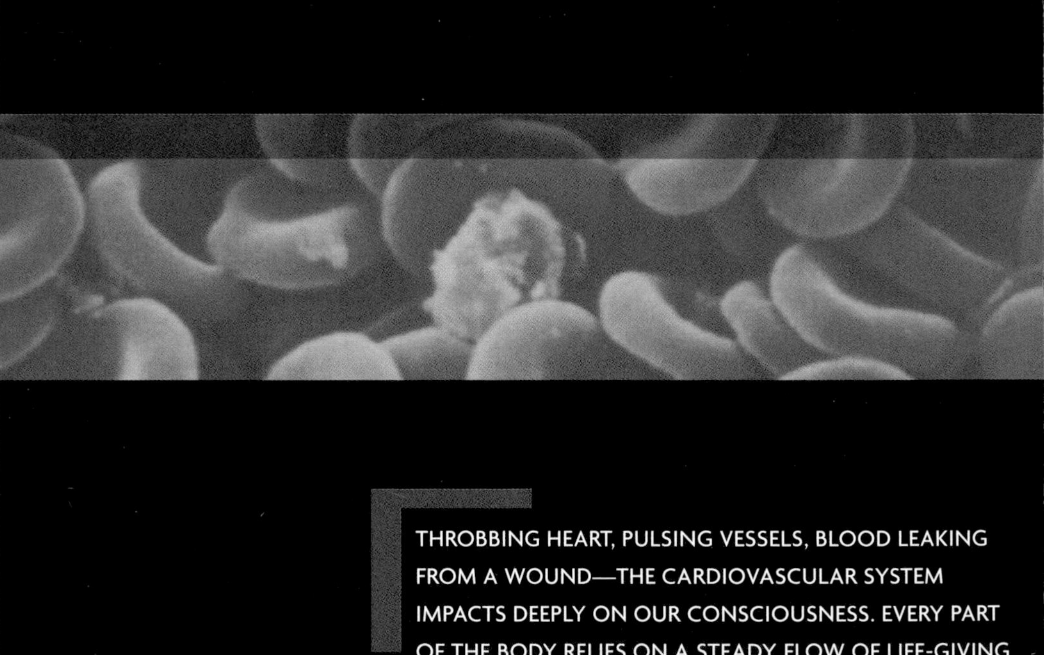

THROBBING HEART, PULSING VESSELS, BLOOD LEAKING FROM A WOUND—THE CARDIOVASCULAR SYSTEM IMPACTS DEEPLY ON OUR CONSCIOUSNESS. EVERY PART OF THE BODY RELIES ON A STEADY FLOW OF LIFE-GIVING

CARDIOVASCULAR SYSTEM

CARDIOVASCULAR ANATOMY

THE CIRCULATORY (OR CARDIOVASCULAR) SYSTEM DELIVERS OXYGEN AND NUTRIENTS TO VIRTUALLY ALL BODY CELLS, AND REMOVES CARBON DIOXIDE AND OTHER WASTES FROM THEM.

The circulatory system comprises the heart, blood vessels, and blood. Working as a pump, the heart beats regularly to send oxygen-rich blood into tough, elastic tubes called arteries, which convey the blood around the body. The arteries divide into tiny capillaries, the walls of which are so thin that oxygen, nutrients, and other substances can pass through to surrounding cells and tissues. Waste products flow from the tissues and cells into the blood for disposal. The capillaries join and enlarge to form veins, which take blood back to the heart. In the figure shown here, vessels carrying oxygenated blood (usually arteries) appear red and those carrying deoxygenated blood (usually veins) are blue. This intricate network is some 90,000 miles (150,000 km) long–almost equivalent to four times the circumference of the Earth.

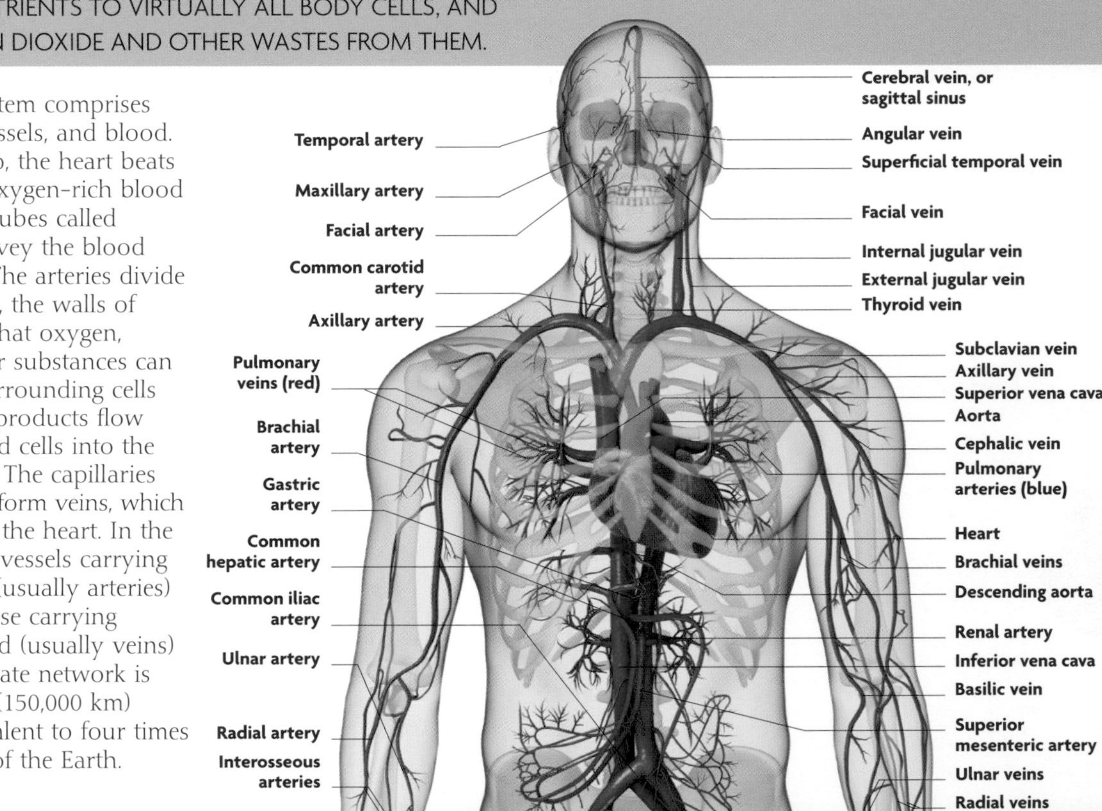
Temporal artery
Maxillary artery
Facial artery
Common carotid artery
Axillary artery
Pulmonary veins (red)
Brachial artery
Gastric artery
Common hepatic artery
Common iliac artery
Ulnar artery
Radial artery
Interosseous arteries
Cerebral vein, or sagittal sinus
Angular vein
Superficial temporal vein
Facial vein
Internal jugular vein
External jugular vein
Thyroid vein
Subclavian vein
Axillary vein
Superior vena cava
Aorta
Cephalic vein
Pulmonary arteries (blue)
Heart
Brachial veins
Descending aorta
Renal artery
Inferior vena cava
Basilic vein
Superior mesenteric artery
Ulnar veins
Radial veins

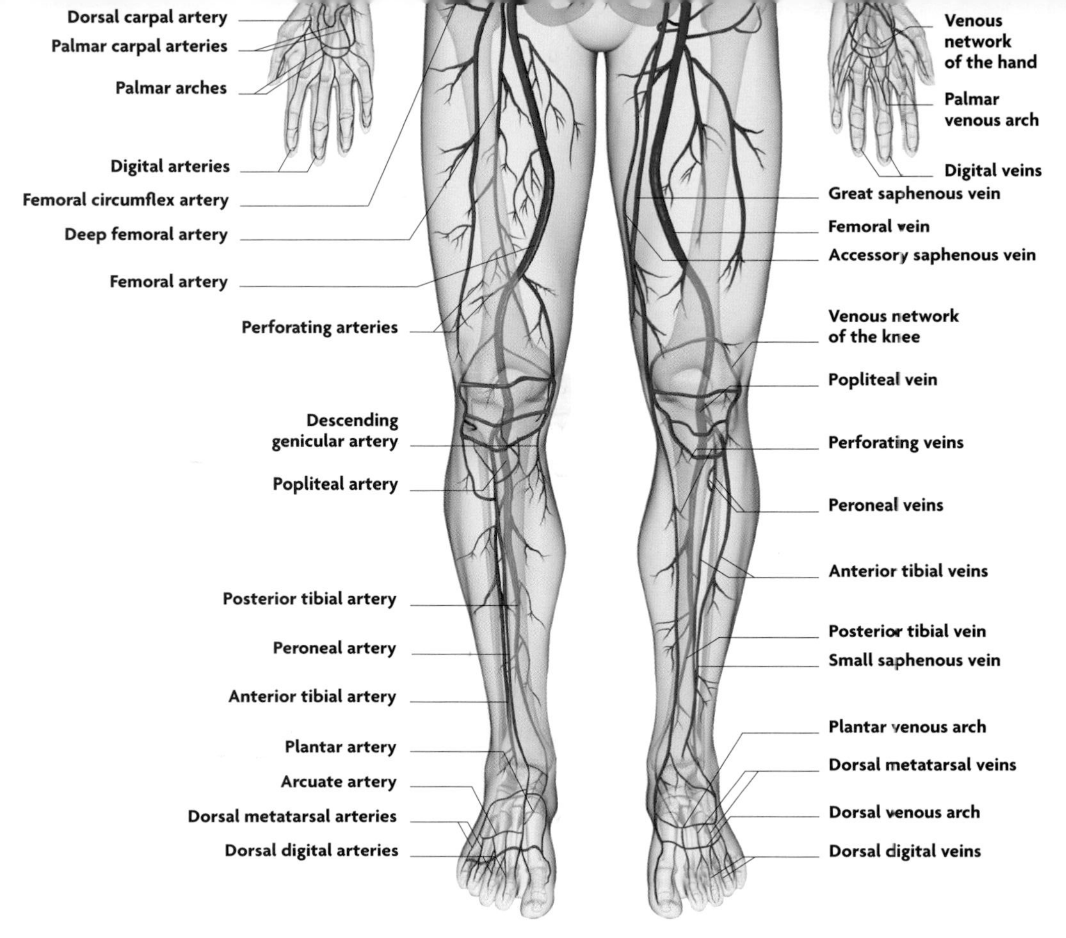
Dorsal carpal artery
Palmar carpal arteries
Palmar arches
Digital arteries
Femoral circumflex artery
Deep femoral artery
Femoral artery
Perforating arteries
Descending genicular artery
Popliteal artery
Posterior tibial artery
Peroneal artery
Anterior tibial artery
Plantar artery
Arcuate artery
Dorsal metatarsal arteries
Dorsal digital arteries
Venous network of the hand
Palmar venous arch
Digital veins
Great saphenous vein
Femoral vein
Accessory saphenous vein
Venous network of the knee
Popliteal vein
Perforating veins
Peroneal veins
Anterior tibial veins
Posterior tibial vein
Small saphenous vein
Plantar venous arch
Dorsal metatarsal veins
Dorsal venous arch
Dorsal digital veins

BLOOD AND BLOOD VESSELS

BLOOD IS A COLLECTION OF SPECIALIZED CELLS SUSPENDED IN A STRAW-COLORED LIQUID CALLED PLASMA. FLOWING AROUND THE BODY, BLOOD CARRIES OXYGEN AND NUTRIENTS, COLLECTS WASTE, DISTRIBUTES HORMONES, AND SPREADS HEAT.

WHAT IS BLOOD?

An adult has about 11 pints (5 liters) of blood. Roughly 50–55 percent of blood is plasma, 90 percent of which is water. Plasma contains dissolved substances such as glucose (blood sugar), hormones, enzymes, and also waste products such as urea and lactic acid. Plasma also contains proteins such as albumins, fibrinogen (important in clotting), and globular proteins or globulins. Alpha and beta globulins help transport lipids, which are fatty substances such as cholesterol. Gamma globulins are mostly the disease-fighting substances known as antibodies. The remaining 45–50 percent of blood is made up of three types of specialized cells. Red blood cells or erythrocytes carry oxygen; various white blood cells, known as leucocytes, are part of the defense system; and platelets or thrombocytes, which are tiny fragments of much larger cells, are involved in the process of clotting.

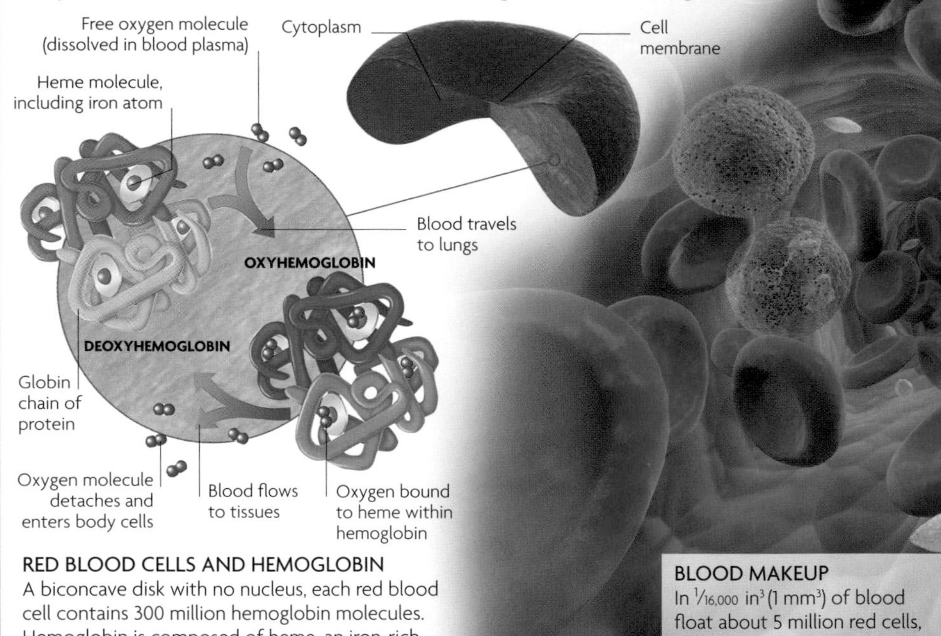
Free oxygen molecule (dissolved in blood plasma)
Cytoplasm
Cell membrane
Heme molecule, including iron atom
Blood travels to lungs
OXYHEMOGLOBIN
DEOXYHEMOGLOBIN
Globin chain of protein
Oxygen molecule detaches and enters body cells
Blood flows to tissues
Oxygen bound to heme within hemoglobin

RED BLOOD CELLS AND HEMOGLOBIN

A biconcave disk with no nucleus, each red blood cell contains 300 million hemoglobin molecules. Hemoglobin is composed of heme, an iron-rich pigment, and globin, ribbonlike protein chains. Oxygen in the lungs latches onto heme to make oxyhemoglobin. In this form, oxygen travels through the bloodstream to all parts of the body.

BLOOD MAKEUP

In $^{1}/_{16,000}$ in^3 (1 mm^3) of blood float about 5 million red cells, 10,000 white cells, and 300,000 platelets. These cells may have to move in single file through the narrowest blood vessels.

ARTERIES

Arteries carry blood away from the heart toward organs and tissues. Apart from the pulmonary arteries, all arteries carry oxygenated blood. Their thick walls and muscular and elastic layers can withstand the high pressure that occurs as the heart pumps blood.

Protective outer coat

Muscle and elastic fibers

Elastic and connective tissue

Inner lining (endothelium)

ARTERY SECTION
Four layers are found in an artery wall. The blood-carrying space, or lumen, is in the center.

VEINS

A vein is more flexible than an artery and its walls are thinner. The blood inside a vein is under relatively low pressure, and flows slowly and smoothly. Many larger veins, particularly the long veins in the legs, contain valves that prevent the backflow of blood, a job helped by muscles around the veins that contract during movement.

Outer layer

Inner lining

Valve cusp, or leaflet

Muscle layer

VEIN SECTION
The muscle layer of a vein is thin and enclosed by two layers; the innermost layer of some veins has valves at regular intervals.

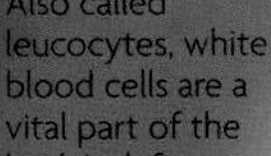

White blood cell
Also called leucocytes, white blood cells are a vital part of the body's defense system

Platelet
Tiny, short-lived cell fragment that has an important role in the clotting of blood

Red blood cell
Red blood cells (erythrocytes) have a lifespan of about 3 months

Blood vessel wall
The thickness of the wall is dependent on the pressure of blood flowing through it

CAPILLARIES

The smallest and most numerous of the blood vessels, capillaries convey blood between arteries and veins. A typical capillary is about $^{1}/_{2,500}$ in (0.01 mm) in diameter, only slightly wider than a red blood cell. Many capillaries enter tissue to form a capillary bed, where oxygen and other nutrients are released, and where waste matter passes into the blood.

CAPILLARY BED
Capillaries link small arteries (arterioles) to veins (venules).

Arteriole
Carries blood rich in oxygen

Capillary

Venule
Contains blood low in oxygen

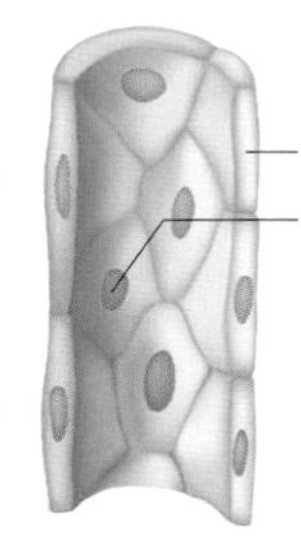
Capillary wall
Cell nucleus

CAPILLARY WALL
The thin capillary wall allows easy movement of substances between surrounding tissues.

HEART STRUCTURE

THE HEART IS A POWERFUL ORGAN ABOUT THE SIZE OF A CLENCHED FIST. LOCATED JUST TO THE LEFT OF CENTER IN BETWEEN THE LUNGS, IT OPERATES AS TWO COORDINATED PUMPS THAT SEND BLOOD AROUND THE BODY.

THE HEART'S BLOOD SUPPLY

The muscular wall, or myocardium, of the heart is constantly active and needs a generous supply of oxygen and energy from blood. To provide this, the heart muscle has its own blood vessels–the right and left coronary arteries. These vessels branch from the main artery, the aorta, just after it leaves the heart, and send smaller blood vessels into the heart muscle. Waste from heart tissue is removed by the coronary veins, in particular by the coronary sinus, a large vein at the back of the heart.

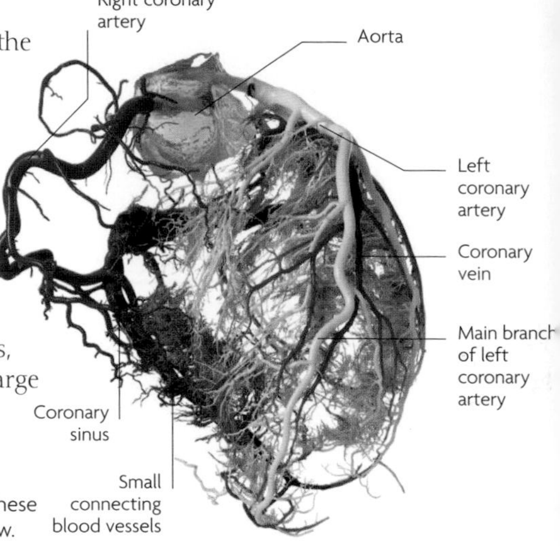
Right coronary artery
Aorta
Left coronary artery
Coronary vein
Main branch of left coronary artery
Coronary sinus
Small connecting blood vessels

CORONARY VESSELS

There are many connecting vessels between the coronary arteries. If an artery becomes blocked, these can provide an alternative route for the blood flow.

CARDIAC SKELETON

In the upper heart, four rigid, cufflike rings, known as the cardiac skeleton, provide points of attachment for the four heart valves and for the heart muscle. The wrap-around muscle fibers in the ventricle walls, and the timing of their contractions, enable the ventricles to squirt blood from the apex (lower end) upward, and out through the pulmonary and aortic valves, rather than squeezing blood down to pool in the apex region.

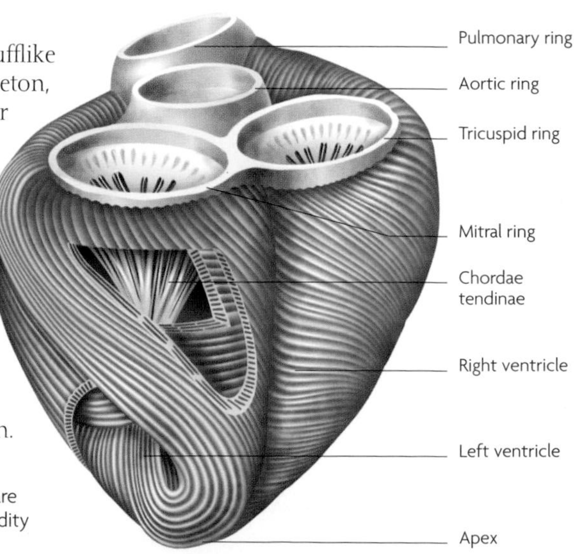
Pulmonary ring
Aortic ring
Tricuspid ring
Mitral ring
Chordae tendinae
Right ventricle
Left ventricle
Apex

FIBROUS FRAMEWORK

Four rings of fibrous tissue in the heart are known as the cardiac skeleton. Their rigidity prevents the valves from deforming.

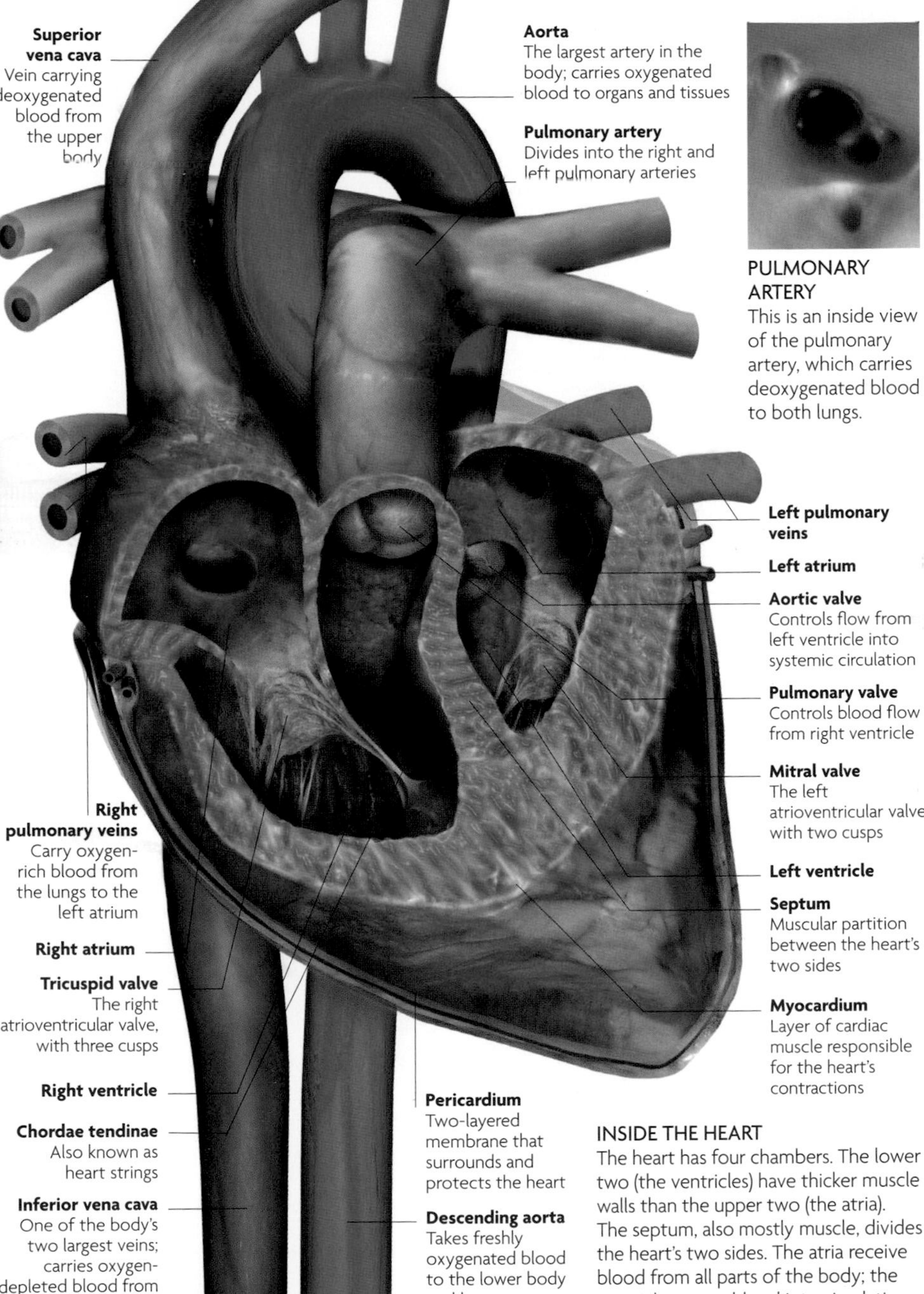
Superior vena cava
Vein carrying deoxygenated blood from the upper body
Aorta
The largest artery in the body; carries oxygenated blood to organs and tissues
Pulmonary artery
Divides into the right and left pulmonary arteries
Left pulmonary veins
Left atrium
Aortic valve
Controls flow from left ventricle into systemic circulation
Pulmonary valve
Controls blood flow from right ventricle
Mitral valve
The left atrioventricular valve, with two cusps
Left ventricle
Septum
Muscular partition between the heart's two sides
Myocardium
Layer of cardiac muscle responsible for the heart's contractions
Right pulmonary veins
Carry oxygen-rich blood from the lungs to the left atrium
Right atrium
Tricuspid valve
The right atrioventricular valve, with three cusps
Right ventricle
Chordae tendinae
Also known as heart strings
Inferior vena cava
One of the body's two largest veins; carries oxygen-depleted blood from the lower body
Pericardium
Two-layered membrane that surrounds and protects the heart
Descending aorta
Takes freshly oxygenated blood to the lower body and legs

PULMONARY ARTERY

This is an inside view of the pulmonary artery, which carries deoxygenated blood to both lungs.

INSIDE THE HEART

The heart has four chambers. The lower two (the ventricles) have thicker muscle walls than the upper two (the atria). The septum, also mostly muscle, divides the heart's two sides. The atria receive blood from all parts of the body; the ventricles pump blood into circulation.

HEART VALVES

The heart has four valves to control blood flow. Each has the same basic structure, although they differ in certain details. The two atrioventricular valves lie between the atria and ventricles. The mitral valve, on the left side, has two cusps, while its right counterpart, the tricuspid valve, has three. The two semilunar valves are at the exits from the ventricles: the pulmonary valve between the right ventricle and the pulmonary artery, and the aortic valve between the left ventricle and the aorta.

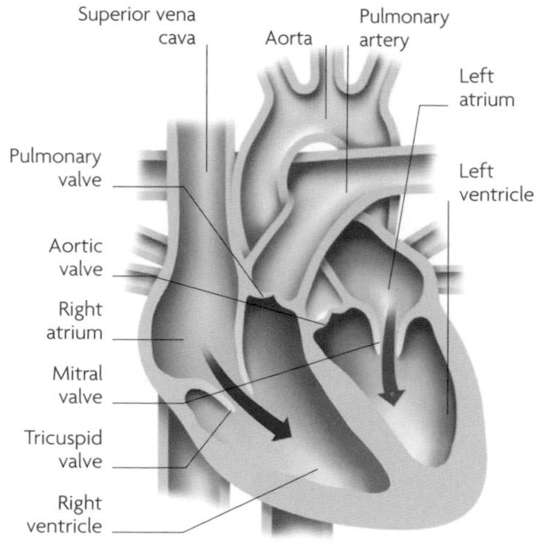
Superior vena cava
Aorta
Pulmonary artery
Left atrium
Left ventricle
Pulmonary valve
Aortic valve
Right atrium
Mitral valve
Tricuspid valve
Right ventricle

FUNCTION OF HEART VALVES

The tricuspid valve controls blood flow from the right atrium to the right ventricle; the pulmonary valve, from the right ventricle into the pulmonary artery; the mitral valve, from the left atrium to the left ventricle; and the aortic valve, from the left ventricle into the aorta.

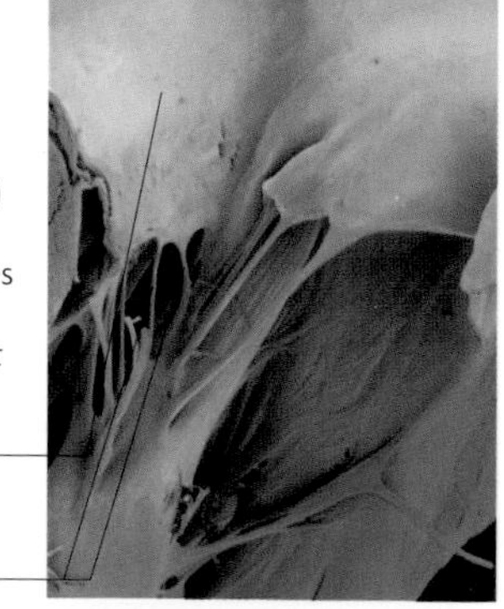
Cusp
Chordae tendinae

MITRAL VALVE

This image of a healthy human heart valve shows the heart strings (chordae tendinae) and valve cusps. The mitral valve lies between the left atrium and the left ventricle.

TWO CUSPS

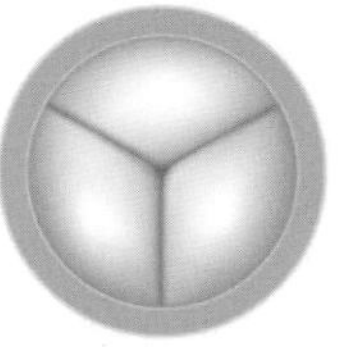

THREE CUSPS

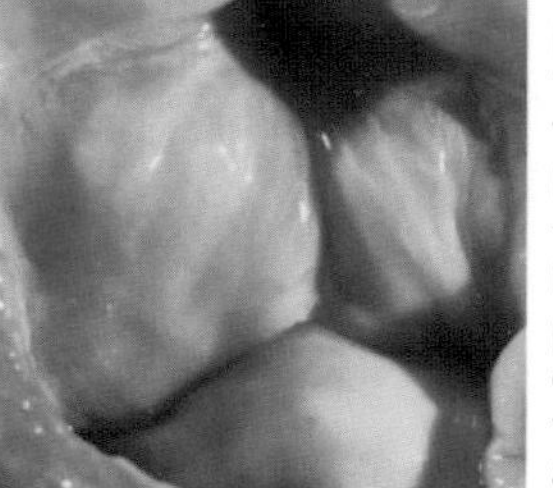
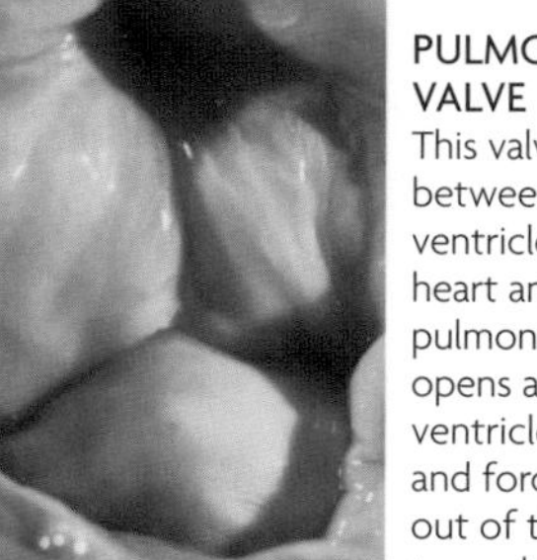

PULMONARY VALVE

This valve lies between the right ventricle of the heart and the pulmonary artery. It opens as the right ventricle contracts and forces blood out of the heart toward the lungs.

Direction of blood flow
Blood pushes against valve
Valve cusp open

HEART VALVE OPEN

The flexible cusps are forced apart by the pressure of blood as the heart contracts.

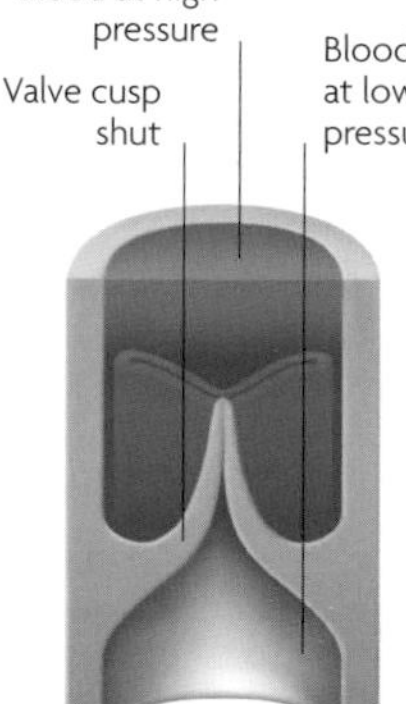
Blood at high pressure
Valve cusp shut
Blood at low pressure

HEART VALVE CLOSED

Back pressure causes the cusps to close and seal at their edges, to stop reverse blood flow.

CARDIAC MUSCLE

The walls of the heart are made of a special type of muscle known as cardiac muscle, which is found only in the heart. Unlike other types of muscle, cardiac muscle can contract repeatedly without becoming tired. However, to maintain this constant activity, the muscle requires a continuous, ample supply of oygenated blood, provided by the coronary arteries.

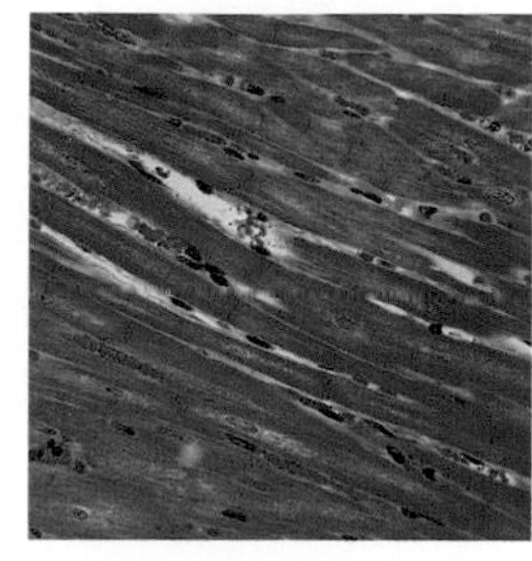

CARDIAC MUSCLE TISSUE
Cardiac muscle is a type of involuntary muscle, with short, branched, striated muscle fibers.

DOUBLE CIRCULATION

The right side of the heart pumps blood to the lungs to be oxygenated, and then back to the left side of the heart (pulmonary circulation). The left side of the heart pumps oxygen-rich blood to all the body's tissues and oxygen-depleted blood back to the right side of the heart (systemic circulation).

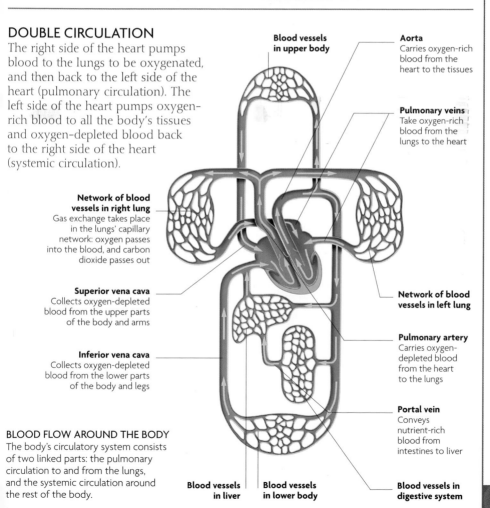

Blood vessels in upper body

Aorta
Carries oxygen-rich blood from the heart to the tissues

Pulmonary veins
Take oxygen-rich blood from the lungs to the heart

Network of blood vessels in right lung
Gas exchange takes place in the lungs' capillary network: oxygen passes into the blood, and carbon dioxide passes out

Superior vena cava
Collects oxygen-depleted blood from the upper parts of the body and arms

Network of blood vessels in left lung

Inferior vena cava
Collects oxygen-depleted blood from the lower parts of the body and legs

Pulmonary artery
Carries oxygen-depleted blood from the heart to the lungs

Portal vein
Conveys nutrient-rich blood from intestines to liver

Blood vessels in liver

Blood vessels in lower body

Blood vessels in digestive system

BLOOD FLOW AROUND THE BODY
The body's circulatory system consists of two linked parts: the pulmonary circulation to and from the lungs, and the systemic circulation around the rest of the body.

HOW THE HEART BEATS

THE HEART IS A DYNAMIC, UNTIRING, PRECISELY ADJUSTABLE DOUBLE-PUMP THAT FORCES BLOOD AROUND THE BODY'S IMMENSE NETWORK OF BLOOD VESSELS—PERHAPS MORE THAN THREE BILLION TIMES DURING A LIFETIME.

The heart's two lower chambers (ventricles) have thick, muscular walls that contract to squeeze blood into the arteries. The upper chambers (atria) have thinner walls and act partly as reservoirs for blood entering from the main veins. Each heartbeat has two main phases: in the first phase (diastole), the heart relaxes and refills with blood; in the second phase (systole), it contracts, forcing the blood out. The whole cycle takes, on average, less than a second. During activity or stress, both the beating rate and the volume of blood pumped out of the heart increase greatly.

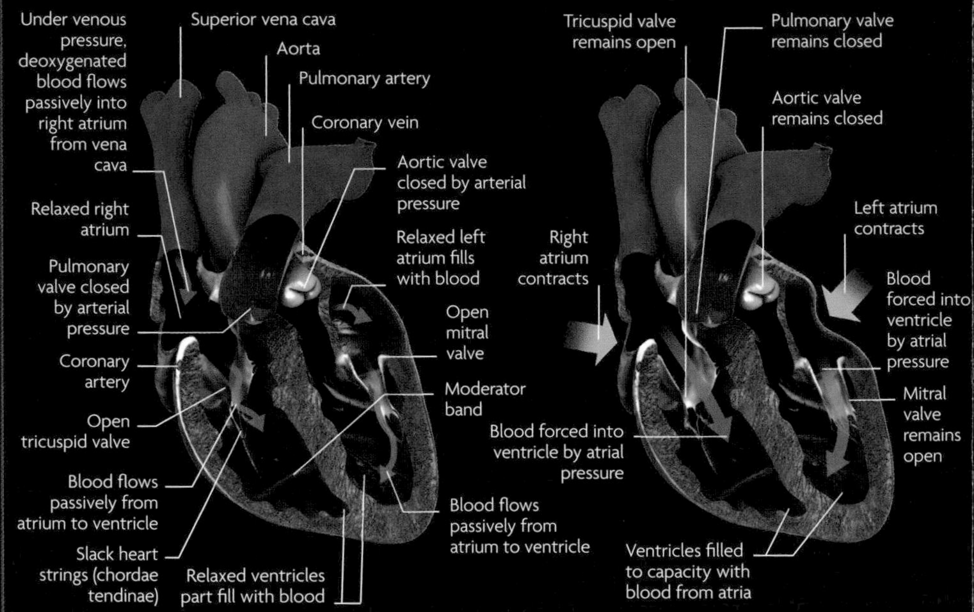
Under venous pressure, deoxygenated blood flows passively into right atrium from vena cava
Superior vena cava
Aorta
Pulmonary artery
Coronary vein
Aortic valve closed by arterial pressure
Relaxed right atrium
Relaxed left atrium fills with blood
Pulmonary valve closed by arterial pressure
Open mitral valve
Coronary artery
Moderator band
Open tricuspid valve
Blood flows passively from atrium to ventricle
Blood flows passively from atrium to ventricle
Slack heart strings (chordae tendinae)
Relaxed ventricles part fill with blood
Tricuspid valve remains open
Pulmonary valve remains closed
Aortic valve remains closed
Left atrium contracts
Right atrium contracts
Blood forced into ventricle by atrial pressure
Mitral valve remains open
Blood forced into ventricle by atrial pressure
Ventricles filled to capacity with blood from atria

1 RELAXATION (LATE DIASTOLE)

During this phase of the heartbeat sequence, the muscular walls of the heart relax. The atrial chambers balloon slightly as they fill with blood coming in under quite low pressure from the main veins. Deoxygenated blood from the body enters the right atrium, while oxygenated blood from the lungs enters the left atrium. Some of the blood in the atria flows down into the ventricles. By the end of this phase, the ventricles are filled to about 80 percent of capacity.

2 CONTRACTION OF THE ATRIA (ATRIAL SYSTOLE)

The heart's natural pacemaker, known as the sinoatrial node, is located in the upper part of the right atrium. It "fires" electrical impulses, similar to those generated by nerves, which set off the contraction phase. Some impulses spread through the atrial walls and stimulate their cardiac muscle to contract. This contraction squeezes the blood inside the atria through the atrioventricular (tricuspid and mitral) valves into the ventricles, whose walls remain relaxed.

HEARTBEAT SYNCHRONIZATION

Contractions of the atria and the ventricles are synchronized by electrical impulses from the sinoatrial node, the heart's natural pacemaker. Toward the end of diastole, the sinoatrial node sends out electrical impulses. These impulses travel through the atria, making them contract (atrial systole). Some impulses travel to the atrioventricular node, which sends them through conducting fibers to the ventricles, which contract in response (ventricular systole). The electrical impulses then travel back toward the atria and the sinoatrial node, which fires again to continue the cycle.

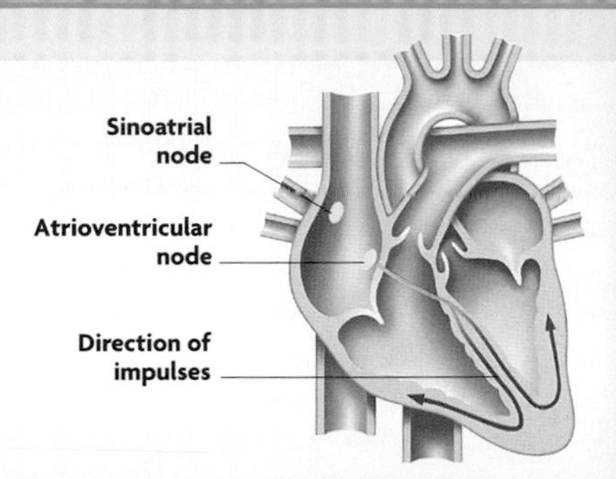

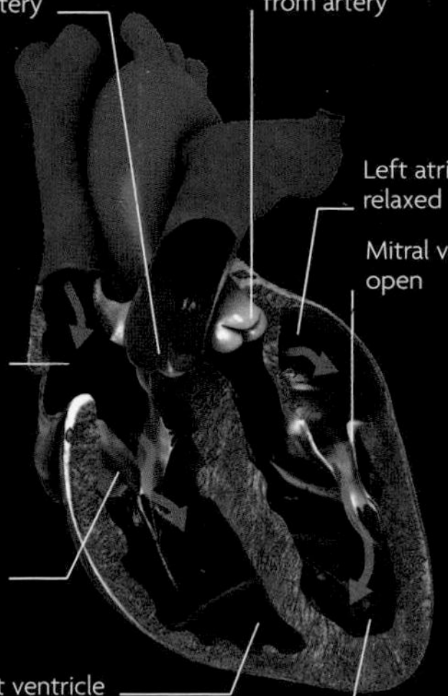

Sinoatrial node
Atrioventricular node
Direction of impulses

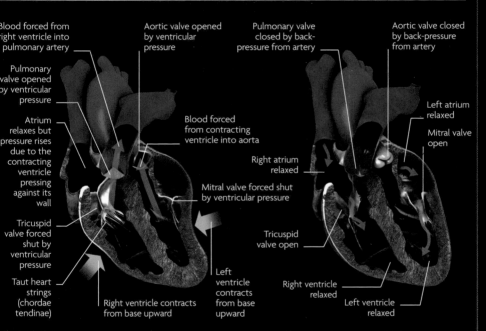

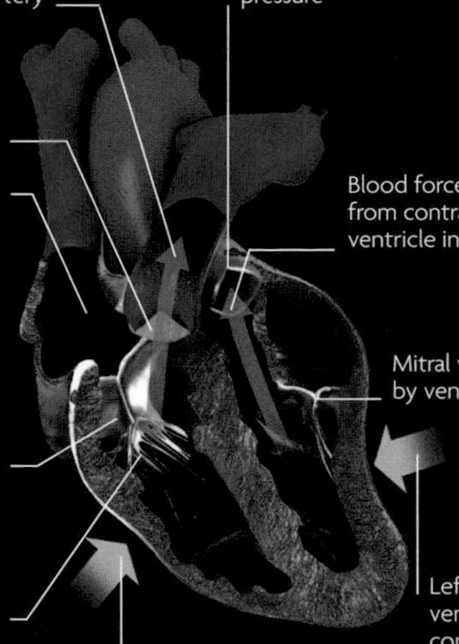

Blood forced from right ventricle into pulmonary artery

Pulmonary valve opened by ventricular pressure

Atrium relaxes but pressure rises due to the contracting ventricle pressing against its wall

Tricuspid valve forced shut by ventricular pressure

Taut heart strings (chordae tendinae)

Right ventricle contracts from base upward

Aortic valve opened by ventricular pressure

Blood forced from contracting ventricle into aorta

Mitral valve forced shut by ventricular pressure

Left ventricle contracts from base upward

Pulmonary valve closed by back-pressure from artery

Aortic valve closed by back-pressure from artery

Left atrium relaxed

Mitral valve open

Right atrium relaxed

Tricuspid valve open

Right ventricle relaxed

Left ventricle relaxed

3 CONTRACTION OF THE VENTRICLES (VENTRICULAR SYSTOLE)

During this most active and powerful stage of the heartbeat, the thick cardiac muscle in the ventricle walls contracts, stimulated by electrical impulses relayed by the atrioventricular node. This causes a rise in ventricular pressure, which opens the aortic and pulmonary valves at the exits of the ventricles. Blood is forced out into the main arteries, making the atrioventricular valves snap shut.

4 RELAXATION (EARLY DIASTOLE)

The walls of the ventricles begin to relax, causing ventricular pressure to reduce. The pressure of the recently ejected blood in the main arteries is now high, so both the aortic and pulmonary valves close. This prevents backflow into the ventricles. As ventricular pressure on the atrioventricular valves relaxes, the valves open. This reduces pressure in the atria, allowing blood to enter once again from the main veins.

CARDIOVASCULAR DISORDERS

DISORDERS OF THE CARDIOVASCULAR SYSTEM MAY AFFECT THE HEART ITSELF, CAUSING STRUCTURAL DAMAGE OR DISRUPTING HEART RHYTHM. BLOCKAGES IN BLOOD VESSELS CAN STARVE TISSUES OF OXYGEN, LEADING TO SERIOUS PROBLEMS ANYWHERE IN THE BODY.

HEART ATTACK

A heart attack (myocardial infarction) is the result of coronary artery disease due to atherosclerosis (see opposite), and the subsequent formation of a blood clot, or thrombus. Once formed, the clot can completely block blood flow to an area of heart muscle, starving it of oxygen and eventually causing tissue death. If possible, the blood flow must be restored to the damaged cells as quickly as possible. A heart attack usually occurs suddenly, with little or no warning. The chest pain is usually severe, is not necessarily brought on by exertion, and persists despite resting. A heart attack can also cause sweating, shortness of breath, nausea, and loss of consciousness.

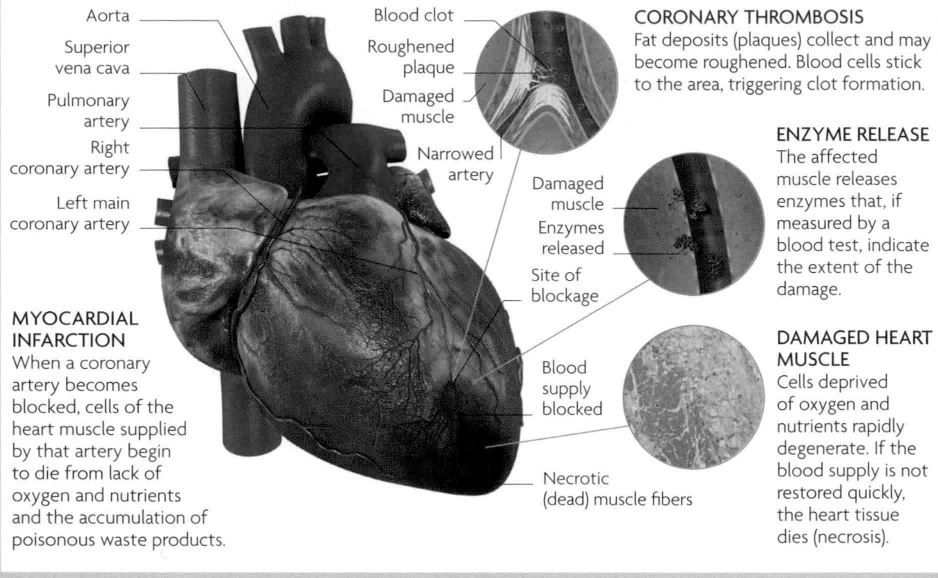
Aorta
Superior vena cava
Pulmonary artery
Right coronary artery
Left main coronary artery
Blood clot
Roughened plaque
Damaged muscle
Narrowed artery
Damaged muscle
Enzymes released
Site of blockage
Blood supply blocked
Necrotic (dead) muscle fibers

MYOCARDIAL INFARCTION
When a coronary artery becomes blocked, cells of the heart muscle supplied by that artery begin to die from lack of oxygen and nutrients and the accumulation of poisonous waste products.

CORONARY THROMBOSIS
Fat deposits (plaques) collect and may become roughened. Blood cells stick to the area, triggering clot formation.

ENZYME RELEASE
The affected muscle releases enzymes that, if measured by a blood test, indicate the extent of the damage.

DAMAGED HEART MUSCLE
Cells deprived of oxygen and nutrients rapidly degenerate. If the blood supply is not restored quickly, the heart tissue dies (necrosis).

ANGINA

Angina is caused by a temporarily inadequate supply of blood to the heart muscle, usually because of arterial narrowing as a result of atherosclerosis (see opposite). The pain most often occurs when the heart's workload is increased, for example with exercise, and fades with rest. Other triggers of angina are stress, cold weather, or a large meal. An attack of angina typically begins with a heavy, constricting pain behind the breastbone. This can spread into the throat and jaw, and down into the arms, especially the left one. The pain usually subsides within about 10–15 minutes. People with angina often take medication to dilate (widen) the coronary arteries.

ATHEROSCLEROSIS

The process that leads to atherosclerosis begins with abnormally high levels of excess fats and cholesterol in the blood. These substances infiltrate the lining of arteries, forming deposits known as atheroma. This can happen in any of the body's arteries, including those supplying the brain with blood, when the result may be a stroke. The atheromatous deposits gradually form raised patches known as plaques. These consist of fatty cores within the arterial wall, covered by fibrous caps. The plaques narrow the space within the artery, restricting the overall flow of blood to tissues beyond the site. They also cause turbulence that disrupts the flow of blood, and the eddies over the plaque surface make the blood more likely to clot. The major risk factors for atherosclerosis include smoking, a diet high in saturated fats, lack of exercise, and excess weight.

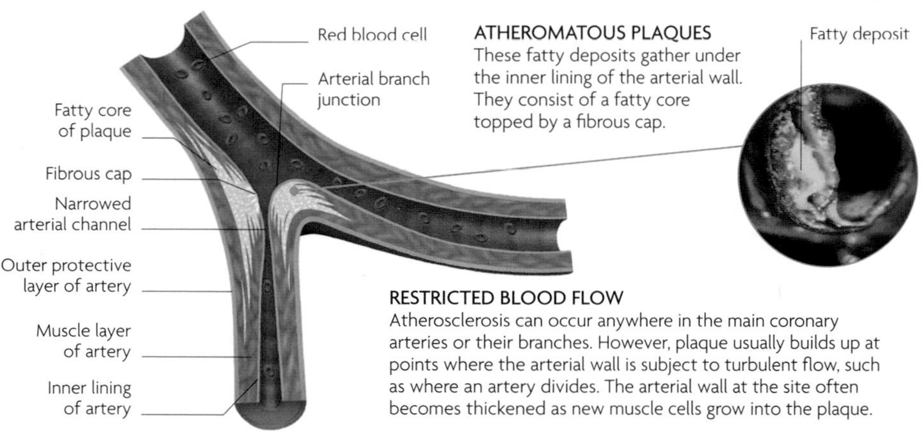
Red blood cell
Arterial branch junction
Fatty core of plaque
Fibrous cap
Narrowed arterial channel
Outer protective layer of artery
Muscle layer of artery
Inner lining of artery
Fatty deposit

ATHEROMATOUS PLAQUES

These fatty deposits gather under the inner lining of the arterial wall. They consist of a fatty core topped by a fibrous cap.

RESTRICTED BLOOD FLOW

Atherosclerosis can occur anywhere in the main coronary arteries or their branches. However, plaque usually builds up at points where the arterial wall is subject to turbulent flow, such as where an artery divides. The arterial wall at the site often becomes thickened as new muscle cells grow into the plaque.

VALVE DISORDERS

There are two main types of heart valve disorder: stenosis and incompetence. In stenosis, the valve outlet is too narrow, restricting blood flow. The condition may be congenital (present at birth), due to an infection such as rheumatic fever, or part of the aging process.

In incompetence, the heart valve does not close fully, allowing backflow of blood. This problem can result from a heart attack or an infection of the valve.

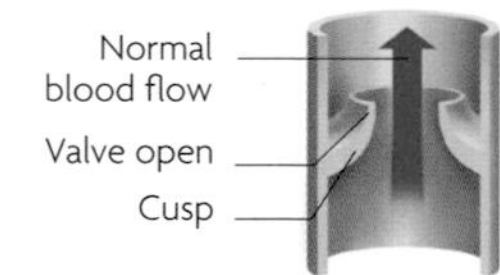
Normal blood flow
Valve open
Cusp

NORMAL VALVE OPEN

As the heart contracts, the high pressure pushes the cusps of the valve open, allowing blood to flow.

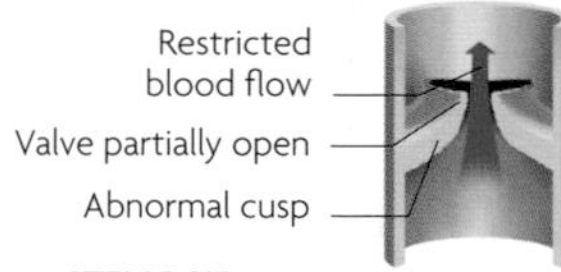
Restricted blood flow
Valve partially open
Abnormal cusp

STENOSIS

The valve stiffens and cannot open fully. Blood flow is restricted, so the heart beats harder to compensate.

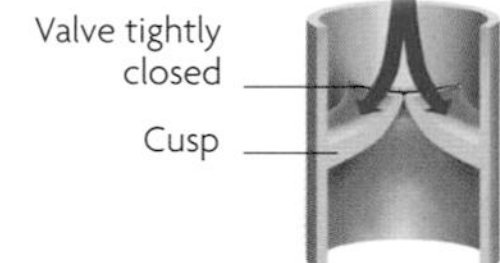
Valve tightly closed
Cusp

NORMAL VALVE CLOSED

The pressure on the other side of the valve increases and the valve cusps snap shut, preventing backflow.

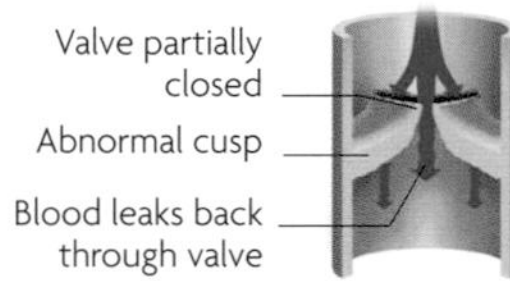
Valve partially closed
Abnormal cusp
Blood leaks back through valve

INCOMPETENCE

The cusps do not close properly, and blood leaks backward. The heart must work harder to circulate blood.

EMBOLISM

Most emboli are fragments of a blood clot (thrombus), or even a whole clot, that has detached from its original site and traveled in the bloodstream to lodge in a blood vessel. An embolus may also be made of fatty material from an atheromatous plaque (see p.157) in an arterial wall, crystals of cholesterol, fatty bone marrow that has entered the circulation following a bone fracture, or an air bubble or amniotic fluid. In a pulmonary embolism, a clot originating elsewhere in the body travels in the veins to the lungs. Clots that form in the heart or arteries can block the blood circulation anywhere in the body. An embolus is most likely to block a blood vessel where it narrows or branches, depriving the tissues of oxygen beyond the site of the blockage. The symptoms of an embolism depend on the site affected.

Embolus traveling to lung

Pulmonary artery

Path of embolus

PULMONARY EMBOLISM
A fragment of blood clot from a leg vein may travel through the veins to the heart's right side, then out along the pulmonary arteries to a lung.

Inferior vena cava

Path of embolus

Thrombotic embolus
A fragment (embolus) composed of blood clot (thrombotic) material may arise anywhere, but veins of the legs and pelvis are common sites

THROMBOSIS

Thrombosis is the blockage of a blood vessel by a blood clot. It is most likely to occur where normal blood flow is disrupted, which may be due to plaques of fatty atheromatous tissue in the walls of an artery (see p.157) or inflammation of the blood vessel. The clot eventually narrows or blocks the passage for blood so that the tissues beyond are deprived of oxygen and nutrients.

THROMBUS FORMATION
Thrombosis can occur in arteries and veins, but commonly happens at a site of atherosclerosis in an artery wall, which disrupts the normal blood flow.

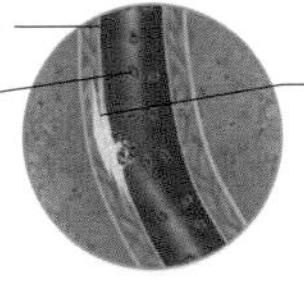
Lining
Platelet
Damage from atheroma (plaque)

1 INTERNAL DAMAGE
When an artery lining is damaged by plaque, platelets clump and release chemicals that begin the clotting process.

Fibrin strands

Thrombus blocking artery; thrombi can also form in veins

2 CLOT FORMATION
The chemicals help convert fibrinogen into insoluble fibrin strands, which trap blood cells, escalating clot formation.

ARRHYTHMIA

An arrhythmia is a heart rate that is unusually slow or fast, or erratic. A normal heartbeat is initiated by specialized cells in the natural "pacemaker," the sinoatrial (SA) node, at the top of the right atrium. They send electrical impulses out through the atrial muscle tissue, stimulating it to contract. These signals are relayed by the atrioventricular (AV) node along nervelike fibers through the septum (central dividing wall) and into the thick muscle tissue of the ventricle walls. A fault in the system can lead to various arrhythmias, including those described here.

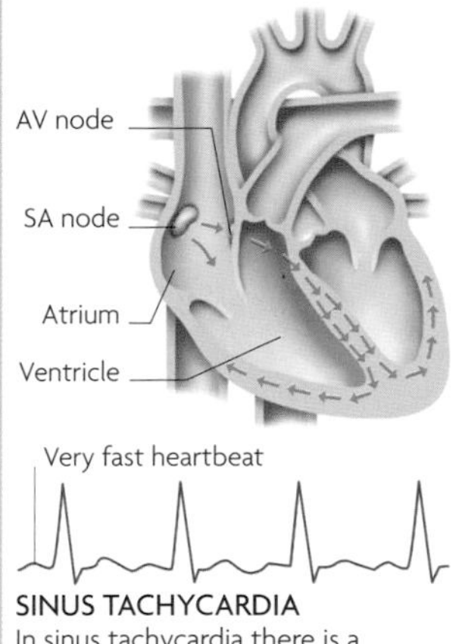
AV node
SA node
Atrium
Ventricle
Very fast heartbeat

SINUS TACHYCARDIA
In sinus tachycardia there is a regular but rapid heart rate, usually more than 100 beats per minute.

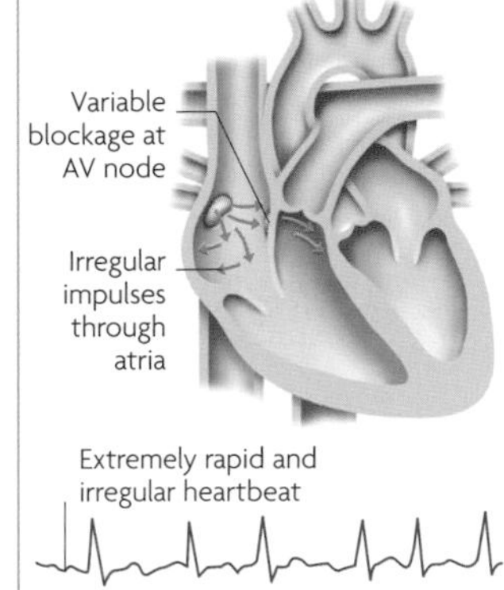
Variable blockage at AV node
Irregular impulses through atria
Extremely rapid and irregular heartbeat

ATRIAL FIBRILLATION
Fibrillations are rapid, disordered, weak contractions with a rate as high as 500 per minute.

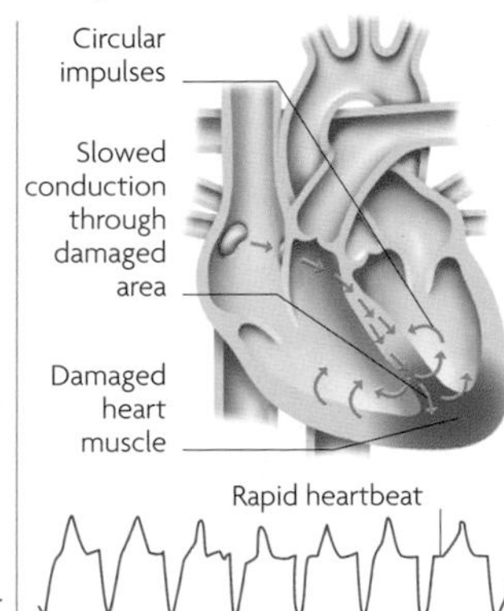
Circular impulses
Slowed conduction through damaged area
Damaged heart muscle
Rapid heartbeat

VENTRICULAR TACHYCARDIA
Very fast ventricular contractions may be caused by heart muscle damage, for example, due to a heart attack.

HYPERTENSION

Blood is under pressure as the heart pumps it around the circulation. In hypertension, this pressure is persistently above normal limits. There are no symptoms of hypertension at first, but despite this, over time it increases the risk of many serious disorders, such as stroke, heart disease, and kidney failure. There is usually no obvious cause for hypertension; however, lifestyle and genetic factors may contribute to the condition, as may being overweight, drinking excessive amounts of alcohol, smoking, and having a high-salt diet. Hypertension is most common in middle-aged and elderly people. A stressful lifestyle may aggravate the condition.

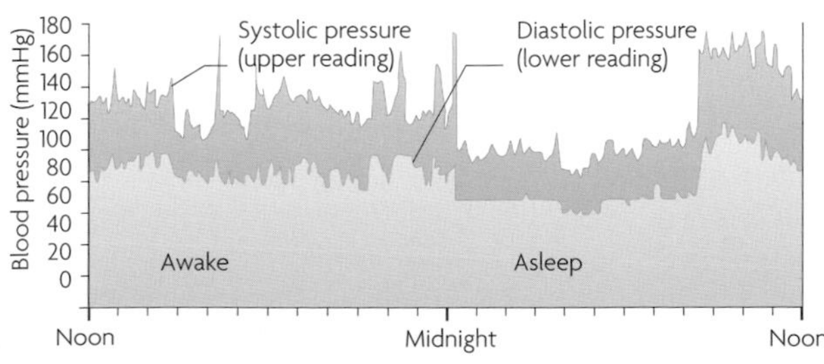
Blood pressure (mmHg)
180
160
140
120
100
80
60
40
20
0
Systolic pressure (upper reading)
Diastolic pressure (lower reading)
Awake
Asleep
Noon
Midnight
Noon

BLOOD PRESSURE GRAPH
Normal blood pressure varies according to activity levels. This graph shows that during sleep, both the systolic and diastolic pressures (see pp.154–155) are much lower.

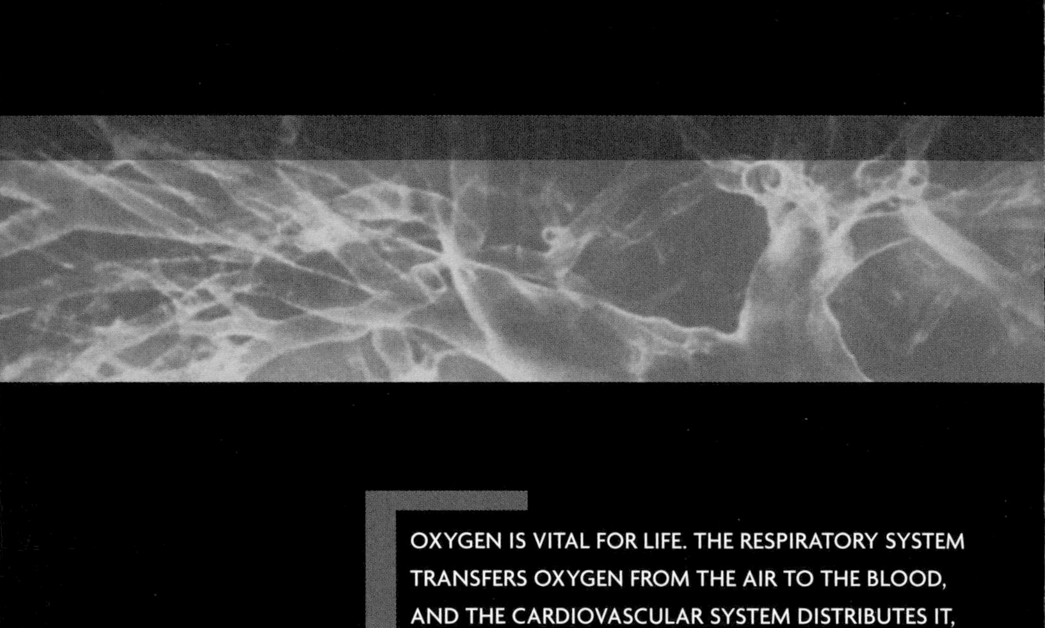

OXYGEN IS VITAL FOR LIFE. THE RESPIRATORY SYSTEM TRANSFERS OXYGEN FROM THE AIR TO THE BLOOD, AND THE CARDIOVASCULAR SYSTEM DISTRIBUTES IT, WHILE THE MUSCULAR AND SKELETAL SYSTEMS DRIVE

RESPIRATORY SYSTEM

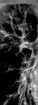

RESPIRATORY ANATOMY

THE RESPIRATORY SYSTEM, IN CLOSE CONJUNCTION WITH THE CIRCULATORY SYSTEM, IS RESPONSIBLE FOR SUPPLYING ALL BODY CELLS WITH ESSENTIAL OXYGEN AND REMOVING POTENTIALLY HARMFUL CARBON DIOXIDE FROM THE BODY.

Air enters the body mainly through the nostrils. The nostrils lead into the nasal cavity, which joins with the pharynx (throat). The larynx, home to the vocal cords, joins the pharynx to the trachea (windpipe). The trachea splits into two airways, known as the primary bronchi, each taking air to one lung. The bronchi divide into secondary and tertiary bronchi, and eventually into minute bronchioles. In the lungs, exchange of gases takes place in tiny sacs called alveoli.

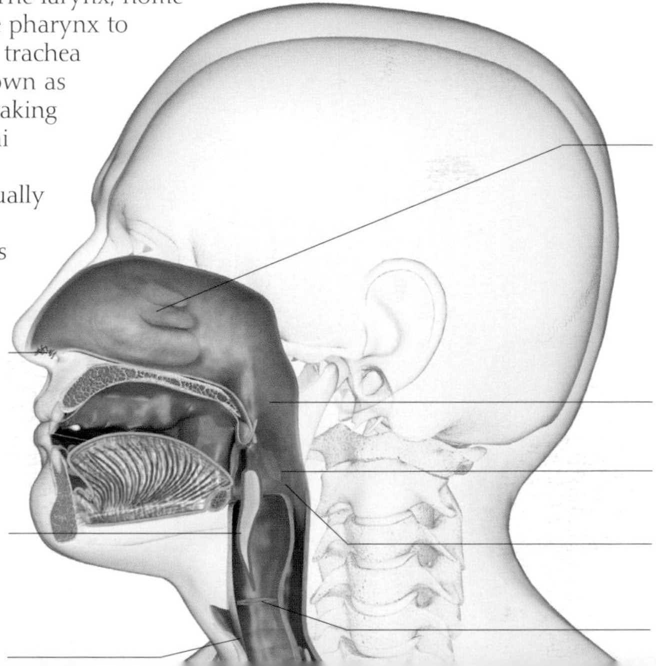

Nose hairs
Situated inside entrance of nostrils; help filter out large particles of dust and debris
Epiglottis
Cartilage flap that tilts over entrance to larynx when swallowing, to keep food, drinks, and saliva from entering trachea
Larynx
Short, cartilaginous tube joining pharynx with trachea; together with vocal cords within it, the larynx plays a vital role in speech production
Nasal cavity
Main route for air to and from lungs; lined with a sticky, mucus-covered membrane that traps dust particles and microorganisms; divided into two by central plate of cartilage (nasal septum); also contains patches called olfactory epithelia in roof of cavity, which are the sensory organs of smell
Nasopharynx
Allows passage of air only
Oropharynx
Permits passage of foods and fluids
Laryngopharynx
Permits passage of foods and fluids
Pharynx
Short tube that begins at rear of nasal cavity and ends at larynx (voice box)
Vocal cord

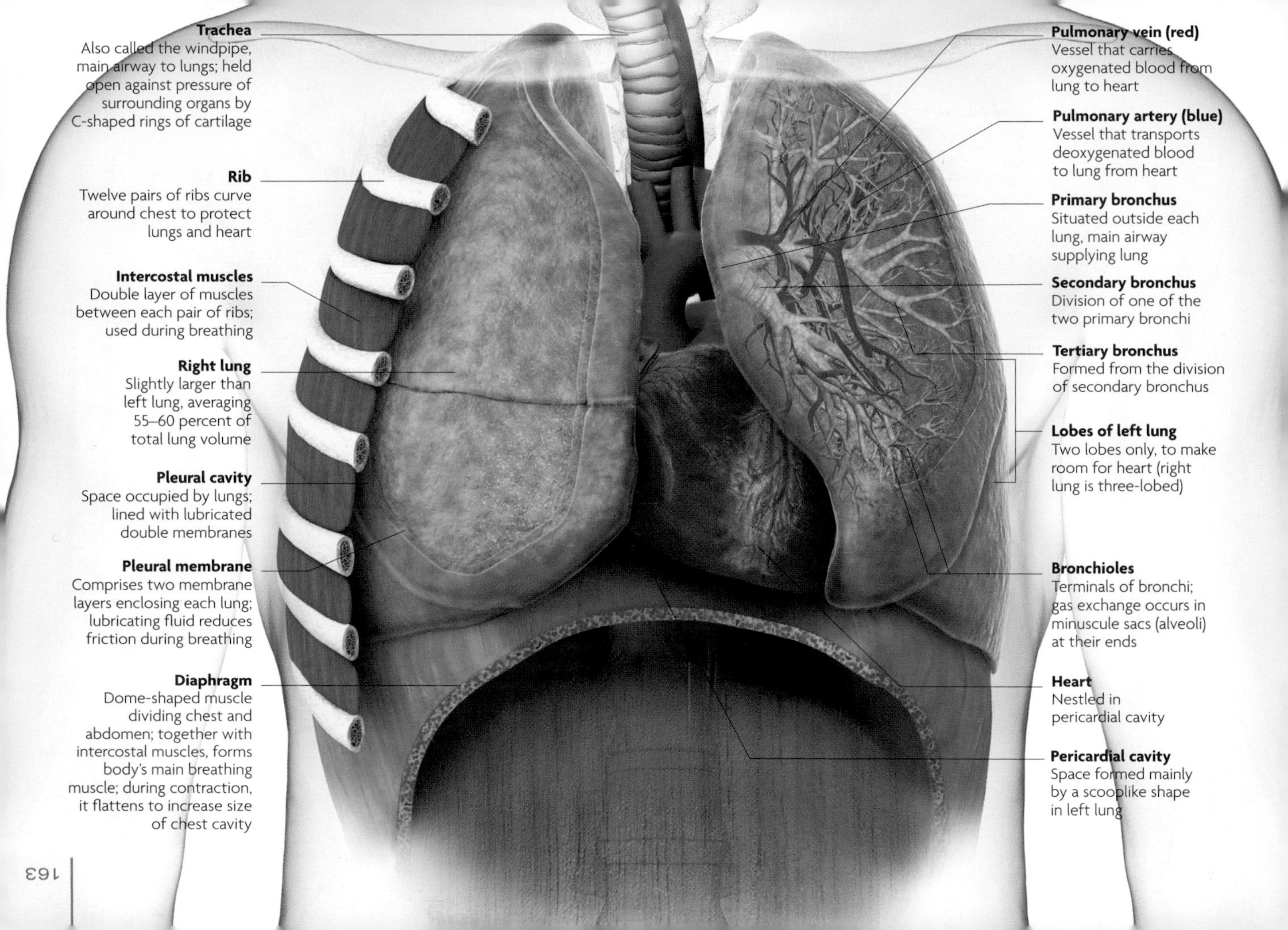

Trachea
Also called the windpipe, main airway to lungs; held open against pressure of surrounding organs by C-shaped rings of cartilage

Rib
Twelve pairs of ribs curve around chest to protect lungs and heart

Intercostal muscles
Double layer of muscles between each pair of ribs; used during breathing

Right lung
Slightly larger than left lung, averaging 55–60 percent of total lung volume

Pleural cavity
Space occupied by lungs; lined with lubricated double membranes

Pleural membrane
Comprises two membrane layers enclosing each lung; lubricating fluid reduces friction during breathing

Diaphragm
Dome-shaped muscle dividing chest and abdomen; together with intercostal muscles, forms body's main breathing muscle; during contraction, it flattens to increase size of chest cavity

Pulmonary vein (red)
Vessel that carries oxygenated blood from lung to heart

Pulmonary artery (blue)
Vessel that transports deoxygenated blood to lung from heart

Primary bronchus
Situated outside each lung, main airway supplying lung

Secondary bronchus
Division of one of the two primary bronchi

Tertiary bronchus
Formed from the division of secondary bronchus

Lobes of left lung
Two lobes only, to make room for heart (right lung is three-lobed)

Bronchioles
Terminals of bronchi; gas exchange occurs in minuscule sacs (alveoli) at their ends

Heart
Nestled in pericardial cavity

Pericardial cavity
Space formed mainly by a scooplike shape in left lung

LUNGS

THE TWO SPONGELIKE LUNGS FILL MOST OF THE CHEST CAVITY AND ARE PROTECTED BY THE RIBS. THEIR ESSENTIAL FUNCTION IS GAS EXCHANGE—TAKING IN VITAL OXYGEN FROM THE AIR AND EXPELLING WASTE CARBON DIOXIDE.

LUNG STRUCTURE

Air enters the lungs from the trachea, which branches at its base into two main airways, the primary bronchi. Each primary bronchus enters its lung at a site called the hilum, which is also where the main blood vessels pass in and out of the lung. The primary bronchus divides into secondary bronchi, and these subdivide into tertiary bronchi, all decreasing in diameter. Many subsequent divisions form the narrowest airways: the terminal and then respiratory bronchioles, which distribute air to the alveoli.

Right lung
Like left lung, has ten bronchopulmonary segments

Superior lobe
Contains three bronchopulmonary segments

Horizontal fissure
Between superior and middle lobes of right lung

Inferior lobe
Contains five segments

Oblique fissure
Between middle and inferior lobes

ALVEOLI

The lungs' microscopic air sacs, known as alveoli, are elastic, thin-walled structures arranged in clumps at the ends of respiratory bronchioles. Around the alveoli are networks of capillaries. Oxygen passes from the air in the alveoli into the blood by diffusion through the alveolar and capillary walls (see p.166). Carbon dioxide diffuses from the blood into the alveoli. There are more than 300 million alveoli in both lungs, providing a huge surface area for gas exchange.

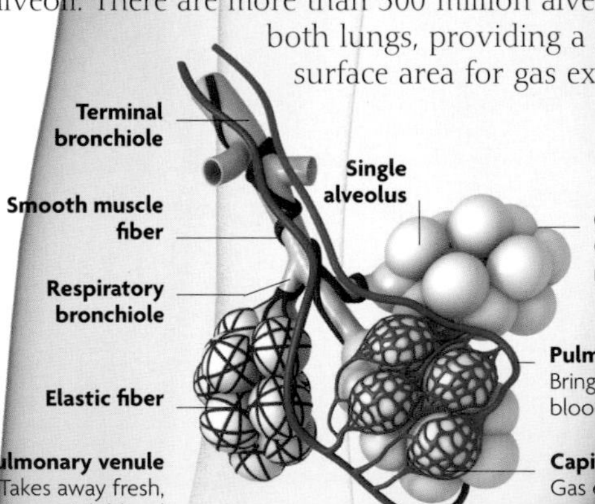
Terminal bronchiole
Single alveolus
Smooth muscle fiber
Group of alveoli
Consists of partly merged alveoli
Respiratory bronchiole
Pulmonary arteriole
Brings used, oxygen-poor blood to alveoli
Elastic fiber
Pulmonary venule
Takes away fresh, oxygen-rich blood
Capillary network
Gas exchange occurs here

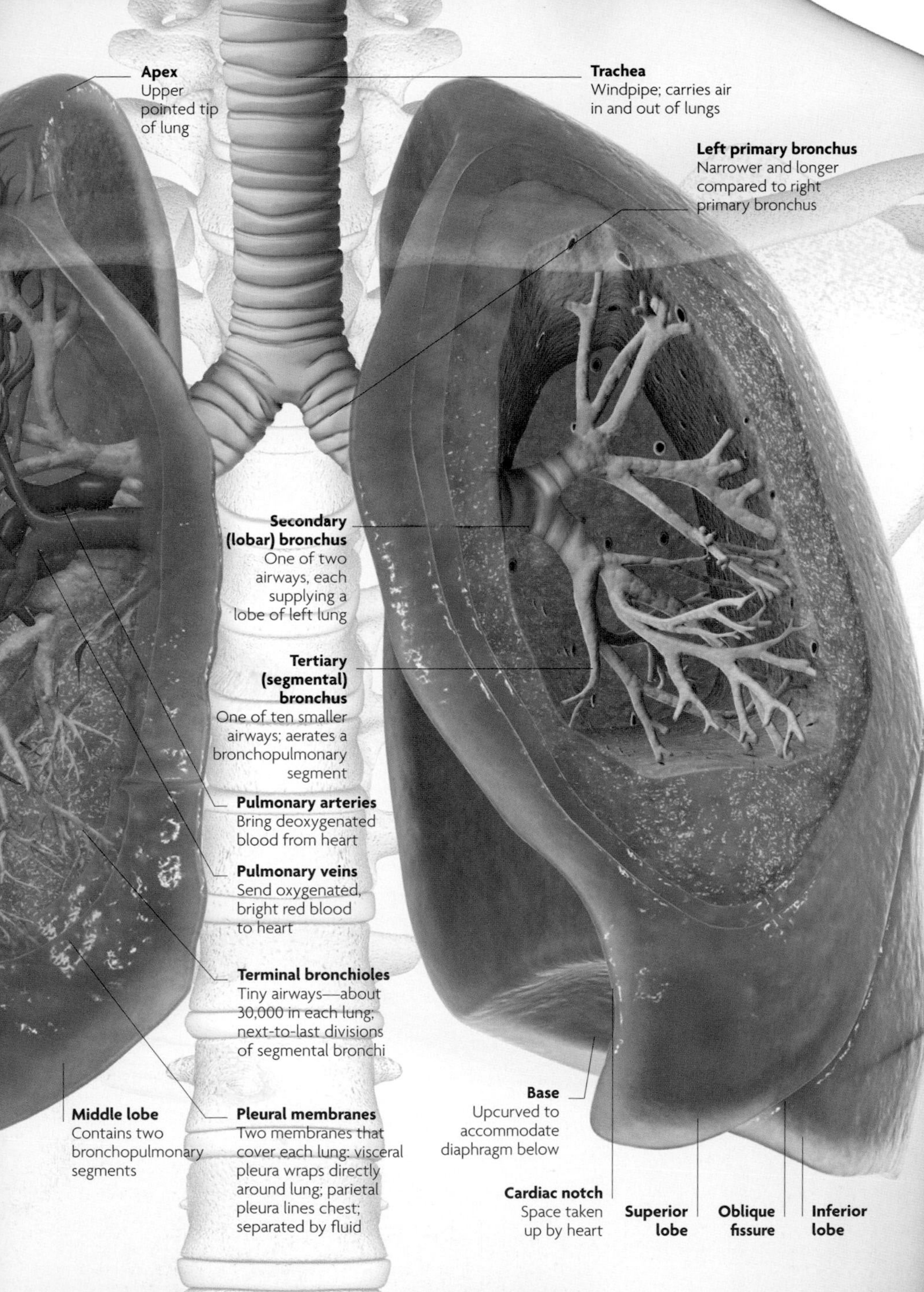

Apex
Upper pointed tip of lung
Trachea
Windpipe; carries air in and out of lungs
Left primary bronchus
Narrower and longer compared to right primary bronchus
Secondary (lobar) bronchus
One of two airways, each supplying a lobe of left lung
Tertiary (segmental) bronchus
One of ten smaller airways; aerates a bronchopulmonary segment
Pulmonary arteries
Bring deoxygenated blood from heart
Pulmonary veins
Send oxygenated, bright red blood to heart
Terminal bronchioles
Tiny airways—about 30,000 in each lung; next-to-last divisions of segmental bronchi
Middle lobe
Contains two bronchopulmonary segments
Pleural membranes
Two membranes that cover each lung: visceral pleura wraps directly around lung; parietal pleura lines chest; separated by fluid
Base
Upcurved to accommodate diaphragm below
Cardiac notch
Space taken up by heart
Superior lobe
Oblique fissure
Inferior lobe

GAS EXCHANGE

THE BODY CANNOT STORE OXYGEN AND NEEDS CONTINUING SUPPLIES. IT ALSO CONSTANTLY PRODUCES CARBON DIOXIDE AS A WASTE PRODUCT. GAS EXCHANGE SWAPS OXYGEN AND CARBON DIOXIDE IN THE LUNGS AND TISSUES.

Oxygen is drawn into the body by the expanding lungs. When it reaches the end of the lungs' airways, the gas dissolves into the fluid lining the alveoli (air sacs). It passes into the blood for distribution to each body cell. Inside cells, the oxygen reacts with glucose to free its energy. Toxic carbon dioxide is a by-product of the process, but gas exchange discharges it into the air. In the lungs and body tissues, gases pass by diffusion: flowing from regions of high to low density.

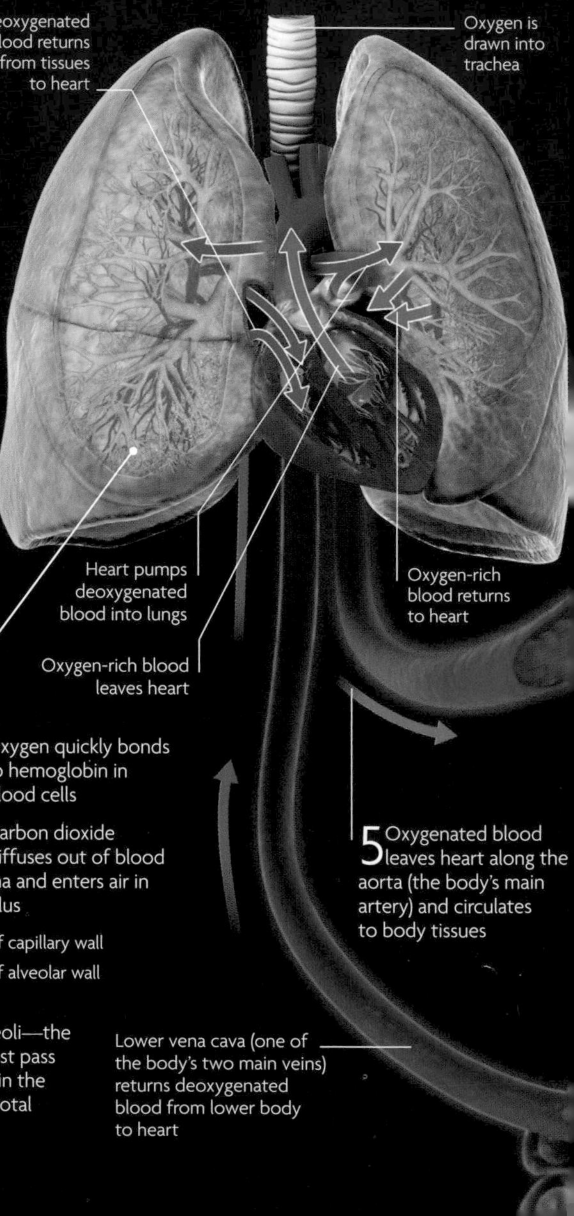

Deoxygenated blood returns from tissues to heart
Oxygen is drawn into trachea
Heart pumps deoxygenated blood into lungs
Oxygen-rich blood returns to heart
Oxygen-rich blood leaves heart
1 Oxygen in air dissolves into fluid lining the alveolus and diffuses through alveolar wall and blood capillary wall
2 Oxygen enters blood plasma inside capillary
Fluid-lined alveolus (air sac)
Blood capillary
3 Oxygen quickly bonds to hemoglobin in red blood cells
4 Carbon dioxide diffuses out of blood plasma and enters air in alveolus
Cell of capillary wall
Cell of alveolar wall
5 Oxygenated blood leaves heart along the aorta (the body's main artery) and circulates to body tissues
Lower vena cava (one of the body's two main veins) returns deoxygenated blood from lower body to heart

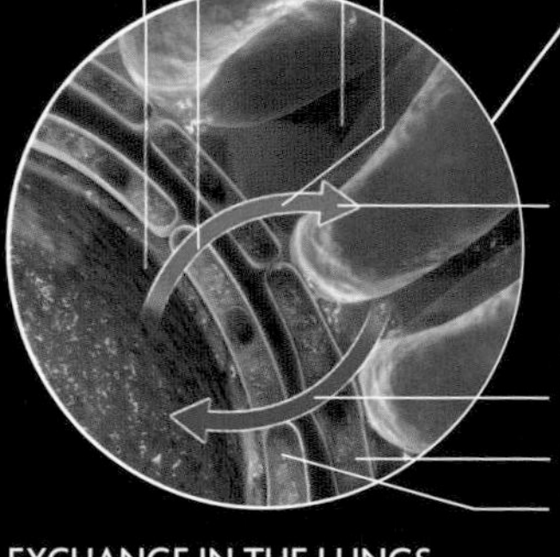

EXCHANGE IN THE LUNGS

When fresh, oxygen-rich air reaches the alveoli—the tiny dead-end air spaces in the lungs—it must pass through several layers to reach the red cells in the blood. But these layers are so thin that the total distance is only $\frac{1}{2500}$ in (0.001 mm).

EXCHANGE IN THE BODY TISSUES

Oxygen levels are higher in the blood than in surrounding tissues. The difference in levels forces oxygen to break its bonds to the hemoglobin in red blood cells and diffuse out of the blood into the adjacent cells. The reverse applies to carbon dioxide, which diffuses from the tissue into the blood plasma.

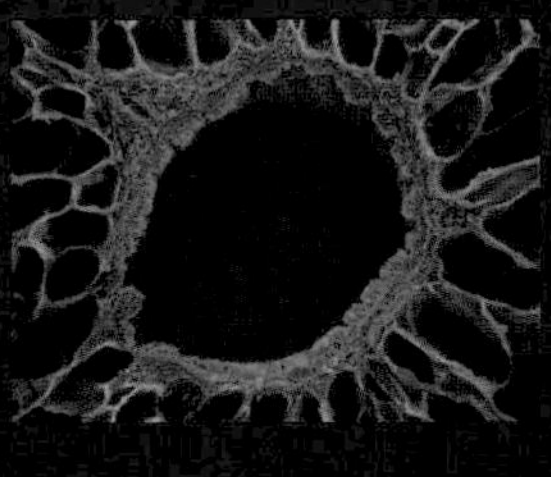

BRONCHIOLE AND ALVEOLI
This microview shows a cross-sectioned bronchiole (red) surrounded by alveoli that have been cut through, so that they resemble air bubbles in a sponge.

6 Oxygenated blood is carried through tissues in capillaries thinner than hair

7 Arriving red blood cells are rich in oxygen, which is bound to hemoglobin in the body of each cell

8 Oxygen leaves the hemoglobin within the blood cells, and diffuses across capillary walls and into tissue cells

9 Carbon dioxide diffuses out of tissue cells, across wall of blood capillary, and into blood plasma

Capillary

Red blood cell

Capillary bed running through tissue

BREATHING AND VOCALIZATION

THE MOVEMENTS OF BREATHING, ALSO KNOWN AS RESPIRATION, BRING FRESH AIR CONTAINING OXYGEN DEEP INTO THE LUNGS AND THEN REMOVE STALE AIR CONTAINING THE WASTE PRODUCT CARBON DIOXIDE.

BREATHING

The movement of air into and out of the lungs is generated by differences in pressure within the lungs compared to the surrounding atmospheric pressure. These differences are produced by forcefully expanding the lungs by muscular action, and then passively allowing them to return to their former size. The rate and depth of breathing can be consciously modified. However, the underlying need to breathe is controlled by an area of the brain where responses to regulate the breathing muscles occur according to the levels of carbon dioxide and oxygen in the blood.

INHALATION

The chief muscles used in respiration at rest are the diaphragm at the base of the chest and the external intercostals between the ribs. For forceful inhalation, additional muscles assist in moving the ribs and sternum to expand the chest further and stretch the lungs even more.

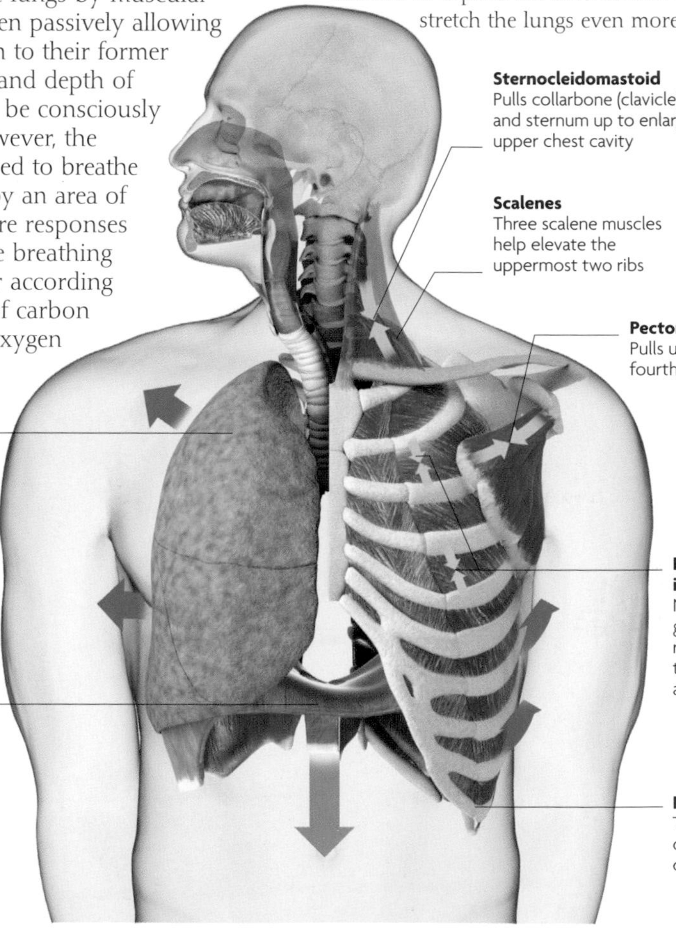

Sternocleidomastoid
Pulls collarbone (clavicle) and sternum up to enlarge upper chest cavity
Scalenes
Three scalene muscles help elevate the uppermost two ribs
Pectoralis minor
Pulls up the third, fourth, and fifth ribs
Lung
Expands as diaphragm pulls down and ribs move up and out
External intercostals
Narrow the gaps between ribs, making them swing up and out
Diaphragm
Contracts and becomes flatter to stretch lungs downward
Ribs
Tilt up and out to expand chest

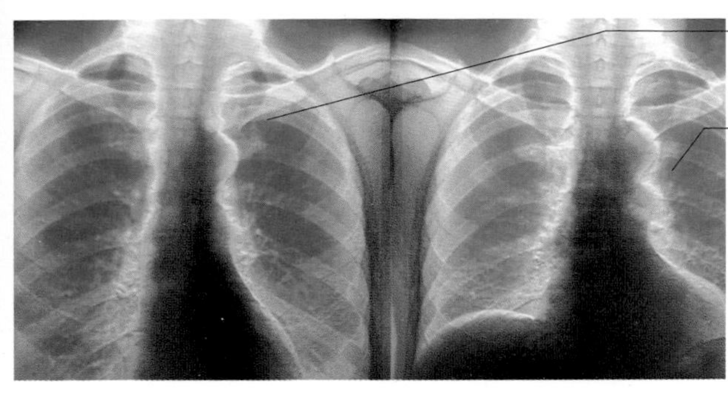

Chest cavity expands as diaphragm and intercostal muscles contract

Chest cavity decreases in size as diaphragm and intercostal muscles relax

DIAPHRAGM MOVEMENT

The abdominal contents (dark area at the bottom of this X-ray) are flattened by the diaphragm muscle during inhalation (left) and then rise up during exhalation (right).

EXHALATION

Breathing out is largely passive. Like a stretched elastic band, the enlarged lungs recoil and shrink when the diaphragm, intercostals, and other inhalation muscles relax, and abdominal pressure pushes the diaphragm upward. Forced expiration brings further muscles into play, to actively compress the lungs beyond their usual resting volume.

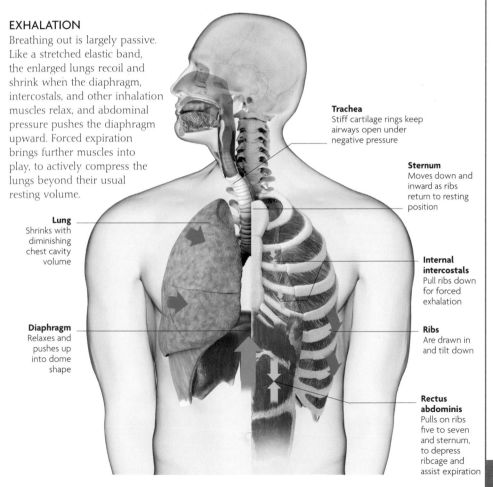

Trachea
Stiff cartilage rings keep airways open under negative pressure

Sternum
Moves down and inward as ribs return to resting position

Lung
Shrinks with diminishing chest cavity volume

Internal intercostals
Pull ribs down for forced exhalation

Diaphragm
Relaxes and pushes up into dome shape

Ribs
Are drawn in and tilt down

Rectus abdominis
Pulls on ribs five to seven and sternum, to depress ribcage and assist expiration

VOLUME AND PRESSURE

Breathing alters the volume of the chest (thoracic cavity). The lungs "suck" on to the inner chest wall, so that as the cavity expands, they also become larger. The main expanding forces are provided by the diaphragm and intercostal muscles. At rest, the diaphragm carries out most of the work, as 17 fl oz (0.5 liters) of air–the tidal volume–shifts in and out with each breath (12 to 17 times every minute). Rate and volume increase automatically if the body needs more oxygen, as during exercise. Then forced inspiration can suck in an extra 70 fl oz (2 liters), and forced expiration expels almost as much, leading to a total air shift, or vital capacity, of more than 150 fl oz (4.5 liters) in a large, healthy adult. The breathing rate can triple, producing a total air exchange more than 20 times greater than at rest.

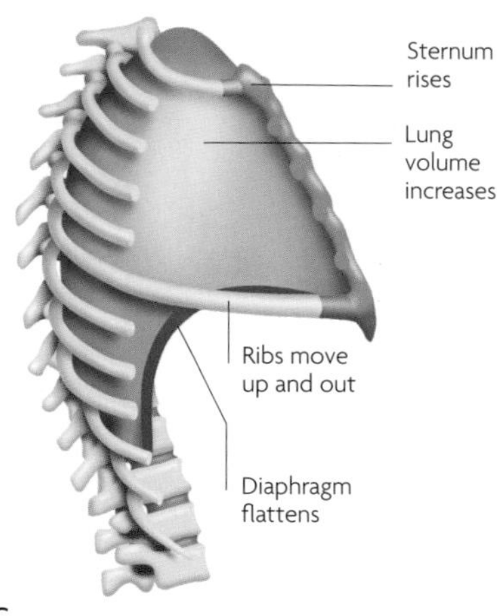
Sternum rises
Lung volume increases
Ribs move up and out
Diaphragm flattens

INHALING

The diaphragm contracts to become less domelike, while the ribs swing upward and outward with a "bucket handle" action to raise the sternum.

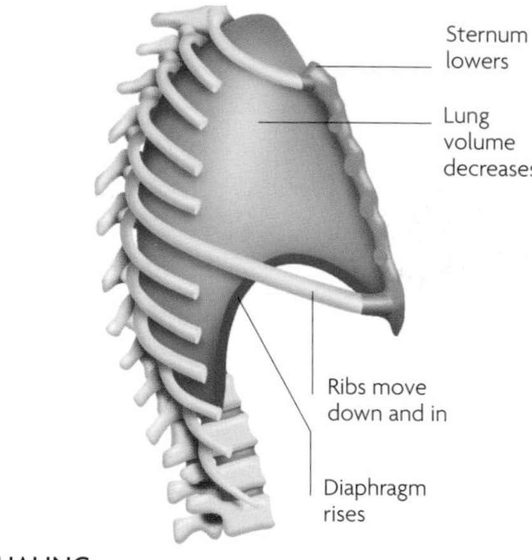
Sternum lowers
Lung volume decreases
Ribs move down and in
Diaphragm rises

EXHALING

The diaphragm relaxes, and the elastic, stretched lungs recoil to become smaller again, allowing the sternum and ribs to move down and inward.

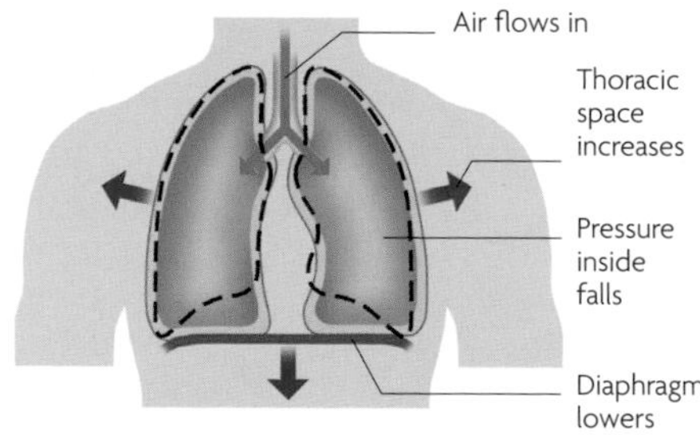
Air flows in
Thoracic space increases
Pressure inside falls
Diaphragm lowers

NEGATIVE PRESSURE

As the lung volume increases, the air pressure within decreases. Atmospheric pressure outside the body is now higher, and air is drawn down the airways and into the lungs—in effect, air is "sucked" in.

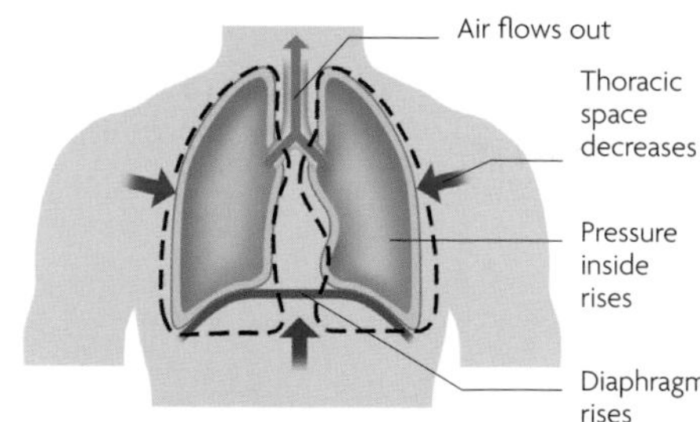
Air flows out
Thoracic space decreases
Pressure inside rises
Diaphragm rises

POSITIVE PRESSURE

As the lung volume diminishes when exhaling, the air is compressed, which raises its pressure within the lungs. So the air is pushed back along the airways and out of the nose and mouth.

VOCALIZATION

The vocal cords are two bands of fibrous tissue within the larynx. During breathing they are separated by a V-shaped gap (the glottis). Sound is produced when the cords tighten together and vibrate as air from the lungs passes between them. Pitch varies according to the tension in the cords. The false vocal cords above help to close off the larynx during swallowing.

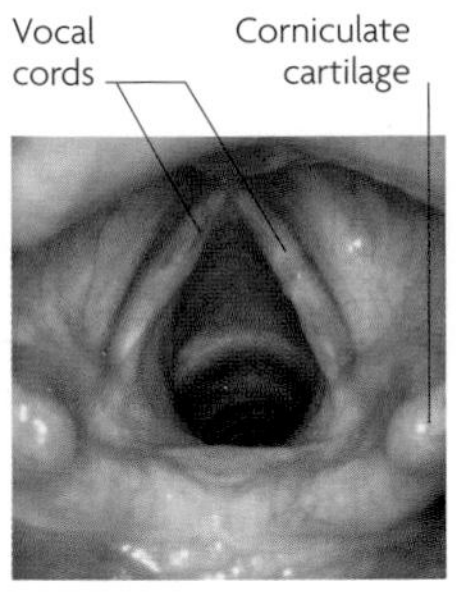
Vocal cords
Corniculate cartilage

CORDS APART
A laryngoscope view shows the vocal cords during normal breathing, when air passes through the gap between them.

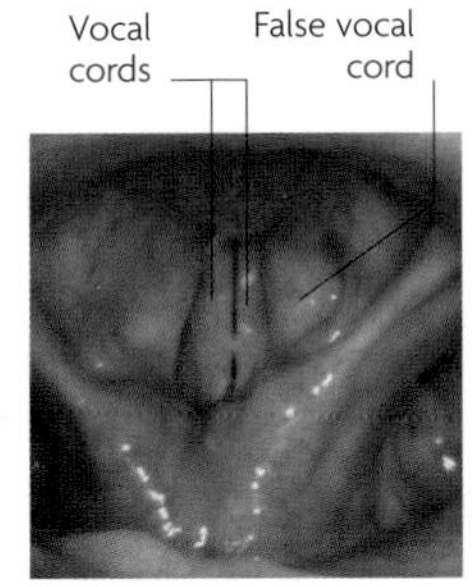
Vocal cords
False vocal cord

CORDS ADJACENT
Laryngeal muscles swing the arytenoid cartilages, to which the vocal cords are attached, and bring them together.

RESPIRATORY REFLEXES

The respiratory reflexes of coughing and sneezing aim to blow out excess mucus, dust, irritants, and obstructions—coughing from the lower pharynx, larynx, trachea, and lung airways, and sneezing from the nasal chambers and nasopharynx. For a cough, the lower pharynx, epiglottis, and larynx close so that air pressure builds up in the lungs, and is released explosively, rattling the vocal cords. In a sneeze, the tongue closes off the mouth, to force air up and out through the nose.

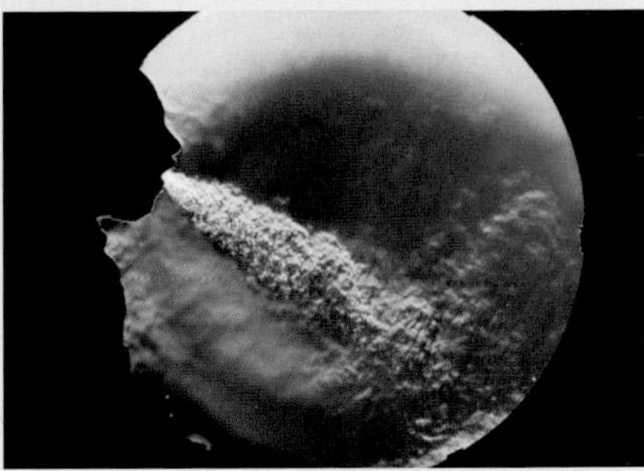

MUCUS SPRAY
Coughs (as shown here) and sneezes propel a spray of tiny mucus droplets from the respiratory airways for distances of up to 10 ft (3 m).

THE LARYNX

The larynx is sited between the pharynx and the trachea. It has a framework of nine cartilages, comprising the paired arytenoids, cuneiforms, and corniculates, and the unpaired epiglottic, thyroid, and cricoid. The thyroid cartilage forms a prominent mound under the skin of the neck, called the "Adam's apple," which is larger and more pronounced in adult males. The cartilages are held in position by numerous muscles and ligaments.

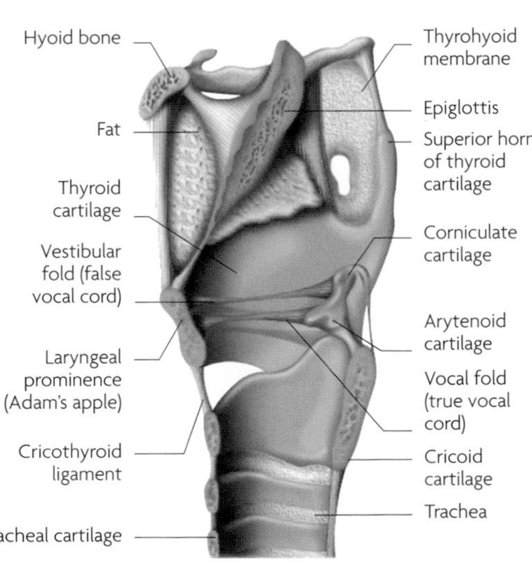
Hyoid bone
Fat
Thyroid cartilage
Vestibular fold (false vocal cord)
Laryngeal prominence (Adam's apple)
Cricothyroid ligament
Tracheal cartilage
Thyrohyoid membrane
Epiglottis
Superior horn of thyroid cartilage
Corniculate cartilage
Arytenoid cartilage
Vocal fold (true vocal cord)
Cricoid cartilage
Trachea

INTERNAL STRUCTURE
The larynx forms a hollow chamber through which air flows silently during normal breathing. The areas of cartilage tilt to bring the vocal cords together for speech.

RESPIRATORY DISORDERS

MILLIONS OF MICROBES FLOAT IN THE AIR, AND EACH BREATH BRINGS THOSE PARTICLES INTO THE RESPIRATORY TRACT, HEIGHTENING THE RISK OF A RESPIRATORY INFECTION. OTHER TYPES OF RESPIRATORY DISORDER INCLUDE DAMAGE CAUSED BY ALLERGIES OR IRRITANTS, AND CANCERS.

COMMON COLD

The common cold is one of the most frequently experienced illnesses but also generally one of the less serious. At least 200 different and highly contagious types of virus can cause the problem. They spread in fluid that floats through air, in tiny droplets of mucus coughed or sneezed out by sufferers, and also in films of moisture transferred from person to person by close contact, such as shaking hands, or via shared objects, such as cups. Symptoms involve frequent sneezing, a runny nose, which at first runs with a clear, thin discharge that may later become thicker and greenish yellow, a headache, slightly raised temperature, and perhaps a sore throat, cough, and reddened eyes. Antibiotic drugs are ineffective because they do not work against viruses. Cold viruses change (mutate) so rapidly that even if antiviral drugs could be made to tackle existing strains, they would be ineffective against the new ones. Most cold remedies, such as decongestants and inhalants, treat the symptoms while the body's immune system attacks the invading microbes.

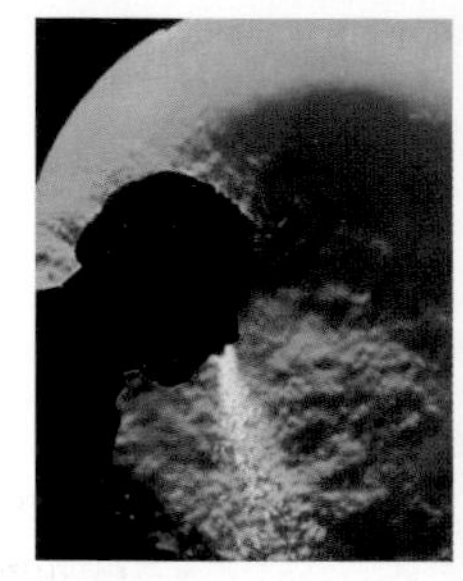

SPREADING INFECTION Coughs and sneezes can rapidly spread common cold viruses by spraying them up to 10 ft (3 m) in mucous droplets.

1 VIRUS INVADES CELLS Virus particles in air land on and invade the cells lining the nose and throat. They rapidly replicate, killing their host cells.

Virus particle

Cell of nasal lining

Released virus particles infect new cells

Multiplied virus particles

2 WHITE CELLS ARRIVE Defensive white blood cells squeeze out of capillaries toward the infected lining cells, which are producing thin mucus.

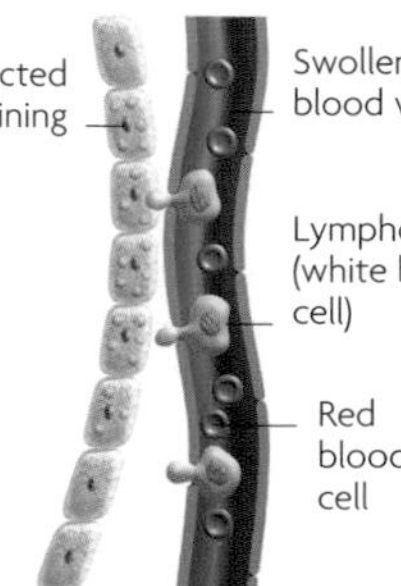
Infected nasal lining
Swollen blood vessel
Lymphocyte (white blood cell)
Red blood cell

3 ANTIBODY PRODUCTION White blood cells known as B cells produce antibodies, which immobilize the virus; other white blood cells destroy infected cells.

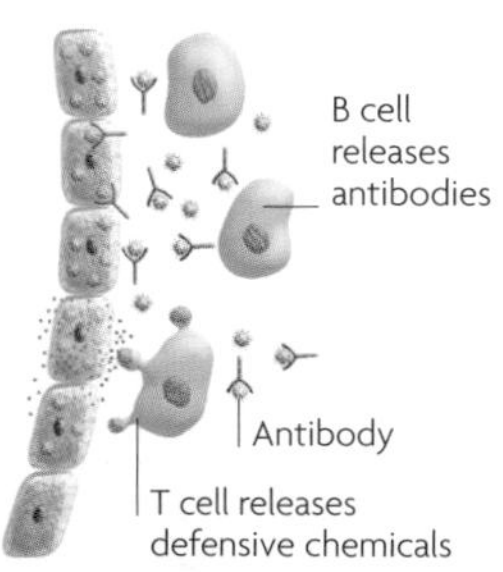
B cell releases antibodies
Antibody
T cell releases defensive chemicals

4 CLEARING UP Other white blood cells called phagocytes engulf virus particles, damaged nasal lining cells, and other debris. The cold subsides.

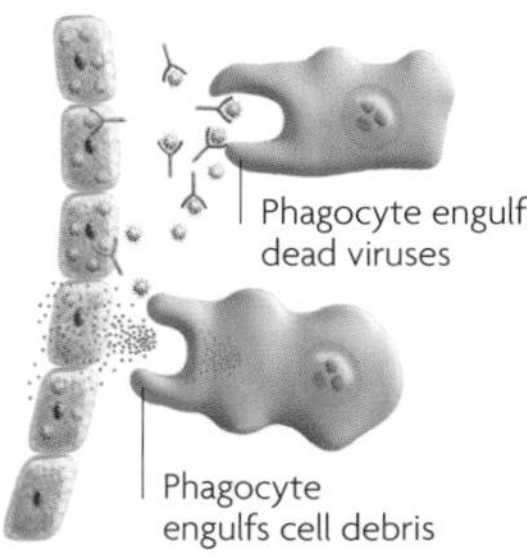
Phagocyte engulfs dead viruses
Phagocyte engulfs cell debris

INFLUENZA

Influenza is primarily an upper respiratory tract infection, but it also has body-wide symptoms: raised temperature, sensations of being hot and sweaty and then cold with shivers, muscle aches, and exhaustion. Even after the main infection has cleared up, there may be lingering depression and fatigue. The influenza viruses are coded A, B, and C and are very contagious. Influenza A tends to produce regular outbreaks and can also affect domestic animals such as pigs, horses, and fowl. Influenza B usually causes more sporadic outbreaks in places where many people gather and interact. Influenza C is less likely to produce serious symptoms. The type A virus is most likely to change or mutate. People at risk of dangerous complications, such as the very young or elderly, can be vaccinated before the main risk time of the winter season. Because the virus can mutate, new vaccines are prepared annually.

ACUTE BRONCHITIS

Bronchitis is inflammation of the larger airways (bronchi) in the lungs. The disorder may be a complication of a respiratory infection. Its acute form develops rapidly, with symptoms including a sputum-producing cough, tight chest, wheezing, and mild fever. Healthy adults usually recover in a few days.

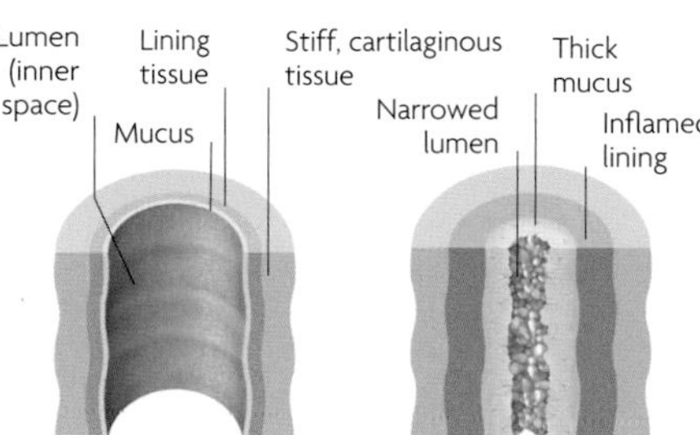
Lumen (inner space)
Lining tissue
Stiff, cartilaginous tissue
Mucus
Narrowed lumen
Thick mucus
Inflamed lining

NORMAL BRONCHUS
The airway lining secretes a thin layer of mucus. The passageway (lumen) allows free air flow.

INFLAMED BRONCHUS
The airway lining swells and produces excess mucus, which may be coughed up.

ASTHMA

Asthma is an inflammatory lung disease that causes recurrent attacks of breathlessness and wheezing due to narrowed airways. Some people have the occasional slight episode; others are prone to severe and even life-threatening attacks. The muscle in the walls of the airways contracts spasmodically, causing constriction of the tubes. The narrowing is worsened by the secretion of excess mucus. Most cases develop in childhood and may be linked to allergy-based problems such as eczema. In many children, the trigger for an attack is an allergic reaction to a foreign substance, such as inhaled particles of pollen, animal hair, or house dust mite droppings.

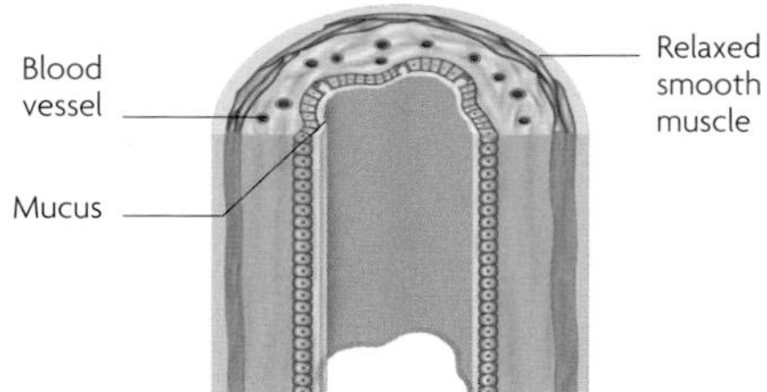
Blood vessel
Relaxed smooth muscle
Mucus

HEALTHY AIRWAY
The bronchiole has relaxed smooth muscle in its walls and a thin coating of protective mucus covering the lining. The passage of air is unrestricted.

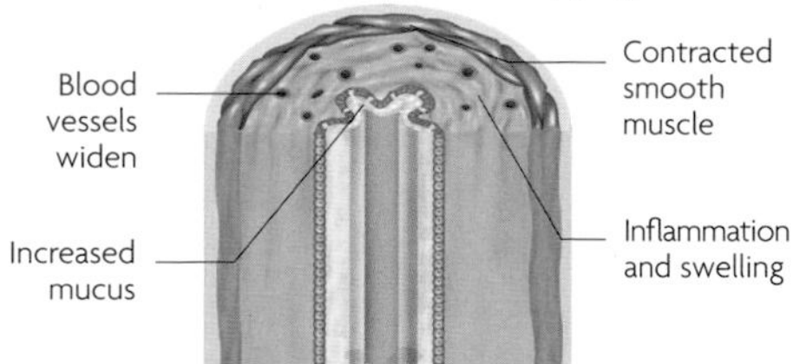
Blood vessels widen
Contracted smooth muscle
Increased mucus
Inflammation and swelling

ASTHMATIC AIRWAY
In an asthma attack, the muscle wall contracts and inflammation causes swelling of the airway lining. The mucus thickens, further narrowing the airway.

LUNG CANCER

The most common cause of lung cancer–responsible for almost 90 percent of all cases–is tobacco smoke. In the past, lung cancer was far more common in men than women, because more men than women smoked. However, the incidence of the disease in women rose rapidly in the final decades of the 20th century. The disease is also becoming increasingly common in developing countries, with the spread of tobacco smoking and growing urban populations. Many inhaled irritants trigger the growth of abnormal cells in the lungs, but cigarette smoke contains thousands of known carcinogenic (cancer-causing) substances. In rare cases, lung cancer is caused by asbestos, toxic chemicals, or the radioactive gas radon. A persistent cough is usually the earliest symptom. Because most people who develop lung cancer are smokers, this is often dismissed as a "smoker's cough." Other symptoms include coughing up blood, wheezing, weight loss, persistent hoarseness, and chest pain. If tests confirm the presence of lung cancer, a lobectomy (removal of a lung lobe) or pneumonectomy (removal of a whole lung) may be performed. This is usually advised only if the tumor is small and has not spread. Chemotherapy and radiation therapy can be given to relieve symptoms, rather than with the aim of curing the disease.

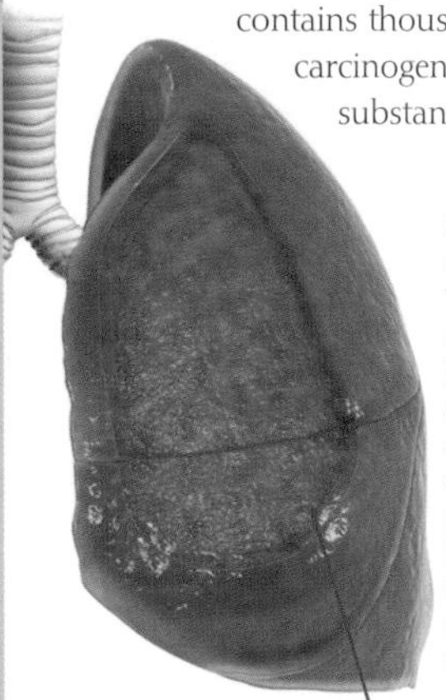

SPREADING CANCER CELLS
Tiny airborne carcinogenic particles lodge in the airways and contribute to the development of cancerous cells. Some of these cells may break away and travel in the blood or lymph to trigger secondary tumors.

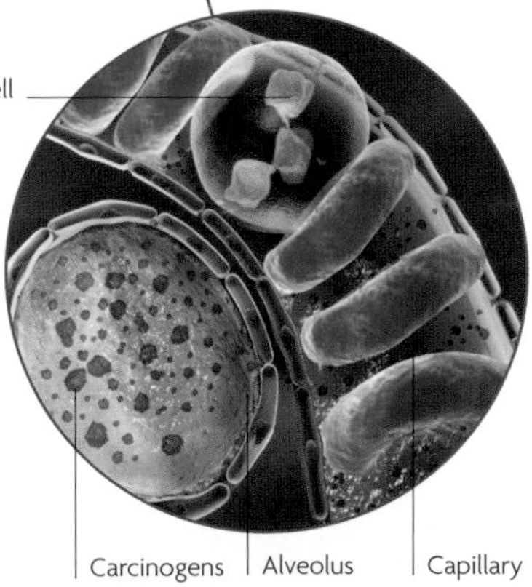

White blood cell

Carcinogens

Alveolus

Capillary

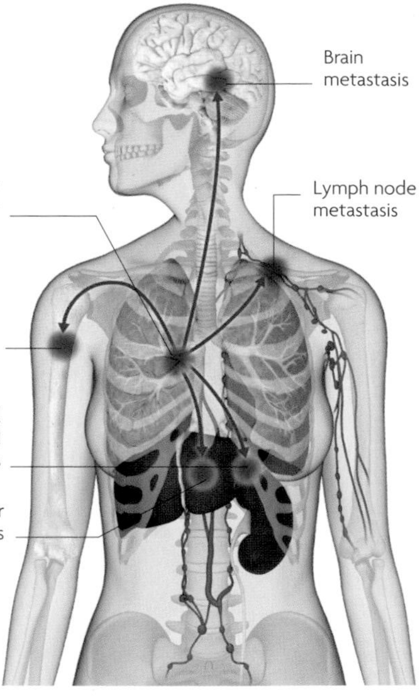

Brain metastasis

Primary tumor

Lymph node metastasis

Bone metastasis

Adrenal gland metastasis

Liver metastasis

THE SPREAD OF LUNG CANCER
Lung cancer can spread (metastasize) to other parts of the body. Metastases in bones can cause pain and fractures; in the brain, headaches and confusion; and in the liver, weight loss and jaundice.

CHRONIC OBSTRUCTIVE PULMONARY DISEASE

Chronic obstructive pulmonary disease (COPD) consists primarily of chronic bronchitis and emphysema, two conditions that usually occur together. In this disorder there is progressive damage to lung tissue, causing restricted airflow in and out of the lungs and shortness of breath.

CHRONIC BRONCHITIS

In chronic bronchitis, the main airways leading to the lungs, the bronchi, become inflamed, congested, and narrowed due to irritation caused by tobacco smoke, frequent infections, or prolonged exposure to pollutants. The inflamed airways begin to produce too much mucus (sputum), resulting in a typical cough that at first is troublesome mostly in damp, cold months but then persists throughout the year. Symptoms such as hoarseness, wheezing, and breathlessness also develop. Eventually a person becomes short of breath even at rest. If a secondary respiratory infection develops, the sputum may change appearance from clear or white to yellow or green.

NORMAL AIRWAY LINING
Glands produce mucus that traps inhaled dust and germs. Tiny surface hairs (cilia) propel the mucus up into the throat, where it is coughed up or swallowed.

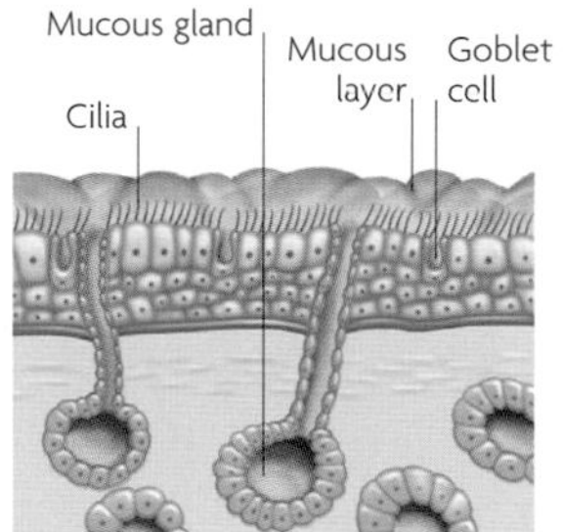
Mucous gland
Cilia
Mucous layer
Goblet cell

AIRWAY IN CHRONIC BRONCHITIS
Inhaled irritants cause glands to produce more mucus. Damaged cilia cannot propel mucus along, so it becomes a bacterial breeding ground.

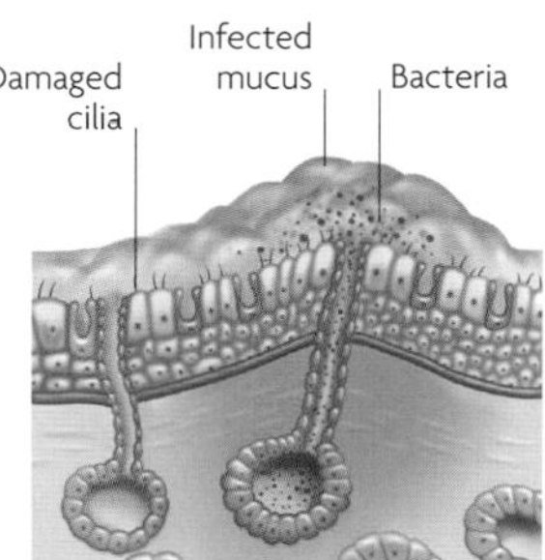
Damaged cilia
Infected mucus
Bacteria

EMPHYSEMA

In emphysema, the air sacs (alveoli) become overstretched. They rupture and merge, which reduces their oxygen-absorbing surfaces and makes gas exchange less efficient. Air also becomes trapped inside them, the lungs over-inflate, and the volume of air moving in and out of the lungs is reduced. Most people affected by emphysema are long-term heavy smokers, although a rare inherited condition called alpha1-antitrypsin deficiency can also cause the disorder. The lung damage is usually irreversible, but giving up smoking may slow the progression of the disease.

HEALTHY TISSUE
The alveoli are grouped, like grapes, and each tiny sac is partly separate from the others. The walls are thin and elastic so they can stretch.

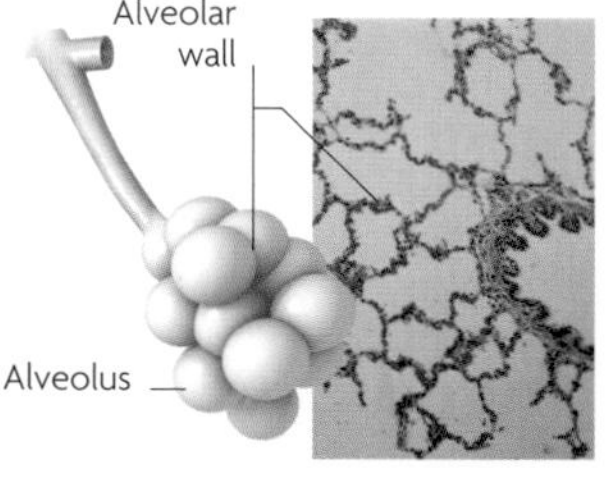
Alveolar wall
Alveolus

DAMAGED TISSUE
Smoke or other pollutants stimulate chemicals that cause the alveolar walls to break down, reducing the area for gas exchange.

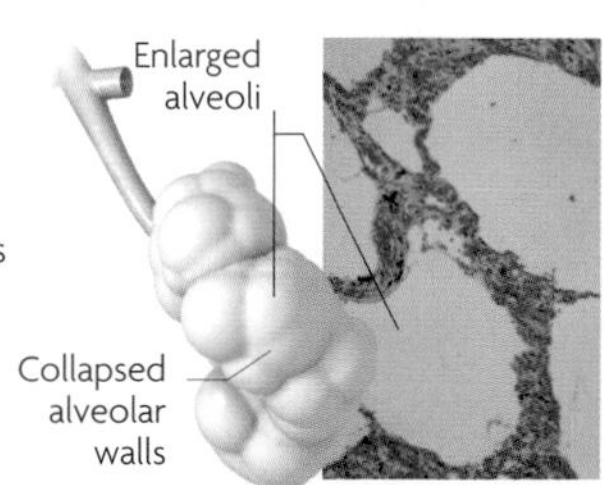
Enlarged alveoli
Collapsed alveolar walls

FEW BODY PARTS RENEW AS RAPIDLY AS THE SKIN. EVERY MONTH THE OUTER LAYER OF EPIDERMIS IS COMPLETELY REPLACED, AT A RATE OF 30,000 FLAKELIKE DEAD CELLS PER MINUTE. THE HAIR AND NAILS ARE LIKEWISE SELF-

SKIN, HAIR, AND NAILS

SKIN, HAIR, AND NAIL STRUCTURE

THE SKIN IS ONE OF THE LARGEST ORGANS OF THE BODY, WEIGHING 6–9LB (3–4KG) AND WITH A SURFACE AREA OF ALMOST 21SQ FT ($2M^2$). IT IS FORMED FROM MANY TYPES OF CELL, SOME OF WHICH PRODUCE HAIR AND NAIL TISSUE.

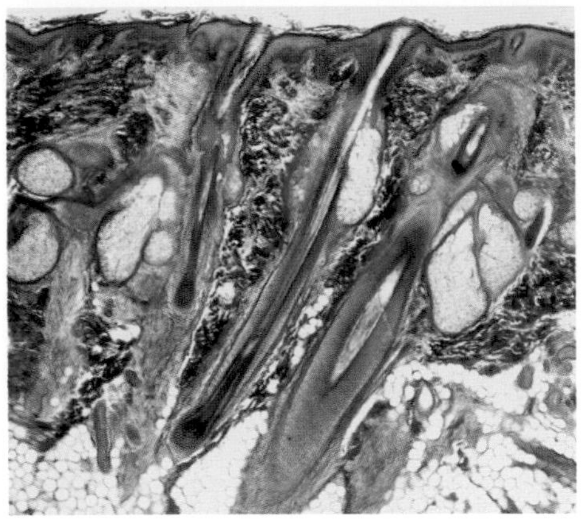

SKIN SECTION
This micrograph shows three hair follicles and globules of sebum in the dermis (blue), with the thin epidermis (purple) on top.

SKIN STRUCTURE

The skin is not just a thin, waterproof covering but a complex organ consisting of a variety of specialized cells. The skin's thickness varies from about 1/50in (0.5mm) on delicate areas such as the eyelids, to 1/5in (5mm) or more on areas of wear and tear, such as the soles of the feet. Skin has two main layers: the outer epidermis, the main function of which is protection, and the underlying dermis. The dermis contains thousands of sensors that are sensitive to touch. It also contains sweat glands and blood vessels, which play a vital role in temperature regulation. Under the dermis is a layer known as subcutaneous fat. This acts as a buffer and provides insulation against extreme heat and cold.

SKIN RENEWAL

The epidermis continually renews and replaces itself. The basal layer consists of boxlike cells that multiply quickly and are gradually pushed up to the surface by new cells forming below. As the cells move upward, they develop tiny prickles that bind them together. They then flatten and fill with a waterproofing protein called keratin. Finally, the cells die, and reach the surface resembling interlocking roof tiles. As they flake away with wear and tear, more cells arrive to replace them.

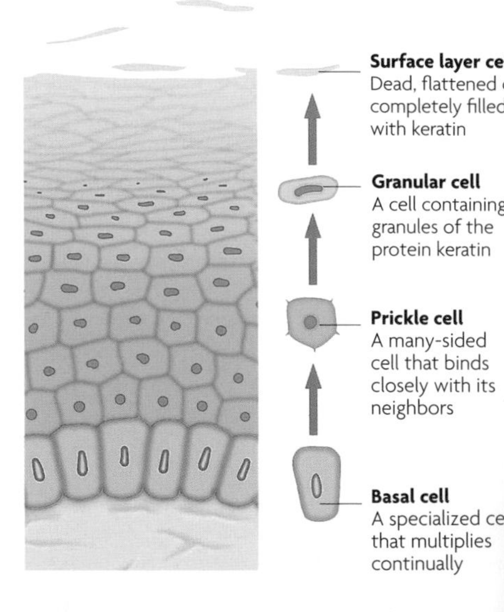
Surface layer cell
Dead, flattened c
completely filled
with keratin
Granular cell
A cell containing
granules of the
protein keratin
Prickle cell
A many-sided
cell that binds
closely with its
neighbors
Basal cell
A specialized cell
that multiplies
continually

EPIDERMAL LAYERS
The procession of skin cells from base to surface creates four layers (five in areas of great friction, such as the palms and soles) in the epidermis. As each cell moves upward, it fills with keratin.

STRUCTURE OF SKIN

A patch of skin the size of a fingernail contains 5 million cells of at least a dozen main kinds, 100 sweat glands, 1,000 touch sensors, 100-plus hairs with sebaceous glands, up to 3⅓ft (1m) of tiny blood vessels, and about 1⅔ft (0.5m) of nerve fibers.

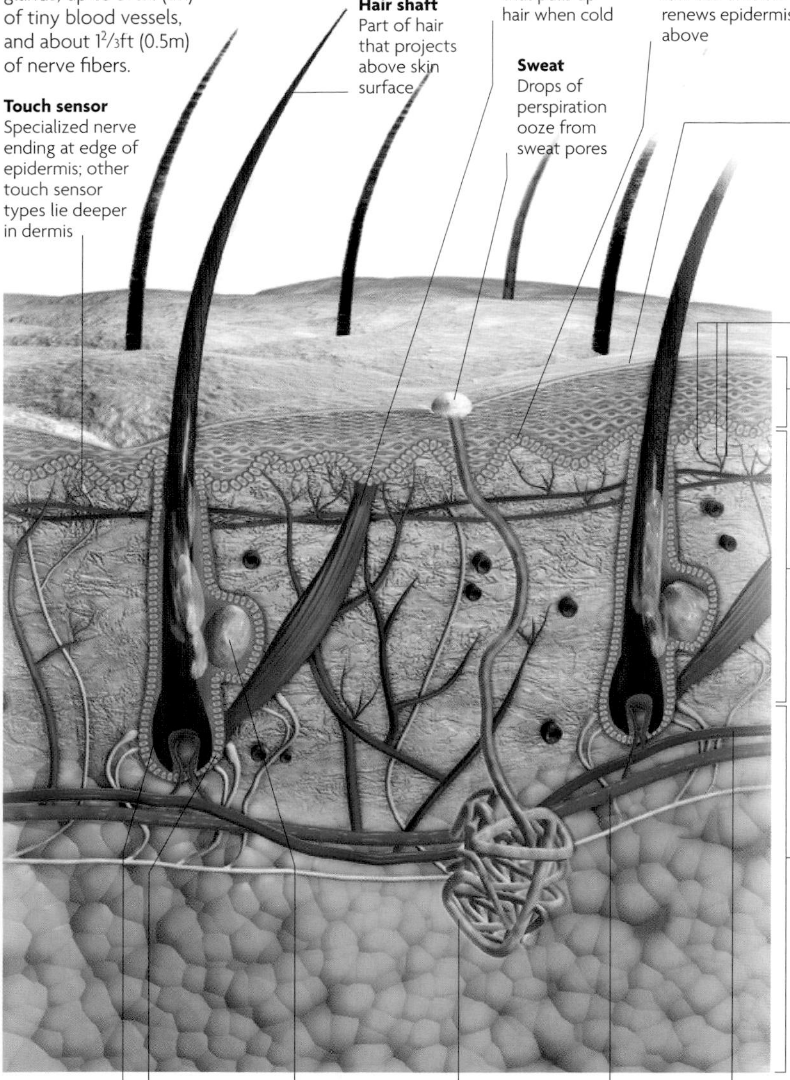

Touch sensor Specialized nerve ending at edge of epidermis; other touch sensor types lie deeper in dermis

Hair shaft Part of hair that projects above skin surface

Arrector pili muscle Tiny muscle that pulls up hair when cold

Sweat Drops of perspiration ooze from sweat pores

Basal epidermal layer Layer in which fast cell division renews epidermis above

Epidermal surface Cornified layer of flat, dead skin cells

Capillaries Tiny blood vessels that supply oxygen and nutrients to tissues and collect waste

Epidermis Outer protective layer consisting of tough, flat cells

Dermis Layer containing blood vessels, glands, and nerve endings

Subcutaneous fat Acts as an insulator, shock absorber, and energy store

Hair bulb Lowest part of hair, where growth occurs

Hair follicle Pouch of epidermis at root of hair

Sebaceous gland Produces sebum that protects hair and lubricates skin

Sweat gland Coiled knot of tubes secreting watery sweat

Arteriole Supplies oxygenated blood

Venule Carries away waste

SKIN REPAIR

If the skin surface is breached, contents leak from damaged cells and stimulate the repair process. Platelets in the blood and the blood-clotting protein fibrinogen together form a mesh of fibers that traps red cells as the beginning of a clot.

Tissue-forming fibroblast cells collect in the area, as do white cells called neutrophils, which ingest cell debris and foreign matter such as dirt and germs. The clot gradually hardens and expels fluid to become a scab, as the skin heals.

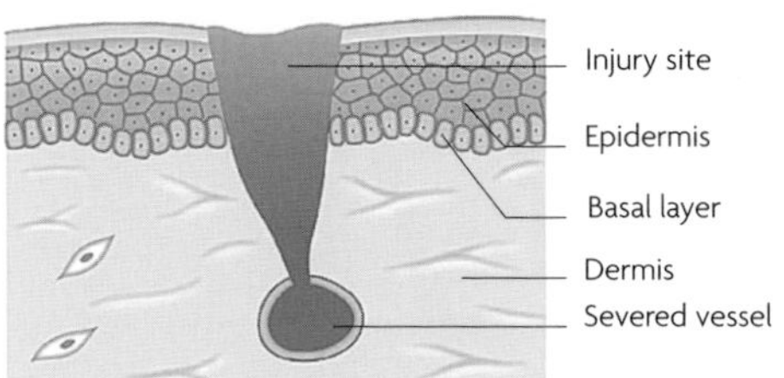
Injury site
Epidermis
Basal layer
Dermis
Severed vessel

1 INJURY The wound breaks open cells and releases their contents. These components attract various defence and repair cells.

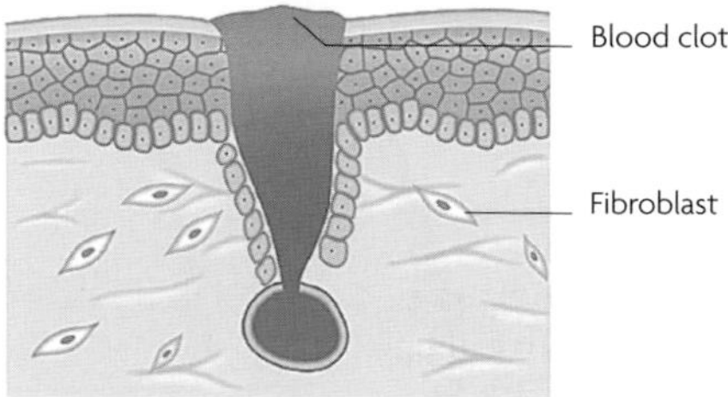
Blood clot
Fibroblast

2 CLOTTING Blood seeps from the vessel and forms a clot. Fibroblasts multiply and migrate to the damaged area.

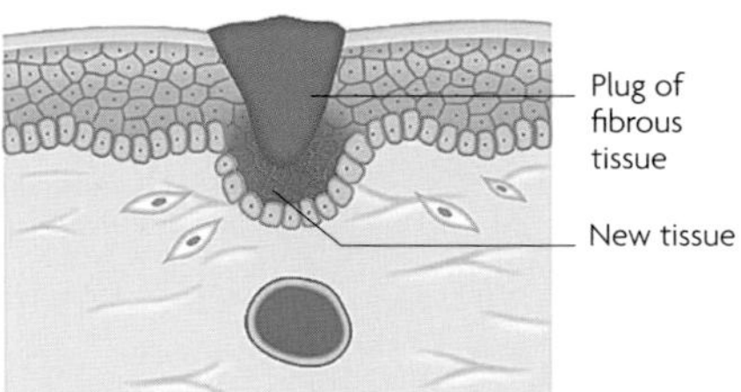
Plug of fibrous tissue
New tissue

3 PLUGGING Fibroblasts produce a plug of fibrous tissue within the clot, which contracts and shrinks. New tissue begins to form beneath.

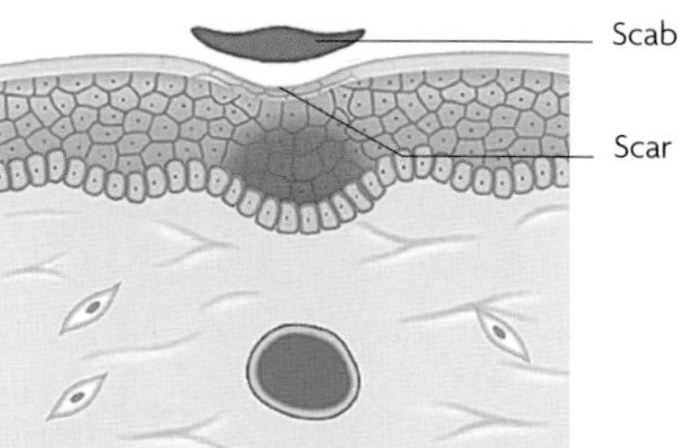
Scab
Scar

4 SCABBING The plug hardens and dries into a scab, which eventually detaches. A scar may remain, but usually fades with time.

NAIL STRUCTURE

Fingernails and toenails are hard plates made of a tough protein called keratin. Growth takes place under a fold of flesh (cuticle) at the nail base. The nail matrix adds keratinized cells to the nail root, and the whole nail is continuously pushed forward along the nail bed toward its free edge. Most nails grow about 1/50in (0.5mm) each week, with fingernails lengthening faster than toenails.

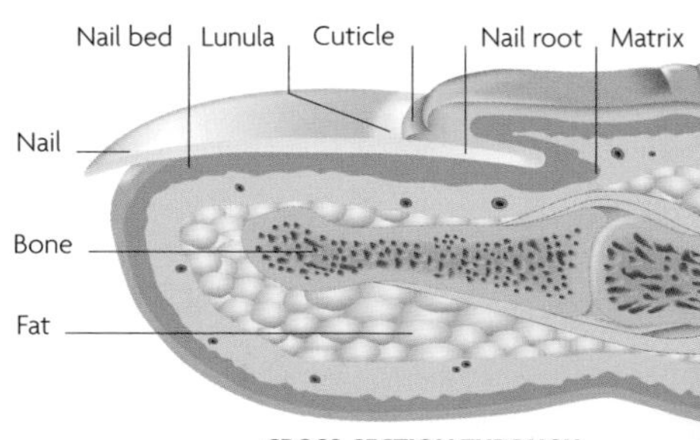
Nail bed
Lunula
Cuticle
Nail root
Matrix
Nail
Bone
Fat

CROSS-SECTION THROUGH NAIL AND FINGER

HAIR GROWTH

Hairs are rods of dead cells filled with keratin. The root, or bulb, is buried in a pit, the follicle. As extra cells add to the root, the hair lengthens. Different kinds of hairs grow at varying rates, with scalp hairs lengthening about 1/100in (0.3mm) each day. Hair does not grow continuously. After three to four years, the follicle goes into a rest phase and the hair may fall out. Three to six months later, the follicle activates again and begins to produce a new hair.

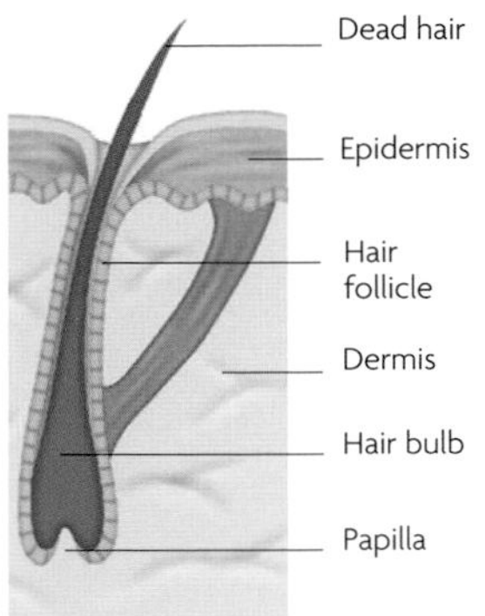
Dead hair
Epidermis
Hair follicle
Dermis
Hair bulb
Papilla

REST PHASE
The hair grows to its maximum length. Activity in the follicle stops and the hair dies.

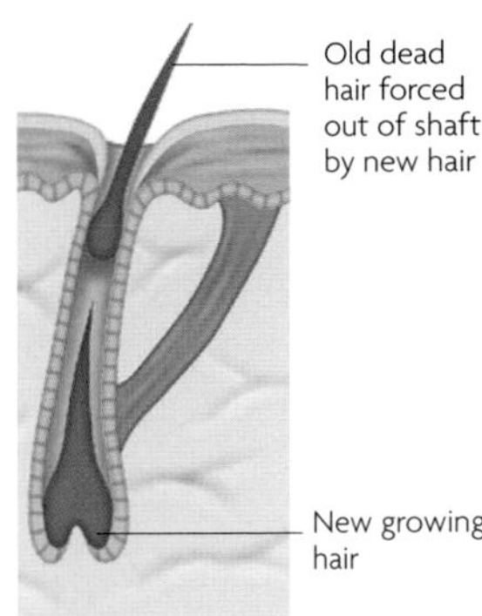
Old dead hair forced out of shaft by new hair
New growing hair

GROWTH PHASE
A new hair sprouts at the base of the follicle. As it grows, the dead hair is shed.

TEMPERATURE REGULATION

One of the skin's functions is to help maintain a constant body temperature. If the body is too hot, blood vessels in the dermis widen (vasodilate) to allow extra blood flow so more heat can be lost from the surface, and sweat glands produce more sweat, which evaporates, drawing away body heat. If the body is cold, the skin's blood vessels narrow (vasoconstrict) to minimize heat loss, and sweating is reduced. Tiny hairs are pulled upright by the arrector pili muscles to trap an insulating layer of air.

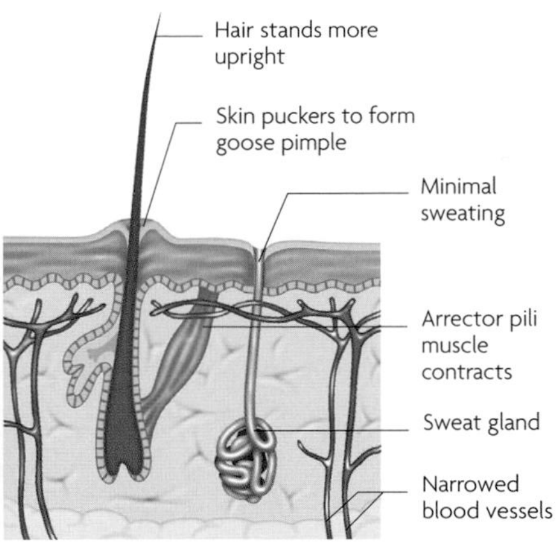
Hair stands more upright
Skin puckers to form goose pimple
Minimal sweating
Arrector pili muscle contracts
Sweat gland
Narrowed blood vessels

FEELING COLD
Tiny body hairs are raised, creating goose pimples at their bases. The skin's blood vessels constrict, reducing blood flow, and sweat glands reduce their activity.

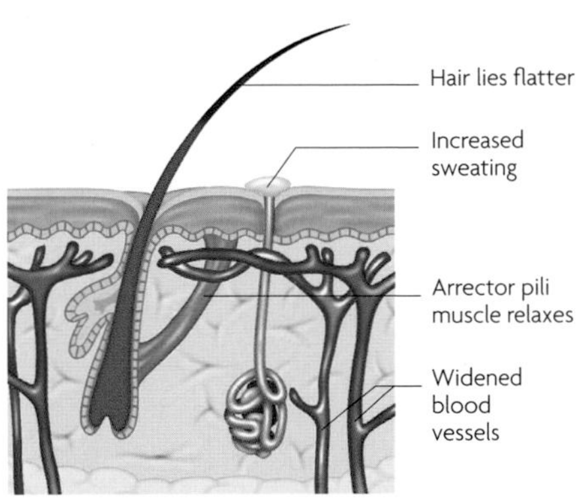
Hair lies flatter
Increased sweating
Arrector pili muscle relaxes
Widened blood vessels

FEELING HOT
Tiny body hairs lie flatter and the goose pimples disappear. The blood vessels widen, increasing blood flow, and sweat glands increase sweat production.

SKIN AND EPITHELIAL TISSUES

SKIN PROTECTS THE UNDERLYING TISSUES AND ALSO PROVIDES THE SENSE OF TOUCH. IT IS A SPECIALIZED TYPE OF EPITHELIUM. EPITHELIAL TISSUES OCCUR THROUGHOUT THE BODY, PROVIDING COVERINGS AND LININGS FOR BODY PARTS AND ORGANS.

COMPLEXITIES OF TOUCH

The sense of touch is based in the lower of the skin layers, the dermis. Microsensors–the endings of tiny nerve cells–in the dermis detect various physical changes, from light contact to heavy, painful pressure. On average, a skin patch the size of a fingernail contains about 1,000 receptors. However, the skin on the fingertips has more than 3,000 receptors that detect light touch for precise feeling. There are also receptor fibers around the bases of hairs, in the follicles (pits) within the dermis. Different types of receptor respond more readily to certain types of stimulation, but almost all respond to most stimuli.

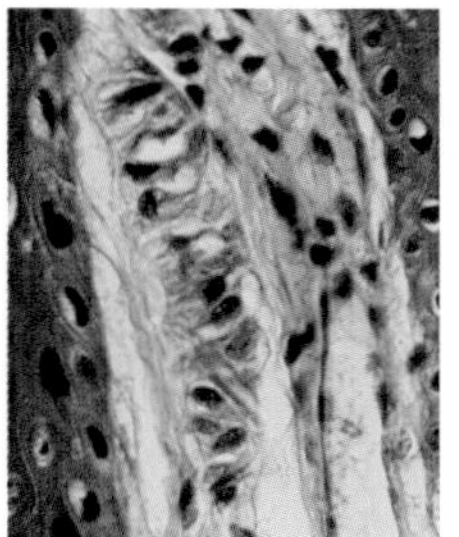

LIGHT-TOUCH SENSOR
This microscope view shows a Meissner's corpuscle (green) in a fingertip. It is important for light, discriminatory touch.

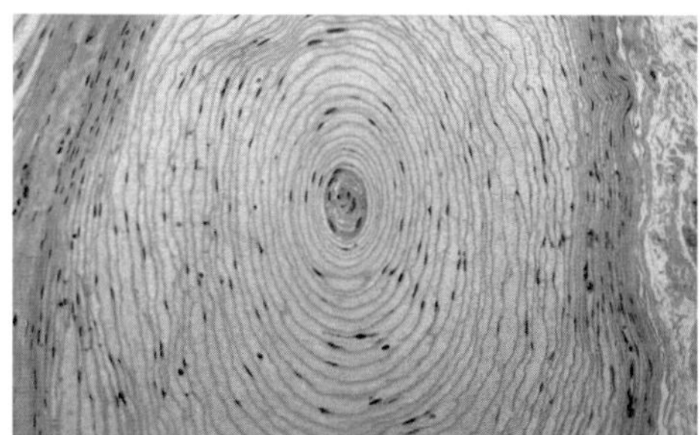

DEEP-PRESSURE SENSOR
Pacinian corpuscles have a multilayered structure and are the largest skin receptors, in some areas being more than 1/25in (1mm) long.

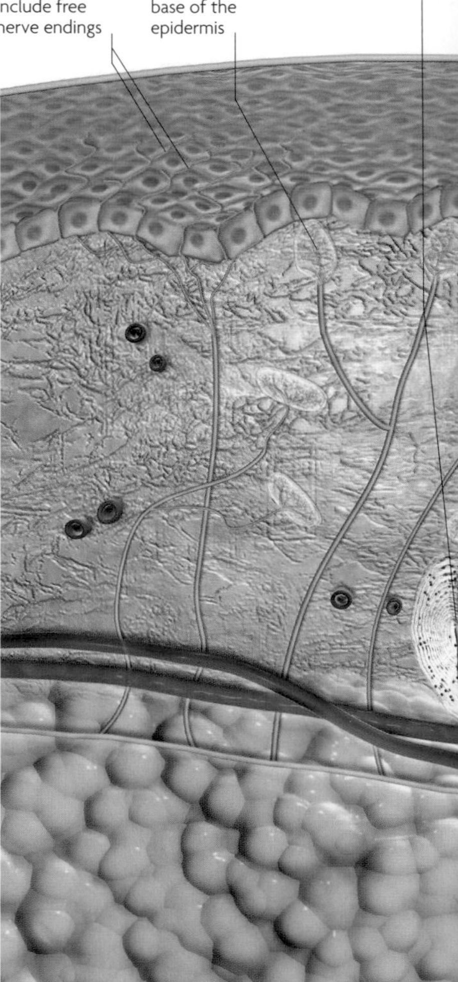
Superficial nerve endings Penetrate the epidermis; occur everywhere in the skin and include free nerve endings
Meissner's corpuscle Upper dermal nerve ending; mostly located just below the base of the epidermis
Pacinian corpuscle Located deep in the dermis

SKIN MICRORECEPTORS

Deformation of the layers within a receptor, and expansion or contraction as a result of temperature changes, generate nerve impulses. The impulses travel along the receptor's nerve fiber, which joins with bundles of other fibers in the deep dermis or tissue layers below. Most receptors "fire" nerve signals infrequently and irregularly when not stimulated, increasing their firing rate when the skin is touched.

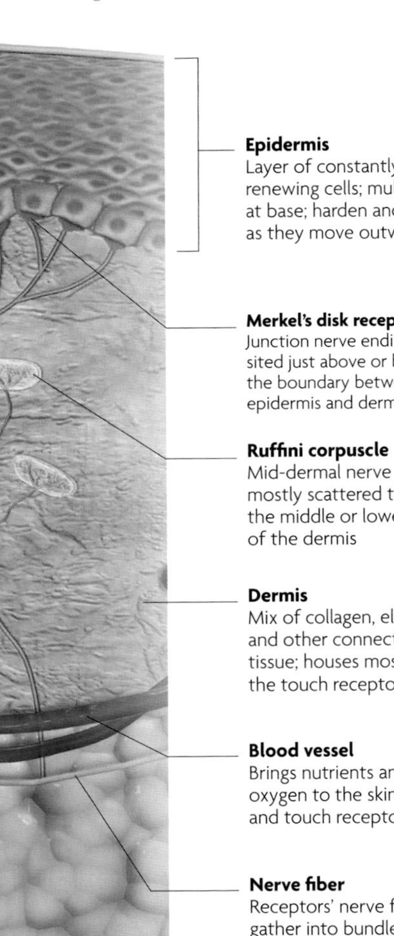

Epidermis
Layer of constantly renewing cells; multiply at base; harden and die as they move outward

Merkel's disk receptor
Junction nerve ending; sited just above or below the boundary between epidermis and dermis

Ruffini corpuscle
Mid-dermal nerve ending; mostly scattered through the middle or lower layers of the dermis

Dermis
Mix of collagen, elastin, and other connective tissue; houses most of the touch receptors

Blood vessel
Brings nutrients and oxygen to the skin layers and touch receptors

Nerve fiber
Receptors' nerve fibers gather into bundles; these convey signals to the main nerves

TYPES OF SENSOR

Each type of microsensor is set at a particular depth in the dermis that best suits its function. The largest receptors, Pacinian corpuscles, are located at the deepest level, near the base of the dermis. Sensors for light touch are located near or just within the epidermal layer.

Free nerve endings
Branching, usually unsheathed sensors of temperature, light touch, pressure, and pain. They are found all over the body and in all types of connective tissue.

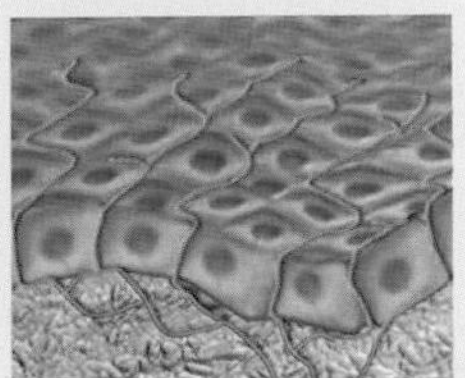

Meissner's corpuscle
Encapsulated nerve ending in the skin's upper dermis, especially on the palms, soles, lips, eyelids, external genitals, and nipples. It responds to light pressure.

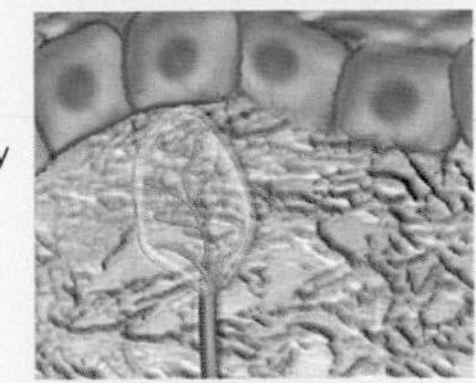

Merkel's disk
Naked (unencapsulated) receptors, usually in the upper dermis or lower epidermis, especially in nonhairy areas. They sense faint touch and light pressure.

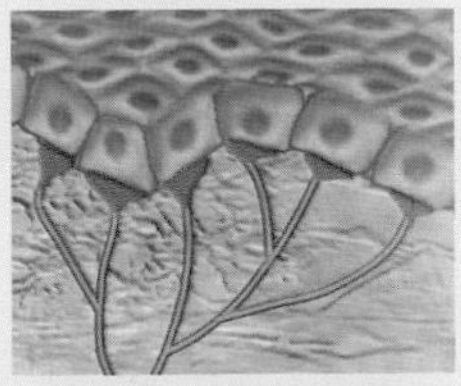

Ruffini corpuscle
Encapsulated receptor in the skin and deeper tissue that reacts to continuous touch and pressure. In joint capsules, it responds to rotational movement.

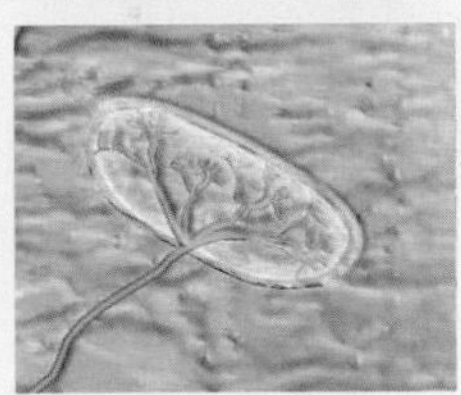

Pacinian corpuscle
Large, covered receptor located deep in the dermis, as well as in the bladder wall, and near joints and muscles. It senses stronger, more sustained pressure.

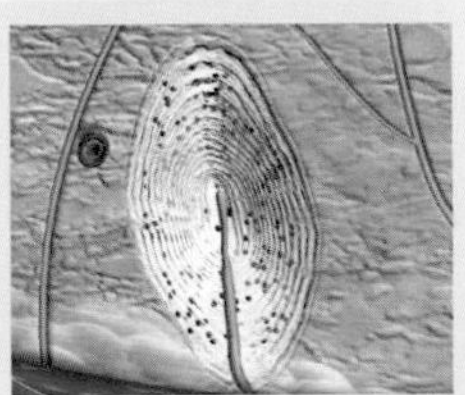

EPITHELIUM

Epithelial tissue, also called epithelium, is an important structural element that acts as a lining or covering for other body tissues. Epithelium can be classified according to the shape and layout pattern of individual cells (see opposite), and also by the arrangement of cells into one or more layers. Most epithelial tissues form membranes and are specialized for protection, absorption, or secretion. They do not contain blood vessels, and their cells are usually anchored to, and stabilized by, a basement membrane. There may be other cell types present, such as goblet cells that secrete blobs of mucus for release onto the surface.

PSEUDOSTRATIFIED EPITHELIUM

This type of columnar epithelium seems to be arranged in vertical layers. However, it actually consists of a single layer of cells of varying shapes and heights. The nuclei (control centers) of the different cell types are also at different levels, creating a layered (stratified) effect. Taller cells may be specialized into mucus-making goblet cells or ciliated cells that trap foreign particles. This type of epithelium occurs in the airway linings, and in the excretory and male reproductive passages and ducts.

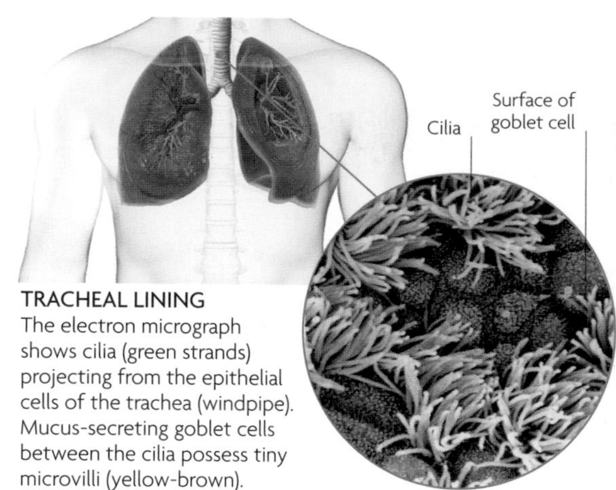
Cilia
Surface of goblet cell

TRACHEAL LINING
The electron micrograph shows cilia (green strands) projecting from the epithelial cells of the trachea (windpipe). Mucus-secreting goblet cells between the cilia possess tiny microvilli (yellow-brown).

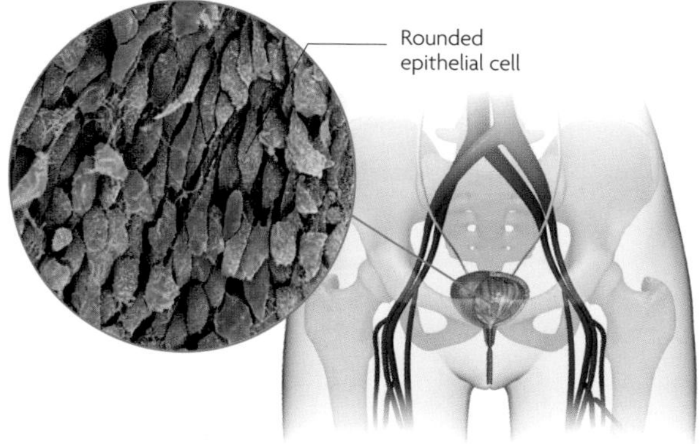
Rounded epithelial cell

BLADDER LINING
The electron micrograph shows the tightly packed epithelial cells of the bladder lining. They are soft and pliable, enabling them to stretch as the bladder fills with urine.

TRANSITIONAL EPITHELIUM

This epithelial tissue is similar to layered (stratified) epithelium, but is able to stretch without tearing. There are usually columnar cells in the basal layer, which become more rounded in the upper layers. As these layers stretch, the cells flatten, or become more squamous. Transitional epithelium is well suited to the urinary system, where it lines areas within the kidneys, ureters, bladder, and urethra. It allows these organs to bulge as urine flows through at pressure. The epithelium also secretes mucus that protects it from acidic urine.

SIMPLE AND LAYERED EPITHELIUM

Simple epithelium is composed of a single layer of cells. This type of tissue is often found in areas where substances need to pass through easily, a single-cell thickness offering minimal resistance. For example, in the air sacs in the lungs, simple epithelium allows the exchange of gases to take place. Layered (stratified) epithelium has two or more layers, and is better for protection in areas such as the mouth or esophagus. Some complex epithelium has more than five layers, but two or three is more usual. The cells may be different shapes in the different layers.

CORNEA STRUCTURE

The epithelium covering the cornea is transparent and about five layers thick. It permits light rays to enter the eye.

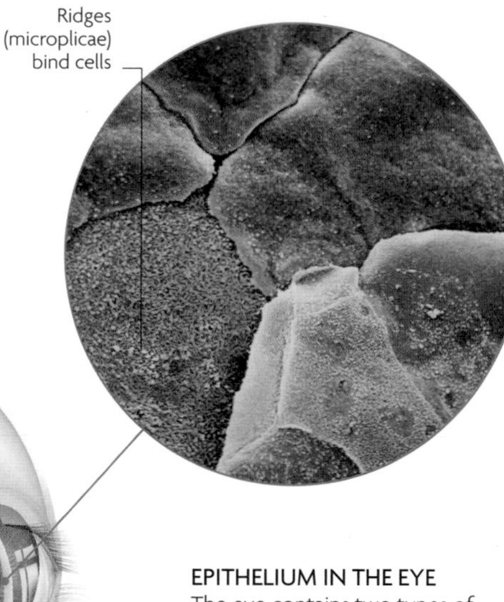
Ridges (microplicae) bind cells

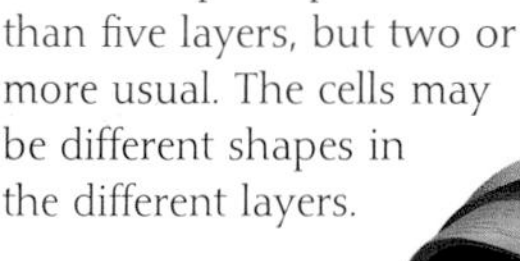

Retina
Cornea

EPITHELIUM IN THE EYE

The eye contains two types of epithelium: simple epithelium in the pigmented layer of the retina, and stratified squamous epithelium in the domed front "window" of the cornea.

TYPES OF EPITHELIAL CELL

The cells that make up the epithelial layers are usually classified according to their shape. Since most epithelial cells, as a consequence of their locations in the body, are subject to friction, compression, and similar physical wear and tear, they divide rapidly to replace themselves.

Squamous
Platelike or flattened cells, wider than deep, resembling paving slabs or random paving; flattened nucleus.

Features: Cells allow selective diffusion, or permeability, allowing certain substances to pass, owing to thinness of the layer.

Cuboidal
Cube- or box-shaped cells, occasionally hexagonal or polygonal; nucleus usually in cell center.

Features: Substances absorbed from one side of the layer can be altered as they pass through the cytoplasm of the cuboidal cells, before leaving.

Columnar
Tall, slim cells, often square, rectangular, or polygonal; large, oval nucleus near cell base.

Features: Protect and separate other tissues; may be topped with cilia for movement of fluid outside the cell or microvilli for absorption.

Glandular
Epithelial cells modified for secretion, usually cuboidal or columnar with secretory granules or vacuoles.

Features: Layers of these cells may be infolded to form pits, pockets, grooves, or ducts, as in sweat glands.

SKIN AND HAIR DEFENSIVE FUNCTIONS

Skin is the body's first line of defense against potential harm. As such, it is well equipped to prevent physical damage due to its supple, cushioned qualities. The epidermal cells that form skin's outermost layers are tightly knit together, but allow a certain amount of pliability. The cells are almost entirely full of the tough protein keratin, which resists attack by many kinds of chemicals. The natural secretion of sebum from the millions of sebaceous glands, each associated with a hair follicle, is slightly oily at body temperature and spreads easily. It furnishes the skin with partially water-repellent and antibiotic qualities, inhibiting the growth of certain microorganisms, and prevents hairs from becoming too brittle.

SCALP HAIR

Head hairs help keep rainwater from the scalp, absorb or deflect some of the energy in knocks and blows, and shield the head from extremes of temperature.

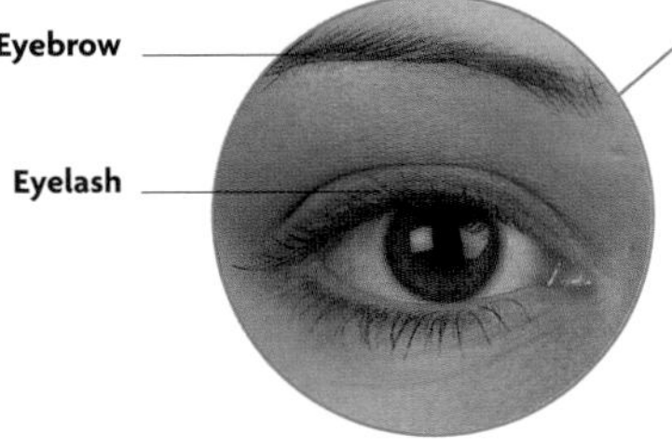

Eyebrow
Eyelash

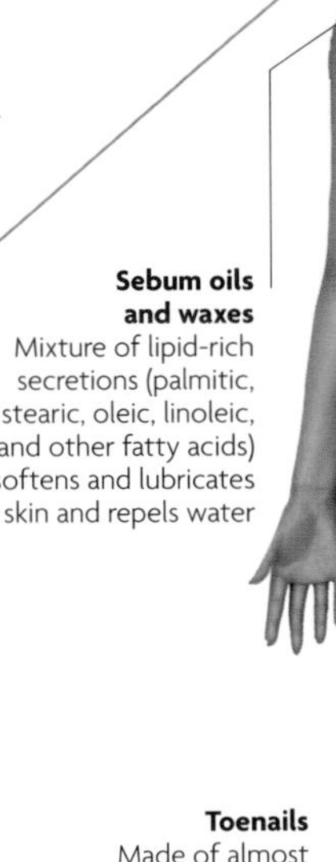

Sebum oils and waxes
Mixture of lipid-rich secretions (palmitic, stearic, oleic, linoleic, and other fatty acids) softens and lubricates skin and repels water
Toenails
Made of almost solid keratin

EYEBROWS AND EYELASHES

The arch of relatively coarse, fast-growing eyebrow hairs helps divert sweat or rainwater on the forehead that might trickle into the eyes. Eyelashes produce swirling air currents when blinking, which push floating particles away from the eye surface.

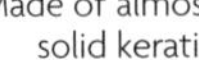

Thick epidermis
Basal cell layer
Dermis

THICKENED SKIN

Areas of skin subjected to regular pressure respond by thickening their epidermis for greater protection and buffering, as in this magnified image of skin from the foot.

ULTRAVIOLET DEFENSES

The Sun's rays include a spectrum of color wavelengths, as well as infrared or IR rays and ultraviolet, UV, rays. Both UV-A and UV-B wavelengths are invisible, but exposure to the latter, in particular, is linked to skin cancers. Skin's defense is its dark pigment, melanin. This forms a screen in the upper epidermis and shields the multiplying cells in the base of the epidermis.

MELANIN PRODUCTION

Melanocytes are melanin-producing cells in the base of the epidermis. They make parcels of melanin granules, known as melanosomes, which pass into surrounding cells.

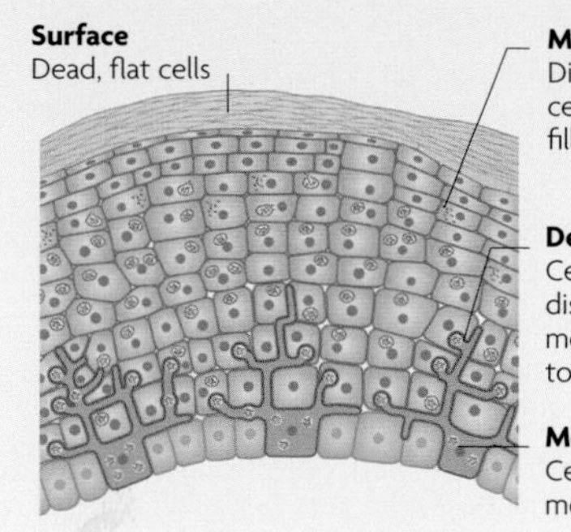
Surface
Dead, flat cells
Melanin granules
Disperse in cell; cell flattens and fills with keratin
Dendrite
Cell projection distributes melanosomes to nearby cells
Melanocyte
Cell body makes melanosomes

SKIN PIGMENTATION

Skin color depends on the type and quantity of two melanin pigments–reddish pheomelanin and brown-black eumelanin–in the epidermis, and on the way the pigment granules are distributed. Each melanocyte has fingerlike dendrites that touch surrounding cells (basal keratinocytes). The melanocyte produces pigment granules within organelles called melanosomes. These move along the dendrites and into nearby cells. Darker skin has larger melanocytes with more melanosomes. Lighter skin has smaller melanocytes and fewer melanosomes. Exposure to UV rays stimulates the melanocytes so that the skin darkens.

COLOR VARIATION

Darker skin tends to have larger melanin-making cells with denser melanosomes, in comparison to lighter skin. The former release their pigment granules, while the latter's granules stay clumped.

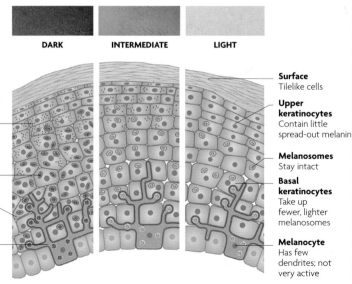

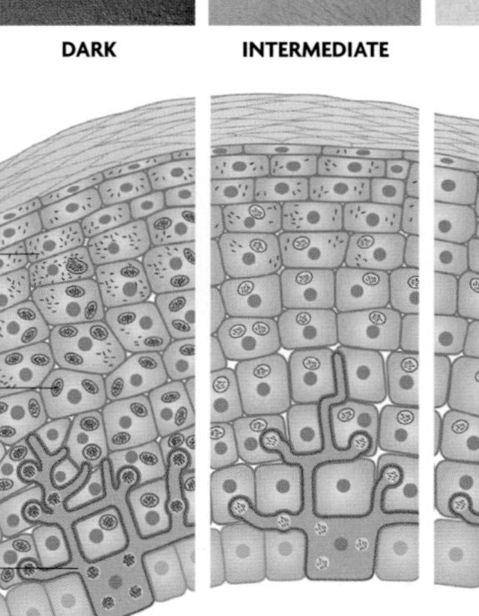
DARK
INTERMEDIATE
LIGHT
Upper keratinocytes
Have evenly spread melanin
Melanosomes
Release melanin granules
Basal keratinocytes
Take up more melanosomes
Melanocyte
Has many dendrites and is active
Surface
Tilelike cells
Upper keratinocytes
Contain little spread-out melanin
Melanosomes
Stay intact
Basal keratinocytes
Take up fewer, lighter melanosomes
Melanocyte
Has few dendrites; not very active

SKIN DISORDERS

SKIN CONTAINS SOME OF THE FASTEST-MULTIPLYING CELLS IN THE BODY. SEVERAL OF ITS DISORDERS, SUCH AS MOLES, RESULT FROM PROBLEMS IN THIS SELF-RENEWAL SYSTEM. THE SKIN IS ALSO SUSCEPTIBLE TO INJURY, ALLERGIC REACTION IN THE FORM OF RASHES, AND INFECTIONS.

RASHES

Some skin rashes are localized, while others are more widespread. Often, the cause of a rash is not clear. The condition may affect quality of life and require long-term control with self-help measures and medication. Psoriasis is a widespread, patchy rash that flares up at intervals. Episodes may be triggered by infection, injury, stress, or as a side-effect of drug treatment. Eczema is one of the most common rashes, especially in children. It is often linked to allergic conditions such as asthma and rhinitis (hay fever). Impetigo is a blistering of the skin caused by bacterial infection, typically through a cut, a cold sore (*Herpes simplex* virus), or scratched, weeping eczema. In vitiligo, the body makes antibodies that attack the skin's pigment-making cells, or melanocytes. It occurs in patchy areas over the body; in about one-third of cases, the pigmentation spontaneously returns.

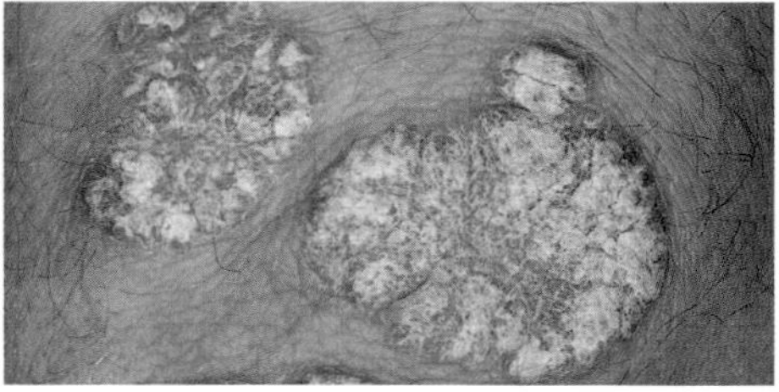

PSORIASIS
There are several types of psoriasis, mostly characterized by intermittently itchy patches of red, thickened, scaly skin, as dead epidermal cells accumulate. Common sites are the knees, elbows, lower back, scalp, and behind the ears.

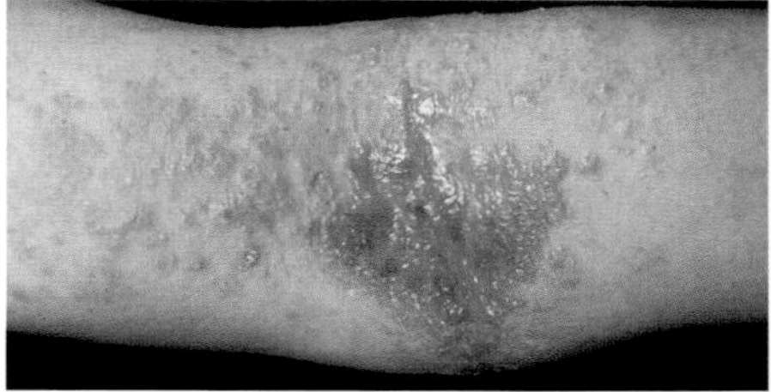

ECZEMA
A typical eczema rash is inflamed and itchy, with small fluid-filled blisters or episodes of dry, scaly, thickened, and cracked skin. Common sites are the hands and creased areas of skin, such as the wrists, elbows, and knees. The condition is also known as dermatitis.

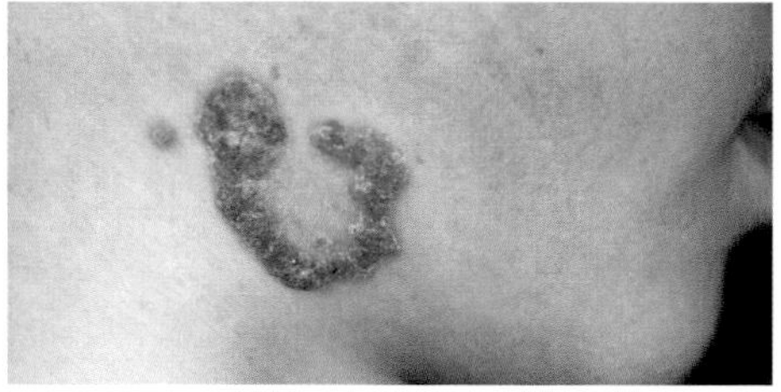

IMPETIGO
This bacterial infection is common on the face, most often around the nose and mouth. The skin develops fluid-filled blisters, which burst. This stage is followed by redness, weeping, and crusting that may itch.

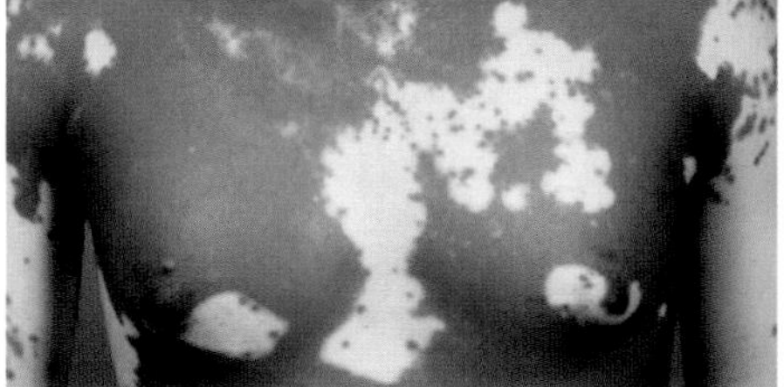

VITILIGO
Depigmented patches of skin develop over months or years, especially on the face and hands. The areas are more distinct in people with dark skin. They do not carry any medical risks to health.

ACNE

In acne vulgaris, the sebaceous glands produce an excessive amount of the oily-waxy secretion, sebum. This reacts in contact with air and forms a plug in the skin pore, which may appear dark with pigmentation (not dirt), as a blackhead or comedone, or pale, as a whitehead. A combination of trapped sebum, dead skin cells, and bacterial infection inflame the area, and a pustule develops. Acne is a common problem at puberty, when hormone surges cause increased sebum output.

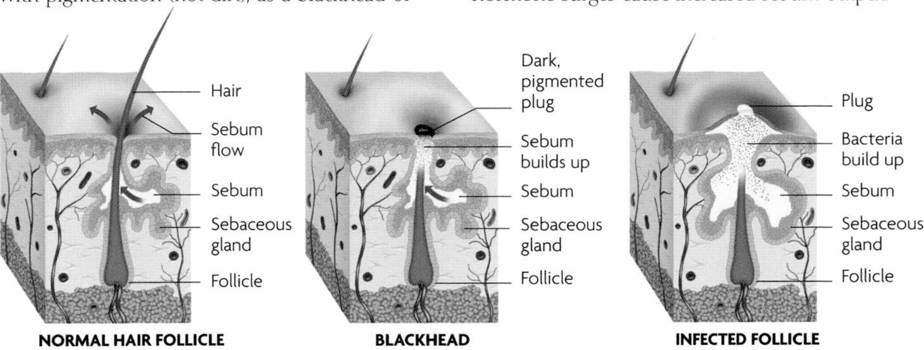
Hair
Sebum flow
Sebum
Sebaceous gland
Follicle
NORMAL HAIR FOLLICLE
Dark, pigmented plug
Sebum builds up
Sebum
Sebaceous gland
Follicle
BLACKHEAD
Plug
Bacteria build up
Sebum
Sebaceous gland
Follicle
INFECTED FOLLICLE

MOLE

A mole, or nevus, is a localized overproduction and aggregation of the skin's pigment cells (melanocytes), with increased amounts of melanin pigment. Moles are very common–most adults have 10–20 moles by the age of 30 years. They can occur almost anywhere on the body and are variable in size, but usually less than ⅖in (1cm) across. Rarely, moles become malignant (cancerous); any change in size or appearance, itching, or bleeding should be discussed with a doctor.

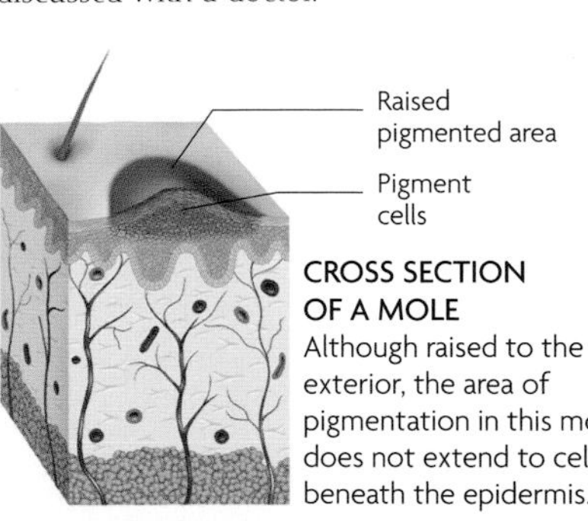
Raised pigmented area
Pigment cells

CROSS SECTION OF A MOLE
Although raised to the exterior, the area of pigmentation in this mole does not extend to cells beneath the epidermis.

CYST

The most common type of cyst is a sebaceous cyst that forms in a hair follicle. A cyst contains sebaceous secretions and dead cells, which are restrained in a strong, baglike capsule. Its surface mound is usually smooth, and some cysts have a paler or darker central region. Common sites include the scalp, face, trunk, and genitals, although they can occur just about anywhere. Treatment may be needed if the cyst becomes enlarged, unsightly, painful, or infected.

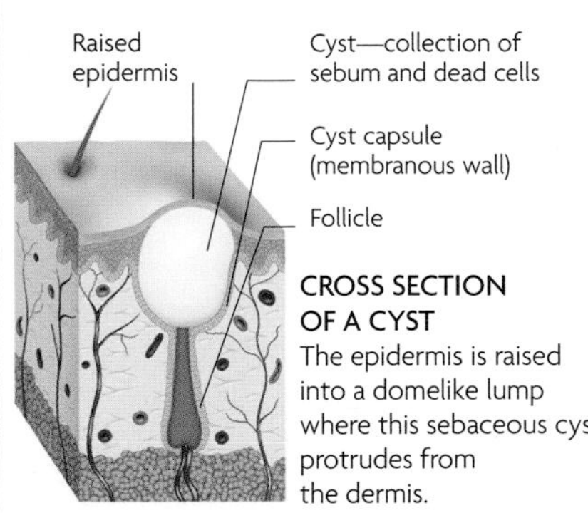
Raised epidermis
Cyst—collection of sebum and dead cells
Cyst capsule (membranous wall)
Follicle

CROSS SECTION OF A CYST
The epidermis is raised into a domelike lump where this sebaceous cyst protrudes from the dermis.

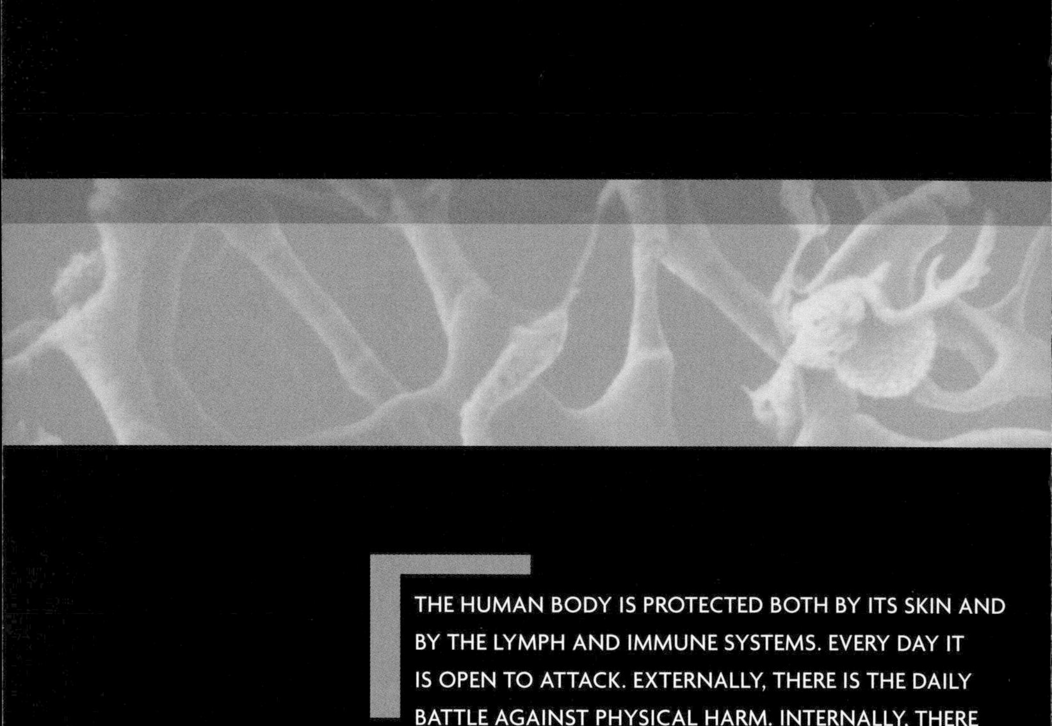

THE HUMAN BODY IS PROTECTED BOTH BY ITS SKIN AND BY THE LYMPH AND IMMUNE SYSTEMS. EVERY DAY IT IS OPEN TO ATTACK. EXTERNALLY, THERE IS THE DAILY BATTLE AGAINST PHYSICAL HARM. INTERNALLY, THERE

LYMPH AND IMMUNITY

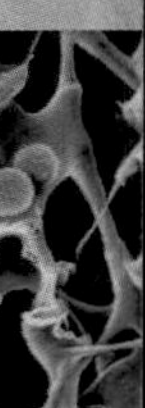

LYMPH AND IMMUNE SYSTEMS

THE IMMUNE SYSTEM, INCORPORATING THE LYMPHATIC SYSTEM, IS THE MAIN MEANS BY WHICH THE BODY IS PROTECTED FROM INVASION BY MICROORGANISMS.

The lymphatic system contains lymph, a fluid that originates in the interstitial spaces between cells. Lymph drains into networks of tiny capillaries in tissue spaces that unite to form larger vessels called lymphatics. Nodes scattered along the length of the lymphatics filter and store the lymph. Organs, such as the thymus and spleen, and lymphoid tissue, such as tonsils and Peyer's patches, complete the system. They contain many specialized white blood cells, which protect the body against non-self material such as microorganisms.

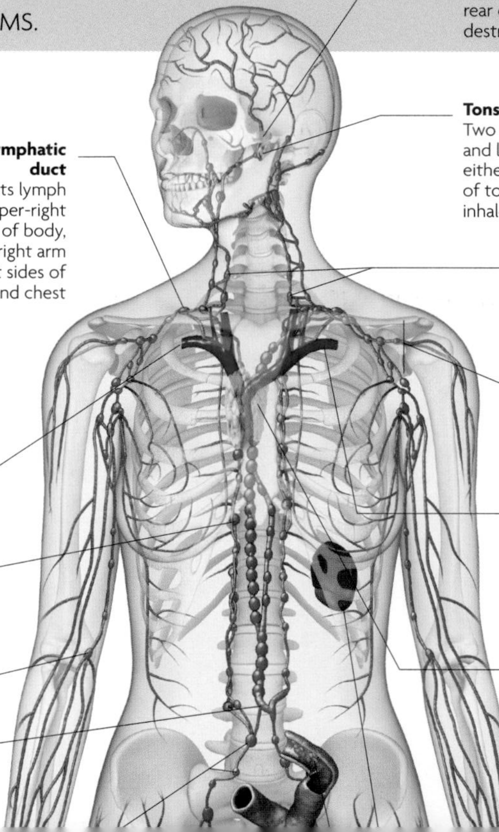

Right lymphatic duct
Collects lymph from upper-right quadrant of body, including right arm and right sides of head and chest
Right subclavian vein
One of two main exit points at which lymph drains into blood system
Thoracic duct
Also called left lymphatic duct; collects lymph from both legs, abdomen, left arm, and left sides of head and chest
Supratrochlear node
Collects lymph from hand and forearm
Cisterna chyli
Enlarged lymph vessel formed from vessels from legs and lower body; eventually narrows into thoracic duct
Adenoids
Also called pharyngeal tonsils; lie at rear of nasal cavity; help filter air and destroy microorganisms
Tonsils
Two pairs of tonsils (palatine and lingual) at back of mouth—on either side of pharynx and at base of tongue—help guard against inhaled microbes
Cervical (neck) nodes
Collect lymph from right or left side of the face, scalp, nasal cavity, and throat
Axillary (armpit) nodes
Drain lymph from arm, breast, chest wall, and upper abdomen
Left subclavian vein
Point at which lymph from left and lower body enters blood after collecting in thoracic duct
Thymus gland
Site of maturation of immune-system T cells (T lymphocytes); T cells develop from stem cells, which migrate here from bone marrow

Lumbar lymph nodes
Drain lymph from abdominal organs

AUXILIARY IMMUNE SYSTEM

Many organs have a role in protecting the body against invading microbes. They form an auxiliary immune system that includes the skin, microscopic hairs, gastric enzymes, and useful bacteria.

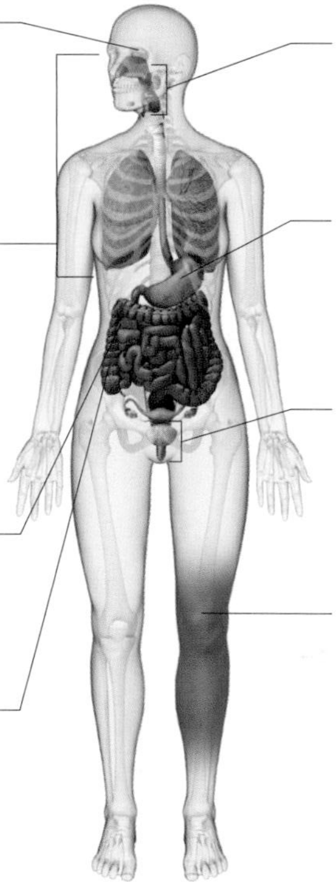

Tear (lacrimal) glands
Tear fluid contains an antibacterial enzyme, lysozyme, that flushes across the eyeball with each blink

Respiratory tract
Nostril hairs trap airborne particles; mucus and cilia in lining of nose and trachea trap and remove dust, microorganisms, and debris

Small intestine
Digestive enzymes, including those in pancreatic juices, attack microbes that survive the stomach

Large intestine
The body's natural gut flora ("friendly" bacteria and other microorganisms) suppress unwanted, harmful microbes

Mouth, and throat
Salivary glands (yellow) produce antibacterial saliva, while mucus and saliva trap airborne particles in throat

Stomach
Powerful hydrochloric acid and digestive enzymes in the gastric juices help destroy ingested organisms

Genitourinary tract
The mucous lining helps trap foreign matter, and harmless bacteria restrict the growth of potentially harmful organisms

Skin
The mechanical barrier formed by skin is the first defense against invading organisms, as well as protecting the body against physical forces such as extremes of temperature, radiation, and various chemicals

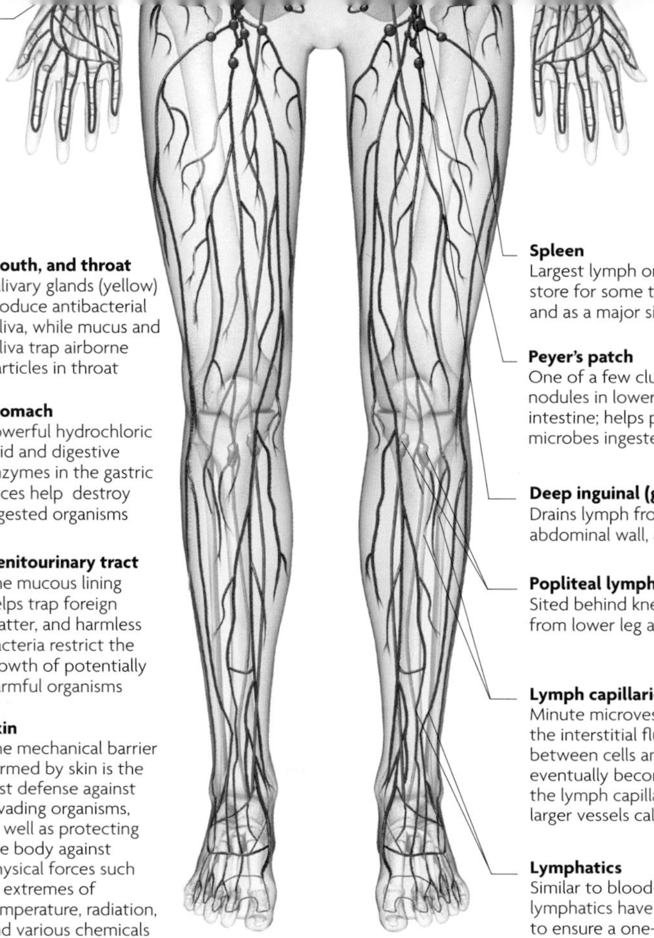

Spleen
Largest lymph organ; spleen acts as store for some types of lymphocyte and as a major site for filtering blood

Peyer's patch
One of a few clusters of lymphoid nodules in lower part of small intestine; helps protect against microbes ingested in food

Deep inguinal (groin) node
Drains lymph from the legs, lower abdominal wall, and external genitals

Popliteal lymph nodes
Sited behind knees; drain lymph from lower leg and foot

Lymph capillaries
Minute microvessels that collect the interstitial fluid, which flows between cells and tissues and eventually becomes lymph fluid; the lymph capillaries unite into larger vessels called lymphatics

Lymphatics
Similar to blood-carrying veins, lymphatics have flap-type valves to ensure a one-way flow of lymph

IMMUNE SYSTEM

THE COMPLEXITIES OF THE IMMUNE SYSTEM CREATE THE CONDITION OF IMMUNITY, IN WHICH, AFTER THE FIRST ATTACK BY A PARTICULAR TYPE OF MICROORGANISM, THE BODY IS PROTECTED OR RESISTANT TO FUTURE INVASIONS.

LYMPH NODES

The lymph nodes produce and harbor lymphocytes that protect the body from disease. They are scattered throughout the body and are also concentrated in groups (see p.192). Small lymphatics (vessels) bring lymph to a node, while a larger vessel carries it away. The nodes filter and clean the lymph, which then drains into the venous bloodstream. Lymph vessels have valves so the fluid flows one way.

INSIDE A NODE

A lymph node, or gland, is a mass of segmented lymphatic tissue covered in a fibrous capsule. It contains sinuses, where many scavenging white blood cells, called macrophages, ingest bacteria as well as other foreign matter. Lymph nodes vary in diameter from 1 to 1/25 to 1 in (25 mm), although they can swell during infection or illness.

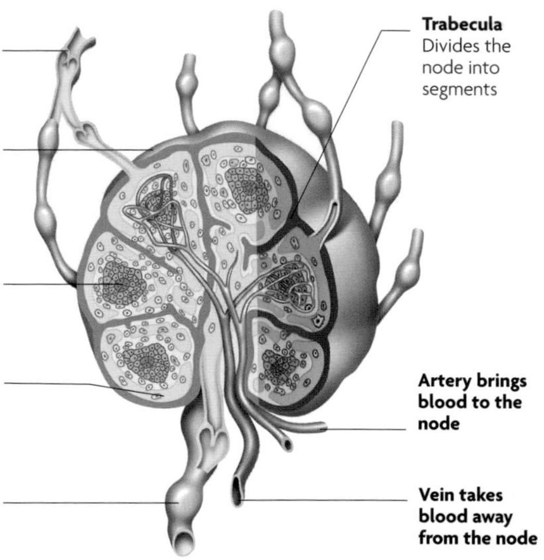
Incoming lymph vessel One of several vessels that transport lymph into the node
Capsule Dense meshwork of fibers that surrounds node
Germinal center Lymphocytes multiply and partly mature here
Sinus Wide channel where lymph slows so macrophages can attack intruders
Outgoing lymph vessel Filtered lymph is carried away by only one vessel
Trabecula Divides the node into segments
Artery brings blood to the node
Vein takes blood away from the node

WHITE CELL TYPES

There are numerous types of white blood cell, which are known by the general name of leucocytes. All white blood cells are derived from the bone marrow. Some of them grow and mature into other types. The largest leucocytes, and the largest cells in the blood, are monocytes. The lymphocytes are the chief immune cells, and they can be either B or T cells, depending on the way the lymphocyte develops.

Monocyte
Has a nucleus that is big and rounded, or indented; engulfs pathogens.

Lymphocyte
Both types (B and T cells) have a large nucleus that almost fills the cell.

Neutrophil
Granulocyte with many particles and multilobed nucleus; engulfs pathogens.

Basophil
Granulocyte with lobed nucleus; involved in allergic reactions.

Eosinophil
Granulocyte with B-shaped nucleus; destroys antigen–antibody complexes.

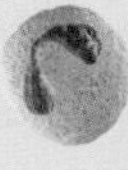

NONSPECIFIC RESPONSE

Any damage, such as burns, extreme cold, corrosive chemicals, or invading organisms, elicits a nonspecific response. The main response is inflammation (see p.198). The damaged tissue releases chemicals that attract white blood cells. Capillary walls become more permeable and porous to let these cells, along with defensive chemicals and fluids, enter and accumulate. The white cells surround, engulf, and destroy any pathogens, and the blood may clot to seal the breach.

INFLAMED TISSUE

The four common signs of inflammation are redness, swelling, increased warmth, and discomfort or pain. They occur after any form of harm in order to limit damage and initiate repair and healing.

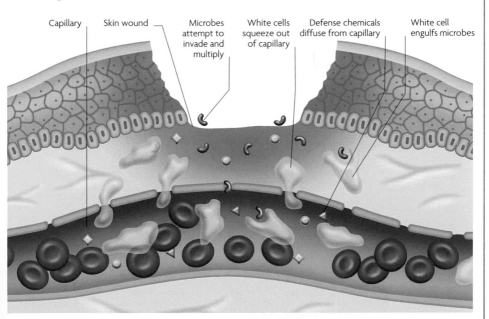

Capillary

Skin wound

Microbes attempt to invade and multiply

White cells squeeze out of capillary

Defense chemicals diffuse from capillary

White cell engulfs microbes

LOCAL INFECTION

If harmful microbes enter body tissues, both the inflammatory and immune responses act swiftly to limit their spread. White blood cells, fluids, microbes, toxins, and debris accumulate as pus. An abscess forms if the pus gathers in a localized area, putting pressure on surrounding structures. This may cause discomfort and pain, especially if the surrounding tissues have no flexibility—for example, in a dental abscess.

DENTAL ABSCESS

Microbes enter through a region of decayed enamel and dentine, infect the pulp, and spread into the root, where pus collects. As pus presses on the pulp nerves, it causes the pain of toothache.

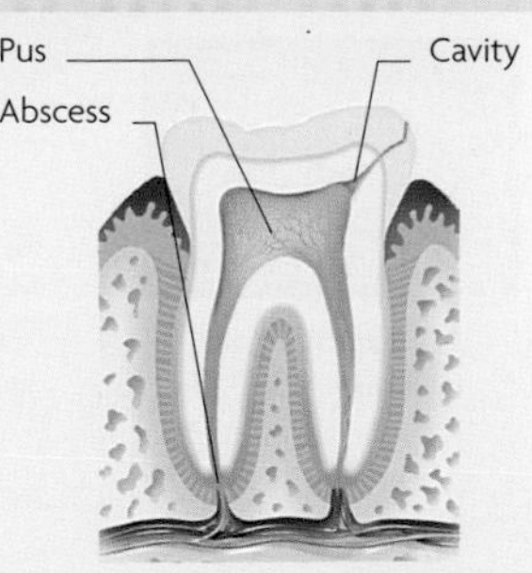
Pus
Cavity
Abscess

SPECIFIC RESPONSE

The two main types of specific defense–cell-mediated and antibody-mediated immunity–may accompany nonspecific reactions such as inflammation, or follow if infection persists. Both depend on the actions of B and T lymphocytes. B cells make protein antibodies known as gammaglobulins, which react against antigens (foreign proteins). Types of T cell multiply and attack the pathogen.

CELL-MEDIATED IMMUNITY

Once a T cell recognizes an antigen, it multiplies rapidly and its offspring form several types. Helper T cells activate B cells to make antibodies and macrophages that engulf the microbes as well as debris. Killer T cells attack the microbes and any body cells that are infected with them, using powerful proteins called lymphokines. Suppressor T cells inhibit the response of other cells to the invading microbes.

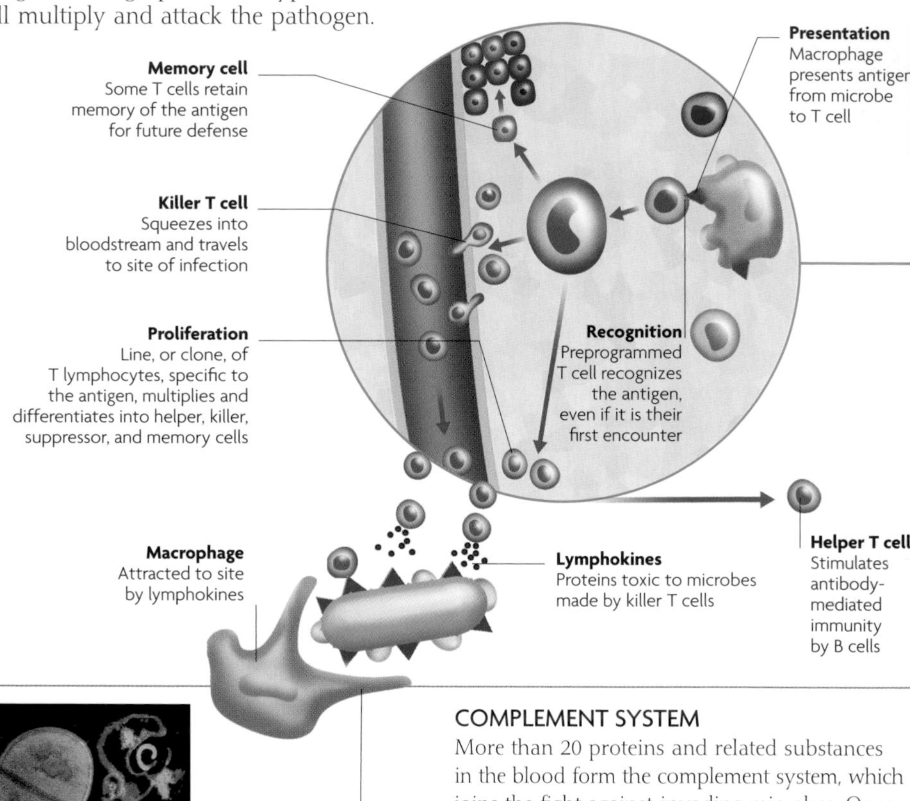
Memory cell
Some T cells retain memory of the antigen for future defense
Killer T cell
Squeezes into bloodstream and travels to site of infection
Proliferation
Line, or clone, of T lymphocytes, specific to the antigen, multiplies and differentiates into helper, killer, suppressor, and memory cells
Presentation
Macrophage presents antigens from microbe to T cell
Recognition
Preprogrammed T cell recognizes the antigen, even if it is their first encounter
Macrophage
Attracted to site by lymphokines
Lymphokines
Proteins toxic to microbes made by killer T cells
Helper T cell
Stimulates antibody-mediated immunity by B cells
Phagocytosis
Macrophage engulfs antigens

LYSED BACTERIUM
Complement dissolves, or lyses, invaders such as bacteria by disrupting their outer membranes (cell shown on right).

COMPLEMENT SYSTEM

More than 20 proteins and related substances in the blood form the complement system, which joins the fight against invading microbes. Once a complement reaction begins, it carries on in a "cascade," with one protein activating the next, and so on. The complement system has several roles. It helps to destroy microbes and stops them from attacking body cells, encourages the activity of white cells, widens blood vessels, and clears away the antigen-antibody complexes.

ANTIBODY-MEDIATED IMMUNITY

B cells produce Y- or T-shaped antibodies, and each type acts against a certain microorganism or "non-self" material by attaching to antigens on its surface. The presence of antigens triggers B cells to multiply. Some develop into plasma cells, which are the main antibody-producing cells. As with cell-mediated immunity, memory cells are made, which can recognize the same antigen and initiate defence many years later.

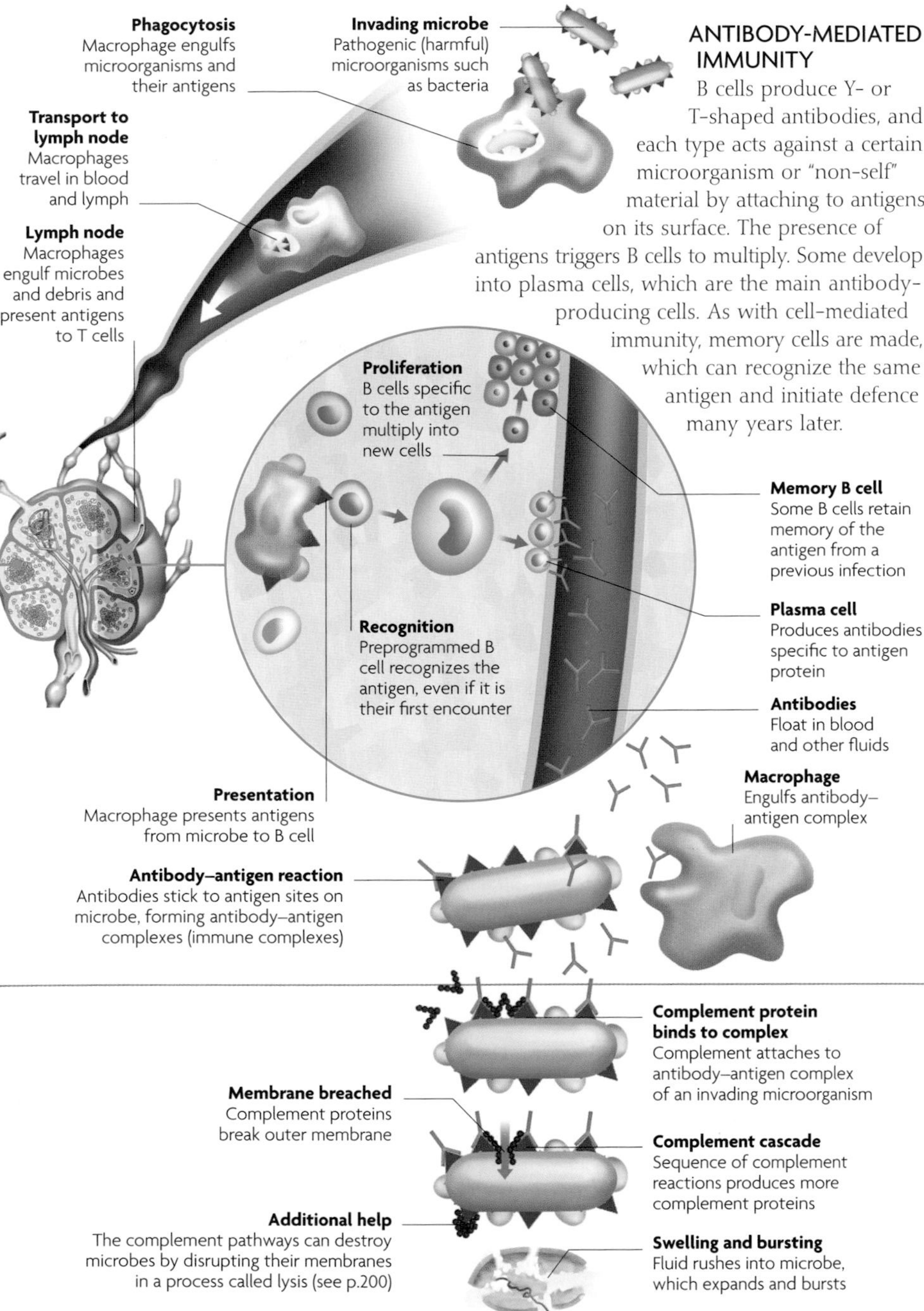
Phagocytosis
Macrophage engulfs microorganisms and their antigens
Invading microbe
Pathogenic (harmful) microorganisms such as bacteria
Transport to lymph node
Macrophages travel in blood and lymph
Lymph node
Macrophages engulf microbes and debris and present antigens to T cells
Proliferation
B cells specific to the antigen multiply into new cells
Memory B cell
Some B cells retain memory of the antigen from a previous infection
Plasma cell
Produces antibodies specific to antigen protein
Recognition
Preprogrammed B cell recognizes the antigen, even if it is their first encounter
Antibodies
Float in blood and other fluids
Presentation
Macrophage presents antigens from microbe to B cell
Macrophage
Engulfs antibody–antigen complex
Antibody–antigen reaction
Antibodies stick to antigen sites on microbe, forming antibody–antigen complexes (immune complexes)
Complement protein binds to complex
Complement attaches to antibody–antigen complex of an invading microorganism
Membrane breached
Complement proteins break outer membrane
Complement cascade
Sequence of complement reactions produces more complement proteins
Additional help
The complement pathways can destroy microbes by disrupting their membranes in a process called lysis (see p.200)
Swelling and bursting
Fluid rushes into microbe, which expands and bursts

INFLAMMATORY RESPONSE

INFLAMMATION IS THE BODY'S RAPID, GENERAL RESPONSE TO ANY KIND OF INSULT OR INJURY, SUCH AS FROM PHYSICAL WOUNDS AND FOREIGN OBJECTS, INFECTING ORGANISMS, CHEMICAL TOXINS, HEAT, OR RADIATION.

The inflammatory response is a nonspecific reaction that passes through defined phases and involves various types of white blood cell and defensive chemicals. The four cardinal signs are redness, swelling, heat, and pain. The process acts to attack, break down, and remove invading material, to dispose of the body's damaged cells and tissues, and to initiate healing.

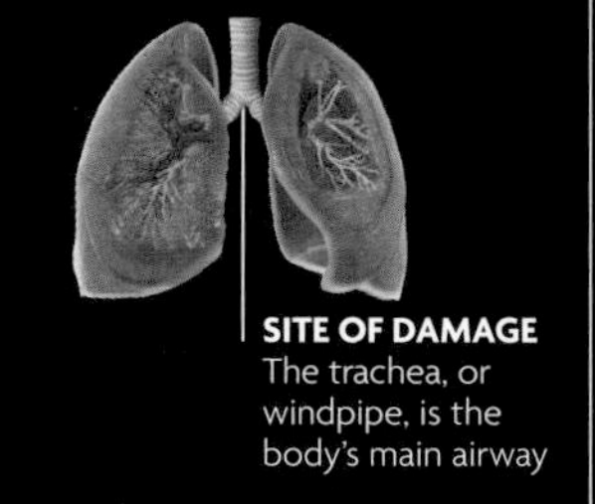

SITE OF DAMAGE
The trachea, or windpipe, is the body's main airway

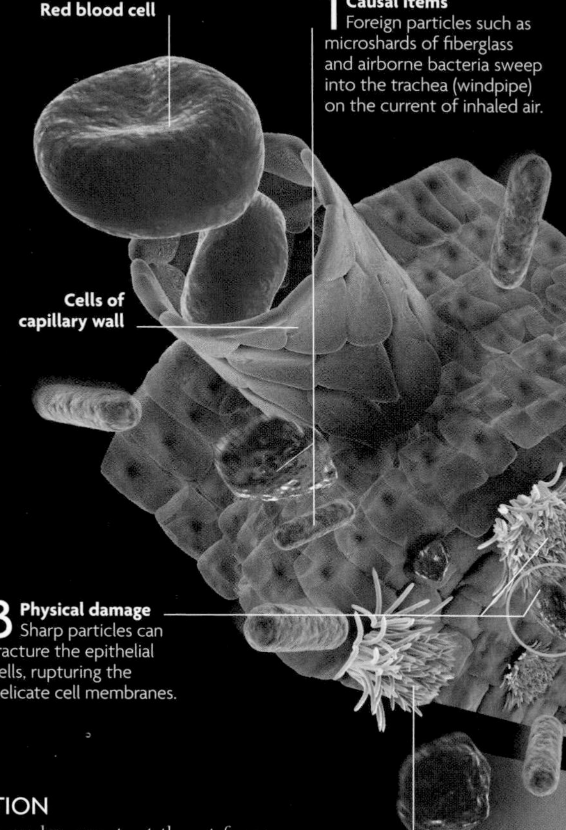

Red blood cell

1 Causal items
Foreign particles such as microshards of fiberglass and airborne bacteria sweep into the trachea (windpipe) on the current of inhaled air.

Cells of capillary wall

3 Physical damage
Sharp particles can fracture the epithelial cells, rupturing the delicate cell membranes.

Tufts of cilia
Hairlike projections borne by some cells of the tracheal lining; the cilia "beat" to remove protective mucus covering the cells

CAUSE OF INFLAMMATION

The respiratory system is under constant threat from tiny inhaled particles of dust and debris and attack by infecting microbes. Here, the lining (epithelium) of the trachea (windpipe) mounts an inflammatory response to dust and bacteria. In reality, this process usually occurs alongside the specific immune response (see p.196), which targets individual foreign substances.

DEFENSIVE CELLS

Various types of white blood cell (leucocytes) become involved in inflammation, including the defensive cells called neutrophils and monocytes (see p.194). The monocytes are immature when they leave the blood vessels and enter the tissues. However, they rapidly develop into active cells called macrophages that replace neutrophils.

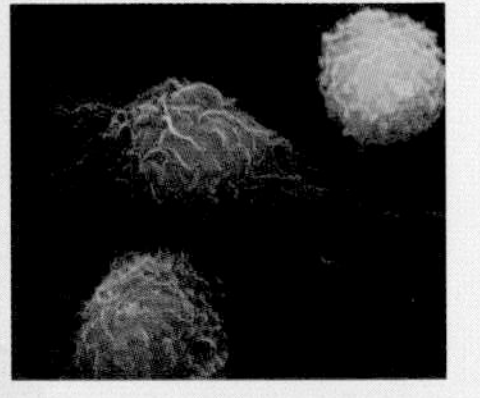

NEUTROPHILS
Among the first cells to take action, these are small but capable of engulfing several pieces of damaged tissue and bacteria.

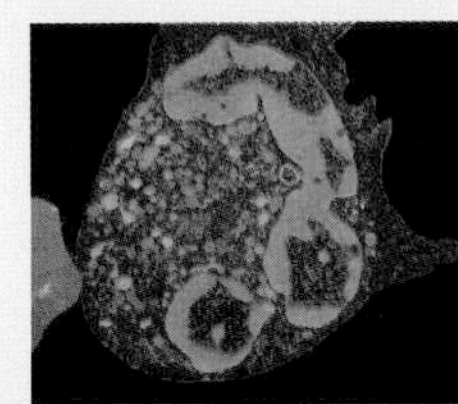

MACROPHAGE
A single macrophage, which means "big eater," can consume up to 100 bacteria or similar-sized items before dying.

2 Physical damage As the air current slows, the particles are trapped in protective mucus secreted by epithelial cells of the tracheal lining.

4 Initial spread Messenger substances, such as histamine and kinins, leak from ruptured cells, especially "mast cells" scattered throughout the tissue.

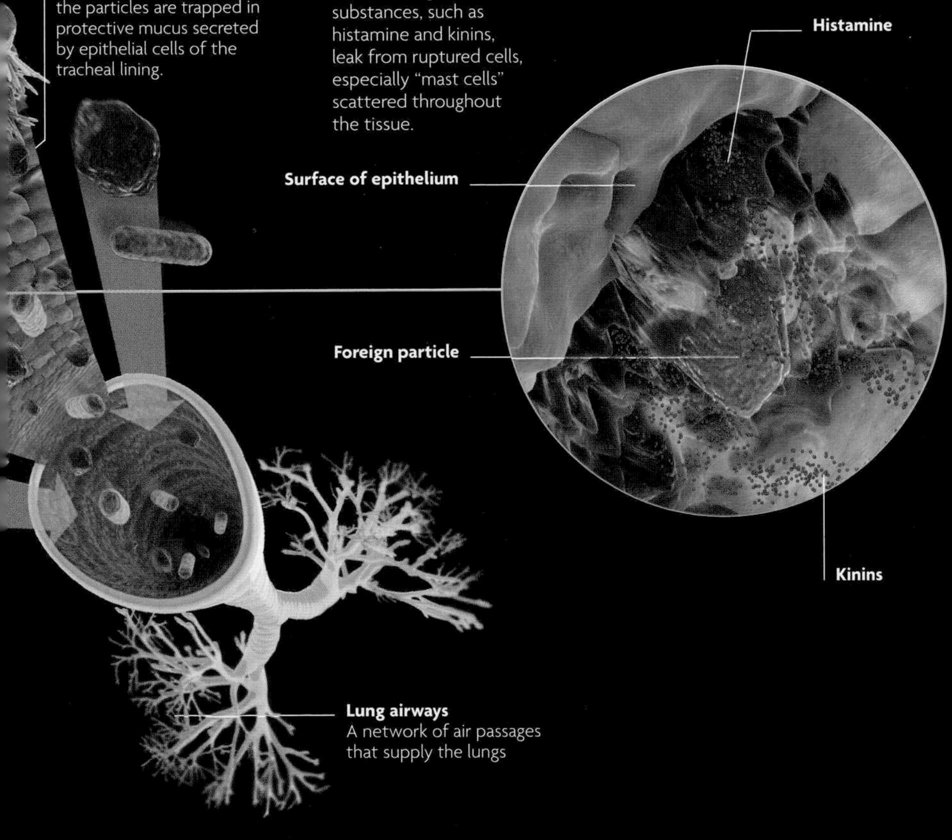

Histamine
Surface of epithelium
Foreign particle
Kinins

Lung airways
A network of air passages that supply the lungs

PHAGOCYTOSIS

Various kinds of white blood cells can surround, engulf, and ingest smaller items, such as bacteria and cellular debris, in a process known as phagocytosis ("cell eating"). The cell exploits its ability to change shape and move, using the intracellular components of microtubules and microfilaments (see p.26) that form its flexible, mobile, internal scaffolding. The ingestion usually takes less than one second, and the consumed material is gradually broken down by enzymes and other chemicals within the cell.

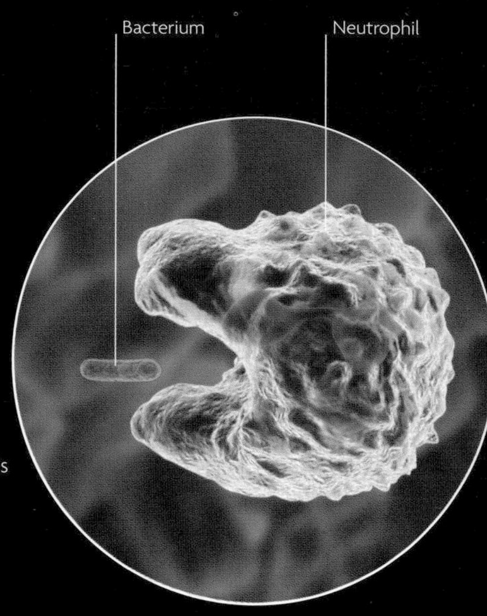
Bacterium
Neutrophil

1 ENGULFING STAGE
The white cell extends pseudopods ("false feet") toward and around the unwanted item—here a bacterium. The pseudopods merge to engulf it.

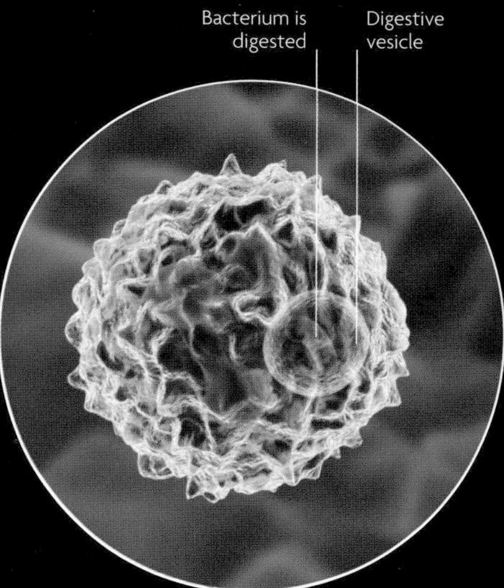
Bacterium is digested
Digestive vesicle

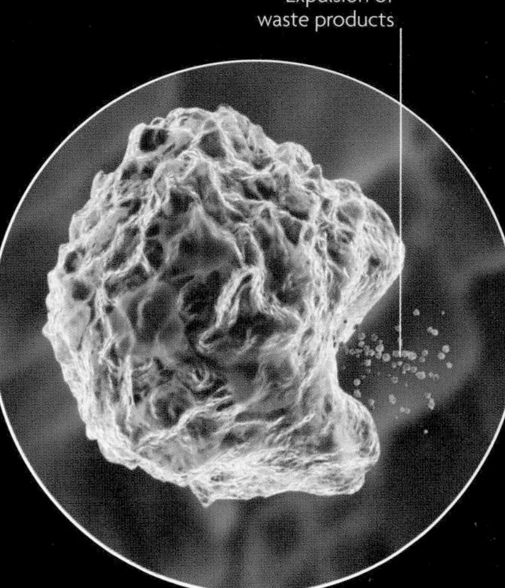
Expulsion of waste products

2 LYSIS STAGE
Any unwanted items are trapped in phagocytic vesicles. Together with enzyme-containing lysosomes, these digestive vesicles form phagolysosomes, in which lysis (breaking down) occurs.

3 EXOCYTOSIS STAGE
Harmless waste products of cell-eating are expelled through the membrane of the white blood cell, or in tiny, membrane-bound, exocytic vesicles, to the extracellular fluid.

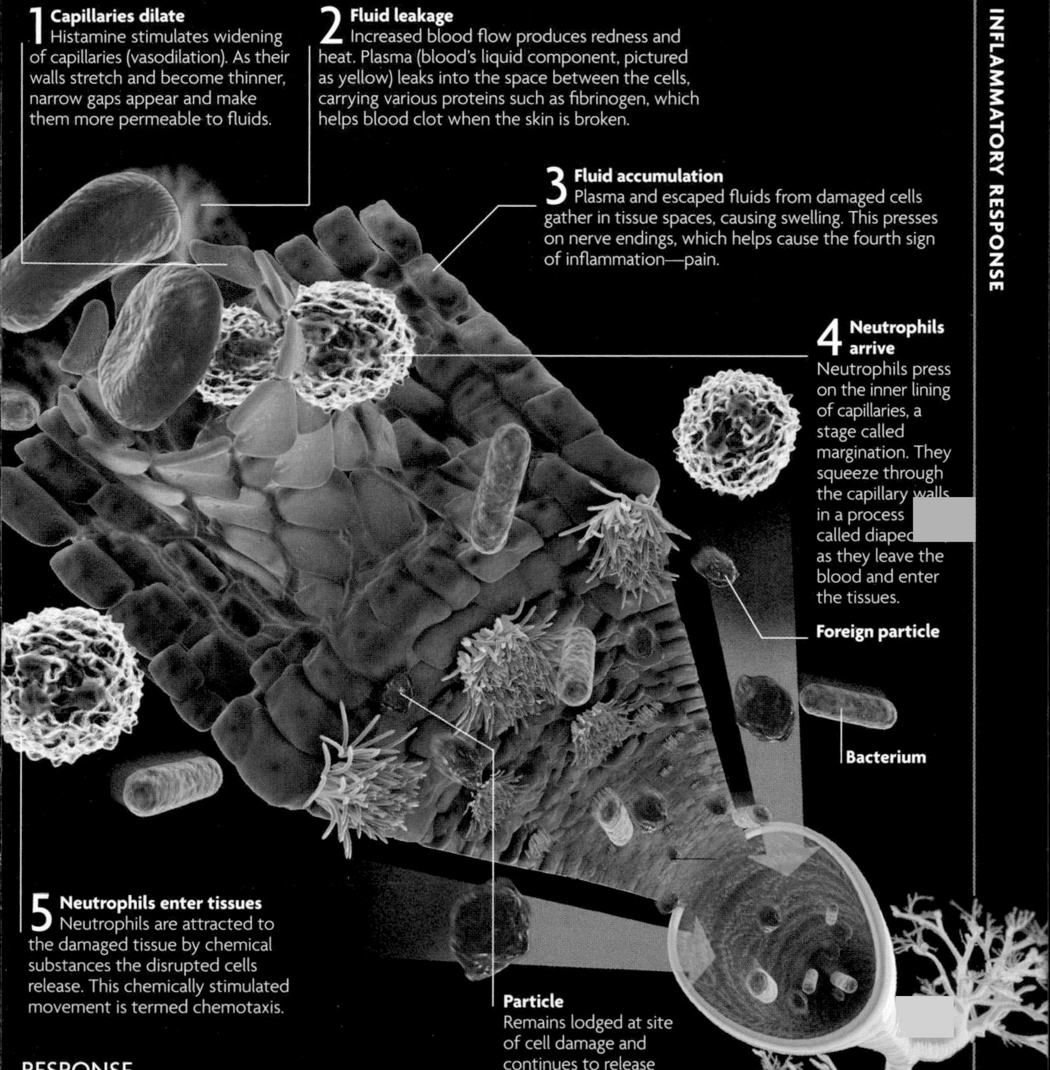

1 Capillaries dilate
Histamine stimulates widening of capillaries (vasodilation). As their walls stretch and become thinner, narrow gaps appear and make them more permeable to fluids.

2 Fluid leakage
Increased blood flow produces redness and heat. Plasma (blood's liquid component, pictured as yellow) leaks into the space between the cells, carrying various proteins such as fibrinogen, which helps blood clot when the skin is broken.

3 Fluid accumulation
Plasma and escaped fluids from damaged cells gather in tissue spaces, causing swelling. This presses on nerve endings, which helps cause the fourth sign of inflammation—pain.

4 Neutrophils arrive
Neutrophils press on the inner lining of capillaries, a stage called margination. They squeeze through the capillary walls in a process called diaped[illegible] as they leave the blood and enter the tissues.

Foreign particle

Bacterium

5 Neutrophils enter tissues
Neutrophils are attracted to the damaged tissue by chemical substances the disrupted cells release. This chemically stimulated movement is termed chemotaxis.

Particle
Remains lodged at site of cell damage and continues to release histamine and kinins (see p.199), which flow into the bloodstream

Bronchial tree
May be affected by inflammation, or the problem may remain restricted to a patch of the trachea

RESPONSE

Once an inflammatory response is triggered, more blood flows to the damaged area. The capillaries widen and become more permeable, allowing plasma and fluid to leak into the space between the cells. White blood cells, such as neutrophils, leave the blood and enter the tissue, drawn to the damaged area by chemicals released from the disrupted cells.

FIGHTING INFECTIONS

AN INFECTION OCCURS WHEN MICROORGANISMS ENTER THE BODY, THEN SURVIVE, MULTIPLY, AND DISRUPT CELL FUNCTION. THE INFECTION MAY BE LOCAL, SUCH AS IN A WOUND, OR SYSTEMIC, IN WHICH MANY PARTS OF THE BODY ARE AFFECTED.

VIRUSES

Viruses are the smallest microbes; millions would cover the head of a pin. Many types of viruses can stay inactive for long periods and survive freezing, boiling, and chemical attack. Yet they can activate suddenly when an opportunity of invading a living cell arises. Viruses are obligate parasites, which means they must have living cells, or host cells, in order to replicate themselves. The typical virus particle has a single or double strand of genetic material (nucleic acid–either DNA or RNA) surrounded by a shell-like coat of protein (capsid), and sometimes a protective outer envelope.

LIFE CYCLE OF A VIRUS

Viruses have very few genes (typically 100–300) and cannot process nutrients. To build copies of itself, a virus takes over a host cell's machinery, causing the cell to die or malfunction.

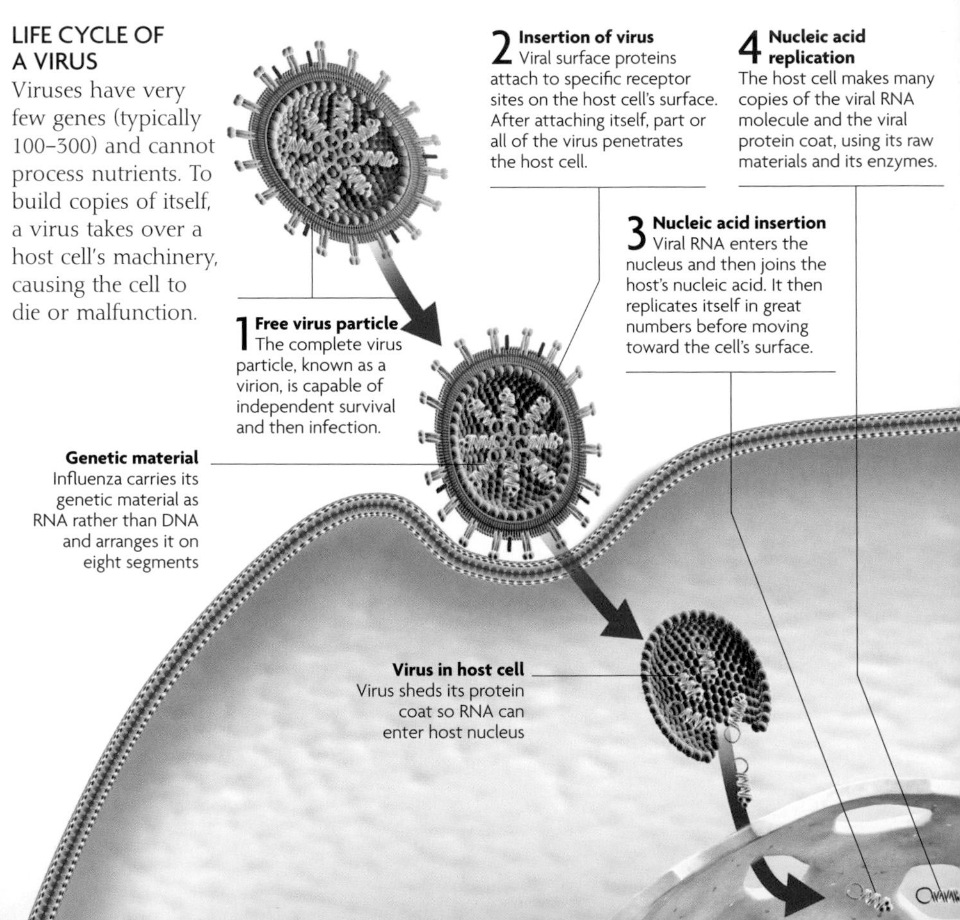
1 Free virus particle The complete virus particle, known as a virion, is capable of independent survival and then infection.
2 Insertion of virus Viral surface proteins attach to specific receptor sites on the host cell's surface. After attaching itself, part or all of the virus penetrates the host cell.
3 Nucleic acid insertion Viral RNA enters the nucleus and then joins the host's nucleic acid. It then replicates itself in great numbers before moving toward the cell's surface.
4 Nucleic acid replication The host cell makes many copies of the viral RNA molecule and the viral protein coat, using its raw materials and its enzymes.
Genetic material Influenza carries its genetic material as RNA rather than DNA and arranges it on eight segments
Virus in host cell Virus sheds its protein coat so RNA can enter host nucleus

VIRUS SHAPES

There are thousands of different types of virus, with various shapes, such as balls, boxes, polygons, sausages, golfballs, spirals, and even tiny "space rockets." Viruses are classified by their size, shape, and symmetry as well as by the disease or groups of diseases they cause. Some, such as the complex virus known as the T4 bacteriophage, attack human pathogenic bacteria.

Spiral (helical)
The protein coat is corkscrewlike, with the genetic material entwined. Examples include myxoviruses and paramyxoviruses.

Protein subunit (capsomer)
Genetic material

Icosahedral
Twenty equal-sided triangles connect to form a faceted container. Examples include adenoviruses and herpes viruses.

Surface protein (antigen)
Triangular face

Complex
Complex viruses resemble a tiny rocket with "landing legs" that settle onto the surface of the host cell. They only attack bacteria.

Head contains genetic material
Helical tail
Leg

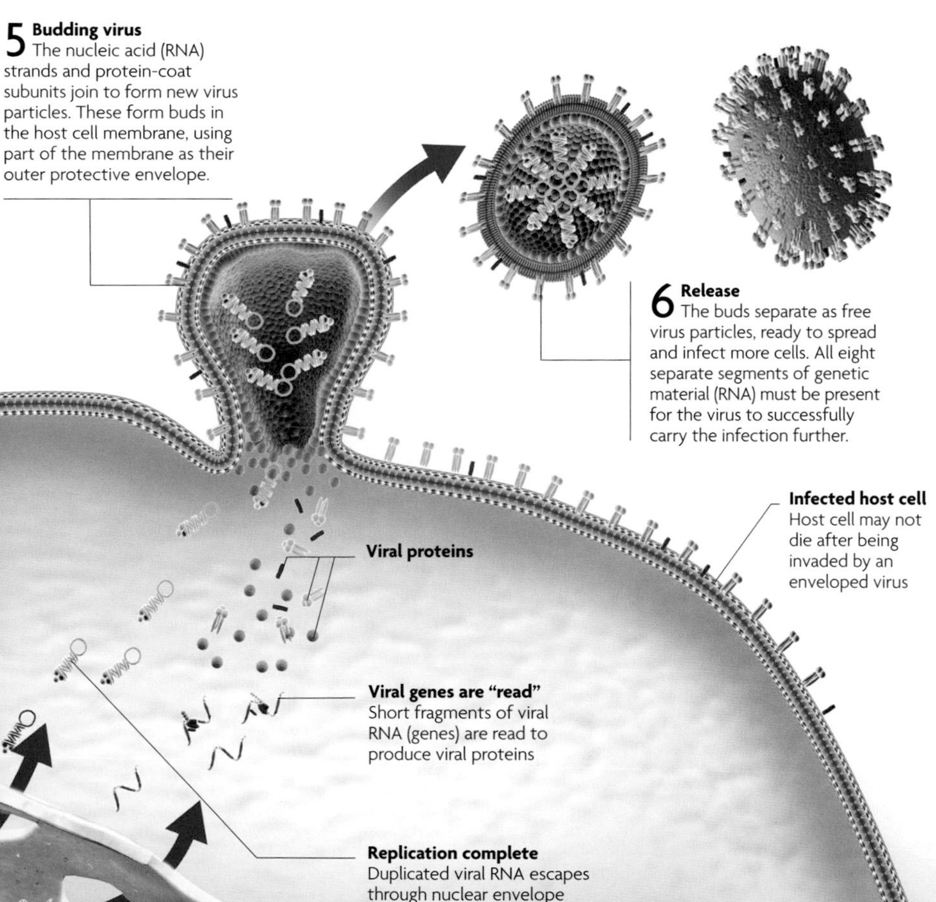
5 Budding virus
The nucleic acid (RNA) strands and protein-coat subunits join to form new virus particles. These form buds in the host cell membrane, using part of the membrane as their outer protective envelope.
6 Release
The buds separate as free virus particles, ready to spread and infect more cells. All eight separate segments of genetic material (RNA) must be present for the virus to successfully carry the infection further.
Infected host cell
Host cell may not die after being invaded by an enveloped virus
Viral proteins
Viral genes are "read"
Short fragments of viral RNA (genes) are read to produce viral proteins
Replication complete
Duplicated viral RNA escapes through nuclear envelope

BACTERIA

The microorganisms known as bacteria are present almost everywhere–in soil, water, air, food, drink, and on and in our own bodies. Many types of bacteria are harmless; indeed, those present naturally in the human intestines, the "gut flora," have a beneficial effect in helping extract nutrients from food. However, hundreds of types of bacteria can cause infections, ranging from mild to lethal. Bacteria are simpler than other single-cell organisms in that their genetic material (DNA) is free in the cell, rather than contained in a membrane-bound nucleus.

STRUCTURE OF A BACTERIUM

A typical rod-shaped bacterium (bacillus) has a cell membrane enclosing cytoplasm and organelles, such as ribosomes, which are distributed in it. Unlike animal cells, it has a semi-rigid cell wall outside its cell membrane.

BACTERIAL SHAPES

There are several typical shapes for bacteria, and these, along with the way they are colored by laboratory stains, are important for classification and determination of their origins and relationships. Many thousands of bacterial types are known, with more discovered each year.

Cocci
Generally spherical. Examples include *Staphylococcus* and *Streptococcus*.

Dividing cocci bacteria

Bacilli
Oval, or rodlike, with or without surface hairs or flagella. Examples include *Streptobacillus*.

Pili (hairs) on surface

Spirilla
Spiral or, more accurately, helical (corkscrewlike) in shape. Examples include *Leptospira* and *Treponema*.

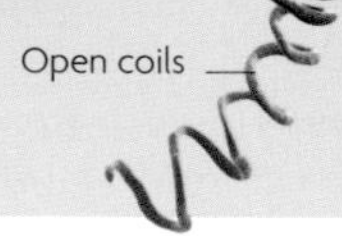
Open coils

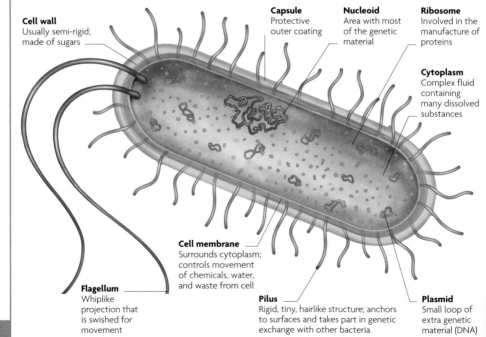

Cell wall
Usually semi-rigid; made of sugars

Capsule
Protective outer coating

Nucleoid
Area with most of the genetic material

Ribosome
Involved in the manufacture of proteins

Cytoplasm
Complex fluid containing many dissolved substances

Cell membrane
Surrounds cytoplasm; controls movement of chemicals, water, and waste from cell

Flagellum
Whiplike projection that is swished for movement

Pilus
Rigid, tiny, hairlike structure; anchors to surfaces and takes part in genetic exchange with other bacteria

Plasmid
Small loop of extra genetic material (DNA)

HOW BACTERIA CAUSE DAMAGE

Disease-causing bacteria can enter the body in several ways: via the airways or digestive tract, during sexual contact, or through broken skin. Some bacteria adhere to and invade body cells. Others produce poisonous substances called bacteriotoxins, or toxins, which may disrupt cell function. For example, the diphtheria toxin from the bacterium *Corynebacterium diphtheriae* damages heart muscle by inhibiting protein production. Some toxins are highly dangerous. A bucket of nerve toxin from *Clostridium botulinum* could kill everyone in the world.

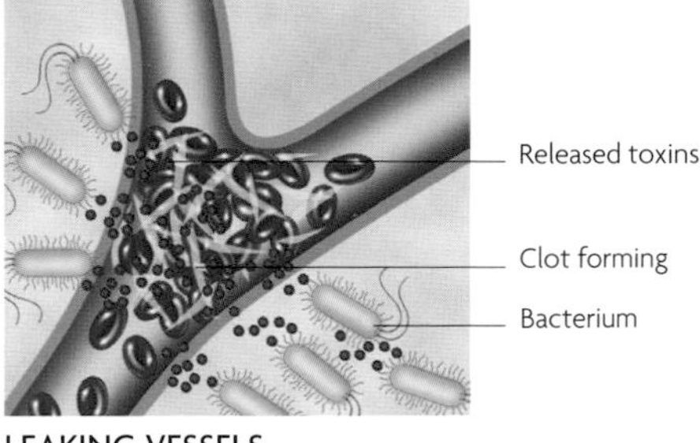
Released toxins
Clot forming
Bacterium

LEAKING VESSELS
Some bacteria release toxins that cause blood to clot in small blood vessels, depriving tissues and organs of their normal blood supply.

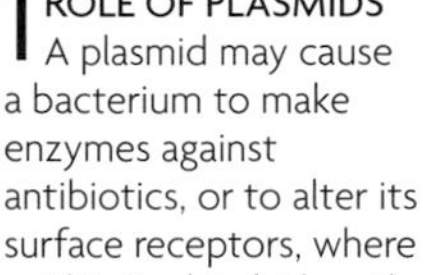

1 ROLE OF PLASMIDS
A plasmid may cause a bacterium to make enzymes against antibiotics, or to alter its surface receptors, where antibiotics bind. Then the plasmid duplicates itself.

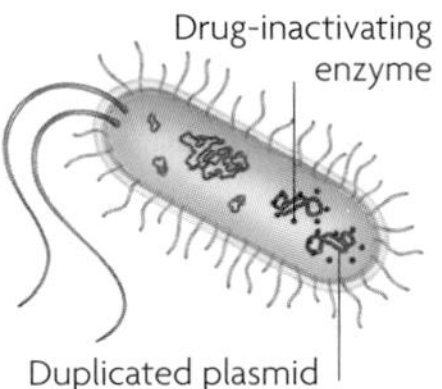
Drug-inactivating enzyme
Duplicated plasmid

RESISTANCE TO ANTIBIOTICS

Many bacteria become resistant to antibiotics by changing (mutating) into new strains. Their most effective mechanism is the transfer of plasmids–fragments of the genetic material DNA –between bacteria. The gene for antibiotic resistance crops up by accident, and the bacterium possessing it can pass it to others.

2 PLASMID TRANSFER
Plasmid transfer takes place during a process known as conjugation. The plasmid copy is passed from the donor, through a pilus, to the recipient bacterium.

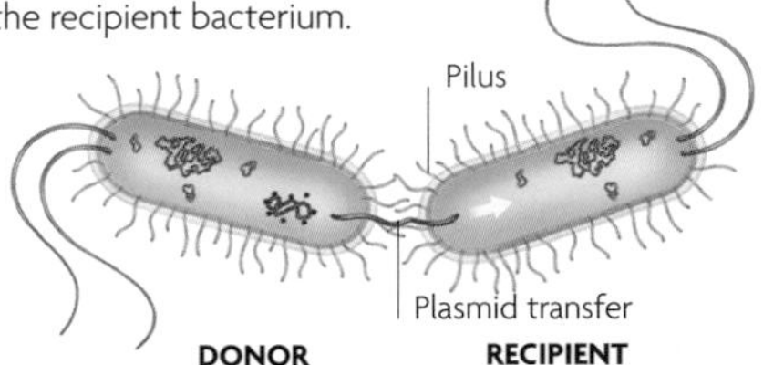
Pilus
Plasmid transfer
DONOR
RECIPIENT

3 DRUG-RESISTANT STRAINS
Recipient bacteria inherit the resistant gene. Plasmid transfer produces populations of bacteria resistant to a range of antibiotics.

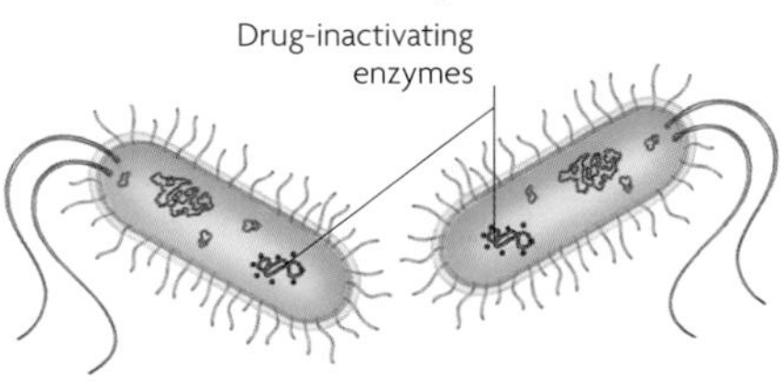
Drug-inactivating enzymes

SUPERBUGS

Some bacteria pass through their life cycle in less than 20 minutes. Fast reproduction, coupled with the incredible numbers of bacteria and rapid transfer of genetic information, gives great scope for mutation (see above). Many strains of bacteria resistant to wide-acting, or broad-spectrum, antibiotics have appeared. These so-called "superbugs" may not be resistant to more specialized, narrow-spectrum, antibiotics.

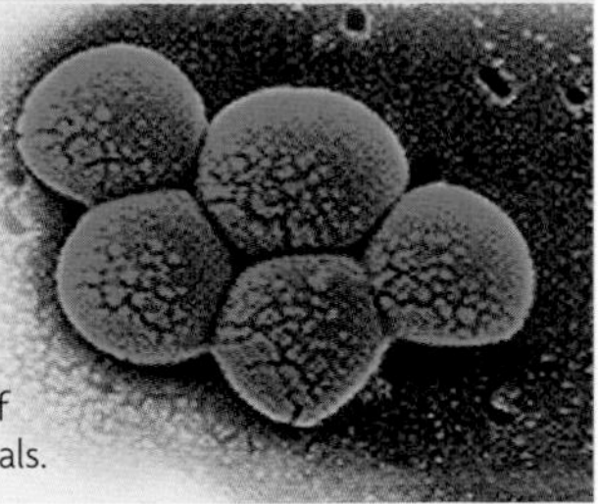

MRSA
Staphylococcus aureus bacteria that are resistant to the antibiotic methicillin are known as MRSA and are a cause of concern in hospitals.

PROTISTS (PROTOZOA)

Protists are single-celled organisms with genetic material contained in a nucleus. Animal-like protists, sometimes called protozoa, are common and usually harmless, although some, such as *Plasmodium*, cause serious diseases.

MALARIAL LIFE CYCLE

Four types of *Plasmodium* cause malaria. They are spread by the female *Anopheles* mosquito. Malaria produces chills and high fever, which can recur and prove fatal if not treated. Most *Plasmodium* have a similar life cycle (see below).

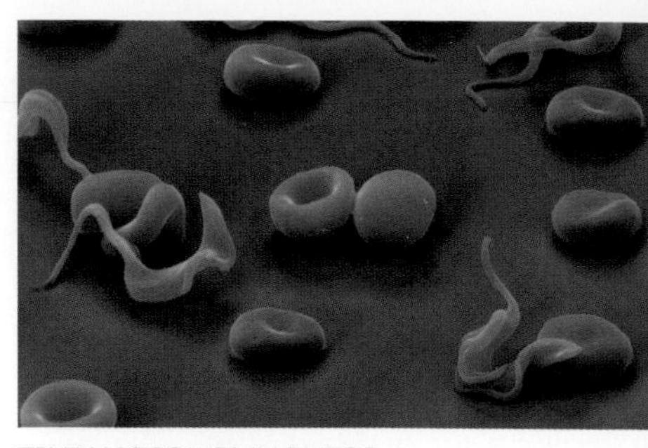

TRYPANOSOMES IN BLOOD
Trypanosomes are wormlike protists (purple), seen here with red blood cells. They cause a disease called trypanosomiasis, or sleeping sickness.

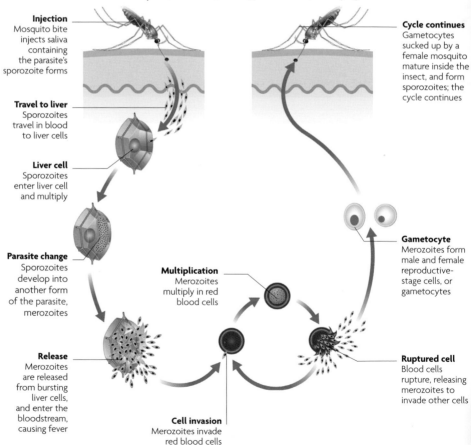

Injection
Mosquito bite injects saliva containing the parasite's sporozoite forms

Travel to liver
Sporozoites travel in blood to liver cells

Liver cell
Sporozoites enter liver cell and multiply

Parasite change
Sporozoites develop into another form of the parasite, merozoites

Release
Merozoites are released from bursting liver cells, and enter the bloodstream, causing fever

Cell invasion
Merozoites invade red blood cells

Multiplication
Merozoites multiply in red blood cells

Ruptured cell
Blood cells rupture, releasing merozoites to invade other cells

Gametocyte
Merozoites form male and female reproductive-stage cells, or gametocytes

Cycle continues
Gametocytes sucked up by a female mosquito mature inside the insect, and form sporozoites; the cycle continues

FUNGI

Disease-causing fungi fall into two main groups: filamentous fungi and single-celled ones. Some types–for example, yeast–cause fairly harmless diseases of the skin, hair, nails, or mucous membranes. Others, such as histoplasmosis, result in potentially fatal infections of vital organs such as the lungs. Some infections may be linked to specific occupations such as farming, while others–ringworm (dermatophytosis), for example–are more likely to affect people with damaged immune systems, such as those with HIV–AIDS.

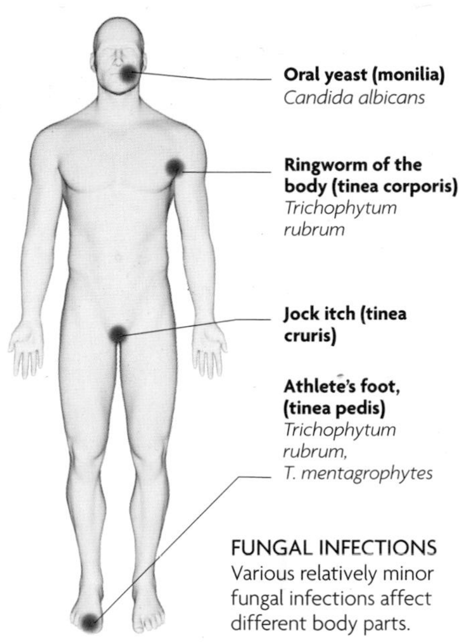
Oral yeast (monilia)
Candida albicans
Ringworm of the body (tinea corporis)
Trichophytum rubrum
Jock itch (tinea cruris)
Athlete's foot, (tinea pedis)
Trichophytum rubrum,
T. mentagrophytes

FUNGAL INFECTIONS
Various relatively minor fungal infections affect different body parts.

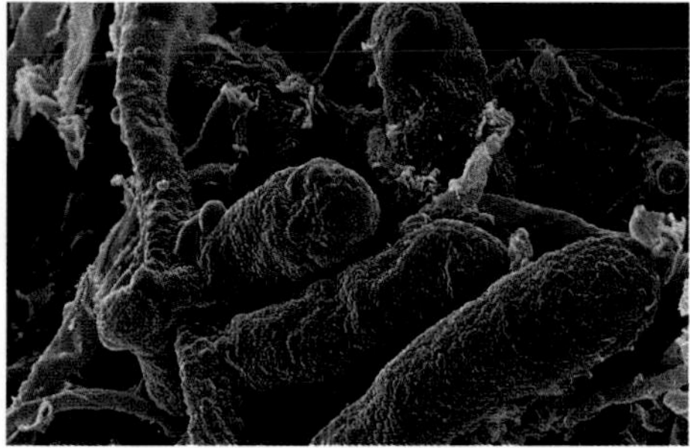

CAUSE OF ATHLETE'S FOOT
Seen here are microscopic threads of the fungus *Epidermophyton floccosum*, which causes the white, itchy skin of athlete's foot.

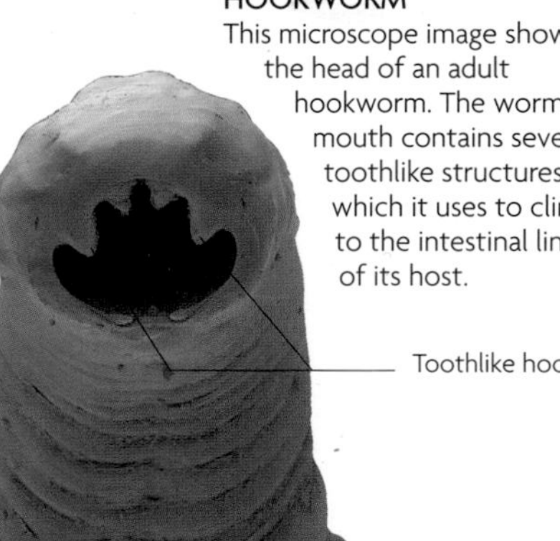
Toothlike hooks

HOOKWORM
This microscope image shows the head of an adult hookworm. The worm's mouth contains several toothlike structures, which it uses to cling to the intestinal lining of its host.

PARASITIC WORMS

Humans, like most other animals, can be infested with parasitic worms that derive all their nutrients from their hosts. At least 20 types of wormlike animals may live in the body as parasites. Most spend at least part of their life cycle in the intestines. A few are members of the annelids, a group of segmented worms that includes common earthworms. Several are roundworms, or nematodes–for example, the hookworm *Ancylostoma duodenale*, which is $^4/_5$ in (1 cm) long and lives in the gut. Another worm-like group is the flatworms; it includes tapeworms, such as *Taenia*, which live in the gut and may reach 30 ft (9 m) in length, and flukes, such as *Schistosoma*, which causes schistosomiasis, or snail fever.

IMMUNE SYSTEM DISORDERS

THE IMMUNE SYSTEM SOMETIMES OVER-REACTS, CAUSING AN ALLERGIC RESPONSE. WHEN THE SYSTEM IS WEAK, IMMUNIZATION CAN HELP BOOST IT. HOWEVER, IT MAY BECOME SO WEAK—BY AN HIV INFECTION, FOR EXAMPLE—THAT EVEN ORDINARY INFECTIONS CAN BE DANGEROUS.

ALLERGIES

When first exposed to an allergen, such as nuts or pollen, the immune system makes antibodies to fight it. The antibodies coat the surface of mast cells in the skin, stomach lining, lungs, and upper airways. If the allergen enters the body again, these cells mount an allergic response.

1 EXPOSURE The first time an allergen enters the body, antibodies bind to the surface of mast cells. These cells contain histamine, which normally causes inflammation.

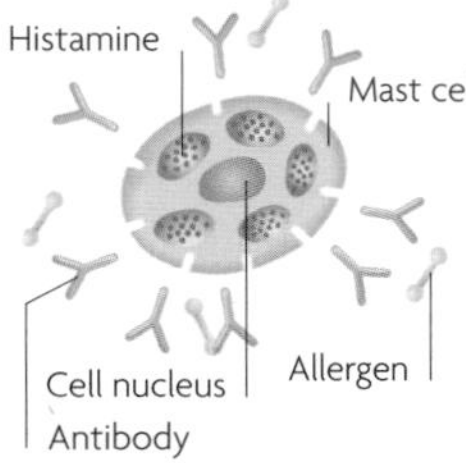
Histamine
Mast cell
Cell nucleus
Allergen
Antibody

2 ANTIBODIES When the allergen returns, the antibodies are triggered into action. If the allergen links two or more antibodies together, the cell bursts.

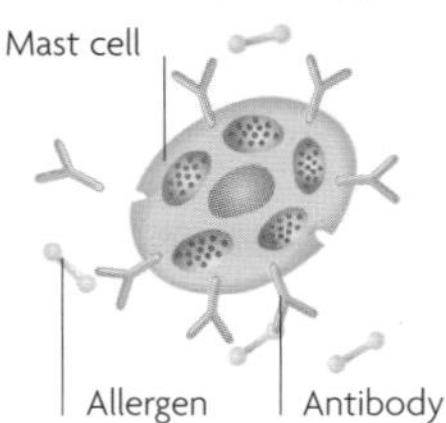
Mast cell
Allergen
Antibody

3 HISTAMINE As the cell bursts, it releases histamine, which causes an inflammatory response. This irritates body tissues and produces all the symptoms of an allergy.

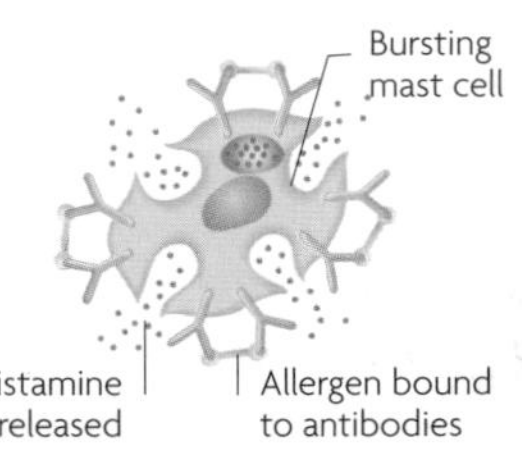
Bursting mast cell
Histamine released
Allergen bound to antibodies

IMMUNIZATION

The process of becoming resistant or immune to a particular microbe as a result of infection is known as natural immunization. Resistance can also be developed artificially. In active immunization, dead or weakened versions of the microbe or its toxic products are injected into the body. The immune response occurs, with the production of antibodies, but the illness does not develop. If urgent protection is needed, or if an immune system is weak, passive immunization can be used by injecting ready-made antibodies. These antibodies provide swift resistance against the microbes, but they gradually degenerate and are not replaced. The body has no memory for making them again.

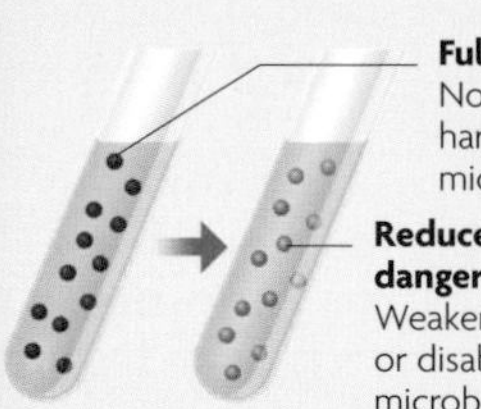
Full strength Normal, harmful microbes
Reduced danger Weakened or disabled microbes

1 VACCINE PRODUCTION A vaccine contains complete or partial microbes, or the toxins they make. It can stimulate the immune response but not cause symptoms.

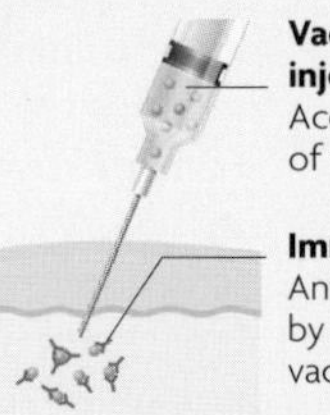
Vaccine injected Accurate dose of vaccine
Immune action Antibodies made by body attack vaccine antigens

2 VACCINE DELIVERY Vaccination stimulates the body's immune system to raise antibodies against the antigens on the disease-carrying organisms.

Trachea (windpipe)

Pathogen attack Invading microbes in mucus droplet in trachea

Invasion Some pathogens invade tissues

Instant response Antibodies launch defense against pathogens

3 IMMUNE RESPONSE A pathogen against which the body has been vaccinated alerts the memory cells, and so the immune system launches an instant defense.

HIV INFECTION

HIV is carried in blood, semen, saliva, vaginal secretions, and breast milk. It is passed on when infected fluids enter the body. HIV infects cells with structures called CD4 molecules on their surface. These CD4+ cells include lymphocytes, which fight infection. The virus multiplies rapidly in CD4+ cells, destroying them in the process. If HIV goes untreated, the number of CD4+ lymphocytes eventually falls so low that the immune system is severely weakened.

AIDS

The onset of AIDS is signaled by the development of disorders called AIDS-defining illnesses. Some of these illnesses are opportunistic infections, caused by organisms that are harmless to healthy people but dangerous to those with reduced immunity; one example is infection by *Candida albicans*, which causes yeast. People who have AIDS may also develop various types of cancer, notably Kaposi's sarcoma.

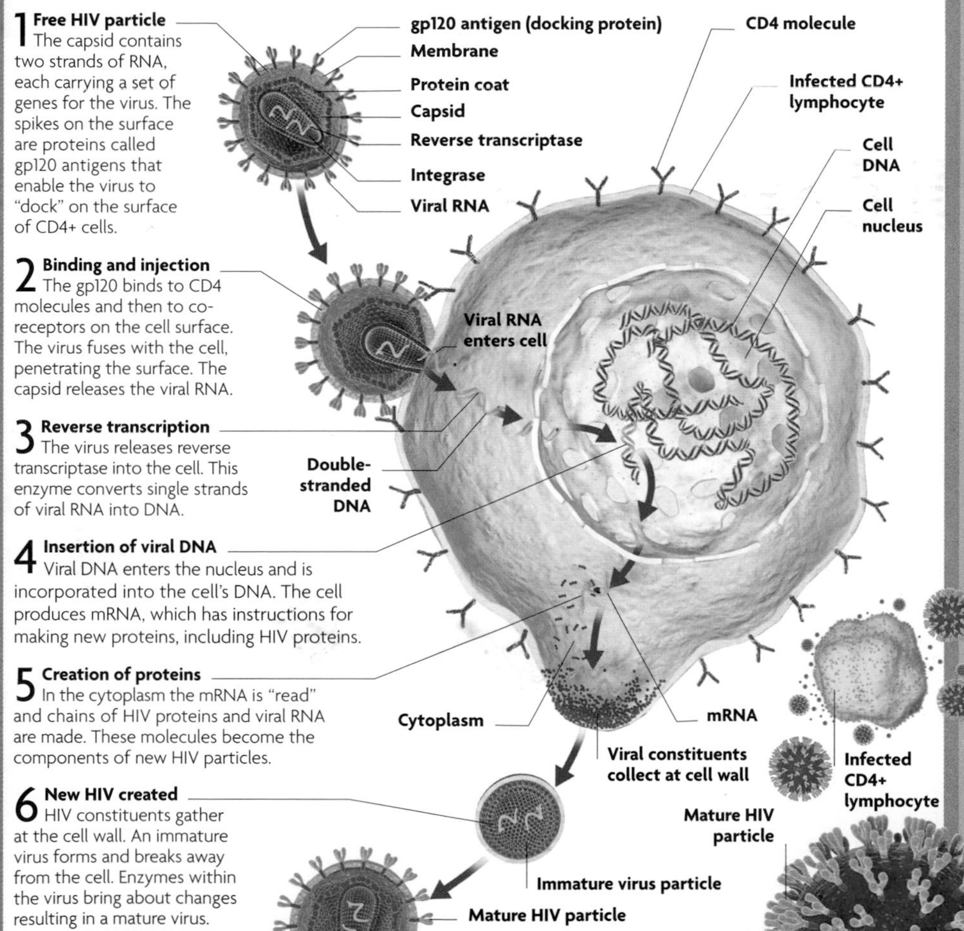
gp120 antigen (docking protein)
Membrane
Protein coat
Capsid
Reverse transcriptase
Integrase
Viral RNA
CD4 molecule
Infected CD4+ lymphocyte
Cell DNA
Cell nucleus
Viral RNA enters cell
Double-stranded DNA
Cytoplasm
mRNA
Viral constituents collect at cell wall
Infected CD4+ lymphocyte
Mature HIV particle
Immature virus particle
Mature HIV particle

1 Free HIV particle The capsid contains two strands of RNA, each carrying a set of genes for the virus. The spikes on the surface are proteins called gp120 antigens that enable the virus to "dock" on the surface of CD4+ cells.

2 Binding and injection The gp120 binds to CD4 molecules and then to co-receptors on the cell surface. The virus fuses with the cell, penetrating the surface. The capsid releases the viral RNA.

3 Reverse transcription The virus releases reverse transcriptase into the cell. This enzyme converts single strands of viral RNA into DNA.

4 Insertion of viral DNA Viral DNA enters the nucleus and is incorporated into the cell's DNA. The cell produces mRNA, which has instructions for making new proteins, including HIV proteins.

5 Creation of proteins In the cytoplasm the mRNA is "read" and chains of HIV proteins and viral RNA are made. These molecules become the components of new HIV particles.

6 New HIV created HIV constituents gather at the cell wall. An immature virus forms and breaks away from the cell. Enzymes within the virus bring about changes resulting in a mature virus.

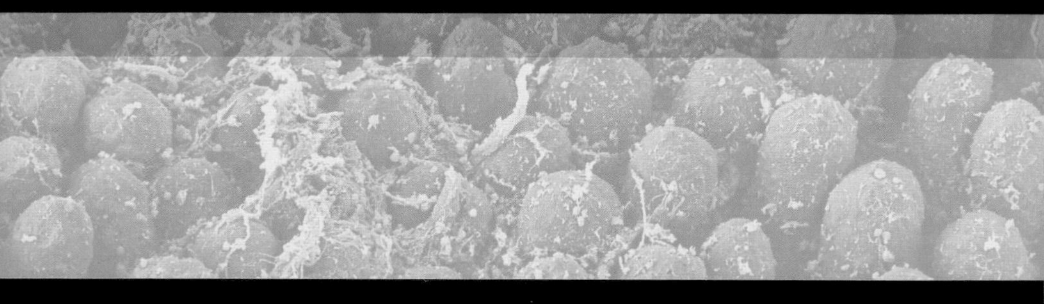

PEOPLE ARE PROBABLY MORE AWARE OF THEIR DIGESTIVE SYSTEM THAN OF ANY OTHER SYSTEM BECAUSE OF ITS FREQUENT MESSAGES. HUNGER, THIRST, GAS, AND BOWEL MOVEMENTS ALL AFFECT DAILY LIFE. EATING WELL AND REGULAR EXERCISE ARE THE BEDROCKS OF GOOD DIGESTIVE HEALTH. PLENTY OF FRESH VEGETABLES AND FRUIT, ADEQUATE FIBER, AND A LOW INTAKE OF ANIMAL FATS AND SALT ARE SIMPLE GUIDELINES FOR MAINTAINING THE WELL-BEING NOT JUST OF THE DIGESTIVE SYSTEM BUT OF THE WHOLE BODY.

DIGESTIVE SYSTEM

DIGESTIVE ANATOMY

THE DIGESTIVE TRACT AND ITS ASSOCIATED ORGANS, INCLUDING THE LIVER, GALLBLADDER, AND PANCREAS, BREAK DOWN FOOD, EXTRACT NUTRIENTS, AND DISPOSE OF WASTE MATERIALS.

After being eaten, or ingested, food embarks on a journey. It can take up to 24 hours to cover a distance of 30 ft (9 m), through various muscular tubes and chambers. The process begins at the mouth, where food is crushed and ground down by the teeth during chewing. The resulting ball, or bolus, of food continues down the throat (pharynx), then travels through the food tube (esophagus) to the stomach, small and large intestine, and anus. In the small intestine, chemicals break down food into molecules small enough to absorb into the blood. What cannot be digested is compacted as feces in the large intestine and eliminated through the anus. Food travels through the system by a process of muscular contraction called peristalsis. The digestive system includes several glands: the spit-making salivary glands; the pancreas, which produces powerful digestive juices; and the body's major nutrient processor, the liver.

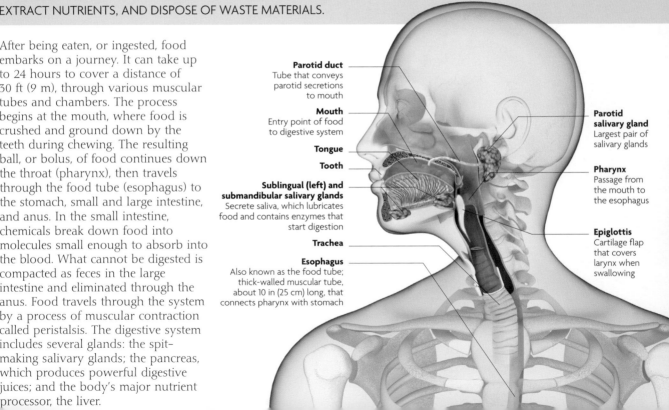

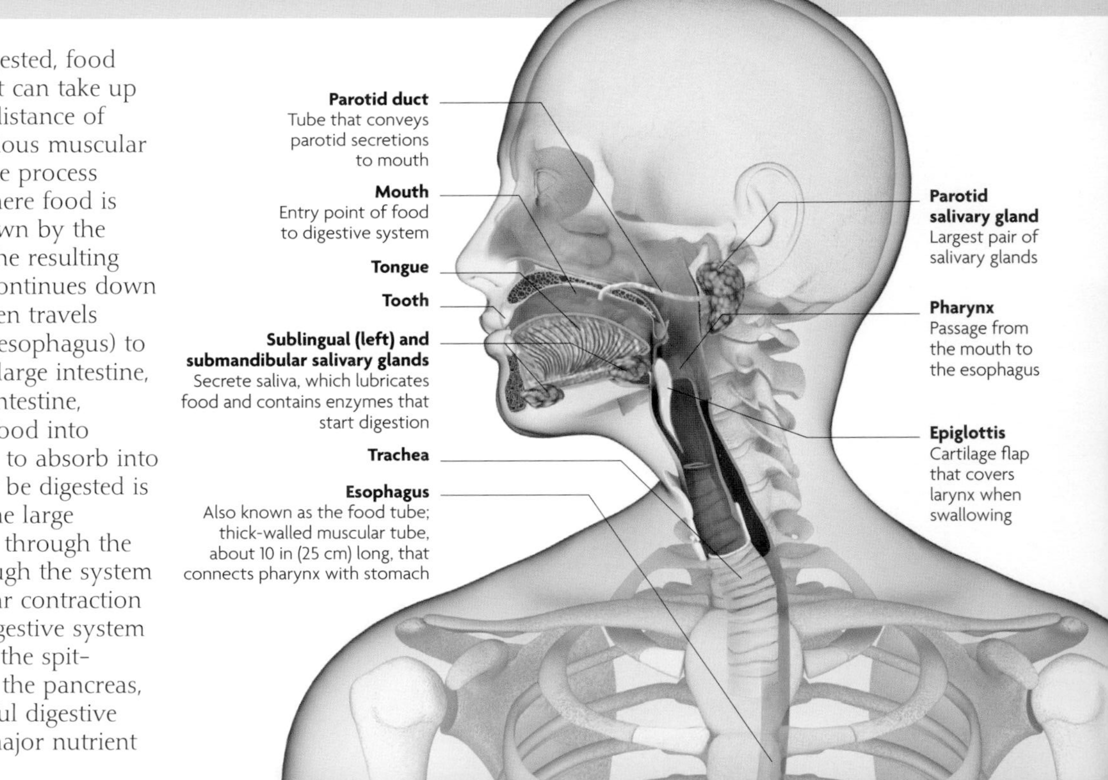
Parotid duct
Tube that conveys parotid secretions to mouth
Mouth
Entry point of food to digestive system
Tongue
Tooth
Sublingual (left) and submandibular salivary glands
Secrete saliva, which lubricates food and contains enzymes that start digestion
Trachea
Esophagus
Also known as the food tube; thick-walled muscular tube, about 10 in (25 cm) long, that connects pharynx with stomach
Parotid salivary gland
Largest pair of salivary glands
Pharynx
Passage from the mouth to the esophagus
Epiglottis
Cartilage flap that covers larynx when swallowing

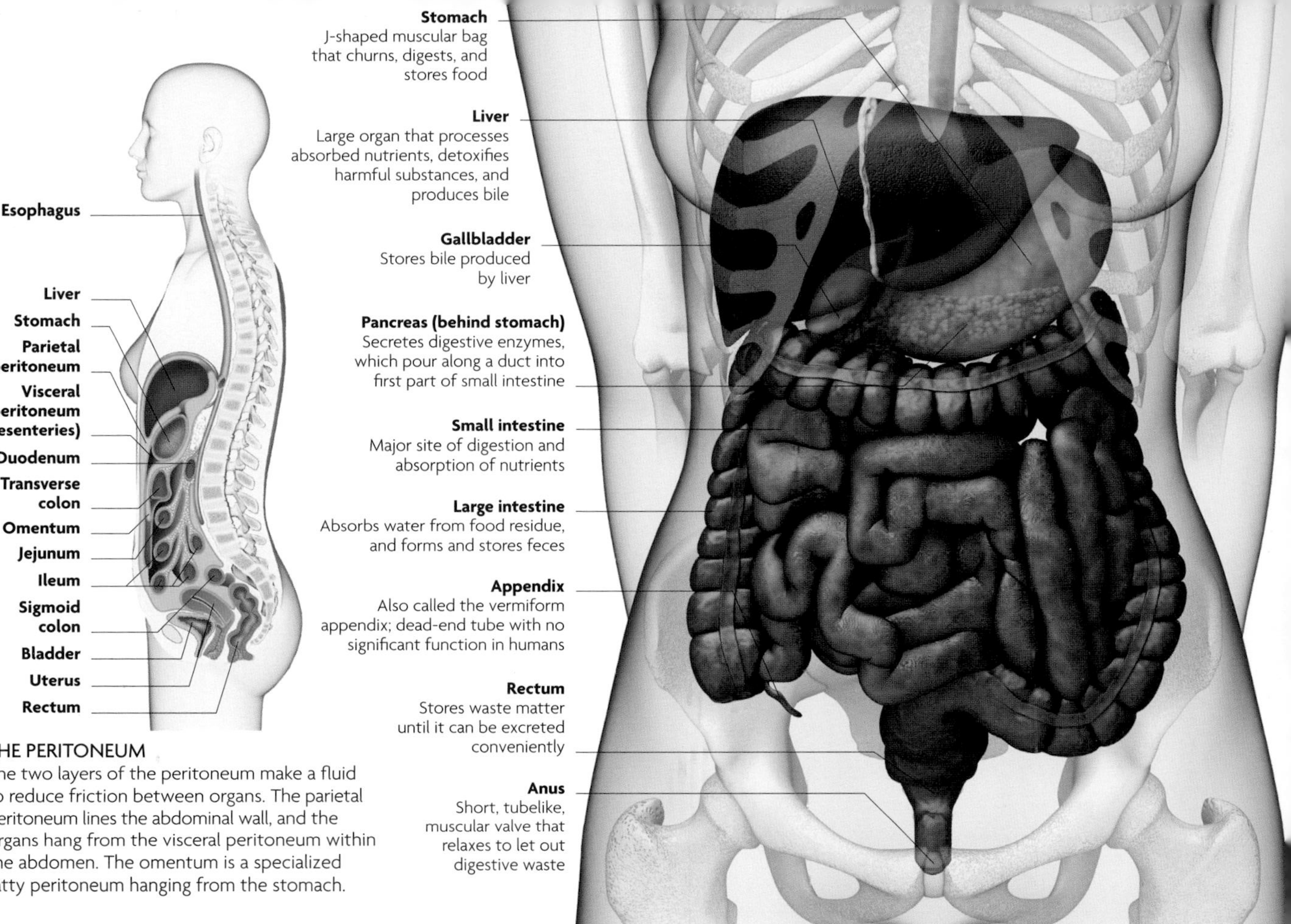
Esophagus
Liver
Stomach
Parietal peritoneum
Visceral peritoneum (mesenteries)
Duodenum
Transverse colon
Omentum
Jejunum
Ileum
Sigmoid colon
Bladder
Uterus
Rectum
Stomach
J-shaped muscular bag that churns, digests, and stores food
Liver
Large organ that processes absorbed nutrients, detoxifies harmful substances, and produces bile
Gallbladder
Stores bile produced by liver
Pancreas (behind stomach)
Secretes digestive enzymes, which pour along a duct into first part of small intestine
Small intestine
Major site of digestion and absorption of nutrients
Large intestine
Absorbs water from food residue, and forms and stores feces
Appendix
Also called the vermiform appendix; dead-end tube with no significant function in humans
Rectum
Stores waste matter until it can be excreted conveniently
Anus
Short, tubelike, muscular valve that relaxes to let out digestive waste

THE PERITONEUM

The two layers of the peritoneum make a fluid to reduce friction between organs. The parietal peritoneum lines the abdominal wall, and the organs hang from the visceral peritoneum within the abdomen. The omentum is a specialized fatty peritoneum hanging from the stomach.

MOUTH AND THROAT

DIGESTION STARTS AT THE MOUTH, WHERE FOOD IS CHEWED, LUBRICATED WITH SALIVA, TURNED INTO A SOFT, MOIST MASS CALLED A BOLUS, AND SWALLOWED.

ANATOMY OF THE MOUTH AND THROAT

The interior of the lips, cheeks, and oral cavity is lined with a tough, firmly anchored mucous membrane and a type of tissue called nonkeratinized squamous epithelium. Cells here multiply rapidly to replace those rubbed away when biting, chewing, and swallowing. The front underside of the tongue has a fleshy central ridge, the frenulum, which connects to the floor of the mouth. The tongue is the body's most flexible muscle. Within it are three pairs of intrinsic muscles; and outside, three pairs of extrinsic muscles run from the tongue to other parts of the throat and neck. The root of the tongue anchors to the lower jaw (mandible) and to the curved hyoid bone in the neck. The rear of the mouth leads to the middle part of the throat, the oropharynx. The whole throat or pharynx, from its nasal to laryngeal regions, is about 5 in (13 cm) long in a typical adult.

NOSE, MOUTH, AND THROAT

The roof of the mouth, or oral cavity, is formed by shelves of the maxillary and palatine bones of the skull (see p.49), together known as the hard palate. This extends rearward as the soft palate, which contains skeletal muscle fibers that allow it to flex when swallowing. The central posterior part of the soft palate extends into a small "finger," the uvula, which can be seen through the open mouth, dangling down from the back, where it helps direct food downward.

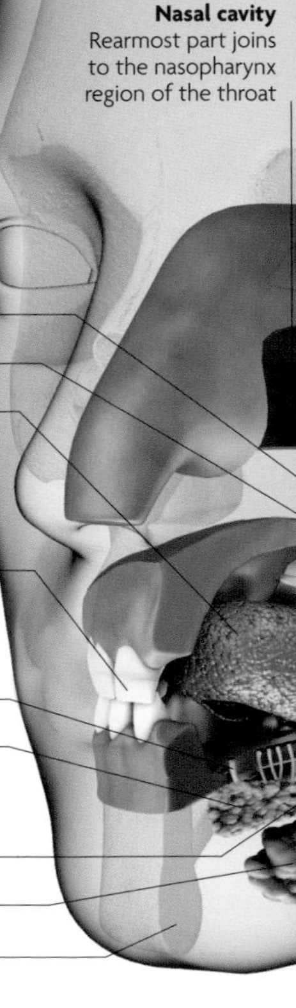
Nasal cavity
Rearmost part joins to the nasopharynx region of the throat
Uvula
Soft palate
Tongue
Moves food around when chewing, contains taste buds, and helps form distinct words in speech
Teeth
Bite off and chew food into a moist, soft pulp, ready to be swallowed
Sublingual duct
Sublingual gland
Produces viscous saliva, which contains enzymes
Submandibular duct
Submandibular gland
Mandible (lower jawbone)
Epiglottis
Cartilaginous flap that blocks off the larynx entrance during swallowing

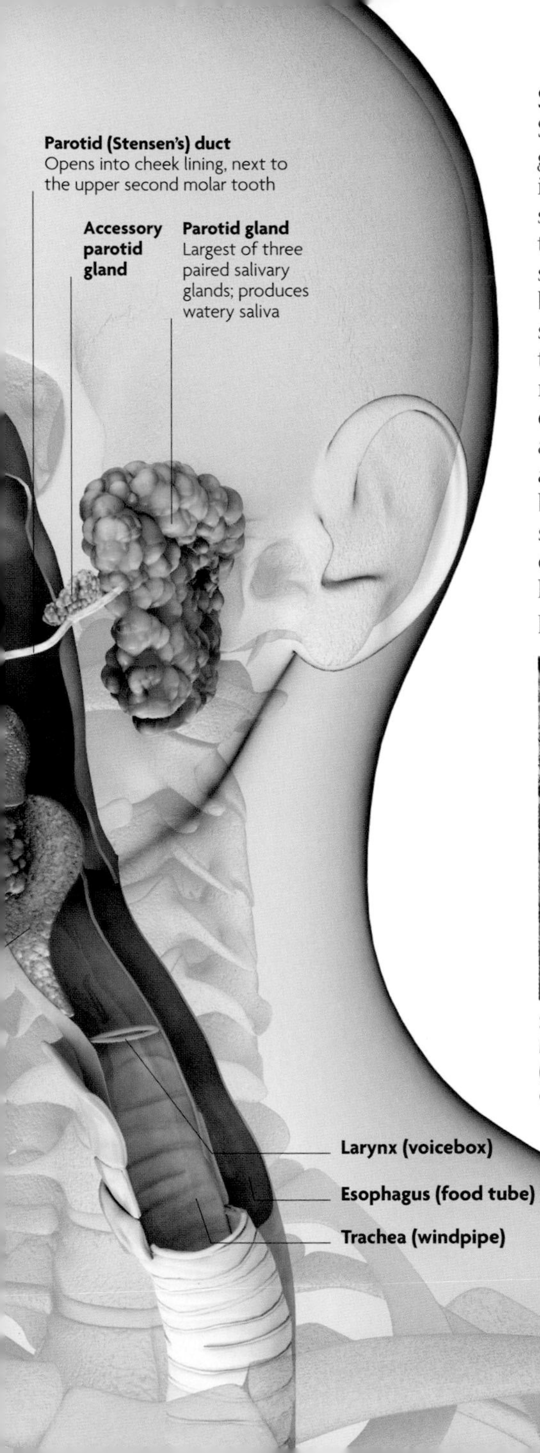

Parotid (Stensen's) duct
Opens into cheek lining, next to the upper second molar tooth
Accessory parotid gland
Parotid gland
Largest of three paired salivary glands; produces watery saliva
Larynx (voicebox)
Esophagus (food tube)
Trachea (windpipe)

SALIVARY GLANDS

Saliva is made by three pairs of salivary glands: the parotid glands, positioned in front of and just below each ear; the submandibulars, on the inner sides of the lower jawbone (mandible); and the sublinguals, in the floor of the mouth, below the tongue. In addition, many small accessory glands are found in the mucous membranes lining the mouth and tongue. Although it is composed of 99.5 percent water, saliva also contains important solutes such as amylase, a digestive enzyme that begins the breakdown of starches, and salts. Saliva lubricates food to make chewing and swallowing easier, and it keeps the mouth moist between periods of eating.

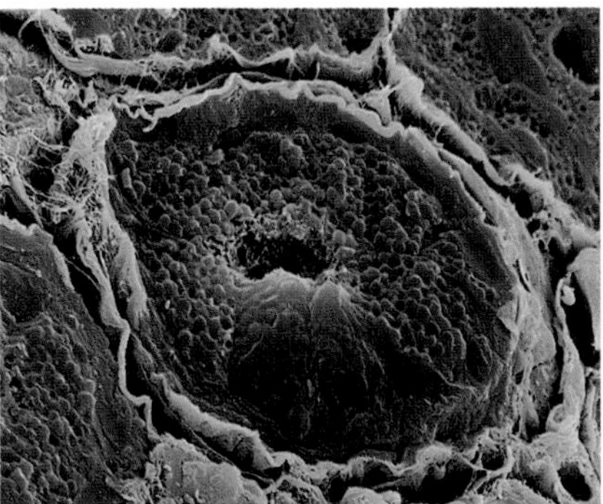

SALIVARY GLAND STRUCTURE
Many small, rounded glandular units called acini (brown), separated by connective tissue (pink), discharge their saliva into tiny central ducts. Acinar ducts converge into the main saliva-carrying glandular ducts.

TEETH

There are four types of tooth, each of which has a different role. The incisors, at the front, are chisel-shaped, with sharp edges for cutting food. The pointed canines, known as "eye teeth," are designed for tearing food. The premolars, with their two ridges, and the flatter molars, which are the largest and strongest teeth, crush and grind food. The crown is the part of the tooth above the gum, while the root is embedded in the jawbone; and where these two meet, at the gum or gingival surface, is the neck of the tooth. The outer layer of the crown is made of a bonelike enamel, which is the hardest substance in the body. Beneath it is a layer of softer but still strong tissue called dentine, which is shock-absorbing. At the center of the tooth, the soft dental pulp contains blood vessels and nerves. Below the gum, bonelike cementum and periodontal ligament tissues secure the tooth in the jawbone.

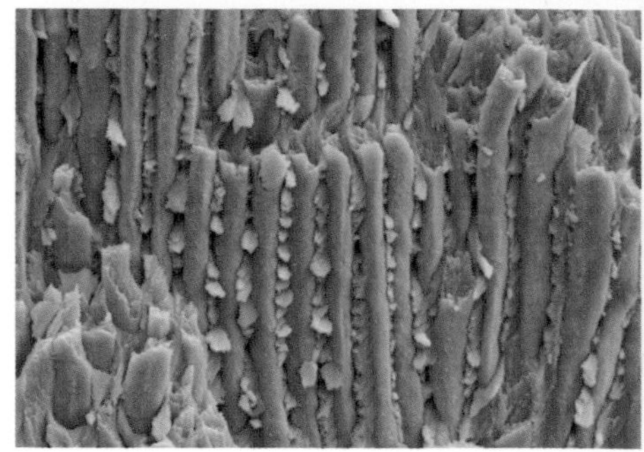

ENAMEL SURFACE
This microscope image of enamel shows U-shaped enamel prisms packed with the crystalline mineral substance hydroxyapatite.

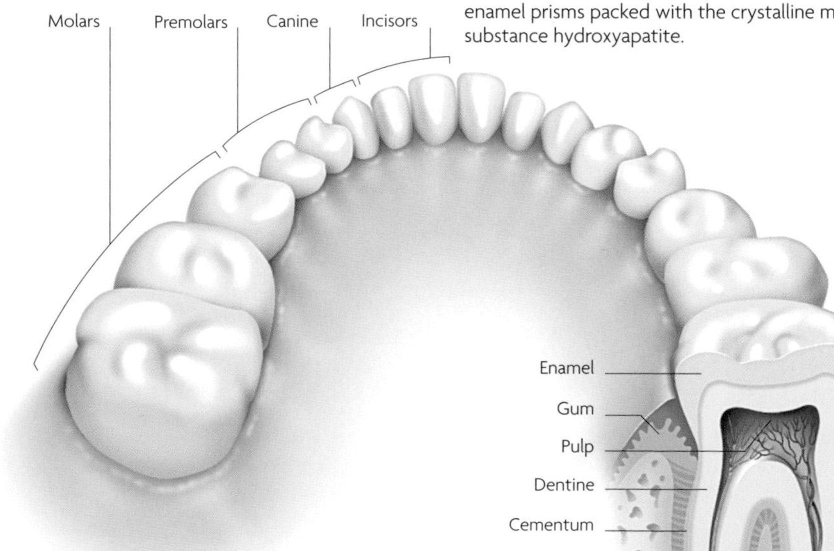

Molars
Premolars
Canine
Incisors

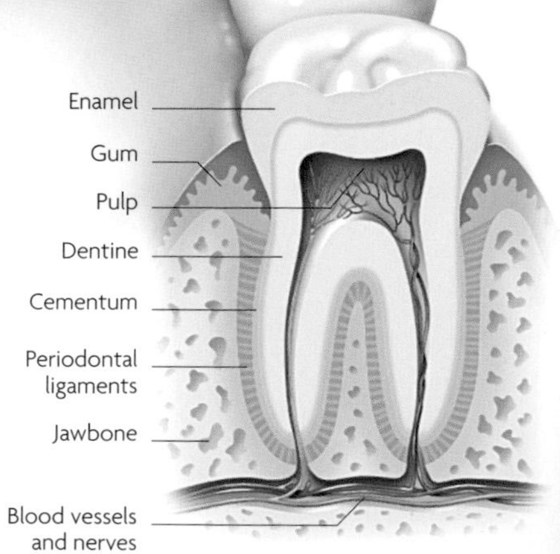

Enamel
Gum
Pulp
Dentine
Cementum
Periodontal ligaments
Jawbone
Blood vessels and nerves

ADULT DENTITION
An adult typically has, in each side of each jaw, two incisors, one canine, two premolars, and three molars. This makes a total of 32 teeth. However, in some individuals certain teeth never grow or erupt out of the gum. This applies particularly to the four rearmost molars, known as "wisdom teeth."

SWALLOWING

The process of swallowing begins as a voluntary action, when the rear of the tongue pushes a bolus of food to the back of the mouth. To swallow a solid item such as a tablet without chewing demands concentration. It is easier to swallow a tablet with water, because drinks are usually gulped down right after entering the mouth. Automatic reflexes control subsequent stages of swallowing, as the muscles of the throat contract and move the bolus rearward and down, and squeeze it into the top of the esophagus. A flap of cartilage known as the epiglottis prevents food from going down "the wrong way" into the larynx and the trachea, where it would cause choking.

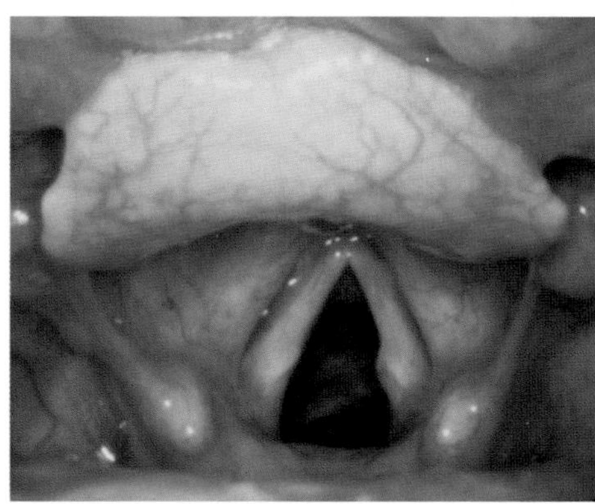

VIEW INTO THE LARYNX
The pale, leaflike flap of the epiglottis is visible at the top of this image. Immediately below it is the inverted "V" of the vocal cords.

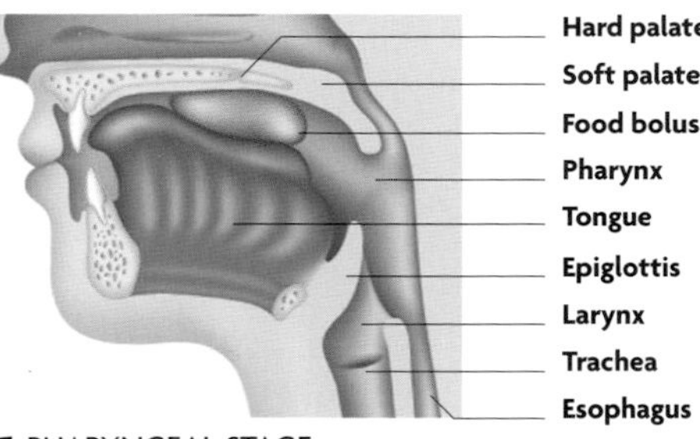
Hard palate
Soft palate
Food bolus
Pharynx
Tongue
Epiglottis
Larynx
Trachea
Esophagus

1 PHARYNGEAL STAGE
Before the food bolus reaches the back of the mouth, the epiglottis is raised in its normal position.

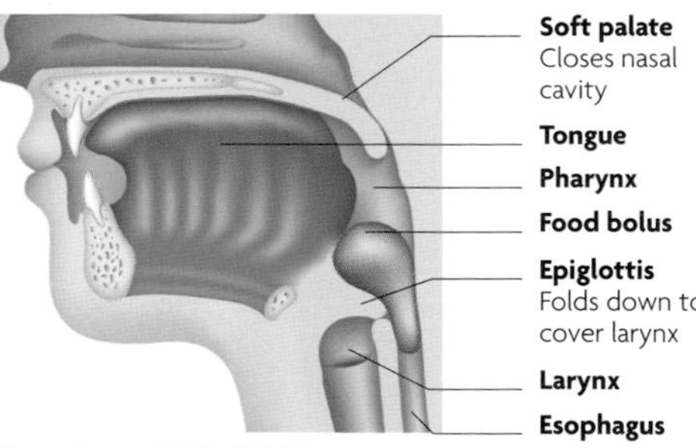
Soft palate
Closes nasal cavity
Tongue
Pharynx
Food bolus
Epiglottis
Folds down to cover larynx
Larynx
Esophagus

2 ESOPHAGEAL STAGE
The larynx rises to meet the tilted epiglottis, closing the trachea. The soft palate lifts to close the nasal cavity. The bolus is pushed down the esophagus.

BREATHE OR SWALLOW

The pharynx is a dual-purpose passageway: for air when breathing, and food, drink, and saliva when swallowing. Nerve signals from the brain operate the muscles of the mouth, tongue, pharynx, larynx, and upper esophagus to prevent food from entering the trachea. If food is inhaled, irritation of the airway triggers the coughing reflex, which expels the inhaled particles and prevents choking. The complex muscle movements involved in swallowing are a voluntary reflex, and they also occur when solid matter contacts touch sensors at the back of the mouth.

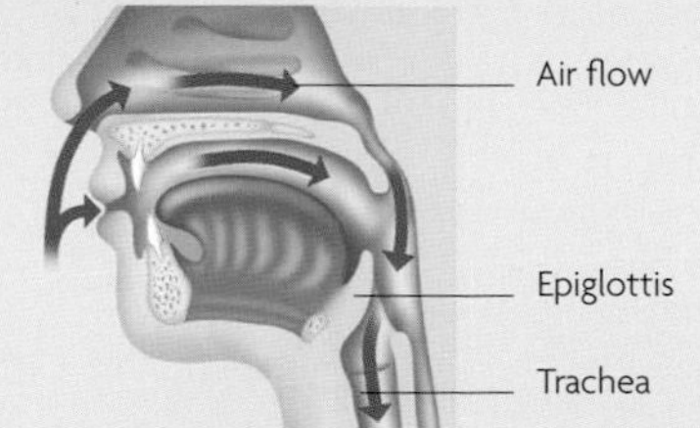
Air flow
Epiglottis
Trachea

DUAL INTAKE
Breathing occurs through the nose or the mouth. The passageways of both meet at the throat, and air flows into the trachea.

STOMACH AND SMALL INTESTINE

THE STOMACH DIGESTS FOOD CHEMICALLY AND PHYSICALLY. THE SMALL INTESTINE CONTINUES THE CHEMICAL BREAKDOWN AND ABSORBS NUTRIENTS.

STOMACH STRUCTURE

The stomach is a muscular-walled, J-shaped sac in which food is stored, churned, and mixed with gastric juices secreted by its lining. The juices include digestive enzymes and hydrochloric acid, which breaks down food and kills potentially harmful microbes. The smooth muscles of the wall contract to combine and squeeze the semiliquid mix of food and gastric juices.

Lymph nodule
Gastric pit
Mucosa
Submucosa
Muscularis (three muscle layers)
Subserous layer
Serosa (outer coating)
Gastric gland
Mucous cell
Parietal cell
Zygomenic cell
Lipase-secreting cell
Enteroendocrine cell

Muscle layers
Longitudinal
Circular
Oblique
The three muscle layers cause the stomach to twist and writhe into almost any shape

LAYERS OF THE STOMACH WALL

The stomach wall has four main layers: the serosa, muscularis, submucosa, and mucosa. The mucosa has deep infolds (gastric pits) that contain the gastric glands, acid-producing parietal cells, zygomenic cells (secreting pepsinogen), and lipase-secreting cells. The enteroendocrine cells secrete the hormone gastrin.

PERISTALSIS

Waves of muscle contraction propel food through the digestive tract. The circular muscle contracts and relaxes in sequence, producing a "traveling wave" known as peristalsis.

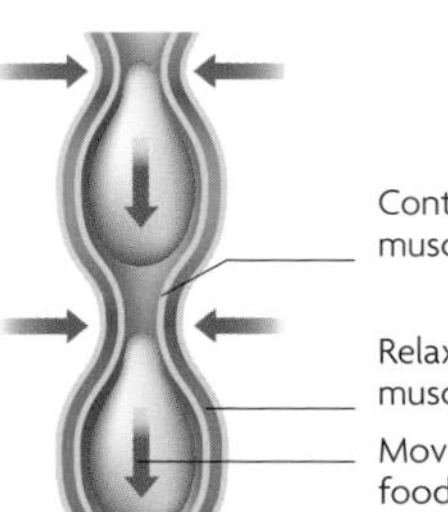
Contracting muscle
Relaxing muscle
Moving food

Duodenum
First and shortest section of the small intestine, about 10 in (25 cm) long

ROLE OF THE SMALL INTESTINE

The duodenum, jejunum, and ileum make up the small intestine. The duodenum receives not only processed food, known as chyme, from the stomach but also digestive secretions from the liver (including bile) and the pancreas. The jejunum and ileum are both long and coiled, but the jejunum is thicker, redder, and slightly shorter. In the small intestine, the chyme is broken down further by pancreatic juices, bile, and the intestine's own secretions, so that nutrients can be absorbed into the blood and lymph circulations.

Serous layer
Clear membrane that covers the stomach exernally

LAYERS OF THE SMALL INTESTINE WALL

The small intestine wall has four layers. The outermost serosa is a protective coat. Next is the muscularis, which has outer longitudinal muscle fibers and inner circular smooth fibers. Inside this is the submucosa, a loose layer carrying vessels and nerves. The innermost mucosa forms ringlike folds, called plicae circulares, that are covered by tiny, fingerlike projections called villi.

Serosa

Muscularis

Submucosa

Mucosa

Villus
Fingerlike extension of mucosa up to 1/25 in (1 mm) long

SECTION OF SMALL INTESTINE

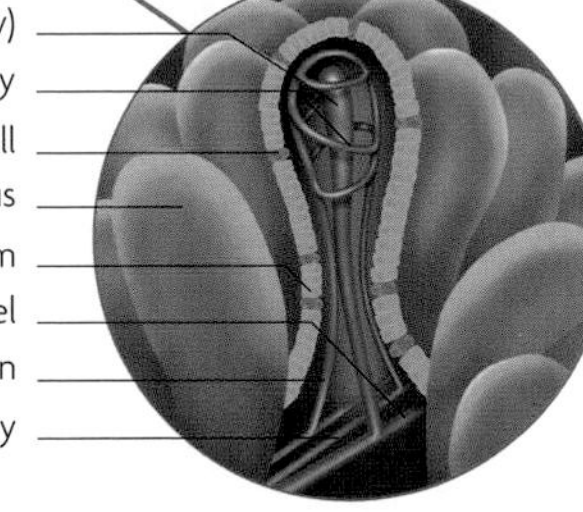
Lacteal (lymph capillary)
Blood capillary
Goblet cell
Villus
Epithelium
Lymph vessel
Vein
Artery

Jejunum
About 6½–8 ft (2–2.5 m) long

Ileum
Third and longest section of the small intestine, up to 11½ ft (3.5 m) in length

THE MIDDLE DIGESTIVE TRACT

The stomach is situated in the upper left abdomen, protected by the lower ribs. The extensive small intestine lies looped and folded beneath it, and occupies most of the lower abdomen.

INTESTINAL VILLI

The epithelium of each villus lets digested nutrients pass into the interior, or lumen. Here, some pass into the lymph via a lacteal, and others enter the blood via a capillary and are carried to the liver. Epithelial cells have tiny microvilli, which increase the surface area of the small intestinal lining. Goblet cells scattered throughout the epithelium secrete mucus that helps the passage of food.

LIVER, GALLBLADDER, AND PANCREAS

THE LIVER IS THE LARGEST INTERNAL ORGAN AND MAKES, PROCESSES, AND STORES MANY IMPORTANT CHEMICALS. THE GALLBLADDER STORES BILE, AND THE PANCREAS SECRETES VITAL DIGESTIVE ENZYMES.

STRUCTURE AND FUNCTION OF THE LIVER

Weighing about 3⅓ lb (1.5 kg), the liver is composed of lobules; these are made up of sheets of liver cells (hepatocytes), tiny branches of the hepatic artery and vein, and bile ducts. Nutrient-rich blood arrives from the intestines via the hepatic portal system (see p.222) and filters through the lobules. The liver has many functions, the most important of which are storing and releasing blood glucose for energy; sorting and processing vitamins and minerals; breaking down toxins into less harmful substances; and recycling old blood cells.

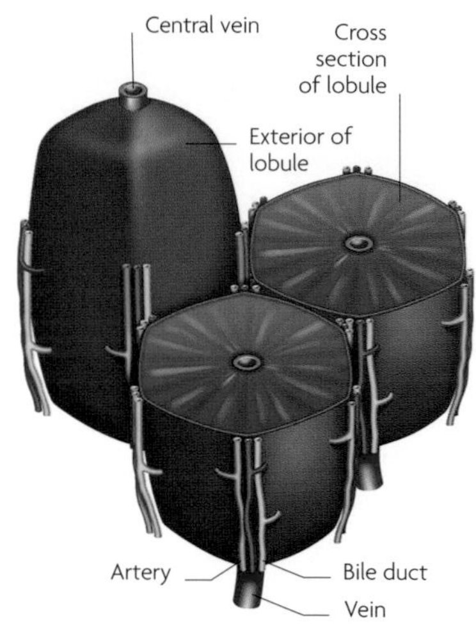
Central vein
Cross section of lobule
Exterior of lobule
Artery
Bile duct
Vein

LIVER LOBULES

The six-sided lobules of the liver nestle together, each supplied with a central vein and blood vessels and bile-collecting vessels around them.

LIVER FUNCTIONS

Most of the liver's tasks are concerned with the process of metabolism. They include breaking down, storing, and circulating vital substances, and constructing complex molecules, such as enzymes.

BILE PRODUCTION	Secretes bile into ducts that lead to the gallbladder.
NUTRIENT PROCESSING	Converts sugars into glycogen and makes amino acids.
GLUCOSE REGULATION	Maintains the level of glucose in the blood.
DETOXIFICATION	Removes harmful substances such as alcohol from the blood.
PROTEIN SYNTHESIS	Makes blood-clotting proteins and proteins for blood plasma.
MINERAL AND VITAMIN STORAGE	Stores iron, copper, and the fat-soluble vitamins.
BLOOD WASTE DISPOSAL	Eliminates bacteria and general foreign particles.
RECYCLING BLOOD CELLS	Breaks down red blood cells and reuses their constituents.

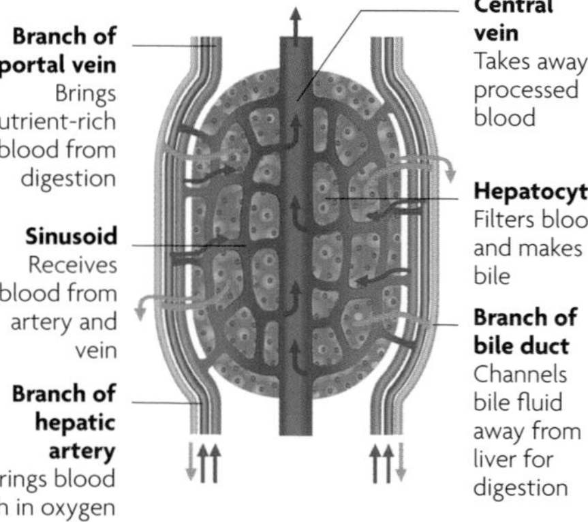
Branch of portal vein Brings nutrient-rich blood from digestion
Sinusoid Receives blood from artery and vein
Branch of hepatic artery Brings blood rich in oxygen
Central vein Takes away processed blood
Hepatocyte Filters blood and makes bile
Branch of bile duct Channels bile fluid away from liver for digestion

INSIDE A LOBULE

Hepatocytes filter the incoming, nutrient-rich blood into constituents that are destined for the bile ducts, storage, or waste disposal.

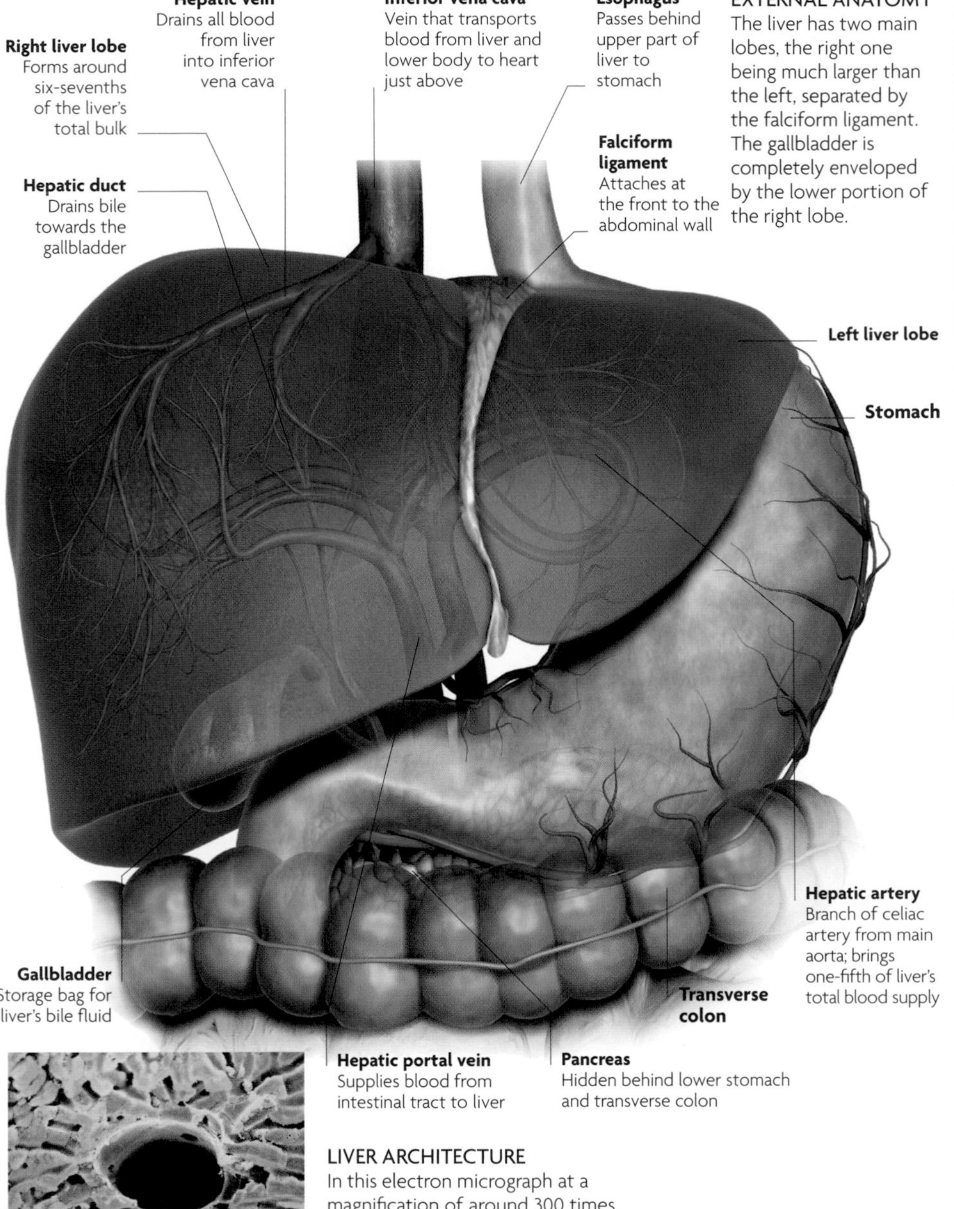
Right liver lobe
Forms around six-sevenths of the liver's total bulk
Hepatic vein
Drains all blood from liver into inferior vena cava
Inferior vena cava
Vein that transports blood from liver and lower body to heart just above
Esophagus
Passes behind upper part of liver to stomach
Falciform ligament
Attaches at the front to the abdominal wall
Hepatic duct
Drains bile towards the gallbladder
Left liver lobe
Stomach
Hepatic artery
Branch of celiac artery from main aorta; brings one-fifth of liver's total blood supply
Gallbladder
Storage bag for liver's bile fluid
Transverse colon
Hepatic portal vein
Supplies blood from intestinal tract to liver
Pancreas
Hidden behind lower stomach and transverse colon

EXTERNAL ANATOMY

The liver has two main lobes, the right one being much larger than the left, separated by the falciform ligament. The gallbladder is completely enveloped by the lower portion of the right lobe.

LIVER ARCHITECTURE

In this electron micrograph at a magnification of around 300 times, sheets of hepatocytes can be seen radiating from the central canal. This canal contains the central vein.

THE HEPATIC PORTAL CIRCULATION

The liver is unusual in that it receives two blood supplies. The hepatic artery delivers oxygen-rich blood to the liver. In addition, the hepatic portal vein supplies the liver with the oxygen-poor, nutrient-rich blood that comes from the digestive tract, before this blood returns to the heart and is pumped throughout the body. This hepatic portal circulation enables the liver to stop toxins absorbed in the intestines from reaching the rest of the body. It also helps regulate the levels of many other substances in the bloodstream. Veins from several organs, including the intestines, pancreas, stomach, and spleen, drain into the hepatic portal vein. It is around 3 in (8 cm) long and supplies up to four-fifths of the blood into the liver. The flow-rate increases after a meal, but falls during physical activity as blood is diverted from the abdominal organs to skeletal muscles.

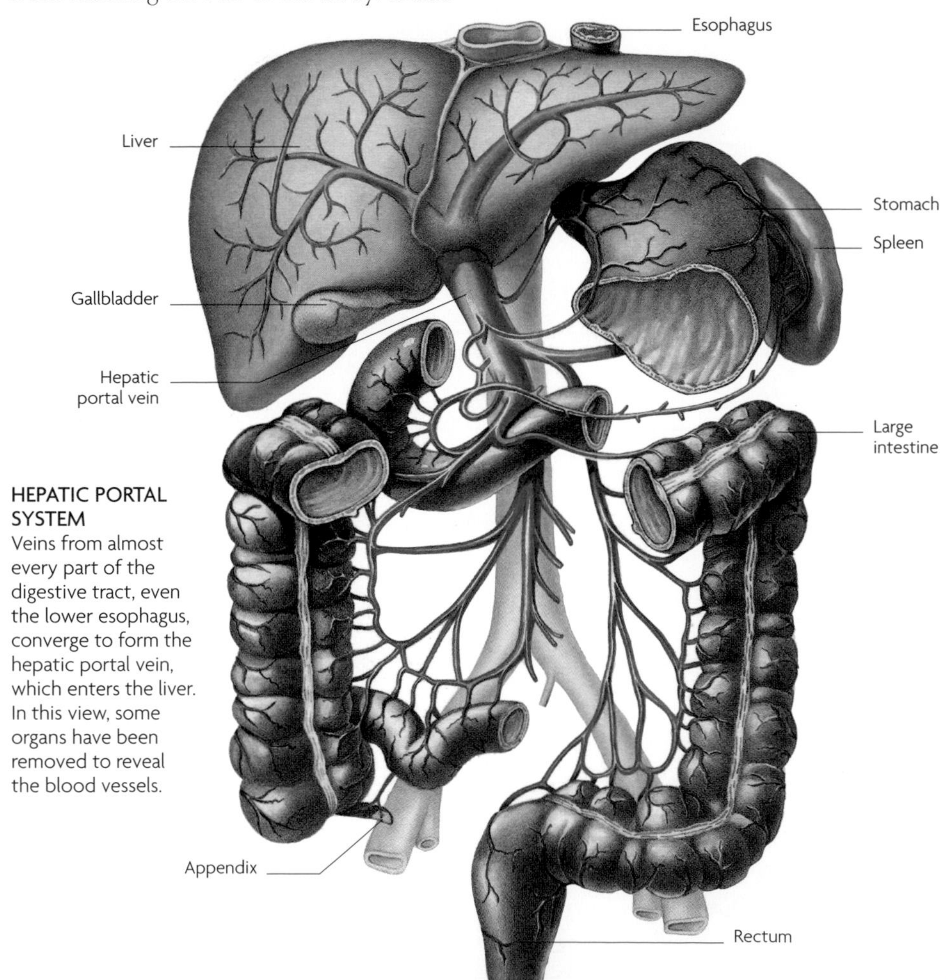
Esophagus
Liver
Stomach
Spleen
Gallbladder
Hepatic portal vein
Large intestine
Appendix
Rectum

HEPATIC PORTAL SYSTEM

Veins from almost every part of the digestive tract, even the lower esophagus, converge to form the hepatic portal vein, which enters the liver. In this view, some organs have been removed to reveal the blood vessels.

BILE TRANSPORT

Bile assists the breakdown of fats (lipids) in the small intestine. The liver secretes up to $1\frac{2}{3}$ pints (1 liter) of bile daily. The bile passes along the left and right hepatic ducts from the liver's two lobes, then along the common hepatic and cystic ducts to the gallbladder. This sac holds around $1\frac{2}{3}$ fl oz (50 ml) of bile and concentrates it, ready for release after a meal. The bile flows along the cystic duct to enter the first part of the small intestine, the duodenum.

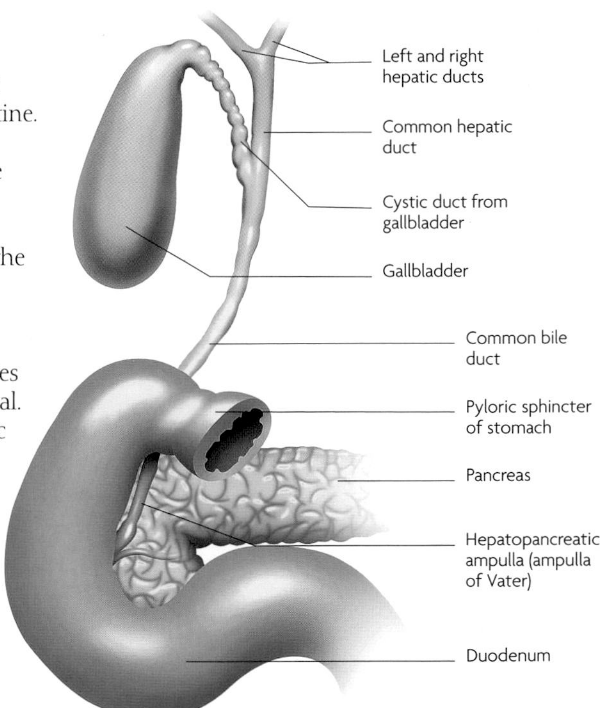
Left and right hepatic ducts
Common hepatic duct
Cystic duct from gallbladder
Gallbladder
Common bile duct
Pyloric sphincter of stomach
Pancreas
Hepatopancreatic ampulla (ampulla of Vater)
Duodenum

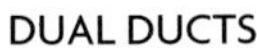

DUAL DUCTS

The common bile duct joins the pancreatic duct at the hepatopancreatic ampulla, which empties into the duodenum.

THE PANCREAS

The head end of this gland nestles in a loop of the duodenum, its main body lies behind the stomach, and its tapering tail sits above the left kidney, below the spleen. Each day, the pancreas produces around $2\frac{2}{3}$ pints (1.5 liters) of digestive juice containing various enzymes that break down proteins, carbohydrates, and lipids. The fluid flows into the main and accessory pancreatic ducts, which empty the juices into the duodenum.

Pancreatic duct

Head of pancreas

Body of pancreas

Tail of pancreas

PANCREATIC STRUCTURE

The pancreas is up to 6 in (15 cm) long, soft and flexible, and gray-pink in color.

LARGE INTESTINE

THE LARGE INTESTINE HAS THREE MAIN REGIONS—THE CECUM, COLON, AND RECTUM. THE 5 FT (1.5 M)-LONG COLON CHANGES LIQUID DIGESTIVE WASTE PRODUCTS INTO A MORE SOLID FORM THAT IS EXCRETED AS FECES.

ROLE OF THE COLON

Once the chemical breakdown of food in the small intestine (see pp.218–19) is complete, almost all the nutrients vital for bodily functions will have been absorbed. The waste product from this process is partially digested, liquefied food (chyme). This passes from the small intestine, through the ileocecal valve, into the cecum. From there, it reaches the first part of the colon, the ascending colon.

The colon's main function is to convert the liquid chyme into semisolid feces for storage and disposal. Sodium, chloride, and water are absorbed through the lining of the colon into blood and lymph, and the feces become less watery. The colon secretes bicarbonate and potassium in exchange for sodium and chloride. There are also billions of symbiotic or "friendly" microorganisms within the colon.

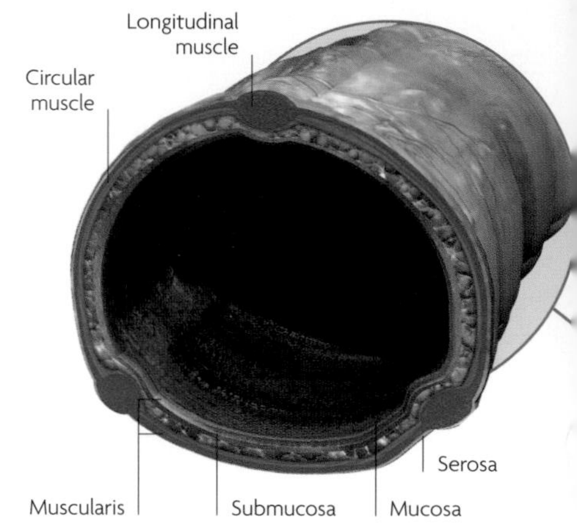
Longitudinal muscle
Circular muscle
Serosa
Muscularis
Submucosa
Mucosa

LAYERS OF THE COLON WALL
Inside the outer coating (serosa), smooth muscle fibers are responsible for colonic movements. The submucosa has many lymphoid nodules, and the mucosa produces lubricating mucus.

GUT FLORA

Billions of microorganisms, mainly bacteria, live in the intestinal tract—chiefly in the large intestine. These "gut flora" are normally harmless, provided they do not spread to other parts of the body and are kept in balance. They produce enzymes that break down certain food components, such as cellulose, which human enzymes cannot digest. The gut flora also produce vitamins K and B, and hydrogen, carbon dioxide, hydrogen sulfide, and methane. They help control harmful microbes in the digestive system and promote the formation of antibodies and the activity of lymphoid tissue in the colonic lining. Overall, the gut flora and the body exist in a mutually beneficial partnership (symbiosis). At least one-third of the weight of excreted feces is composed of these bacteria.

BACTERIA IN THE COLON
This electron microscope image (magnified over 2,000 times) shows clusters of rodlike bacteria (purple) on the lining of the colon.

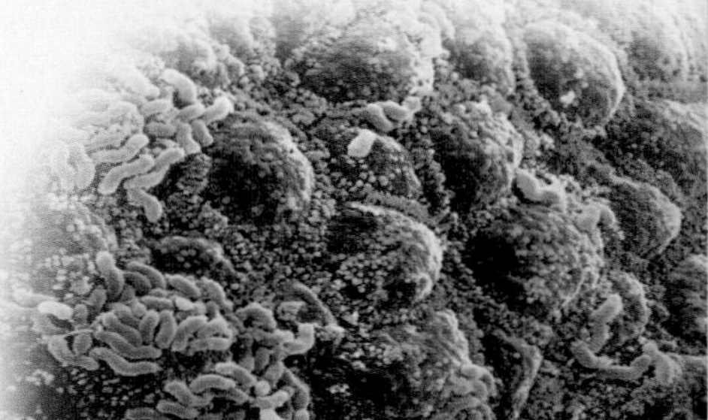

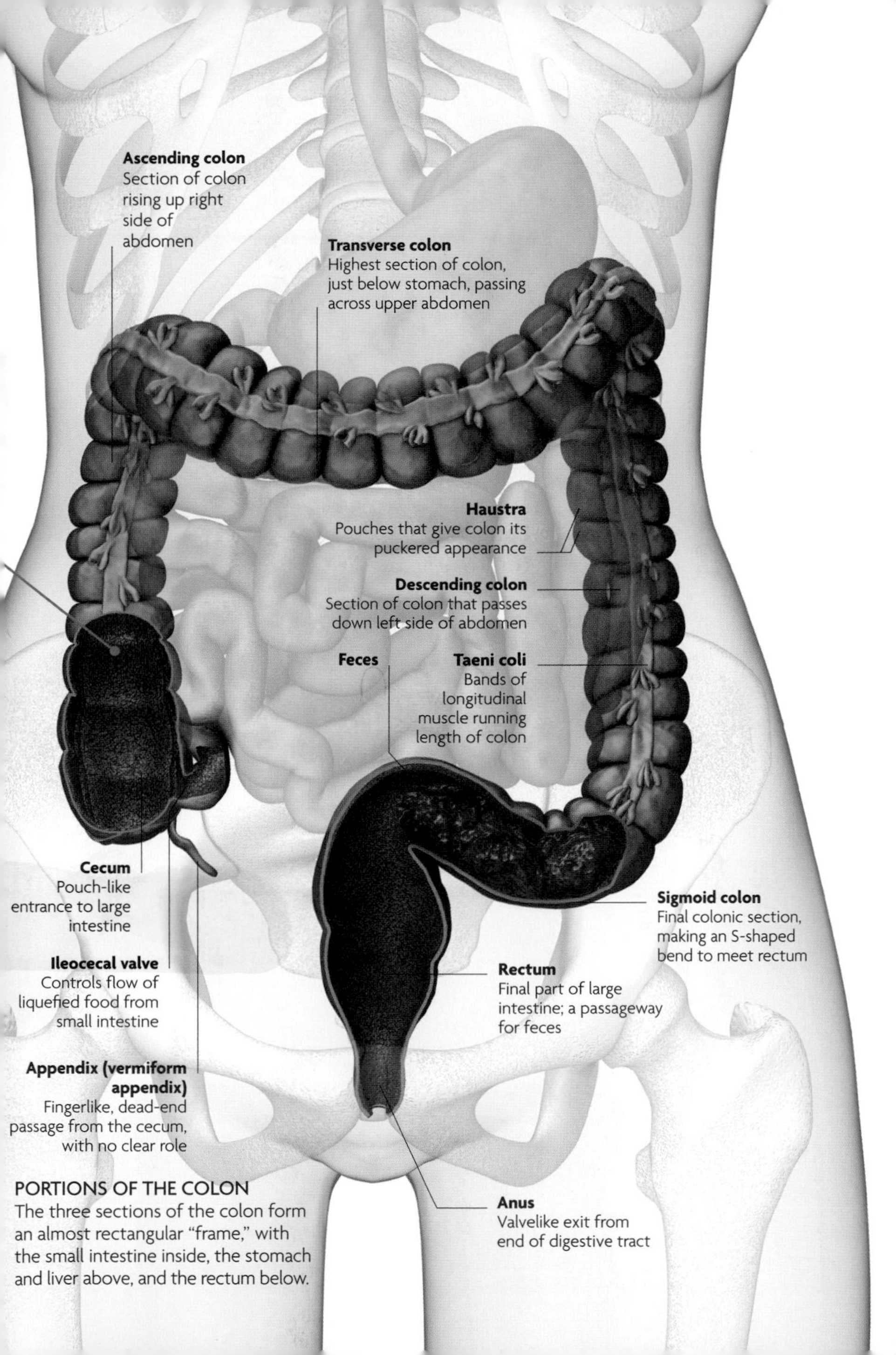

Ascending colon
Section of colon rising up right side of abdomen
Transverse colon
Highest section of colon, just below stomach, passing across upper abdomen
Haustra
Pouches that give colon its puckered appearance
Descending colon
Section of colon that passes down left side of abdomen
Feces
Taeni coli
Bands of longitudinal muscle running length of colon
Cecum
Pouch-like entrance to large intestine
Sigmoid colon
Final colonic section, making an S-shaped bend to meet rectum
Ileocecal valve
Controls flow of liquefied food from small intestine
Rectum
Final part of large intestine; a passageway for feces
Appendix (vermiform appendix)
Fingerlike, dead-end passage from the cecum, with no clear role
Anus
Valvelike exit from end of digestive tract

PORTIONS OF THE COLON
The three sections of the colon form an almost rectangular "frame," with the small intestine inside, the stomach and liver above, and the rectum below.

COLONIC MOVEMENT

The colon has three bands of muscle called taenia coli, which form pouches called haustra (see p.225). Muscular movements mix and propel feces toward the rectum. The motion of feces varies in rate, intensity, and nature, depending mainly on the stage of digestion of the contents. The three main types of motion are known as segmentation, peristaltic contractions, and mass movements. Fecal material passes more slowly through the colon than through the small intestine, enabling the reabsorption into the blood of up to 4¼ pints (2 liters) of water every day.

SEGMENTATION

A series of ringlike contractions occurs at regular intervals. These churn and mix feces but do not propel them along the colon.

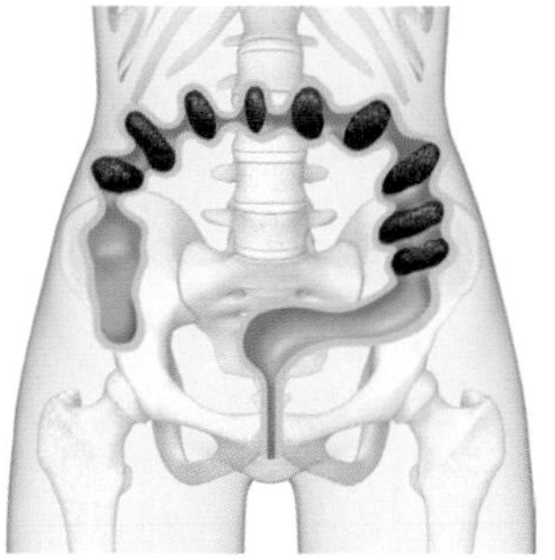

PERISTALTIC CONTRACTIONS

Small waves of movement called peristaltic contractions (see p.218) propel feces toward the rectum. The muscles behind the contents contract, while those in front relax.

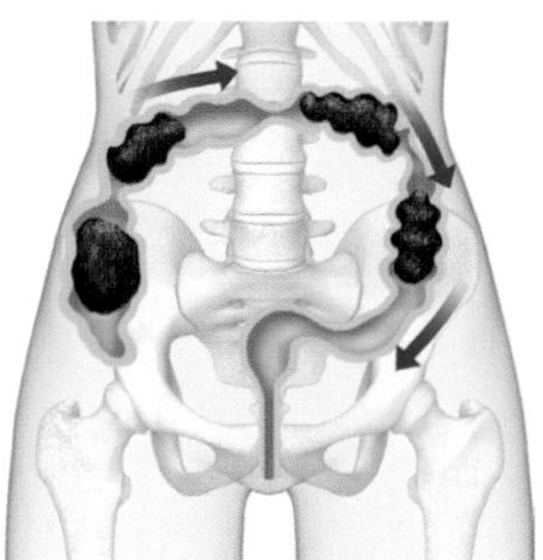

MASS MOVEMENTS

These extra-strong peristaltic waves move from the middle of the transverse colon. They happen two or three times a day and drive feces into the rectum.

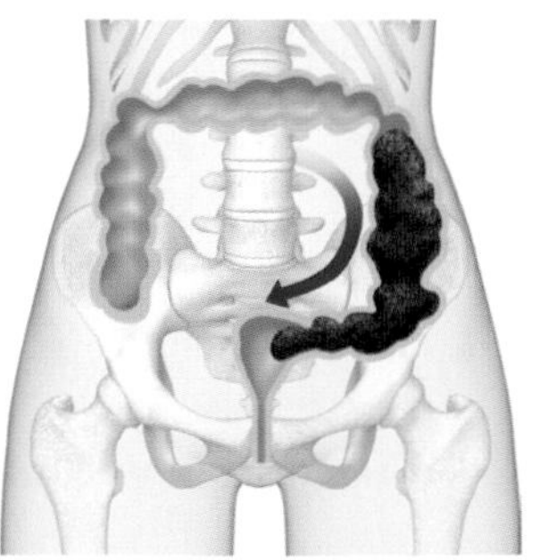

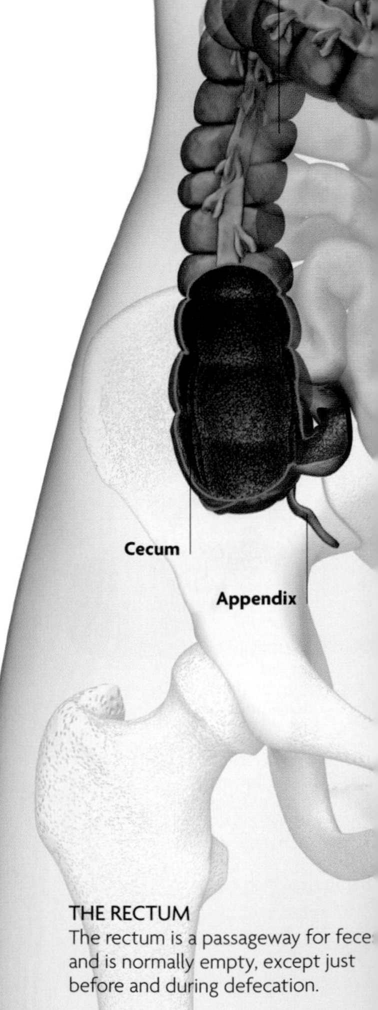
Ascending colon
Cecum
Appendix

THE RECTUM

The rectum is a passageway for fece[s] and is normally empty, except just before and during defecation.

RECTUM, ANUS, AND DEFECATION

The rectum is around 5 in (12 cm) long. Below it is the anal canal, about 1½ in (4 cm) long. In the walls of the anal canal, there are two sets of strong muscles–the internal and external sphincters. During defecation, peristaltic waves in the colon push feces into the rectum, which triggers the defecation reflex. Contractions push the feces along, and the anal sphincters relax to allow them out of the body through the anus. The defecation reflex may be aided by voluntary contraction of the abdominal muscles, or overridden by conscious control.

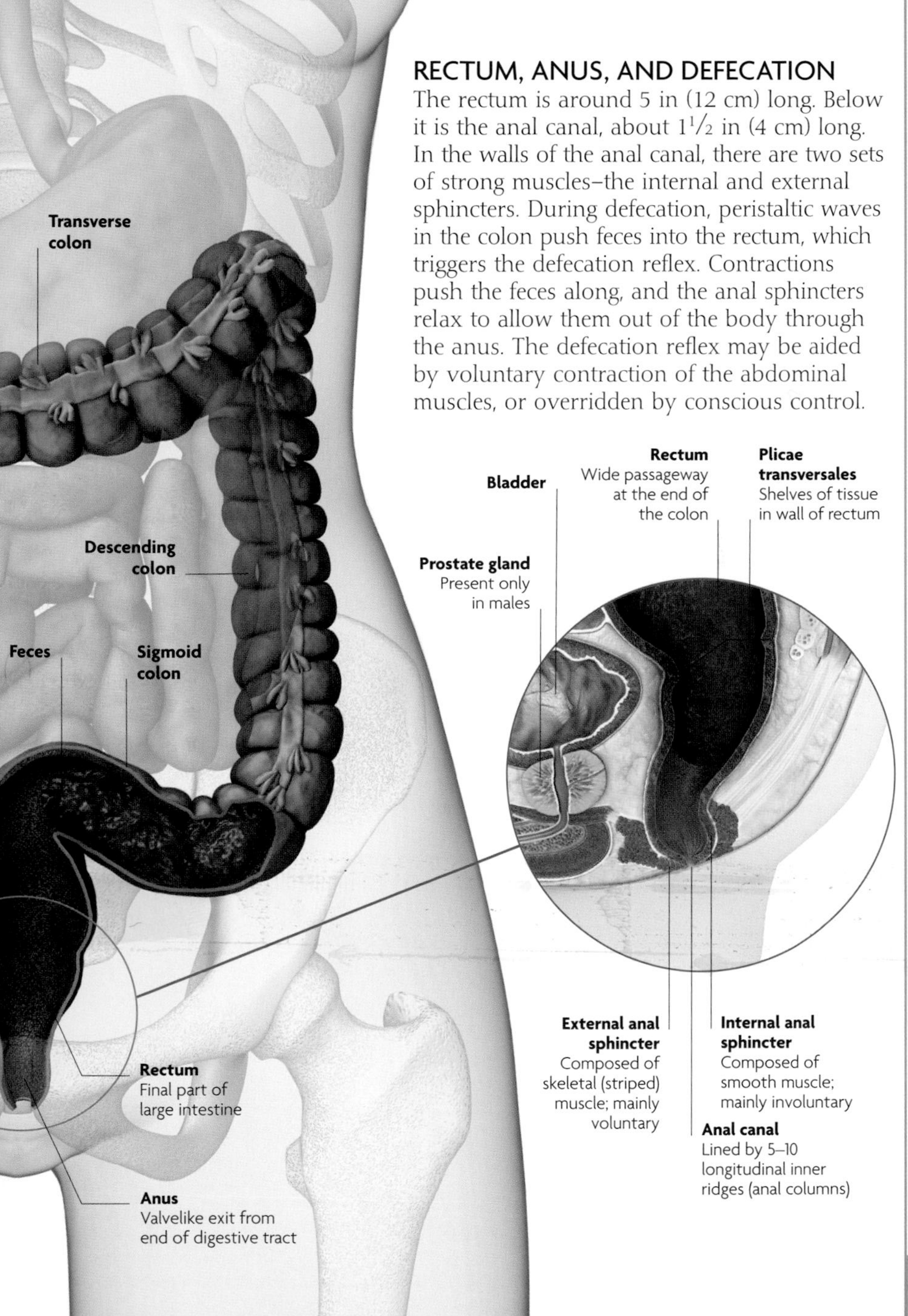

Transverse colon
Descending colon
Feces
Sigmoid colon
Rectum Final part of large intestine
Anus Valvelike exit from end of digestive tract
Bladder
Rectum Wide passageway at the end of the colon
Plicae transversales Shelves of tissue in wall of rectum
Prostate gland Present only in males
External anal sphincter Composed of skeletal (striped) muscle; mainly voluntary
Internal anal sphincter Composed of smooth muscle; mainly involuntary
Anal canal Lined by 5–10 longitudinal inner ridges (anal columns)

DIGESTION

THE DIGESTIVE PROCESS INVOLVES A SERIES OF PHYSICAL AND CHEMICAL ACTIONS THAT BREAK DOWN FOOD INTO NUTRIENT PARTICLES SMALL ENOUGH FOR ABSORPTION.

Vigorous physical digestion of food–mashing and churning–starts in the mouth and continues in the stomach using muscular movement. The stomach and the mouth secrete digestive chemicals (enzymes), too. By the time the pulverized food and enzymes (chyme) reach the duodenum, many food particles are already microscopically small, yet not small enough to pass across cell membranes into the body tissues. Chemical digestion in the small intestine splits large molecules into even smaller, absorbable particles that can enter the blood.

1 IN THE STOMACH

The stomach lining has gastric pits that make various substances: hydrochloric acid to kill microbes in swallowed food; the enzyme gastric lipase to begin breaking down fat; mucus to protect the stomach from digestive enzymes; and pepsin to digest proteins. Pepsin on its own could digest the stomach wall, so it is first released in an inactive form (pepsinogen), then activated by the stomach's acid.

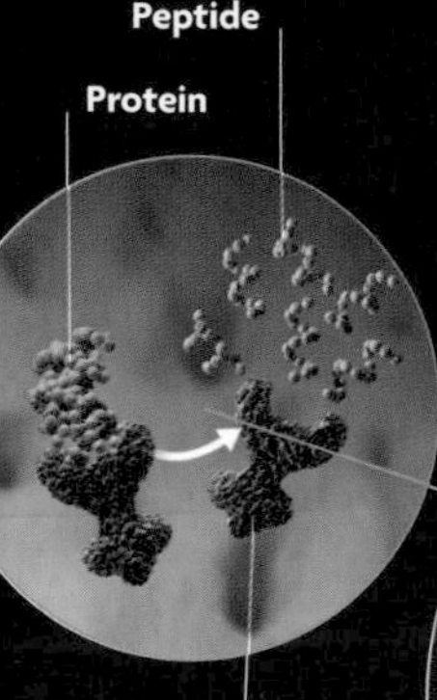
Peptide
Protein
Pepsin enzyme

PEPSIN IN ACTION
Pepsin is activated when it meets the acid of the stomach's interior. It splits protein molecules into shorter amino-acid chains called peptides.

Pepsin Protein-digesting enzyme

Gastric mucosa Stomach lining

Gastric lipase Fat-digesting enzyme

Hydrochloric acid

Mucus

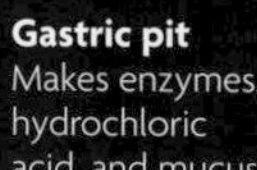

Gastric pit Makes enzymes, hydrochloric acid, and mucus

DIGESTIVE JOURNEY
As it travels from the mouth to the small intestine, food is broken down into smaller and smaller particles.

2 IN THE DUODENUM

Chyme is squirted into the duodenum and mixed with bile and secretions from the pancreas. These include alkalis, such as bicarbonates, that neutralize stomach acid, and about 15 enzymes, which work on carbohydrates, proteins, and fats (lipids).

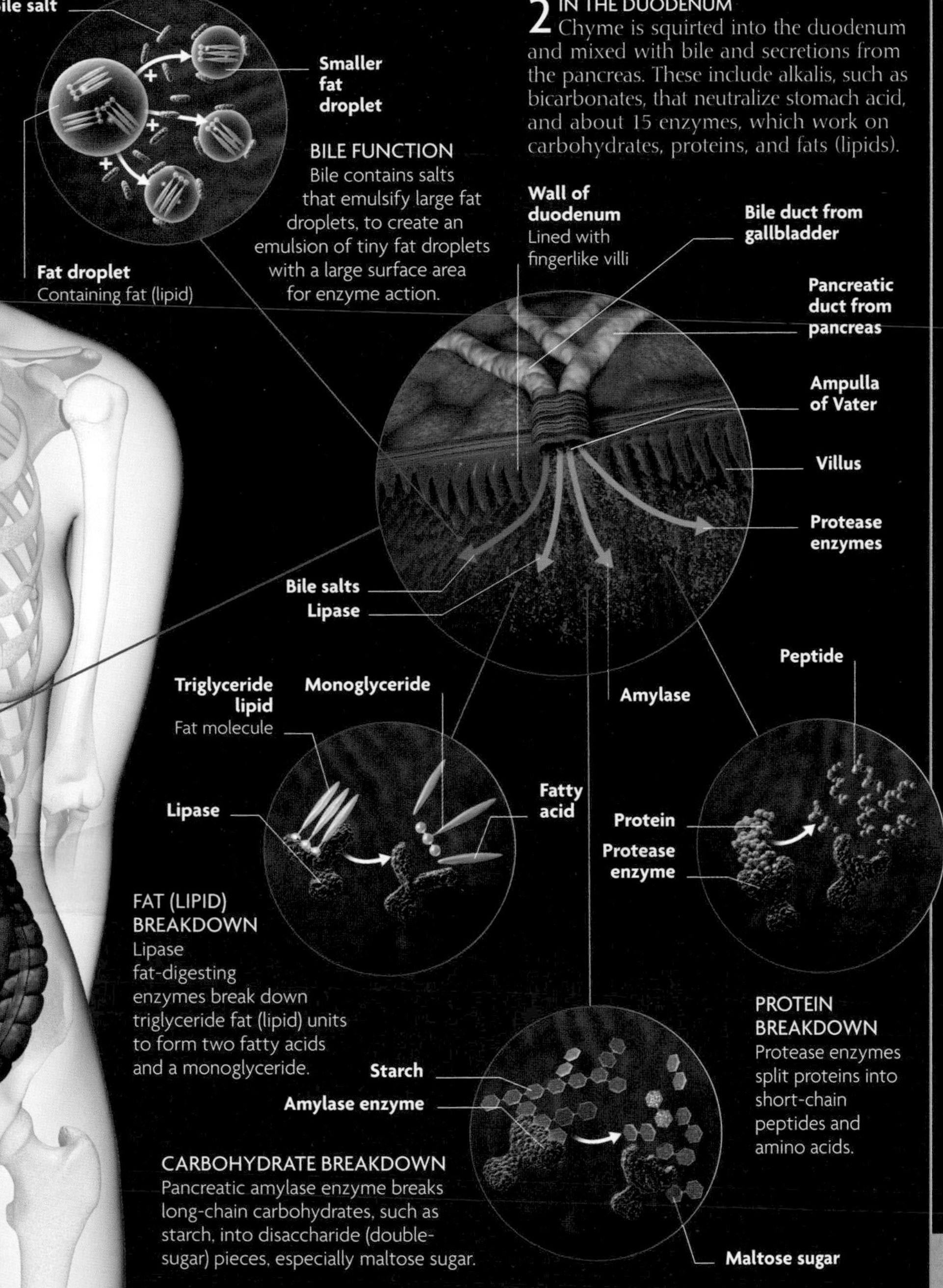

Bile salt
Smaller fat droplet
Fat droplet
Containing fat (lipid)
Wall of duodenum
Lined with fingerlike villi
Bile duct from gallbladder
Pancreatic duct from pancreas
Ampulla of Vater
Villus
Protease enzymes
Bile salts
Lipase
Amylase
Peptide
Triglyceride lipid
Fat molecule
Monoglyceride
Lipase
Fatty acid
Protein
Protease enzyme
Starch
Amylase enzyme
Maltose sugar

BILE FUNCTION

Bile contains salts that emulsify large fat droplets, to create an emulsion of tiny fat droplets with a large surface area for enzyme action.

FAT (LIPID) BREAKDOWN

Lipase fat-digesting enzymes break down triglyceride fat (lipid) units to form two fatty acids and a monoglyceride.

PROTEIN BREAKDOWN

Protease enzymes split proteins into short-chain peptides and amino acids.

CARBOHYDRATE BREAKDOWN

Pancreatic amylase enzyme breaks long-chain carbohydrates, such as starch, into disaccharide (double-sugar) pieces, especially maltose sugar.

3 IN THE SMALL INTESTINE

After the duodenum, the remainder of the small intestine is the site for the final breakdown of food substances and their absorption into the blood and lymphatic fluids. The pancreatic juices and bile fluids continue to work, but the small intestine releases few further enzymes into its inner passage, the lumen. Instead, its enzymes act within the lining cells, and on their surfaces. These enzymes include lactase and maltase, which break down the double (disaccharide) sugars, lactose and maltose, into single-unit glucose and galactose. Intestinal peptidases convert short peptide chains (originally from proteins) into their subunits, amino acids. The fingerlike villi of the intestine lining have surface cells bearing smaller projections of their own (microvilli), where some of these final changes occur.

HOW ENZYMES WORK

An enzyme is a biological catalyst—a substance that boosts the rate of a biochemical reaction, but remains unchanged itself. Most enzymes are proteins. They affect the reactions of digestive breakdown, and also the chemical changes that release energy and build new materials for cells and tissues. Each enzyme has a specific shape due to the way its long chains of subunits (amino acids) fold and loop. The substance to be altered (the substrate) fits into a part of the enzyme known as the active site. In the case of digestion, the enzyme may undergo a slight change in 3-D configuration that encourages the substrate to break apart at specific bonds between its atoms.

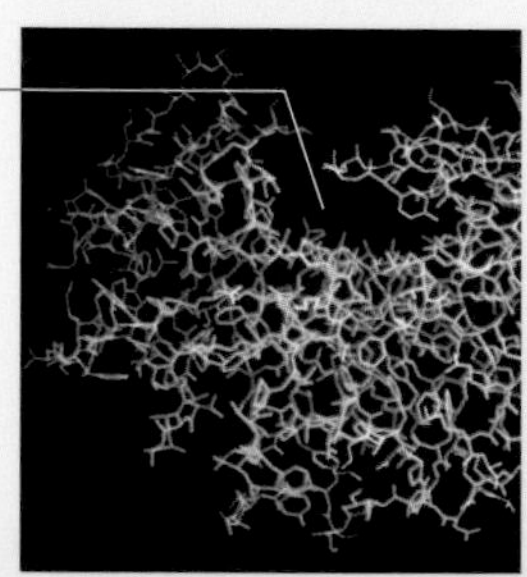
Active site

PEPSIN
A computer model of this digestive enzyme shows the active site as the gap at the top. A protein molecule will slot in here and break apart.

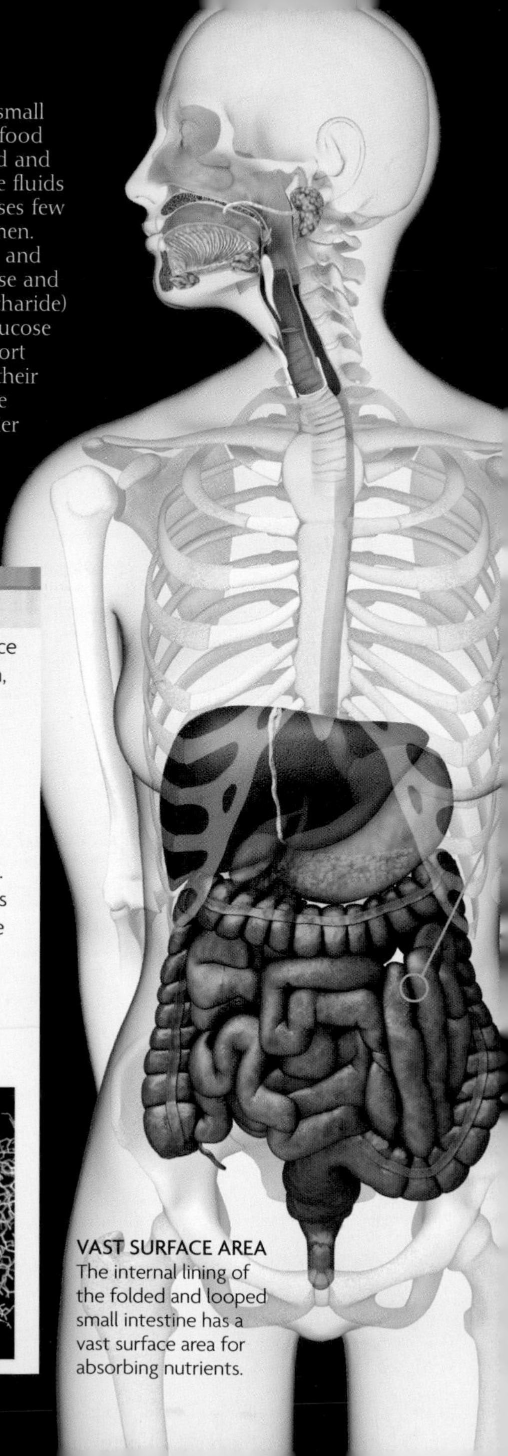

VAST SURFACE AREA
The internal lining of the folded and looped small intestine has a vast surface area for absorbing nutrients.

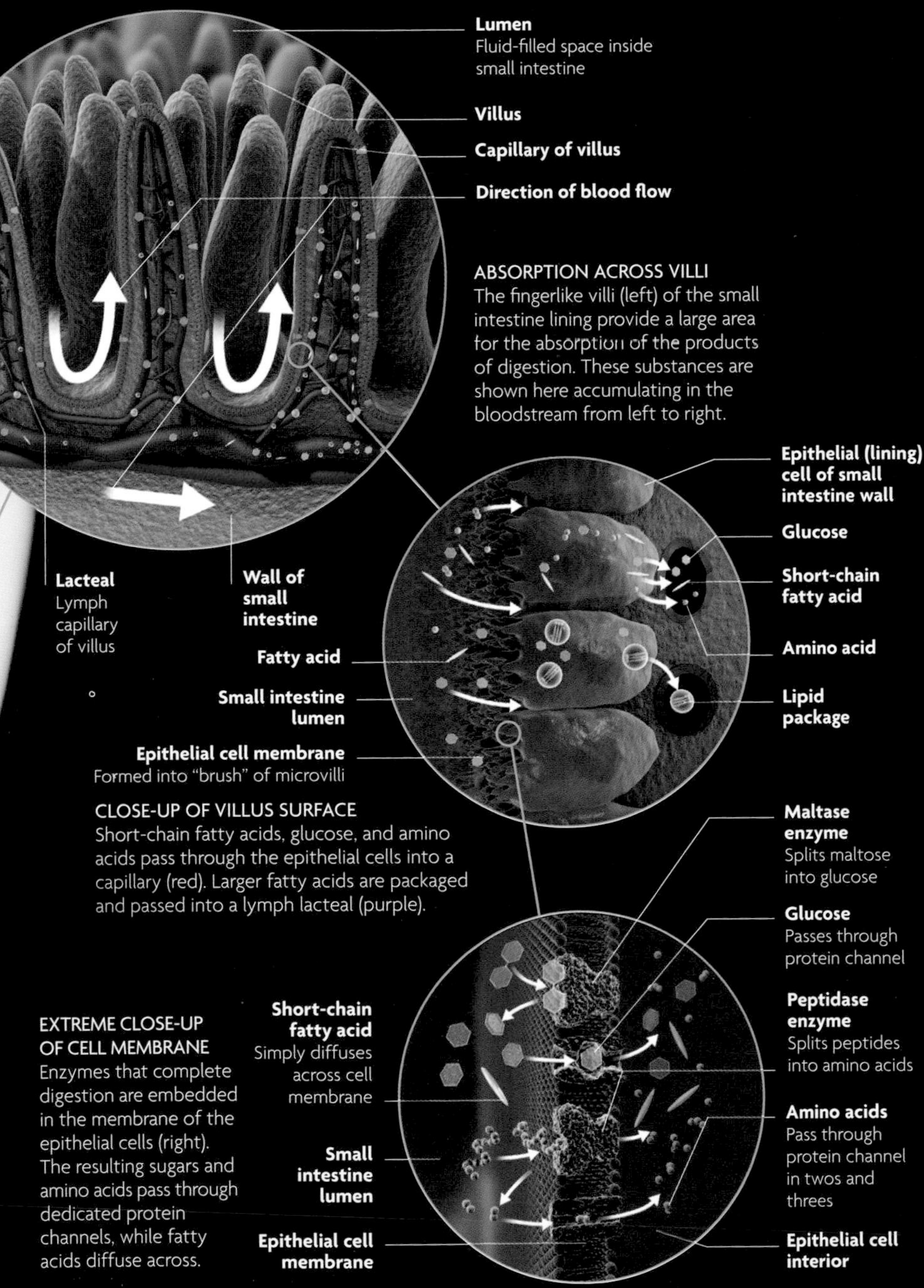

Lumen
Fluid-filled space inside small intestine
Villus
Capillary of villus
Direction of blood flow
Lacteal
Lymph capillary of villus
Wall of small intestine
Epithelial (lining) cell of small intestine wall
Glucose
Short-chain fatty acid
Amino acid
Lipid package
Fatty acid
Small intestine lumen
Epithelial cell membrane
Formed into "brush" of microvilli
Maltase enzyme
Splits maltose into glucose
Glucose
Passes through protein channel
Peptidase enzyme
Splits peptides into amino acids
Amino acids
Pass through protein channel in twos and threes
Short-chain fatty acid
Simply diffuses across cell membrane
Small intestine lumen
Epithelial cell membrane
Epithelial cell interior

ABSORPTION ACROSS VILLI
The fingerlike villi (left) of the small intestine lining provide a large area for the absorption of the products of digestion. These substances are shown here accumulating in the bloodstream from left to right.

CLOSE-UP OF VILLUS SURFACE
Short-chain fatty acids, glucose, and amino acids pass through the epithelial cells into a capillary (red). Larger fatty acids are packaged and passed into a lymph lacteal (purple).

EXTREME CLOSE-UP OF CELL MEMBRANE
Enzymes that complete digestion are embedded in the membrane of the epithelial cells (right). The resulting sugars and amino acids pass through dedicated protein channels, while fatty acids diffuse across.

NUTRIENTS AND METABOLISM

THE BODY'S INTERNAL BIOCHEMICAL REACTIONS, CHANGES, AND PROCESSES ARE TERMED METABOLISM. DIGESTION PROVIDES THE NUTRIENTS AS RAW MATERIALS, WHICH ENTER METABOLIC PATHWAYS IN ALL CELLS AND TISSUES.

TAKING IN NUTRIENTS

Carbohydrates, fats, proteins, vitamins, minerals, and other nutrients are absorbed at different stages along the digestive tract. Blood from the major absorption sites of the intestines flows along the hepatic portal vein (see p.222) to the liver. Here, nutrients are broken down, stored, and released according to the body's needs.

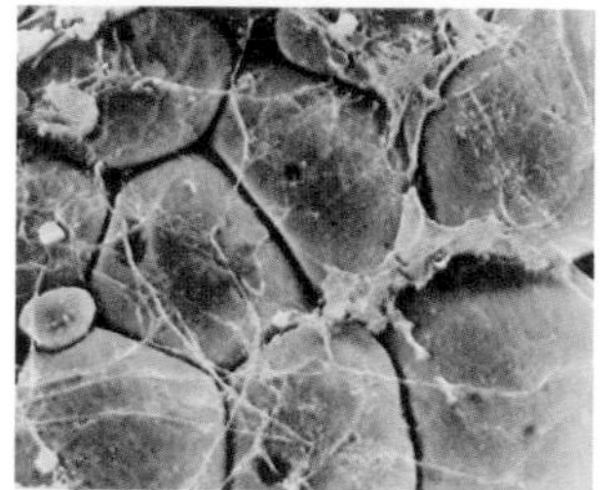

FAT TISSUE
Adipose tissue consists of cells replete with fat droplets that can be used as a concentrated energy store in times of need.

FINAL STAGES OF DIGESTION

The colon absorbs minerals, salts, and some vitamins, and reabsorbs a large amount of water, too. Fiber, such as pectin and cellulose, bulks up the digestive remnants as they are made into feces. Fiber helps delay the absorption of some molecules, including sugars, and spreads their uptake through time rather than in one short "rush." In addition, fiber binds with some fatty substances, such as cholesterol, and helps prevent their overabsorption.

CECUM
Each day about 3½–17 fl oz (100–500 ml) of digestive fluids, undigested leftovers, rubbed-off intestinal linings, and other matter enters the first chamber of the large intestine, the cecum.

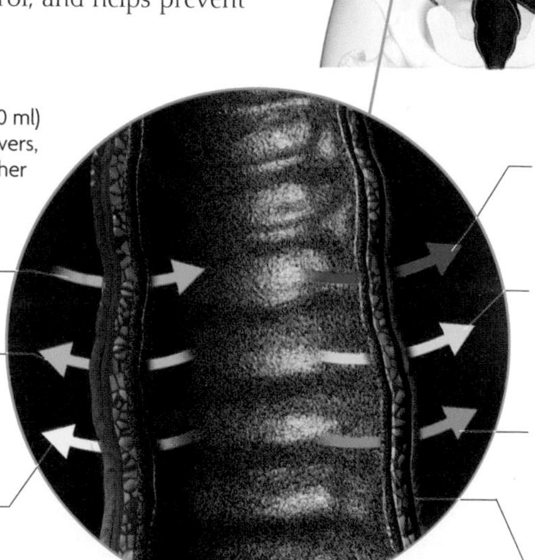

Bicarbonate and potassium
Secreted into lumen to replace recovered sodium

Chloride
Recovered from feces; with sodium, it maintains acid-alkali balance in tissues

Sodium
Also recovered from feces

Vitamin K
Manufactured by symbiotic bacteria

B vitamins
Some types released by bacterial fermentation

Water
Large intestine reabsorbs ⅔ of water in feces

Colon

BREAKDOWN AND BUILDING UP

Catabolism is the breaking down of complex molecules into simpler ones during energy production. Anabolism is the building up of complex molecules from simpler ones–for example, amino acids make peptide chains, which combine to form proteins.

INTERPLAY

Metabolism is a complex interplay of construction and destruction, with many molecules being recycled as they pass between the two processes.

SIMPLE MOLECULES FROM DIGESTED FOOD

CATABOLIC PROCESSES
Molecules with high-energy bonds, such as fats and glucose, are broken down

ANABOLIC PROCESSES
Joining of small molecules into new, larger molecules, such as proteins and DNA

ENERGY

COMPLEX MOLECULES

HOW THE BODY USES FOOD

The three major food components yield different breakdown products. Carbohydrates (starches and sugars) can be reduced to glucose; proteins to single amino acids; and fats (lipids) to fatty acids and glycerol. Glucose is the body's most adaptable and readiest source of energy. Amino acids are remade into the body's own proteins, both structural (collagen, keratin, and similar tough substances) and functional (enzymes). Fatty acids form the bi-lipid membranes around and inside cells (see p.28). However, the body can divert nutrients to different uses as conditions dictate.

ENERGY PRODUCTION

The simple sugar glucose is the energy source used by all cells to power their life processes. Fats, or in starvation circumstances, proteins, also suffice, either from the liver or mobilized from storage in tissues.

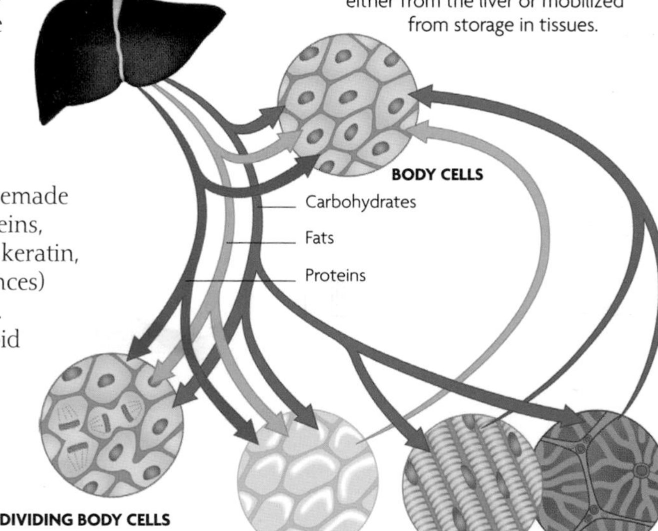
LIVER
BODY CELLS
Carbohydrates
Fats
Proteins
DIVIDING BODY CELLS
FAT CELLS
MUSCLE CELLS
LIVER LOBULES

GROWTH, RENEWAL, AND REPAIR

Cells are maintained using amino acids to build up different protein structures, fats to form membranes, and glucose to provide the energy. Cells that are dividing for growth or repair require increased supplies of these nutrients.

ENERGY STORAGE

Surplus glucose is converted into glycogen, which is stockpiled in the liver and muscle cells. Fatty acids are a concentrated energy store, and they can be derived from dietary fats, excess amino acids, or glucose.

DIGESTIVE TRACT DISORDERS

LIFESTYLE FACTORS, SUCH AS EXCESSIVE ALCOHOL CONSUMPTION, A POOR OR LOW-FIBER DIET, AND FOOD SENSITIVITIES, CONTRIBUTE TO MANY DIGESTIVE TRACT DISORDERS, ALTHOUGH SOME PROBLEMS ARE RELATED TO BACTERIA OR TO A COMPROMISED IMMUNE SYSTEM.

PEPTIC ULCERS

Most peptic ulcers are associated with *Helicobacter pylori* bacteria. These damage the mucous lining that normally protects against the powerful acidic juices in the stomach and first part of the duodenum. Other contributory factors include alcohol consumption, smoking, certain medications, family history, and diet. Upper abdominal pain is a common symptom. With a duodenal ulcer, this is often worse before a meal and relieved by eating; in a gastric ulcer, eating tends to aggravate the pain.

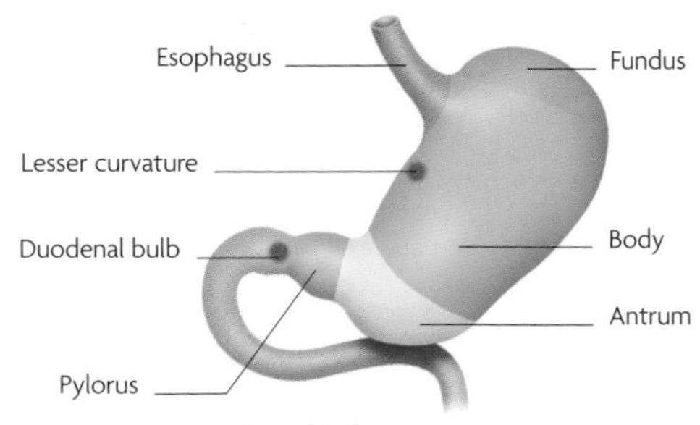
Esophagus
Fundus
Lesser curvature
Body
Duodenal bulb
Antrum
Pylorus

SITES OF PEPTIC ULCERS
A common site for ulcers is in the first part of the duodeum (duodenal bulb). In the stomach, most ulcers develop in the lesser curvature.

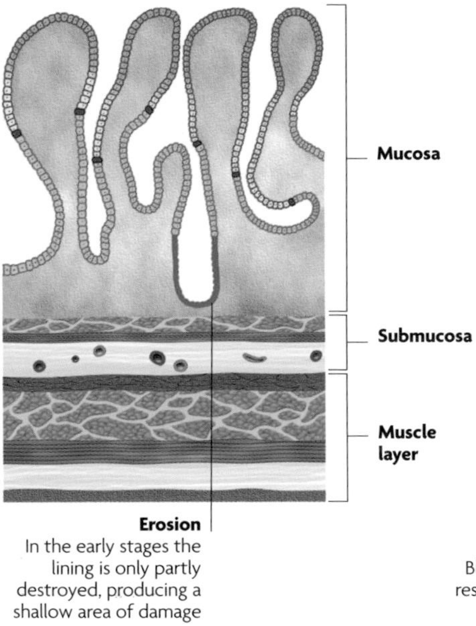

Mucosa

Submucosa

Muscle layer

Erosion
In the early stages the lining is only partly destroyed, producing a shallow area of damage

EARLY ULCER
If the protective mucous barrier coating the stomach lining breaks down, gastric juices containing strong acid and enzymes come into contact with mucosal cells.

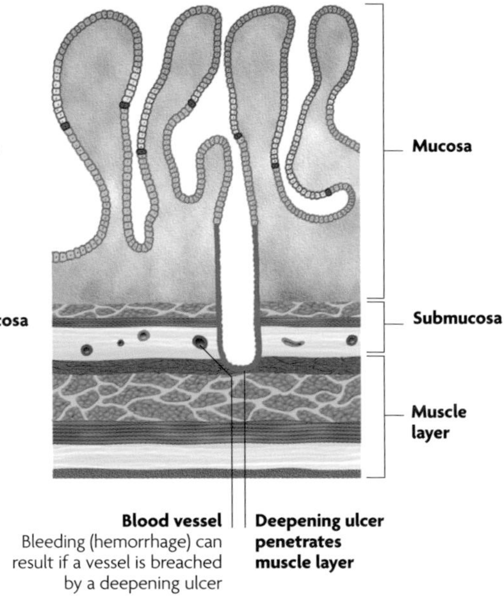

Mucosa

Submucosa

Muscle layer

Blood vessel
Bleeding (hemorrhage) can result if a vessel is breached by a deepening ulcer

Deepening ulcer penetrates muscle layer

PROGRESSIVE ULCERATION
A true ulcer penetrates the entire lining (mucosal layer) as well as the submucosa and muscle layers. In severe cases, it can perforate the stomach or duodenal wall.

GASTRITIS

Inflammation of the stomach lining, called gastritis, causes discomfort or pain, as well as nausea and vomiting. Gastritis that comes on suddenly, known as sudden onset (acute) gastritis, may be caused by overindulging, especially in alcohol consumption, or by taking medications known for their effect on the stomach lining, such as aspirin. Chronic gastritis develops over the longer term and may be due to repeated insult to the lining by alcohol, tobacco, or drugs. Another common cause is the bacterium *Helicobacter pylori*. Gastritis usually gets better with medication and by removing the underlying cause.

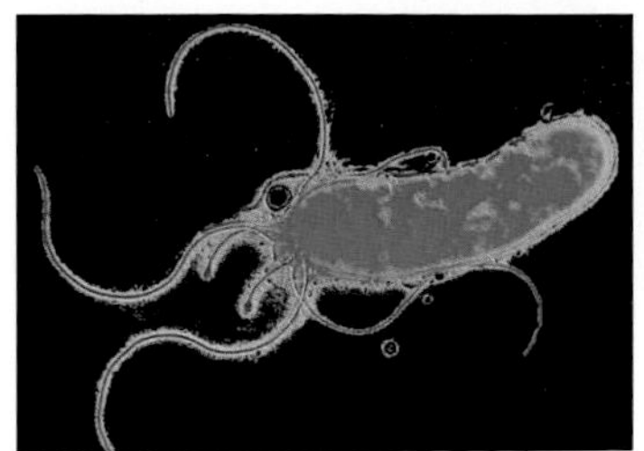

COMMON CULPRIT
At least 50 percent of people have *H. pylori* in their stomach lining. If the bacteria cause symptoms, antibiotics can eradicate them.

HIATUS HERNIA

The esophagus passes through a taut gap (hiatus) in the muscular sheet of the diaphragm, which lies between the abdomen and the chest cavity. The hiatus helps the esophageal sphincter (ring of muscle at the lower end of the esophagus) to prevent acidic stomach contents from passing up into the lower esophagus. In a hiatus hernia, the upper section of the stomach protrudes up through this gap. Any symptoms of a hiatus hernia are those of heartburn (gastric reflux). There are two types of hiatus hernia: sliding and para-esophageal. Sliding hernias usually have no symptoms, and it is estimated that they are present in around a third of all people over 50. In rare cases, however, para-esophageal hernias can cause severe pain and require surgery.

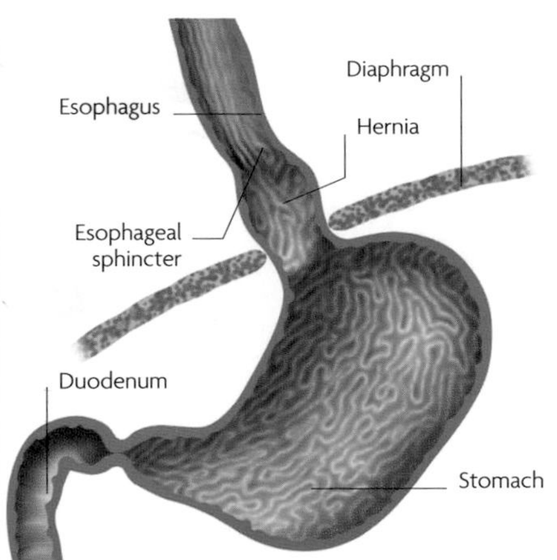
Diaphragm
Esophagus
Hernia
Esophageal sphincter
Duodenum
Stomach

SLIDING HIATUS HERNIA
This is the most common type of hiatus hernia, and occurs when the junction between the esophagus and the stomach slides up through the diaphragm.

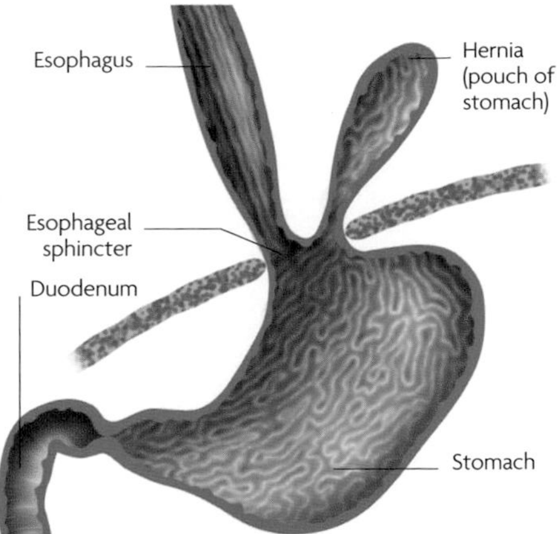
Esophagus
Hernia (pouch of stomach)
Esophageal sphincter
Duodenum
Stomach

PARA-ESOPHAGEAL HIATUS HERNIA
In about 1 in 10 hernias, a pouchlike part of the stomach is pushed upward through the diaphragm and lies adjacent to the lower esophagus.

ALCOHOLIC LIVER DISEASE

Regular, excessive alcohol consumption over many years can lead to serious liver damage. Women do not metabolize alcohol as efficiently as men and are more vulnerable to its side effects. The toxic effects of chemicals in alcohol can damage the liver in different ways and may, in some people, increase the risk of liver cancer. Almost all long-term, heavy drinkers develop a "fatty liver" because alcohol produces fat when it is broken down. If a person stops drinking, the fat disappears and the liver may eventually return to normal. However, continued heavy drinking can lead to alcoholic hepatitis, or inflammation of the liver. Symptoms vary from none at all to acute illness and jaundice. The final stage of alcoholic liver damage is cirrhosis, which can be fatal. Often, the only treatment option at this stage is a liver transplant.

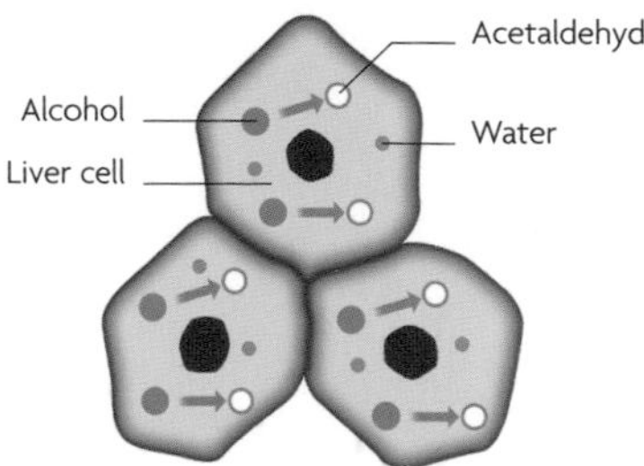
Acetaldehyde
Alcohol
Water
Liver cell

1 HOW DAMAGE OCCURS
Alcohol (ethanol) breaks down into acetaldehyde, which is thought to bind with proteins in liver cells and cause damage, inflammation, and fibrosis.

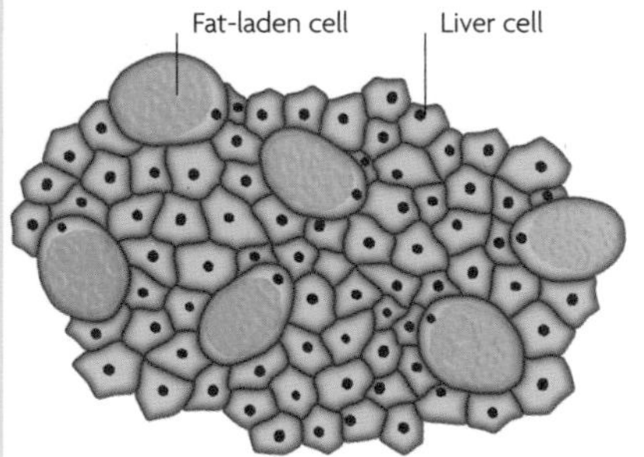
Fat-laden cell
Liver cell

2 FATTY LIVER
One by-product of alcohol metabolism is fat. Liver cells in excessive drinkers swell with globules of fat that are clearly visible as yellow or white patches if the liver is cut open. The condition is reversible if drinking stops.

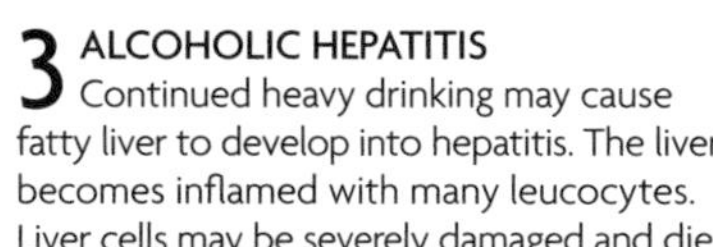

3 ALCOHOLIC HEPATITIS
Continued heavy drinking may cause fatty liver to develop into hepatitis. The liver becomes inflamed with many leucocytes. Liver cells may be severely damaged and die.

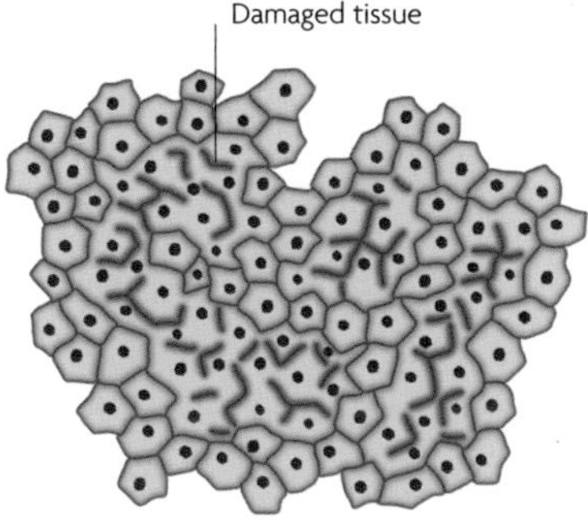
Damaged tissue

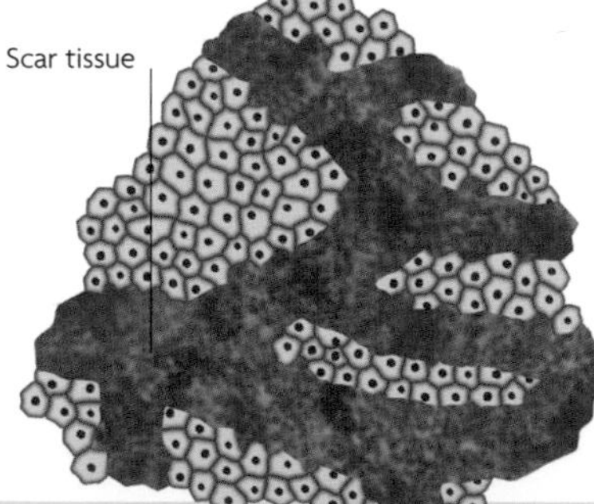
Scar tissue

4 CIRRHOSIS
In this final stage of alcoholic liver disease, the permanent fibrosis and scarring of the liver tissue becomes life-threatening. Because the cells are permanently damaged, the liver is unable to carry out its normal functions.

PORTAL HYPERTENSION

One of the complications of liver cirrhosis is portal hypertension. As the tissue becomes progressively scarred and fibrosed, it obstructs the flow of blood into the liver from the portal vein, a large vessel carrying blood from the digestive tract. Pressure builds up in the vein, and can cause other vessels "upstream" to become distended. Among these are veins in the abdomen and rectum, and those that supply the esophagus with blood. The swollen veins, or varices, protrude into the esophagus and may bleed. In some cases, only slight oozing occurs. In others, a major hemorrhage causes massive vomiting of blood. Not everyone who has liver cirrhosis develops portal hypertension and esophageal varices. In those who do develop the condition, the varices can be treated with drugs to reduce the blood pressure or injected with a sclerosing (hardening) agent, much like that used to treat varicose veins.

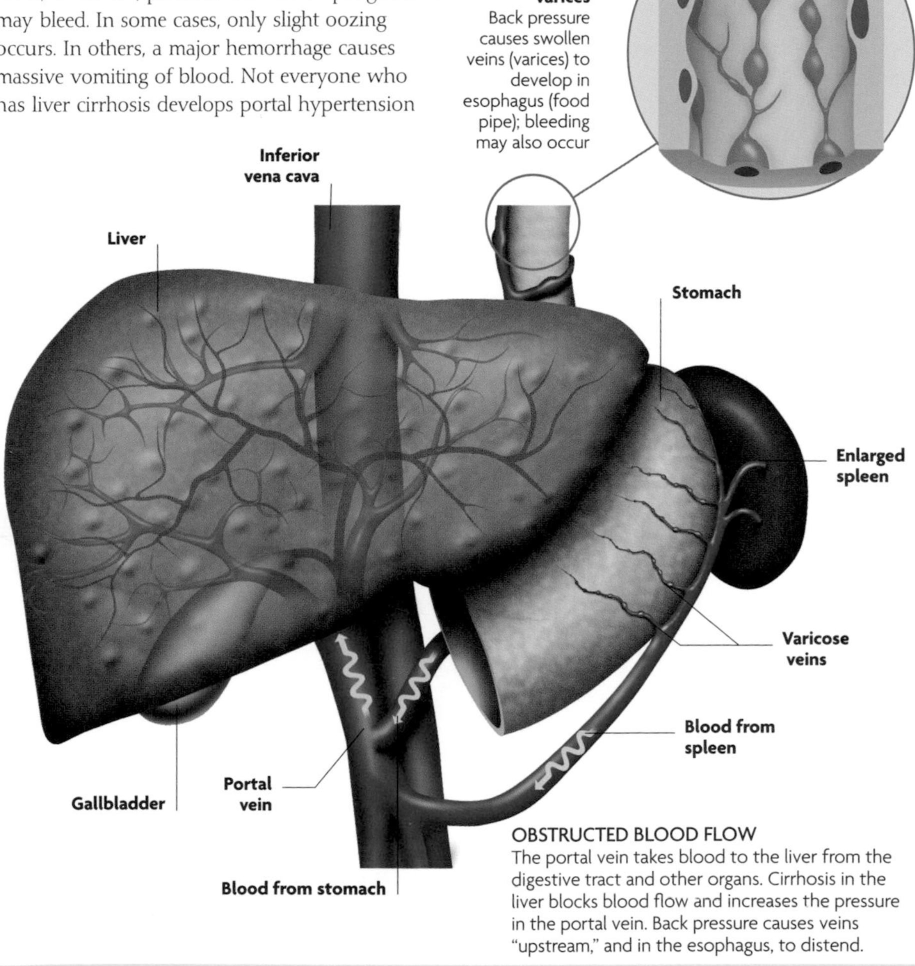
Esophageal varices
Back pressure causes swollen veins (varices) to develop in esophagus (food pipe); bleeding may also occur
Inferior vena cava
Liver
Stomach
Enlarged spleen
Varicose veins
Blood from spleen
Portal vein
Gallbladder
Blood from stomach

OBSTRUCTED BLOOD FLOW
The portal vein takes blood to the liver from the digestive tract and other organs. Cirrhosis in the liver blocks blood flow and increases the pressure in the portal vein. Back pressure causes veins "upstream," and in the esophagus, to distend.

GALLSTONES

In developed countries, the majority of gallstones consist primarily of cholesterol, a fatty substance processed in the liver and stored in the gallbladder as one of the constituents of bile. Gallstones can develop if the normal "mix" of bile is altered and the cholesterol content is high. They are far more common in women and are unusual before the age of 30. Most people have no symptoms. If there are symptoms, they begin when a stone lodges in one of the ducts leaving the gallbladder. The main symptom is pain, which varies in intensity and often develops after a fatty meal. For symptomatic gallstones, the treatment is usually a cholecystectomy, or keyhole surgery to remove the gallbladder.

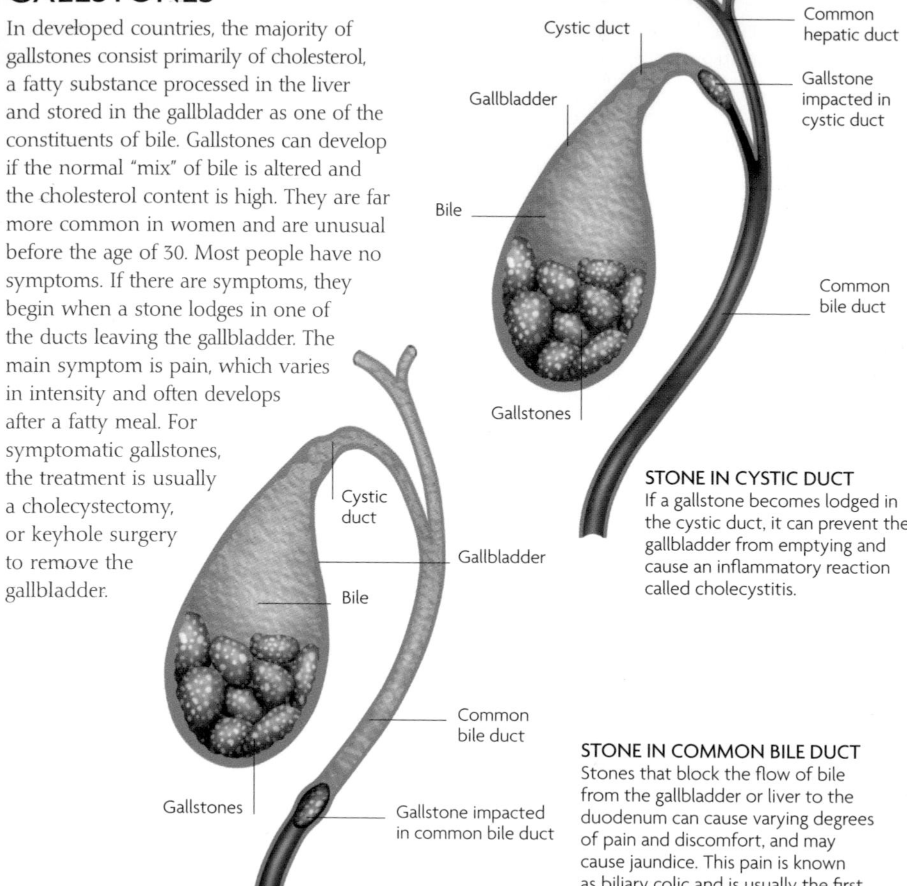
Cystic duct
Common hepatic duct
Gallbladder
Gallstone impacted in cystic duct
Bile
Common bile duct
Gallstones
Cystic duct
Gallbladder
Bile
Common bile duct
Gallstones
Gallstone impacted in common bile duct

STONE IN CYSTIC DUCT
If a gallstone becomes lodged in the cystic duct, it can prevent the gallbladder from emptying and cause an inflammatory reaction called cholecystitis.

STONE IN COMMON BILE DUCT
Stones that block the flow of bile from the gallbladder or liver to the duodenum can cause varying degrees of pain and discomfort, and may cause jaundice. This pain is known as biliary colic and is usually the first sign that a gallstone is present.

PANCREATITIS

Pancreatitis is a serious inflammation of the pancreas, and can be either acute or chronic. In both types, the inflammation is triggered by the enzymes that the pancreas itself normally manufactures to aid the digestion of food when it enters the duodenum. In pancreatitis, these enzymes become activated while they are still inside the pancreas, and begin to digest the tissue.

There are many causes of acute pancreatitis, the most common of which are gallstones (see above), excessive intake of alcohol, some drugs, and certain infections, such as mumps. Chronic pancreatitis is usually associated with long-term alcoholism. In both types, the main feature is pain. In acute pancreatitis, this is particularly severe and may be accompanied by nausea and vomiting.

DIVERTICULAR DISEASE

In diverticular disease, patches of the colon wall bulge outward into pouches called diverticula. Most people with diverticular disease are aged over 50 and have eaten a low-fiber diet for many years, with consequent straining as they pass hard stools. The problem becomes more common with increasing age. The lowest part of the colon, known as the sigmoid colon (see p.227), is most commonly affected, but the whole colon can be involved. About 95 percent of people with diverticular disease do not show symptoms, but some people have abdominal pain and irregular bowel habits. In diverticulitis, the pouches become inflamed, causing severe pain, fever, and constipation. The pain is often in the lower left abdomen, and may fade after passing gas or stools.

1 HARD FECES Soft, bulky feces are able to pass easily along the colon. If feces are hard and dry, usually due to lack of fiber or "roughage" in the diet, the contractions of the smooth muscle layers of the colon must increase in force, putting pressure on the walls of the colon.

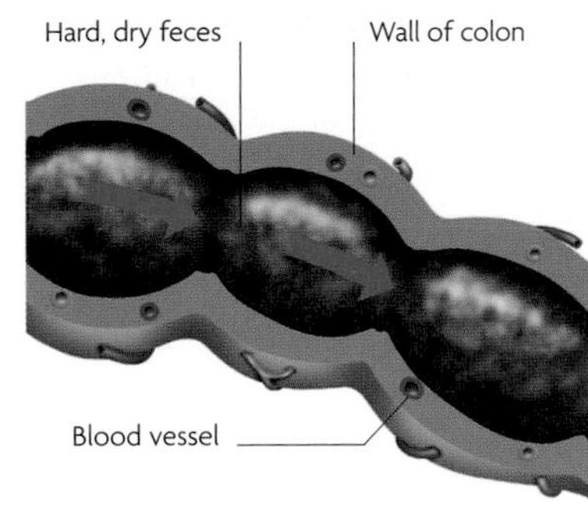
Hard, dry feces
Wall of colon
Blood vessel

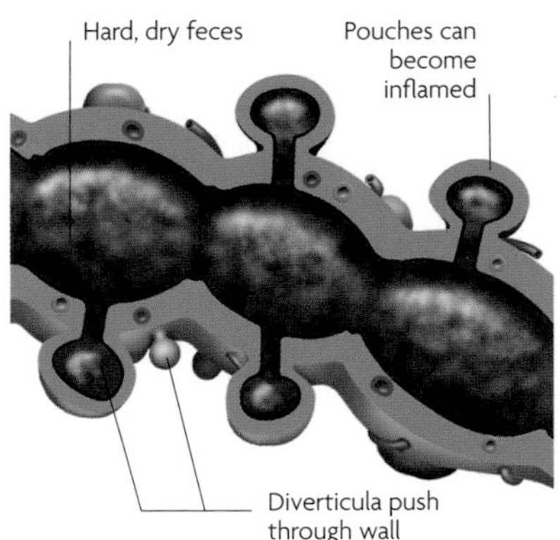
Hard, dry feces
Pouches can become inflamed
Diverticula push through wall

2 POUCHES FORM Eventually, the increased pressure pushes small areas of colon lining through points of weakness in the muscle of the wall, often near a blood vessel. The pea- to grape-sized pouches that form easily trap bacteria and may become inflamed.

COLORECTAL CANCER

Cancer of the colon, rectum, or both is one of the most common cancers in the industrialized world. Risk factors include family history and aging. A malignant tumor in the intestinal wall can start as a polyp in the lining. A high-fat, low-fiber diet, excess alcohol, lack of exercise, and obesity can make this cancer more likely to develop. Symptoms are a change in bowel habits and stool consistency, abdominal pain, loss of appetite, fecal blood, and a sensation of not fully emptying the bowels. Colorectal cancer can be detected by screening programs, such as fecal tests for blood and colonoscopy. If it is detected and treated early, the chances of survival for five years or longer are high.

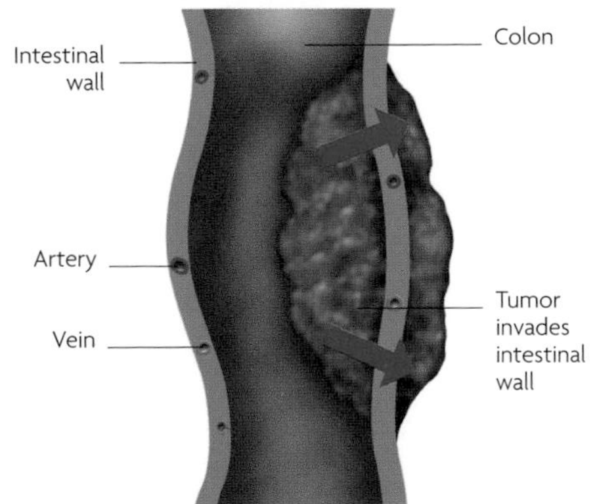
Intestinal wall
Colon
Artery
Vein
Tumor invades intestinal wall

COLONIC TUMOR
Over time, malignant tumors grow and invade the intestinal wall, from where the cancer can spread to other parts of the body via the bloodstream.

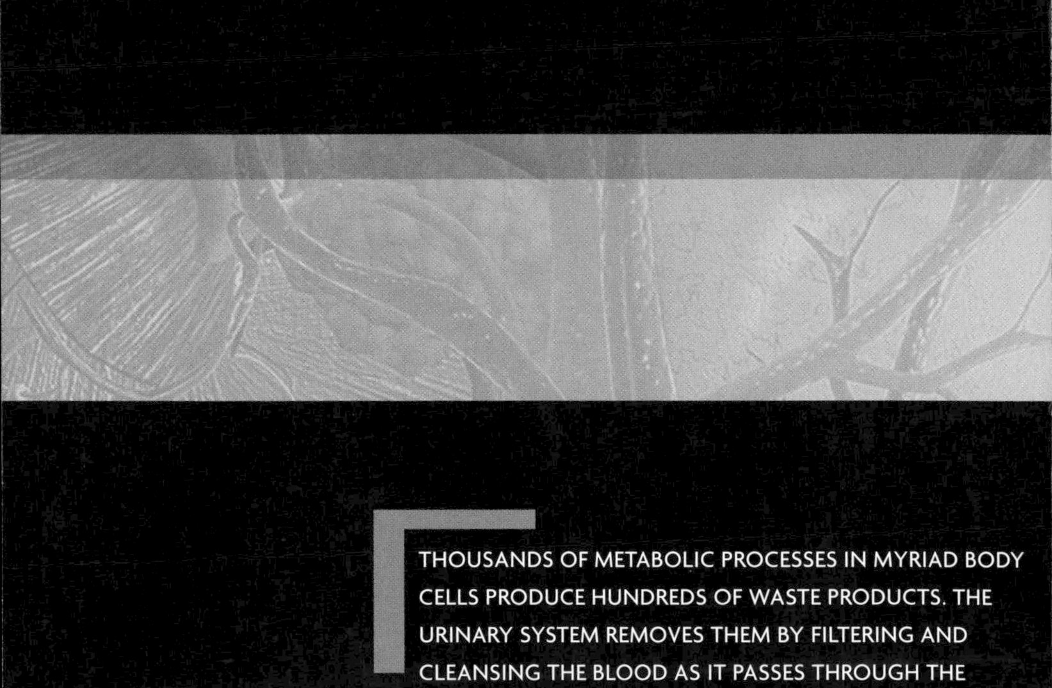
THOUSANDS OF METABOLIC PROCESSES IN MYRIAD BODY
CELLS PRODUCE HUNDREDS OF WASTE PRODUCTS. THE
URINARY SYSTEM REMOVES THEM BY FILTERING AND
CLEANSING THE BLOOD AS IT PASSES THROUGH THE

URINARY SYSTEM

URINARY ANATOMY

THE URINARY SYSTEM IS COMPOSED OF TWO KIDNEYS, TWO URETERS, A BLADDER, AND A URETHRA. IT REGULATES THE VOLUME AND COMPOSITION OF BODY FLUIDS AND EXPELS WASTE AND EXCESS WATER FROM THE BODY.

The two kidneys are reddish organs resembling beans in shape. They are situated on either side of the abdomen, just above the waist and toward the back of the body. Each kidney contains many microscopic filtering units that remove unwanted waste, minerals, and excess water from the blood as urine. A ureter transports the urine to the bladder, which gradually becomes spherical, then pear-shaped, as it fills up. Eventually, stretch receptors in the bladder wall initiate a conscious desire to urinate. The urethra then conducts urine from the bladder to the outside.

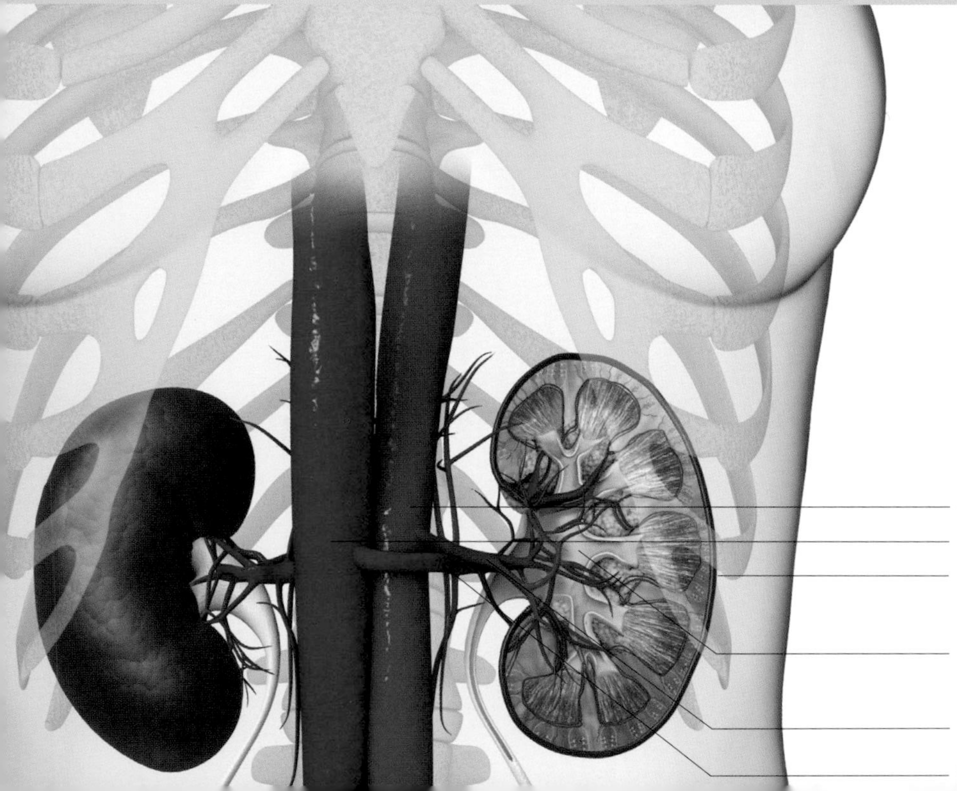

Aorta

Inferior vena cava

Kidney
Each is about 4–5 in (10–12.5 cm) long, and contains about 1 million filtering units

Renal pelvis
Funnel-shaped chamber in which urine collects before passing down the ureter

Renal artery

Renal vein

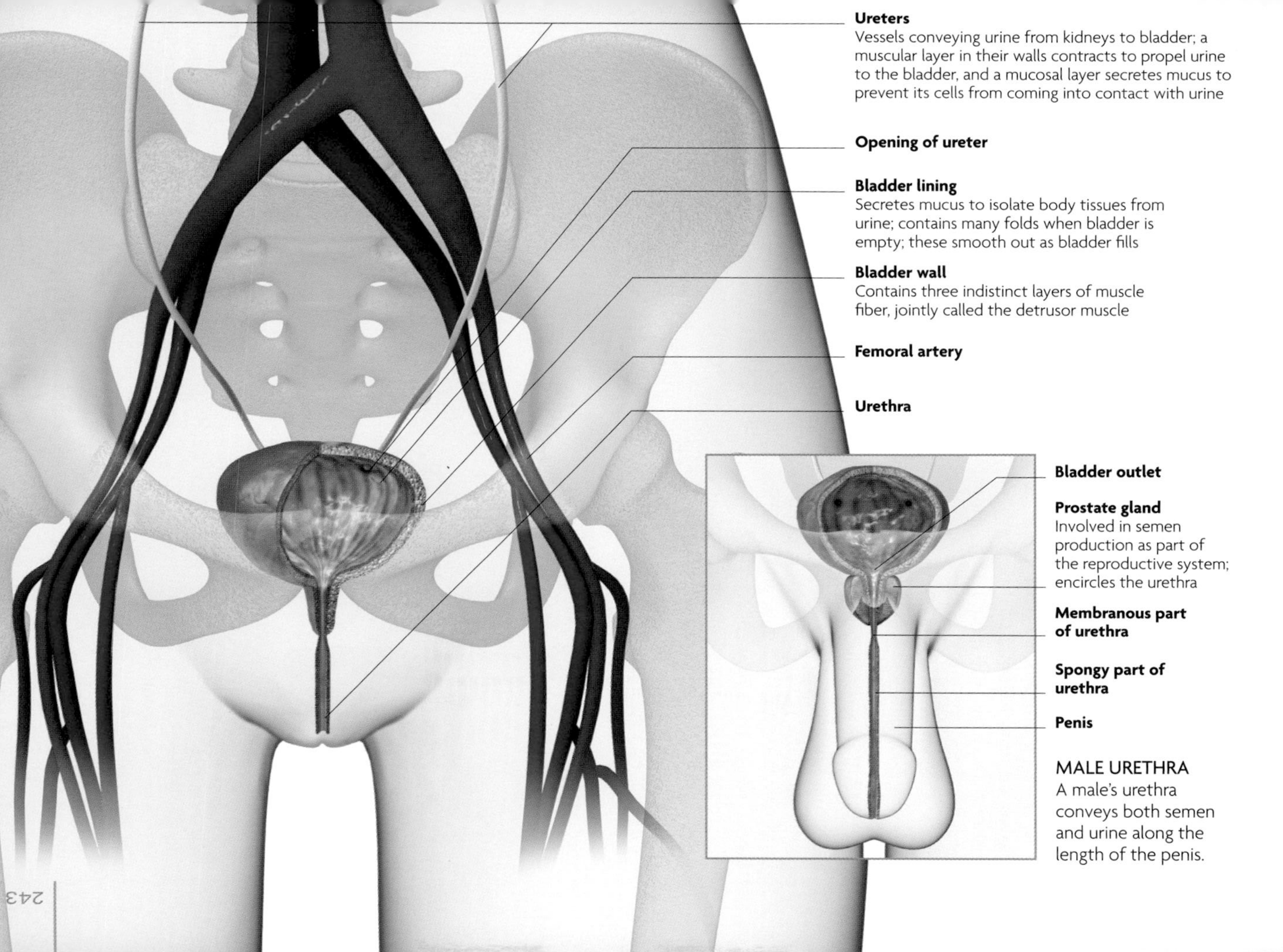
Ureters
Vessels conveying urine from kidneys to bladder; a muscular layer in their walls contracts to propel urine to the bladder, and a mucosal layer secretes mucus to prevent its cells from coming into contact with urine
Opening of ureter
Bladder lining
Secretes mucus to isolate body tissues from urine; contains many folds when bladder is empty; these smooth out as bladder fills
Bladder wall
Contains three indistinct layers of muscle fiber, jointly called the detrusor muscle
Femoral artery
Urethra
Bladder outlet
Prostate gland
Involved in semen production as part of the reproductive system; encircles the urethra
Membranous part of urethra
Spongy part of urethra
Penis

MALE URETHRA
A male's urethra conveys both semen and urine along the length of the penis.

KIDNEY STRUCTURE

THE KIDNEYS ARE A PAIR OF ORGANS SITUATED EITHER SIDE OF THE SPINAL COLUMN AND AT THE UPPER REAR OF THE ABDOMINAL CAVITY. THEY FILTER WASTE PRODUCTS FROM THE BLOOD AND EXCRETE THEM, ALONG WITH EXCESS WATER, AS URINE.

INSIDE THE KIDNEY

Each kidney is protected by three outer layers: a tough external coat of fibrous connective tissue, the renal fascia; a layer of fatty tissue, the adipose capsule; and inside this, another fibrous layer, the renal capsule. The main body of the kidney also has three layers: the renal cortex, which is packed full of knots of capillaries known as glomeruli and their capsules; next, the renal medulla, which contains capillaries and urine-forming tubules; and a central space where the urine collects, known as the renal pelvis. The glomeruli, capsules, and tubules are the constituent parts of the kidney's million-plus microfiltering units, called nephrons.

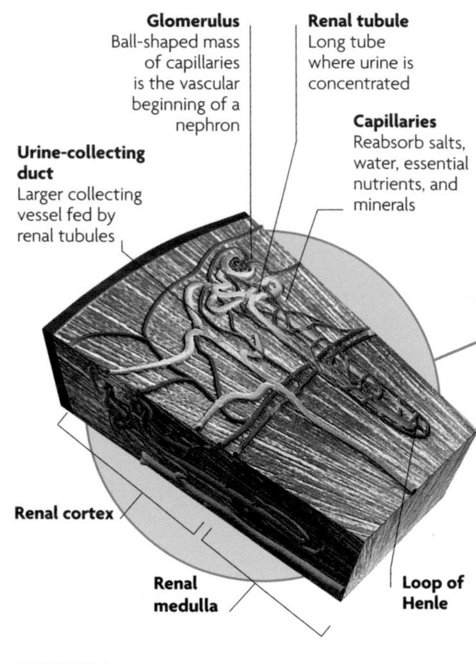

Glomerulus Ball-shaped mass of capillaries is the vascular beginning of a nephron
Renal tubule Long tube where urine is concentrated
Capillaries Reabsorb salts, water, essential nutrients, and minerals
Urine-collecting duct Larger collecting vessel fed by renal tubules
Renal cortex
Renal medulla
Loop of Henle

NEPHRON

Each microfiltering unit, or nephron, spans the cortex and medulla. The glomerulus, capsule, proximal and distal tubules, and the smaller urine-collecting ducts are in the cortex. The medulla contains mainly the long tubule loops of Henle and the larger urine-collecting ducts.

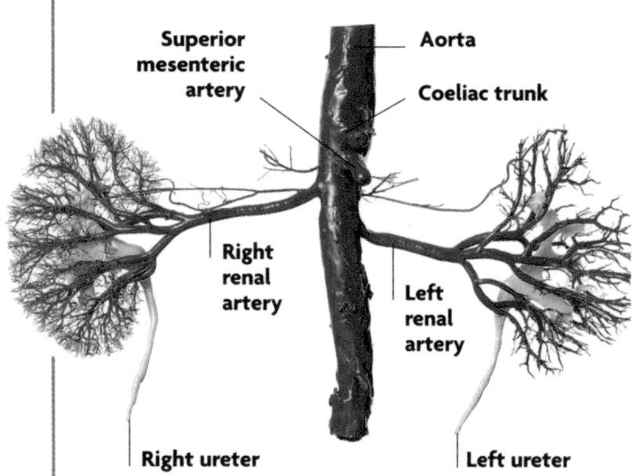

Superior mesenteric artery
Aorta
Coeliac trunk
Right renal artery
Left renal artery
Right ureter
Left ureter

BLOOD SUPPLY TO THE KIDNEYS

The left and right renal arteries are branches of the aorta, which carries blood directly from the heart. The arteries leave the celiac trunk of the aorta just below the superior mesenteric artery. The renal arteries form a branching network that supplies blood to the kidneys.

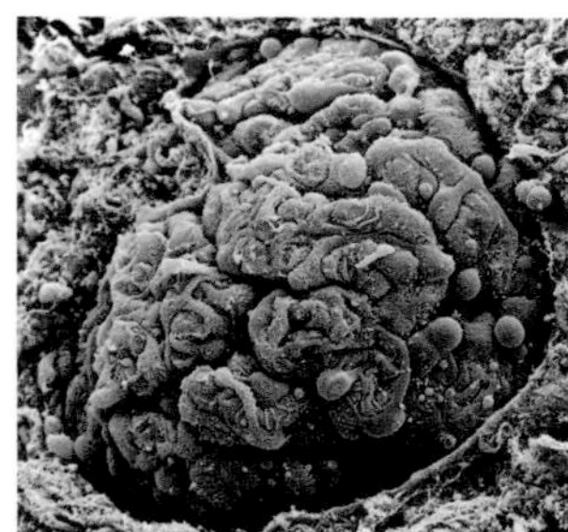

GLOMERULUS

This microscope image shows the tangled system of a glomerulus (pink). A filtrate fluid oozes from the glomerulus and is collected by the cuplike Bowman's capsule (brown).

KIDNEY CROSS SECTION

This cutaway shows the kidney's main layers: the cortex and the medulla, which forms segments known as renal pyramids. The renal artery and vein circulate huge amounts of blood—about 2½ pints per min (1.2 liters per min) at rest, which is up to one-quarter of the heart's total output.

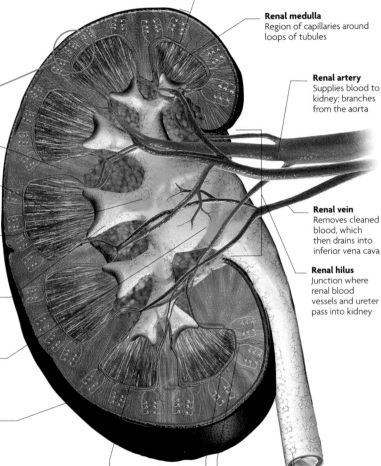

Renal cortex
Outer region of kidney; packed with microscopic structures called glomeruli, which make it look granular

Renal medulla
Region of capillaries around loops of tubules

Renal artery
Supplies blood to kidney; branches from the aorta

Renal column
Tissue between renal pyramids

Renal pyramid
Cone-shaped region of renal medulla

Major calyx
Several minor calyces merge to form a major calyx

Minor calyx
Cuplike cavity that receives urine from the renal papilla

Renal pelvis
Funnel-shaped tube into which the major calyces merge

Renal capsule
Thin covering of white fibrous tissue around the whole kidney

Renal vein
Removes cleaned blood, which then drains into inferior vena cava

Renal hilus
Junction where renal blood vessels and ureter pass into kidney

Renal papilla
Apex of the renal pyramid

Arcuate arteries and veins
Vessels forming arch-like links between the cortex and medulla

Interlobular arteries and veins
Branches of the renal artery and vein

Ureter
Tube for urine to pass to the bladder

STRUCTURE OF A NEPHRON

Each nephron consists of two tubes: one for carrying blood and one for forming urine. Both have convoluted routes between the renal cortex and medulla. The blood vessel starts as the afferent arteriole and finishes as a venule that carries the blood away.

The renal tube starts at Bowman's capsule around the glomerulus, and leads into the proximal convoluted tubule, which dips into and out of the medulla as the loop of Henle. Eventually, it feeds urine into a large urine-collecting duct.

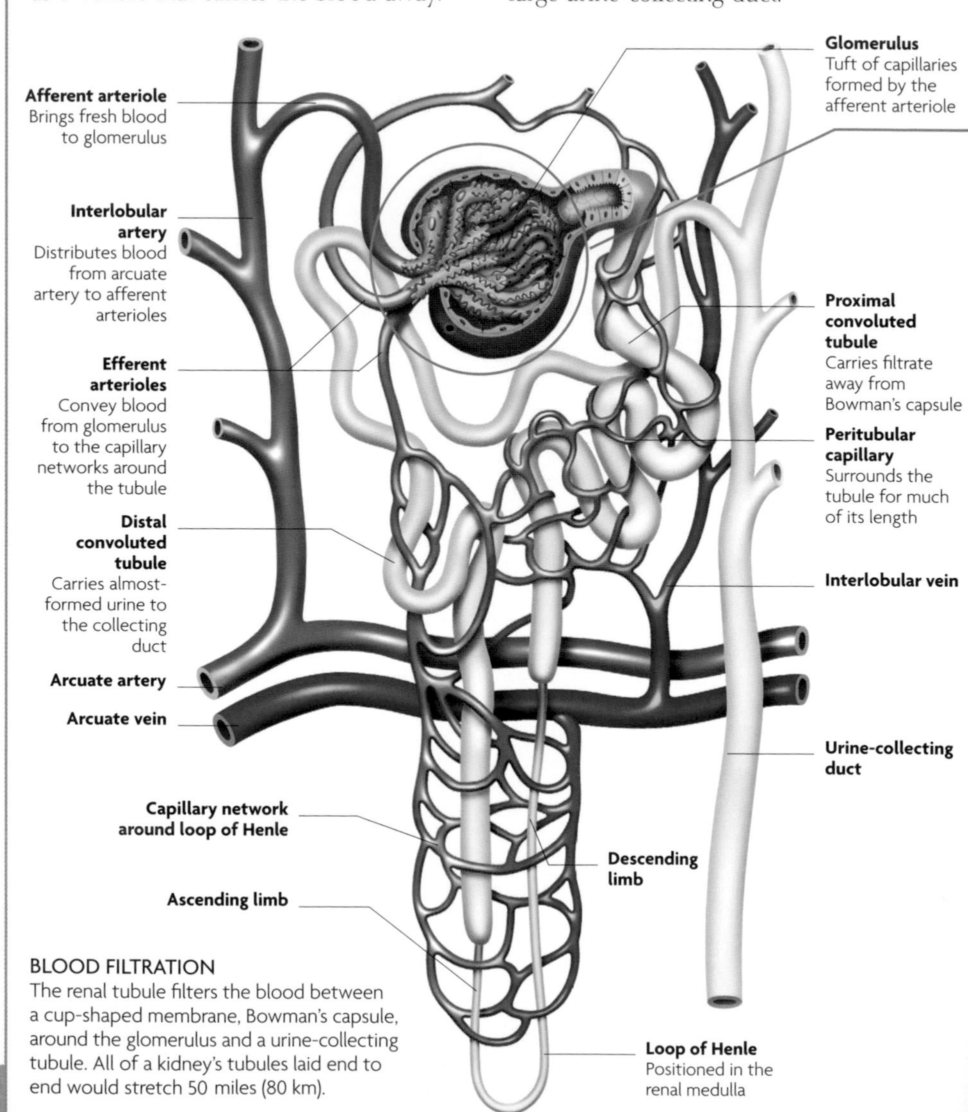
Afferent arteriole
Brings fresh blood to glomerulus
Interlobular artery
Distributes blood from arcuate artery to afferent arterioles
Efferent arterioles
Convey blood from glomerulus to the capillary networks around the tubule
Distal convoluted tubule
Carries almost-formed urine to the collecting duct
Arcuate artery
Arcuate vein
Capillary network around loop of Henle
Ascending limb
Glomerulus
Tuft of capillaries formed by the afferent arteriole
Proximal convoluted tubule
Carries filtrate away from Bowman's capsule
Peritubular capillary
Surrounds the tubule for much of its length
Interlobular vein
Urine-collecting duct
Descending limb
Loop of Henle
Positioned in the renal medulla

BLOOD FILTRATION

The renal tubule filters the blood between a cup-shaped membrane, Bowman's capsule, around the glomerulus and a urine-collecting tubule. All of a kidney's tubules laid end to end would stretch 50 miles (80 km).

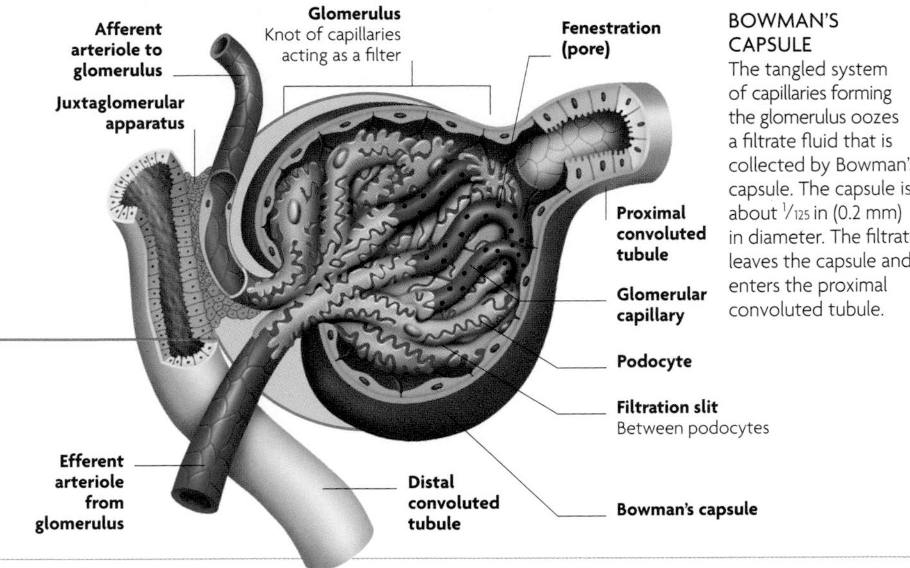
Glomerulus
Knot of capillaries acting as a filter
Afferent arteriole to glomerulus
Juxtaglomerular apparatus
Fenestration (pore)
Proximal convoluted tubule
Glomerular capillary
Podocyte
Filtration slit
Between podocytes
Efferent arteriole from glomerulus
Distal convoluted tubule
Bowman's capsule

BOWMAN'S CAPSULE

The tangled system of capillaries forming the glomerulus oozes a filtrate fluid that is collected by Bowman's capsule. The capsule is about $^1/_{125}$ in (0.2 mm) in diameter. The filtrate leaves the capsule and enters the proximal convoluted tubule.

MAKING URINE

The nephrons of both kidneys filter the body's total blood plasma about twice every hour. They produce about 317 pints (150 liters) of glomerular filtrate per day, 99 percent of which is reabsorbed to leave an average $3^1/_5$ pints (1.5 liters) of urine.

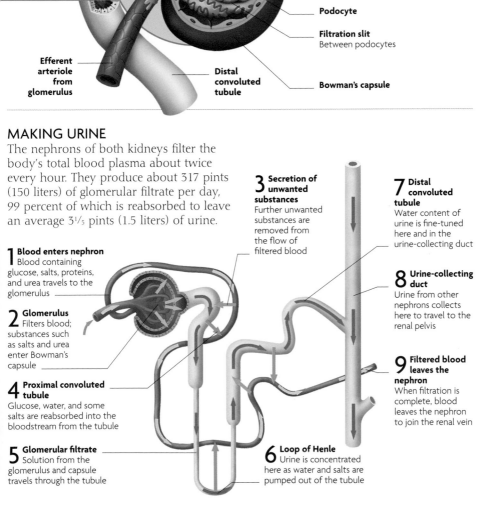

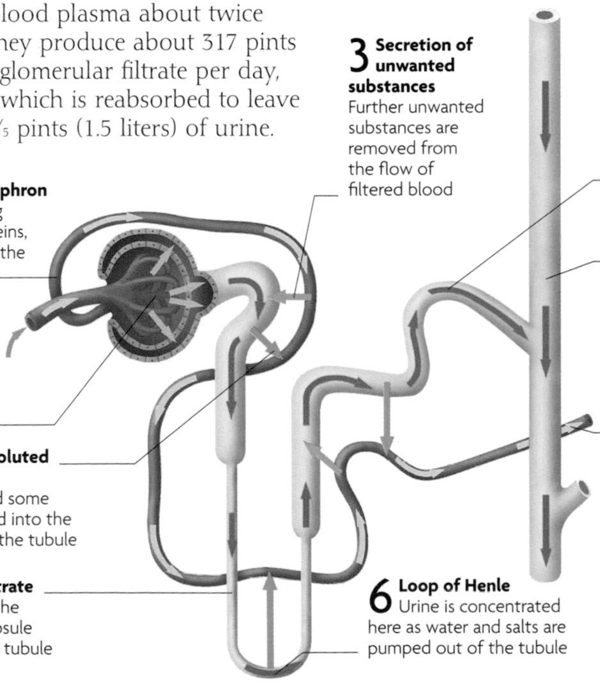

1 Blood enters nephron Blood containing glucose, salts, proteins, and urea travels to the glomerulus

2 Glomerulus Filters blood; substances such as salts and urea enter Bowman's capsule

3 Secretion of unwanted substances Further unwanted substances are removed from the flow of filtered blood

4 Proximal convoluted tubule Glucose, water, and some salts are reabsorbed into the bloodstream from the tubule

5 Glomerular filtrate Solution from the glomerulus and capsule travels through the tubule

6 Loop of Henle Urine is concentrated here as water and salts are pumped out of the tubule

7 Distal convoluted tubule Water content of urine is fine-tuned here and in the urine-collecting duct

8 Urine-collecting duct Urine from other nephrons collects here to travel to the renal pelvis

9 Filtered blood leaves the nephron When filtration is complete, blood leaves the nephron to join the renal vein

URINARY DISORDERS

PARTS OF THE URINARY TRACT ARE SUSCEPTIBLE TO INFECTIONS, RESULTING IN CONDITIONS SUCH AS CYSTITIS. SOME CHRONIC KIDNEY DISEASES ARE ALSO CAUSED BY INFECTION. COMMON SYMPTOMS, SUCH AS INCONTINENCE, CAN BE VERY TROUBLESOME.

URINARY TRACT INFECTIONS

The urine flowing through the urinary tract moves in one direction—from the kidneys through the ureters to the bladder, and then through the urethra to leave the body. During urination, the flow from the bladder is rapid and copious, but for long periods urine remains stagnant in the bladder. Infections can enter the body through the urethra and spread to the bladder, and sometimes up the ureters to the kidneys. The adult female urethra is 1½in (4 cm) long, compared to the male's of 8 in (20 cm). This short length and the proximity of its outlet to the anus (allowing bacteria from the anal area to enter the urethra) together account for females' greater susceptibility to urinary infection. One of the most common urinary infections is inflammation of the bladder, known as cystitis. The main symptoms of cystitis are burning pain and a frequent need to urinate, but often with little urine on each occasion.

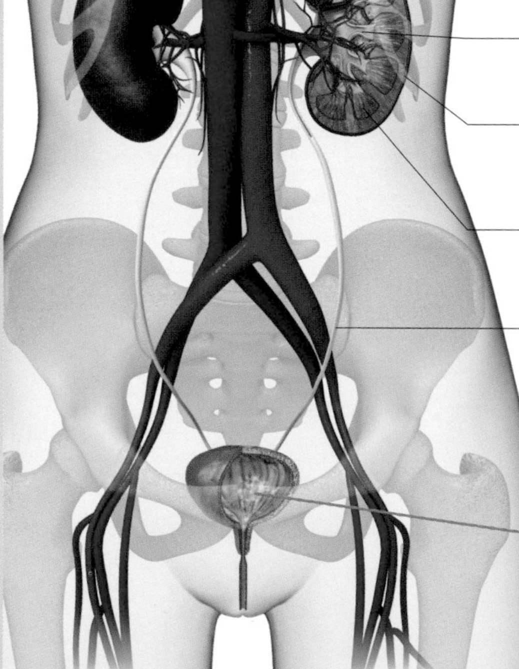

SITES OF DISORDERS

Each of the urinary organs is affected by its own characteristic diseases. However, a disorder of any single organ can affect other parts of the system.

Pyelonephritis
An acute infection of the urine-collecting system of the kidney

Diabetic nephropathy
Changes to capillaries in the kidneys, which may lead to kidney failure; caused by long-term diabetes mellitus

Glomerulonephritis
Inflammation of the filtering units of the kidney (glomeruli); often related to an autoimmune process

Reflux
Forcing of urine up the ureters by back pressure; can be caused by a blockage of the urethra

CYSTITIS
This micrograph shows cystitis affecting a bladder lining. Bacteria (yellow rods) colonize the lining's inner surface (blue), causing inflammation.

INCONTINENCE

A tendency to leak urine, incontinence most commonly occurs in women, elderly people, and those with brain or spinal cord damage. Women after childbirth are susceptible because their pelvic floor muscles may be weak. There are different types, such stress incontinence (see below). In urge incontinence, irritable bladder muscle causes the bladder to contract and expel all its urine. In total incontinence, a nervous system disorder such as dementia causes total loss of bladder function.

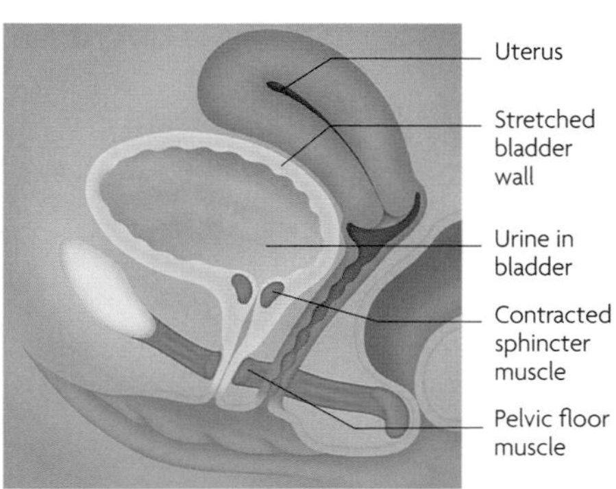
Uterus
Stretched bladder wall
Urine in bladder
Contracted sphincter muscle
Pelvic floor muscle

NORMAL BLADDER

A healthy bladder expands like a balloon as it fills with urine. The sphincter muscles and surrounding pelvic floor muscles keep the exit closed. Nerve signals from stretch sensors in the bladder wall travel to the brain, signaling the need for emptying.

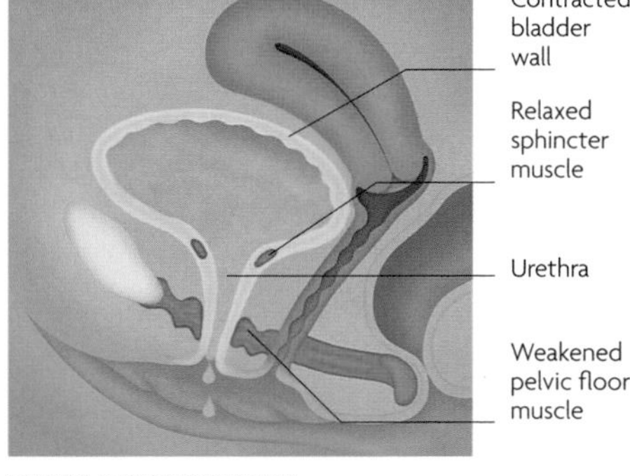
Contracted bladder wall
Relaxed sphincter muscle
Urethra
Weakened pelvic floor muscle

STRESS INCONTINENCE

To empty the bladder, the sphincter and pelvic floor muscles relax, and the detrusor muscle in the bladder wall contracts, forcing urine along the urethra. In incontinence, weak muscles may allow this to happen without proper control, so urine leaks out.

KIDNEY STONES

Kidney stones are solid, mineral-rich objects formed from chemicals, such as calcium salts, in urine. They can take years to form, and grow in various shapes and sizes. A stone may stay in the kidney and cause few problems, but it can increase the risk of urinary tract infection.

CRYSTALS

Kidney stones are usually formed from the mineral salt calcium oxalate, when it crystallizes from the urine. Crystals of this salt are shown here.

WHERE KIDNEY STONES FORM

Kidney stones can occur in any of the urine-collecting parts of the kidneys, such as the calyx or the renal pelvis.

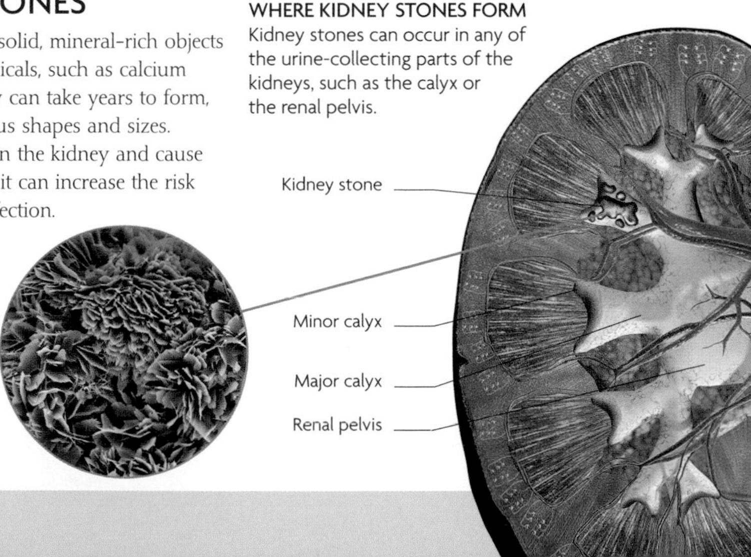
Kidney stone
Minor calyx
Major calyx
Renal pelvis

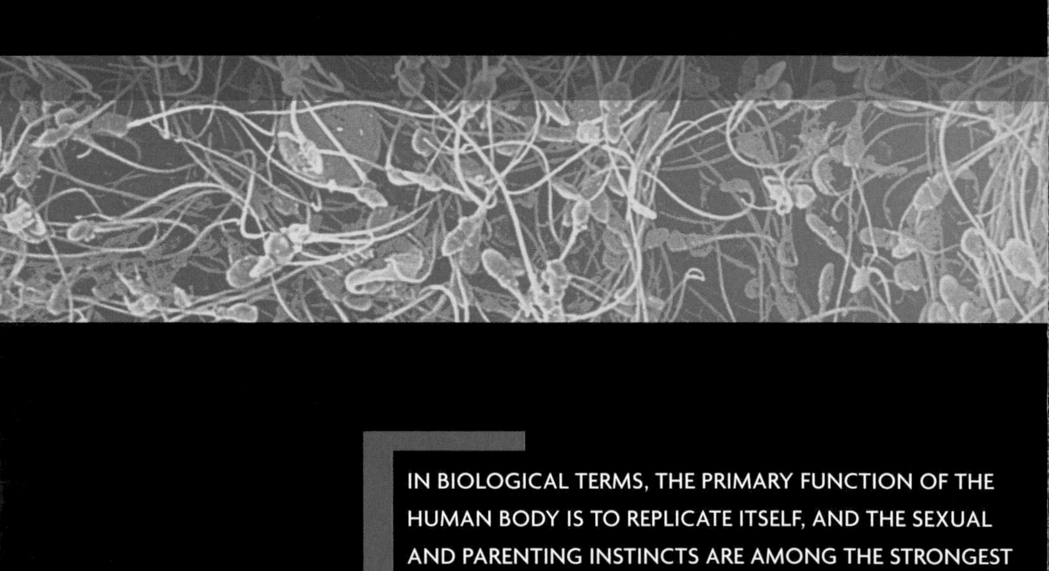

IN BIOLOGICAL TERMS, THE PRIMARY FUNCTION OF THE HUMAN BODY IS TO REPLICATE ITSELF, AND THE SEXUAL AND PARENTING INSTINCTS ARE AMONG THE STRONGEST OF OUR BASIC DRIVES. AS SCIENCE WIDENS THE GAP

REPRODUCTION AND LIFE CYCLE

MALE REPRODUCTIVE SYSTEM

THE MALE SYSTEM PRODUCES SEX CELLS (GAMETES) CALLED SPERM. UNLIKE FEMALE EGG MATURATION, WHICH OCCURS IN CYCLES AND CEASES AT MENOPAUSE, SPERM PRODUCTION IS CONTINUOUS, DECREASING GRADUALLY WITH AGE.

THE REPRODUCTIVE ORGANS

The male reproductive organs include the penis, two testes, several storage and transport ducts, and supporting structures. The oval-shaped testes lie outside the body in a pouch of skin called the scrotum, where they maintain the optimum temperature for making sperm—approximately 5°F (3°C) lower than body temperature. They also secrete the sex hormone testosterone. From each testis, sperm pass into a coiled tube–the epididymis–for the final stages of maturation. They are stored in the epididymides until they are either broken down and reabsorbed, or ejaculated–forced by movement of seminal fluid from the accessory glands (see 254) down a duct called the vas deferens.

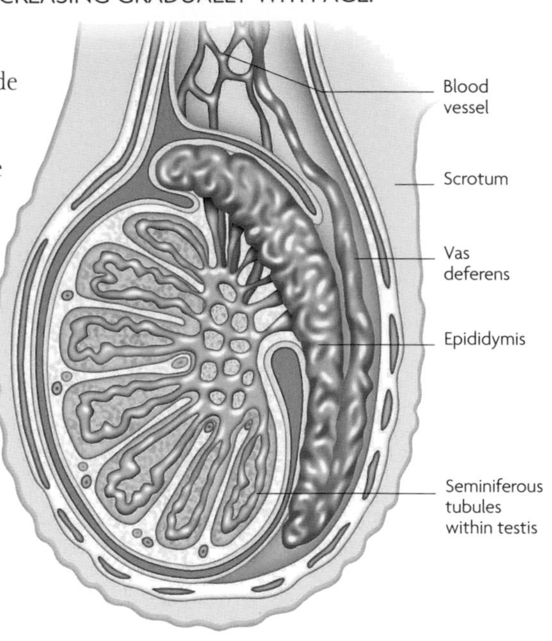
Blood vessel
Scrotum
Vas deferens
Epididymis
Seminiferous tubules within testis

INSIDE THE SCROTUM

The scrotum contains two testes, where sperm are manufactured within tubes called seminiferous tubules, and the two epididymides, where sperm are stored. Each epididymis is a tube about 20 ft (6 m) long, which is tightly coiled and bunched into a length of just 2 in (4 cm).

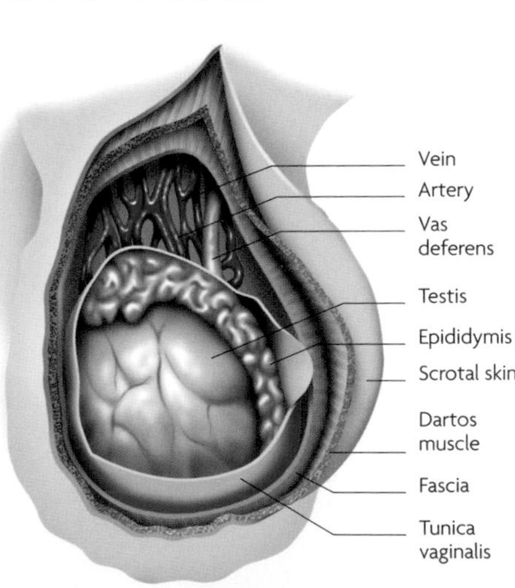
Vein
Artery
Vas deferens
Testis
Epididymis
Scrotal skin
Dartos muscle
Fascia
Tunica vaginalis

SCROTAL LAYERS

Each testis is covered by a thin tissue layer, the tunica vaginalis, and a layer of connective tissue called fascia. An outer layer called the dartos muscle relaxes in hot weather, dropping the testes to keep them cool, and draws them up in cold weather so they do not become too chilled. The spermatic cord suspends each testis within the scrotum: it contains the testicular artery and vein, lymph vessels, nerves, and the vas deferens.

MALE REPRODUCTIVE ORGANS

A midline section through the male lower abdomen shows how the penis and scrotum hang outside the body. Inside the body is a complex system of ducts, tubes, and glands where sperm mature and are stored before being ejaculated in semen.

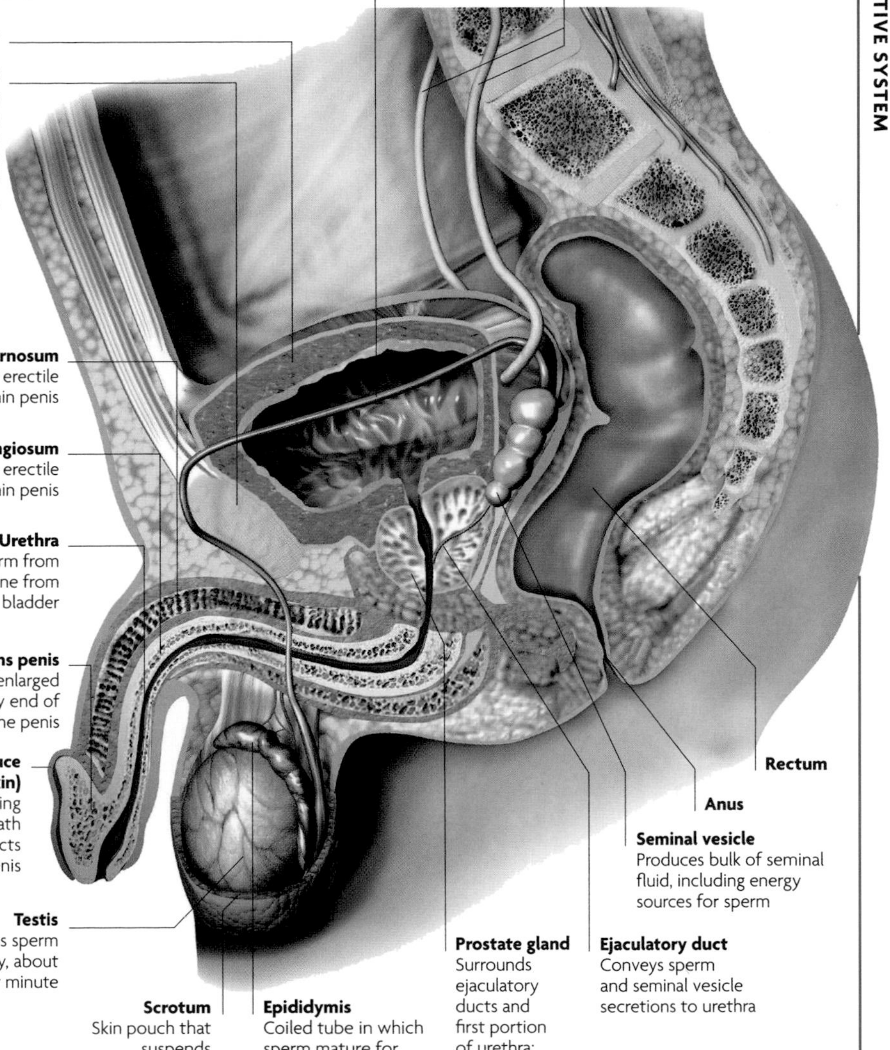

Vas deferens
Thick-walled duct, with narrow central space (lumen) that carries sperm

Ureters
Carry urine from kidneys to bladder; part of the urinary system

Bladder

Cartilage of pubic symphysis
Joint at midpoint of front of pelvis

Corpus cavernosum
Spongy erectile tissue within penis

Corpus spongiosum
Spongy erectile tissue within penis

Urethra
Conveys sperm from testes, or urine from bladder

Glans penis
Sensitive, enlarged fleshy end of the penis

Prepuce (foreskin)
Loosely fitting skin sheath that protects the glans penis

Testis
Produces sperm continuously, about 50,000 per minute

Scrotum
Skin pouch that suspends testes away from body and keeps them cool

Epididymis
Coiled tube in which sperm mature for about 1–3 weeks

Prostate gland
Surrounds ejaculatory ducts and first portion of urethra; produces fluid for semen

Ejaculatory duct
Conveys sperm and seminal vesicle secretions to urethra

Seminal vesicle
Produces bulk of seminal fluid, including energy sources for sperm

Anus

Rectum

ACCESSORY GLANDS

The seminal vesicles and the prostate and bulbourethral glands are together termed the accessory glands. Their secretions are added to sperm during ejaculation. Fluid from the seminal vesicles makes up about 60 percent of semen by volume, and contains sugar (fructose), vitamin C, and prostaglandins. Prostate secretions account for about 30 percent of semen, and include enzymes, fatty acids, cholesterol, and salts to adjust the semen's acid–alkali balance. Secretions from the bulbourethral glands make up 5 percent of semen, and neutralize the acidity of urine traces in the urethra.

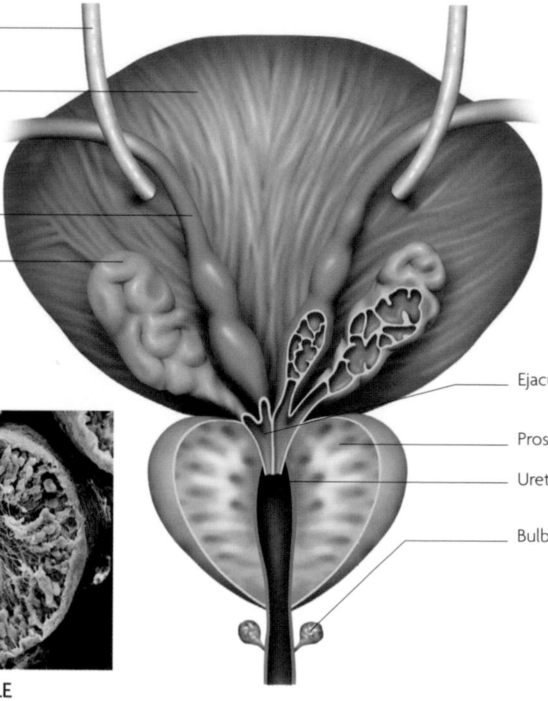

Ureter
Bladder
Vas deferens
Seminal vesicle
Ejaculatory duct
Prostate gland
Urethra
Bulbourethral gland

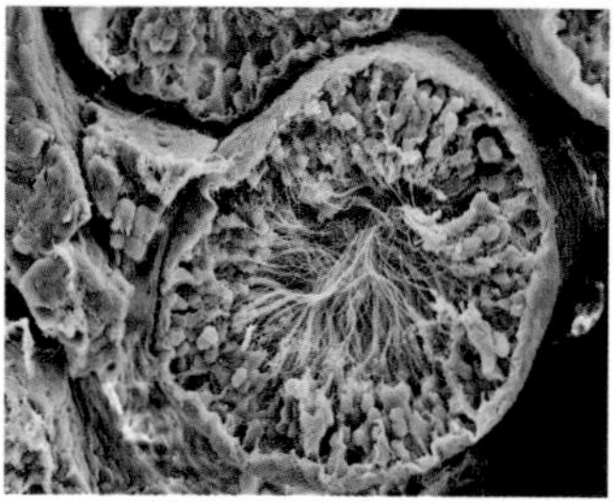

REAR VIEW OF GLANDS

SEMINIFEROUS TUBULE
This cross section of a seminiferous tubule shows sperm and their long tails as they move toward the center.

MAKING SPERM

Each testis is a mass of more than 800 tightly looped and folded seminiferous tubules. Here, sperm begin as bloblike cells called spermatogonia lining the inner wall. As they mature, they develop tails and move steadily toward the middle of the tubule. Thousands of sperm are produced every second, each taking about two months to mature.

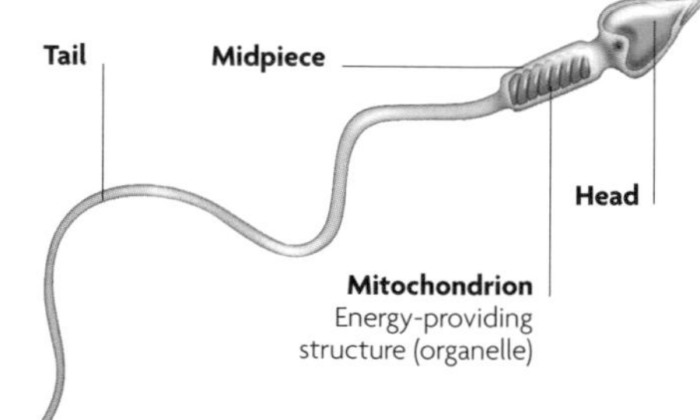
Tail
Midpiece
Head
Mitochondrion
Energy-providing structure (organelle)

PATHWAY FOR SPERM

During ejaculation, waves of muscle contraction squeeze the sperm in their fluid from the epididymis along the vas deferens. This tube is joined by a duct from the seminal vesicle to form the ejaculatory duct. The left and right ejaculatory ducts join the urethra within the prostate gland.

In the male, the urethra is a dual-purpose tube that carries urine from the bladder during urination and sperm from the testes. During ejaculation, however, the sphincter at the base of the bladder is closed because of high pressure in the urethra, preventing the passage of urine.

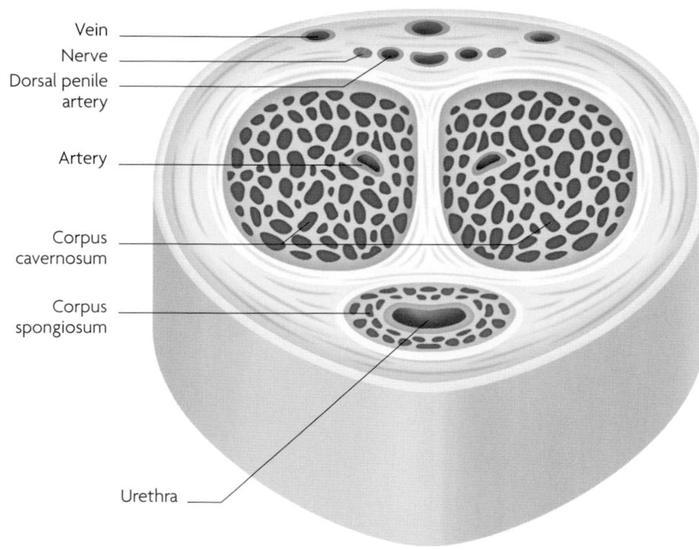

Vein
Nerve
Dorsal penile artery
Artery
Corpus cavernosum
Corpus spongiosum
Urethra

PENILE ERECTION

During arousal, large quantities of arterial blood enter the corpus spongiosum and corpus cavernosum, compressing the veins. As a result, blood cannot drain from the penis and it becomes hard and erect.

SEMEN

Seminal fluid, or semen, is sperm mixed with fluid added by the accessory glands (see opposite), including the prostate gland. The prostate secretes fluid through tiny ducts to mix with sperm as they are ejaculated down the urethra. The final mix has around 300–500 million sperm in $\frac{1}{15}$–$\frac{1}{6}$ fl oz (2–5 ml) of fluid.

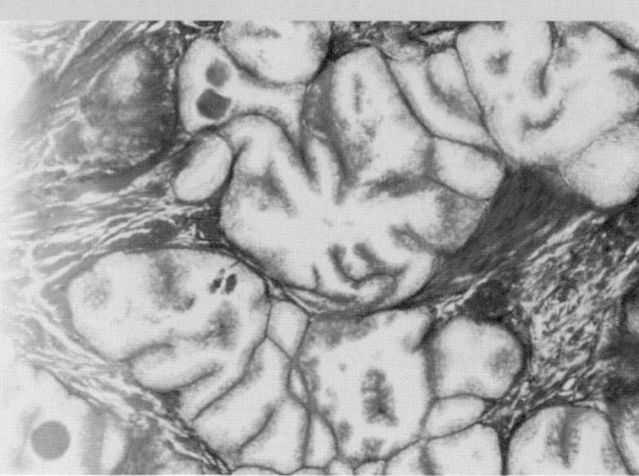

PROSTATE GLAND

This microscopic view of a section of prostate gland tissue shows a number of secretory ducts (orange and white).

FEMALE REPRODUCTIVE SYSTEM

THE FEMALE REPRODUCTIVE ORGANS RELEASE AN EGG AT REGULAR INTERVALS AND, IF IT IS FERTILIZED, PROTECT AND NOURISH THE EMBRYO AND FETUS.

REPRODUCTIVE TRACT

The female reproductive glands (ovaries) lie within the abdomen. From puberty, they mature and release the female sex cells (gametes), known as egg cells or ova. This release occurs roughly once a month as part of the menstrual cycle (see p.283). The ripe egg travels along the fallopian tube to the uterus, the muscular sac in which it develops into an embryo and then a fetus. Unfertilized eggs and the uterine lining are shed via the vagina. The ovaries also make the female sex hormone estrogen.

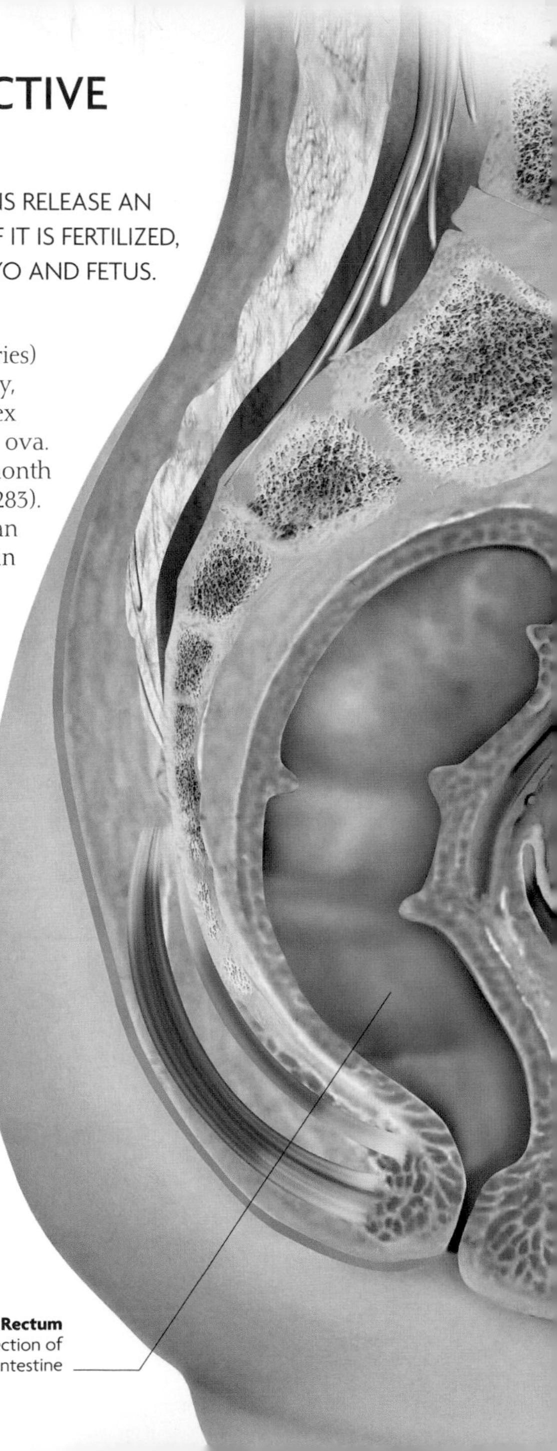

REPRODUCTIVE ORGANS

A cross section through the female lower abdomen reveals the reproductive structures and organs. The ovaries sit against the abdominal wall. The fallopian tubes arch from them, opening into the muscular, thick-walled womb (uterus).

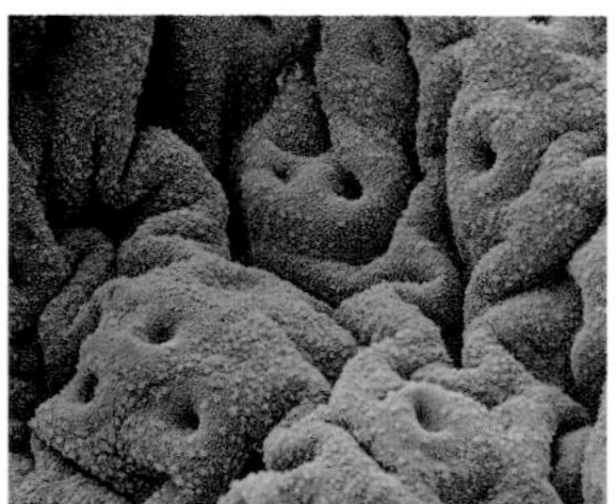

ENDOMETRIUM

This electron micrograph shows the thick, folded, glandular endometrium (the lining of the uterus). The tissue shown is very rich in blood and ready to receive a fertilized egg.

Rectum
Last section of large intestine

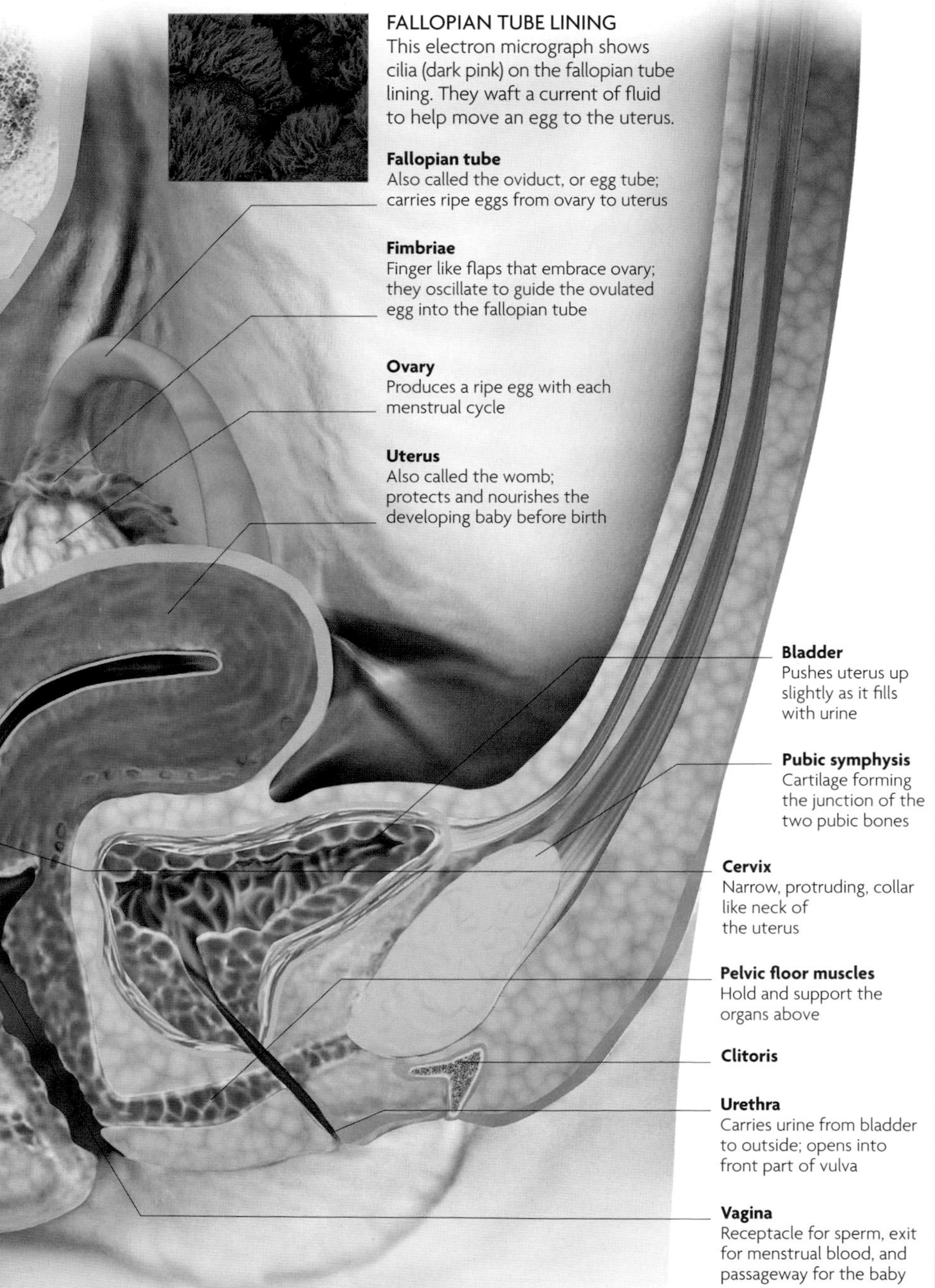

FALLOPIAN TUBE LINING
This electron micrograph shows cilia (dark pink) on the fallopian tube lining. They waft a current of fluid to help move an egg to the uterus.
Fallopian tube
Also called the oviduct, or egg tube; carries ripe eggs from ovary to uterus
Fimbriae
Finger like flaps that embrace ovary; they oscillate to guide the ovulated egg into the fallopian tube
Ovary
Produces a ripe egg with each menstrual cycle
Uterus
Also called the womb; protects and nourishes the developing baby before birth
Bladder
Pushes uterus up slightly as it fills with urine
Pubic symphysis
Cartilage forming the junction of the two pubic bones
Cervix
Narrow, protruding, collar like neck of the uterus
Pelvic floor muscles
Hold and support the organs above
Clitoris
Urethra
Carries urine from bladder to outside; opens into front part of vulva
Vagina
Receptacle for sperm, exit for menstrual blood, and passageway for the baby

VULVA

The external genital parts of the female are together known as the vulva. They are sited under the mons pubis, a mound of fatty tissue that covers the junction of the two pubic bones, the pubic symphysis. On the outside are the flaplike labia majora, and the smaller, foldlike labia minora lie within them. The labia majora contain sebaceous glands, smooth muscle, and sensory nerve endings. At puberty, their exposed surfaces begin to grow hairs. Within the vulva are the openings to the vagina and the urethra. At the front end of the labia minora is the clitoris. Like the male penis, it is sensitive and engorges with blood when aroused.

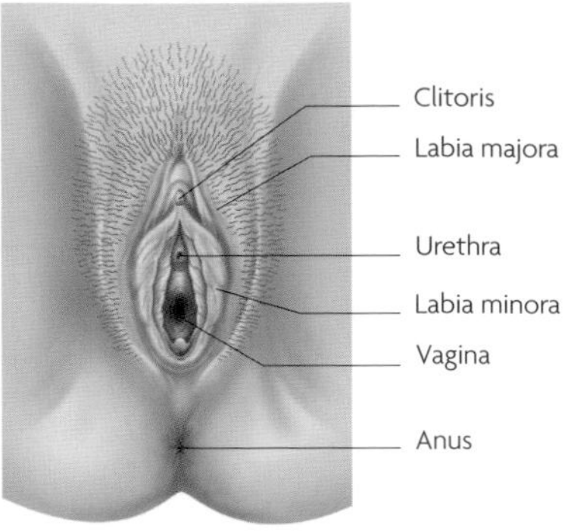

Clitoris
Labia majora
Urethra
Labia minora
Vagina
Anus

EXTERNAL GENITALS
The external genitals have a protective role, preventing infection from reaching the urethra or vagina, but allowing urine to exit.

OVULATION

An ovary contains thousands of immature egg cells. During a menstrual cycle, follicle-stimulating hormone (FSH) causes one egg to develop inside a primary follicle. As the the follicle enlarges, it moves to the ovary's surface and produces more estrogen. At ovulation, a surge of luteinizing hormone (LH) causes this secondary follicle to rupture and release the ripe egg. The lining of the empty follicle thickens into a corpus luteum—a temporary source of hormones.

EGG RELEASE
This colored electron micrograph shows an egg (red) as it is being released from its follicle into the abdominal cavity. Tendrils (fimbriae) at the end of each fallopian tube guide the egg into the tube.

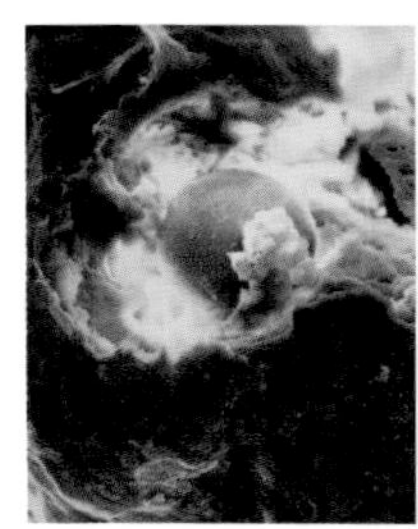

INSIDE AN OVARY
The ovary contains eggs that are undeveloped, eggs inside follicles at various stages of maturation, and empty follicles forming corpora lutea. Most of the glandular tissue around these follicles is known as the stroma.

Primary follicle
Early development; contains primary oocyte (unripe egg cell)

Secondary follicle
Mature stage of development, containing secondary oocyte (ripened egg)

Ovarian ligament
Stabilizes position of ovary within abdomen

Egg

Corpus luteum
An empty follicle, filled with hormone-producing cells

BREASTS

Breasts contain modified sweat glands called mammary glands, which produce milk at childbirth. A breast contains 15–20 lobes of compound areolar glands, each lobe resembling a bunch of grapes on a long stalk. The cells of the glands secrete milk, which flows along merging lactiferous ducts toward the nipple. The breast also contains a widespread drainage system of lymph vessels (see p.192).

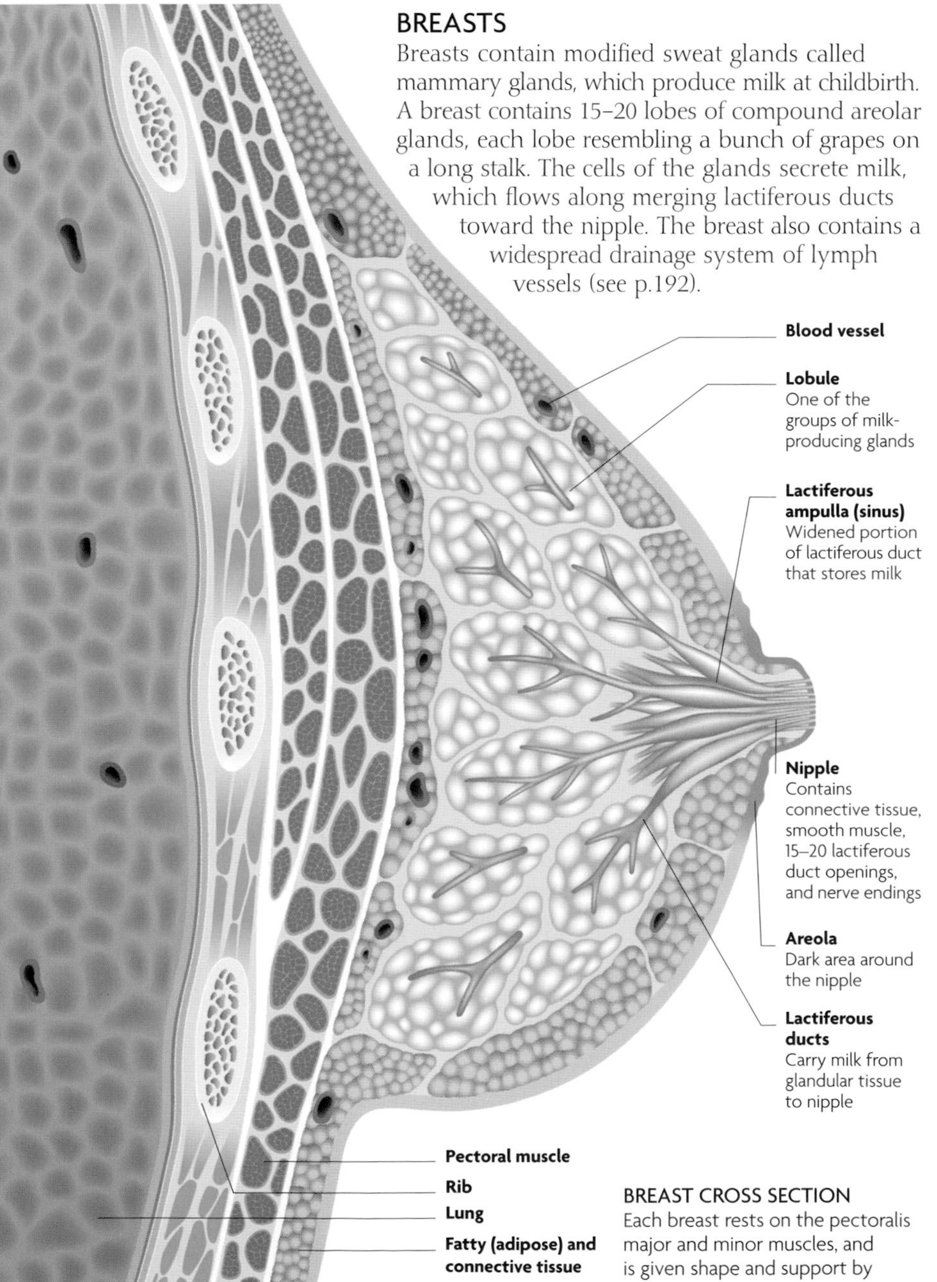

Blood vessel
Lobule
One of the groups of milk-producing glands
Lactiferous ampulla (sinus)
Widened portion of lactiferous duct that stores milk
Nipple
Contains connective tissue, smooth muscle, 15–20 lactiferous duct openings, and nerve endings
Areola
Dark area around the nipple
Lactiferous ducts
Carry milk from glandular tissue to nipple
Pectoral muscle
Rib
Lung
Fatty (adipose) and connective tissue

BREAST CROSS SECTION

Each breast rests on the pectoralis major and minor muscles, and is given shape and support by internal suspensory ligaments.

CONCEPTION TO EMBRYO

THE EMBRYONIC CELLS REPEATEDLY DIVIDE, AND BECOME IMPLANTED INTO THE UTERUS LINING.

The first eight weeks in the uterus are known as the embryo stage, in which the fertilized egg becomes a tiny human body, no larger than a thumb. The fertilized egg develops into an enlarging cluster of cells, the blastocyst. Some cells will form the baby's body, while others become the protective membranes or the placenta, which nourishes the embryo and removes waste products.

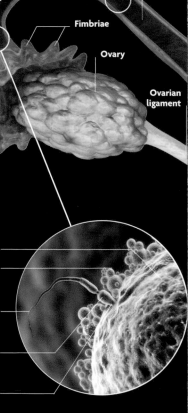

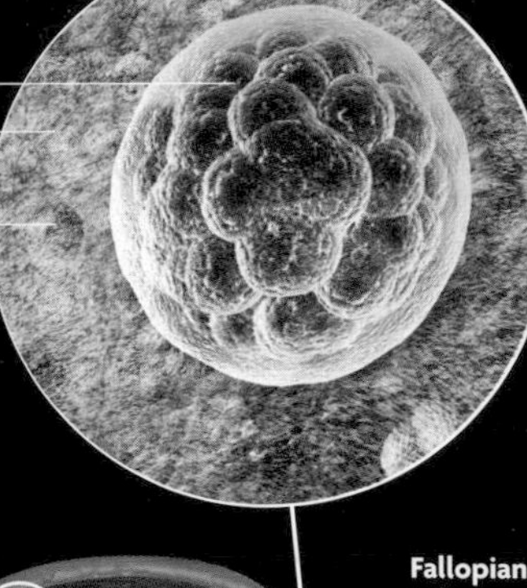
Morula
Fallopian tube lining
Cilia

3 MORULA A cluster of 16–32 cells, the morula leaves the fallopian tube and enters the uterus about 3–4 days after fertilization.

Fallopian tube

Fimbriae

Ovary

Ovarian ligament

Fallopian tube
Conveys zygote toward uterus

First cleavage
Large zygote splits itself into two cells

Cilia
Microhairs waft the zygote along

Goblet cells
Secrete fluid into tube

2 ZYGOTE The fertilized egg passes along the fallopian tube. Within 24–36 hours it has divided into two cells, then 12 hours later into four cells, and so on. This process is known as cleavage. At each stage the resulting cells become smaller, gradually approaching normal body cell size.

1 FERTILIZATION Fertilization takes place in the fallopian tube, when the head of the sperm cell, or spermatozoon, penetrates the much larger ripe egg cell, or mature ovum. This forms a single cell—the fertilized egg, or zygote, which contains 23 pairs of chromosomes (see p.286).

Ovum (egg cell)
Up to 1/250 in (0.1 mm) across (huge compared to other cells); contains 23 maternal chromosomes

Corona cell
Secretes chemicals to aid egg development

Tail of sperm
Lashes to propel sperm toward egg

Sperm head
Contains 23 paternal chromosomes

Acrosome
"Cap" of sperm head, which penetrates egg cell membrane

4 BLASTOCYST About six days after fertilization, the cell cluster forms a hollow cavity known as a blastocyst. It floats within the uterus for around 48 hours before landing on the thick uterus lining (endometrium), which softens to aid implantation (burrowing of the blastocyst into the endometrium). The inner group of cells will become the embryo itself.

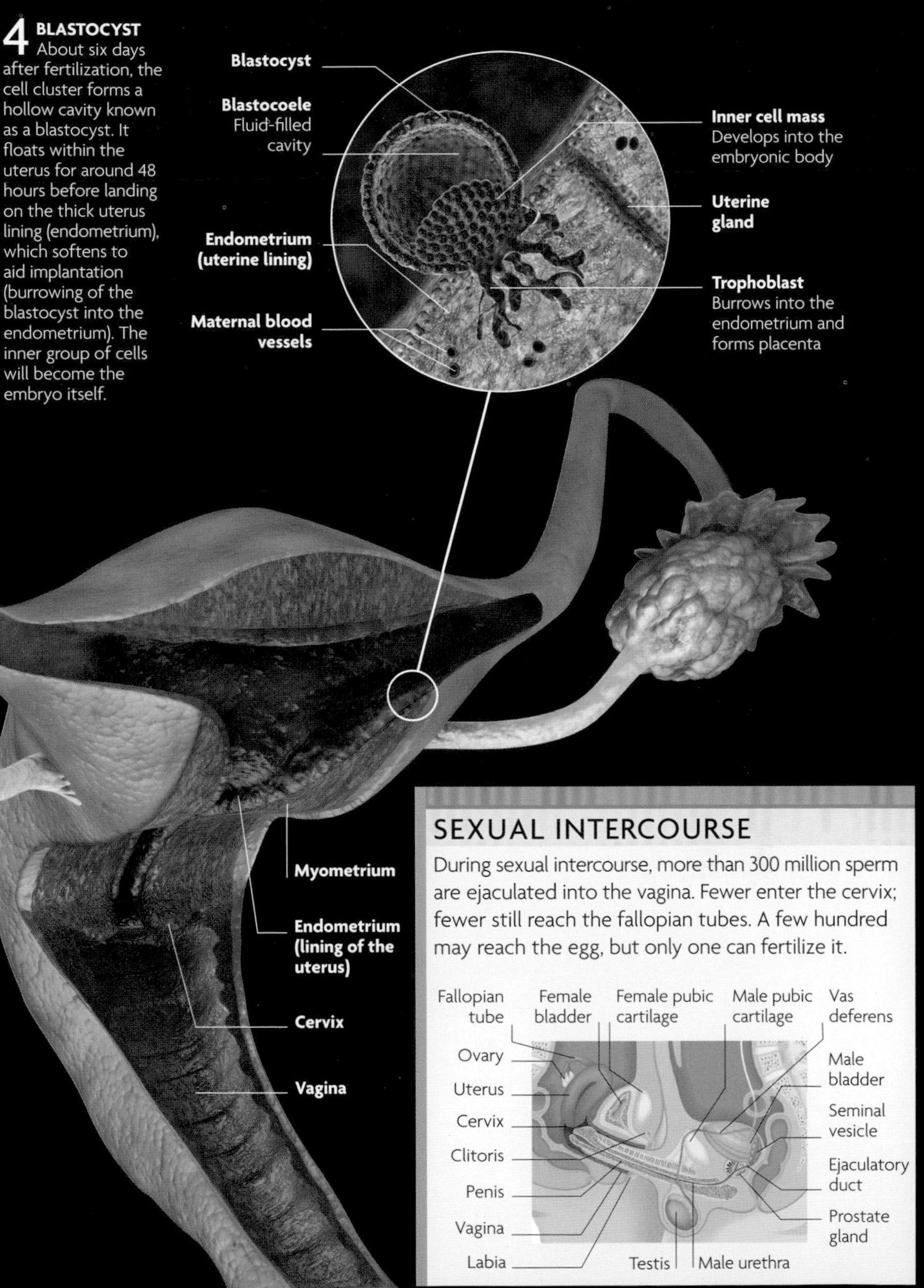
Blastocyst
Blastocoele Fluid-filled cavity
Endometrium (uterine lining)
Maternal blood vessels
Inner cell mass Develops into the embryonic body
Uterine gland
Trophoblast Burrows into the endometrium and forms placenta
Myometrium
Endometrium (lining of the uterus)
Cervix
Vagina

SEXUAL INTERCOURSE

During sexual intercourse, more than 300 million sperm are ejaculated into the vagina. Fewer enter the cervix; fewer still reach the fallopian tubes. A few hundred may reach the egg, but only one can fertilize it.

Fallopian tube
Female bladder
Female pubic cartilage
Male pubic cartilage
Vas deferens
Ovary
Uterus
Cervix
Clitoris
Penis
Vagina
Labia
Testis
Male urethra
Male bladder
Seminal vesicle
Ejaculatory duct
Prostate gland

EMBRYONIC DISK

Within the inner cell mass, an embryonic disk forms. This separates the cell cluster into the amniotic cavity, which develops into a sac that will fill with fluid and fold around to cover the embryo, and the yolk sac, which helps transport nutrients to the embryo during the second and third weeks. The disk develops three circular sheets called the primary germ layers—ectoderm, mesoderm, and endoderm—from which all body structures will derive.

EARLY DEVELOPMENT

As soon as implantation has taken place in the lining of the uterus, development begins. The embryonic disk forms the three germ layers, and the placenta starts to form from the trophoblast.

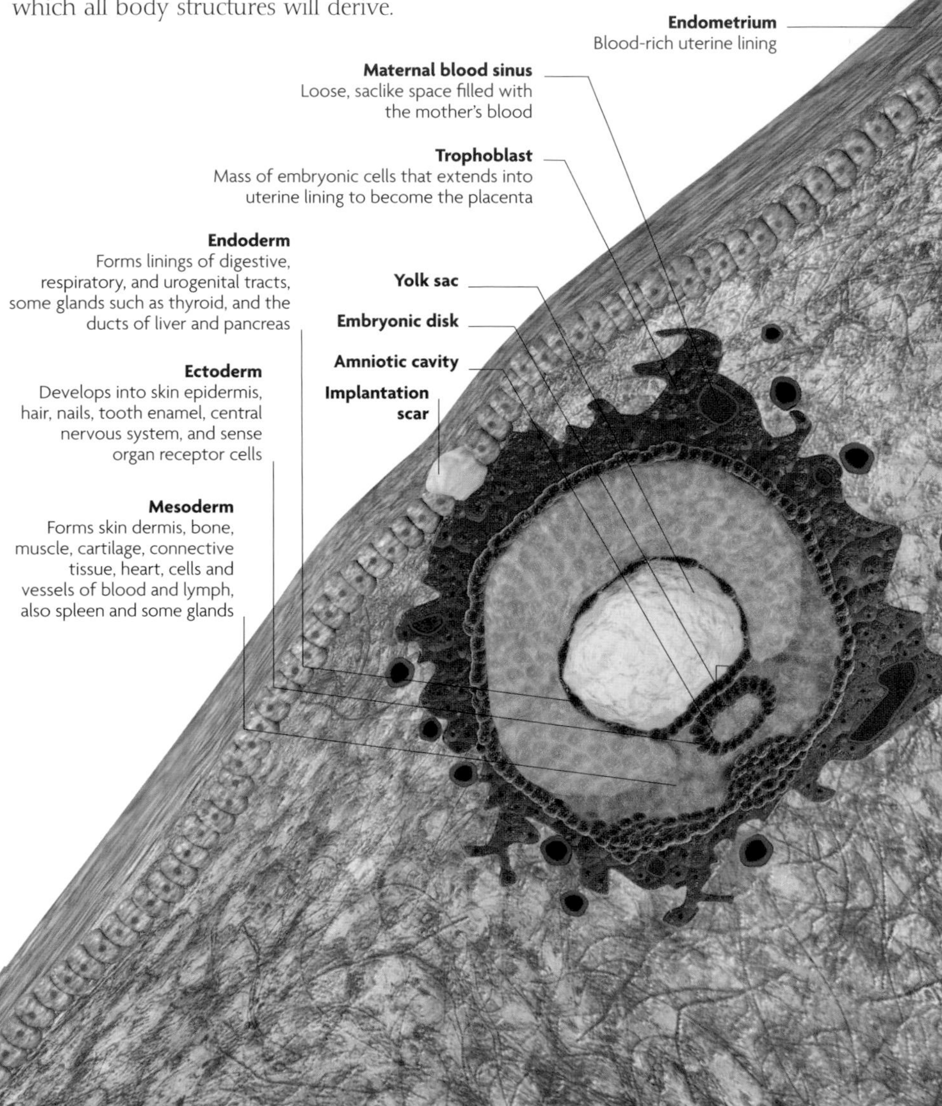
Endometrium
Blood-rich uterine lining
Maternal blood sinus
Loose, saclike space filled with the mother's blood
Trophoblast
Mass of embryonic cells that extends into uterine lining to become the placenta
Endoderm
Forms linings of digestive, respiratory, and urogenital tracts, some glands such as thyroid, and the ducts of liver and pancreas
Yolk sac
Embryonic disk
Amniotic cavity
Implantation scar
Ectoderm
Develops into skin epidermis, hair, nails, tooth enamel, central nervous system, and sense organ receptor cells
Mesoderm
Forms skin dermis, bone, muscle, cartilage, connective tissue, heart, cells and vessels of blood and lymph, also spleen and some glands

GROWING EMBRYO

In general, development is from the head down: the brain and head take shape early, then the body, arms, and lastly the legs. Eight weeks after fertilization, all major organs and body parts have formed. From this time on, the baby is known as a fetus.

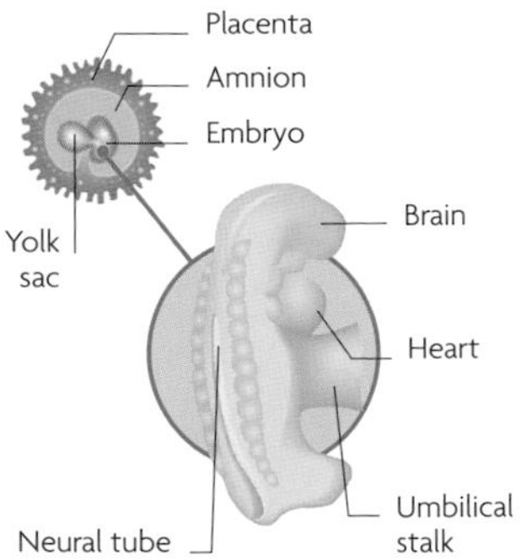
Placenta
Amnion
Embryo
Yolk sac
Brain
Heart
Umbilical stalk
Neural tube

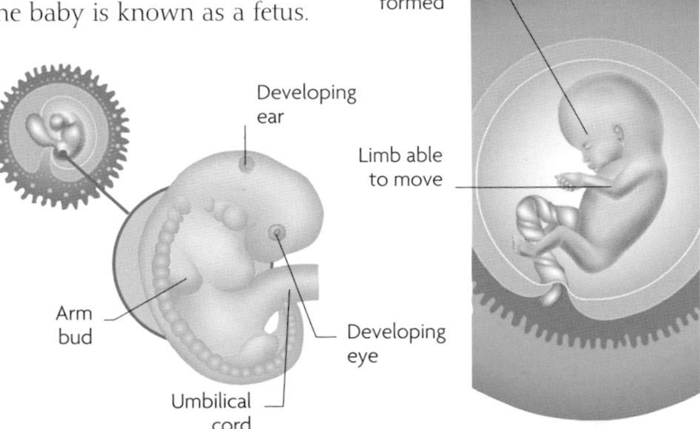
Developing ear
Arm bud
Developing eye
Umbilical cord
Face formed
Limb able to move

THREE WEEKS

The embryo is $^4/_{50}$–$^5/_{50}$ in (2–3 mm) long. The neural tube forms. It will become the spinal cord, with a brain at one end. A tubelike heart pulsates.

FOUR WEEKS

The embryo is about $^1/_5$ in (4–5 mm) long. A four-chambered heart beats, sending blood through simple vessels. Intestines, liver, lungs, and limb buds can be seen.

EIGHT WEEKS

The embryo is around 1–1 $^1/_5$ in (25–30 mm) long. The face, neck, fingers, and toes can be seen.

DEVELOPMENT OF THE PLACENTA

The placenta derives from the trophoblast—the outer layer of the blastocyst (the mass of cells that results from the fusion of egg and sperm). It begins to form soon after the fertilized egg implants in the uterine lining (see opposite), and becomes almost fully developed by the fifth month of pregnancy.

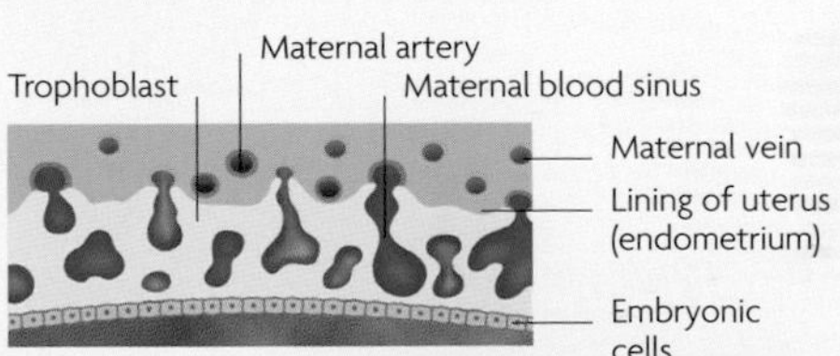
Trophoblast
Maternal artery
Maternal blood sinus
Maternal vein
Lining of uterus (endometrium)
Embryonic cells

1 Embryonic cells extend into uterine blood vessels, so that maternal blood flows into spaces (sinuses) within the trophoblast.

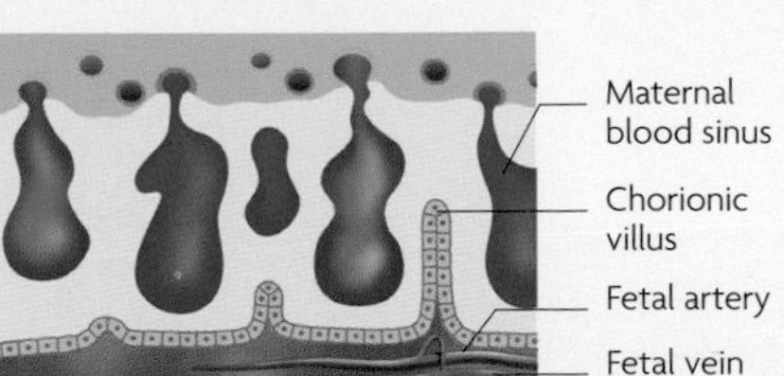
Maternal blood sinus
Chorionic villus
Fetal artery
Fetal vein

2 Fingerlike projections, called chorionic villi, grow and are surrounded by maternal blood sinuses. Later, fetal blood vessels grow into the villi.

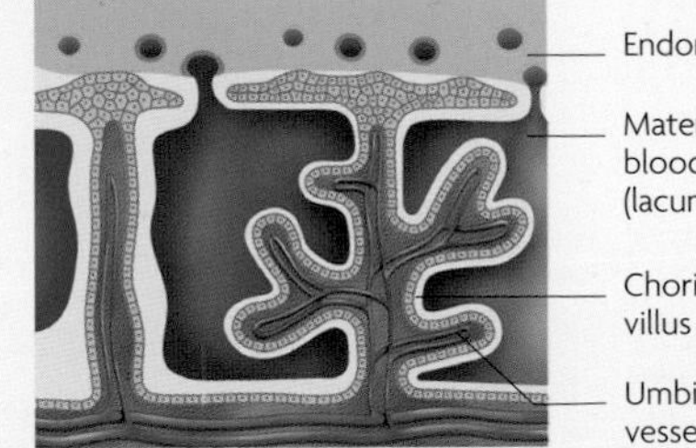
Endometrium
Maternal blood chamber (lacuna)
Chorionic villus
Umbilical cord vessel

3 The villi branch further and the maternal blood sinuses enlarge into lacunae ("lakes"), supplying the placenta with oxygen and nutrients.

FETAL DEVELOPMENT

FROM THE EIGHTH WEEK, WHEN THE BABY STARTS TO BE KNOWN AS A FETUS, ITS BODY GROWS LARGER AND STRONGER.

CHANGES IN THE FETUS

By 12 weeks, the fetus has a large head compared with the rest of its body, and all major internal organs have developed. By around 16 weeks, the fetus can move its limbs vigorously. As its growth continues, the fetus becomes leaner, but by the seventh to eighth month, it starts to accumulate fat and to assume the "chubby" appearance of the newborn.

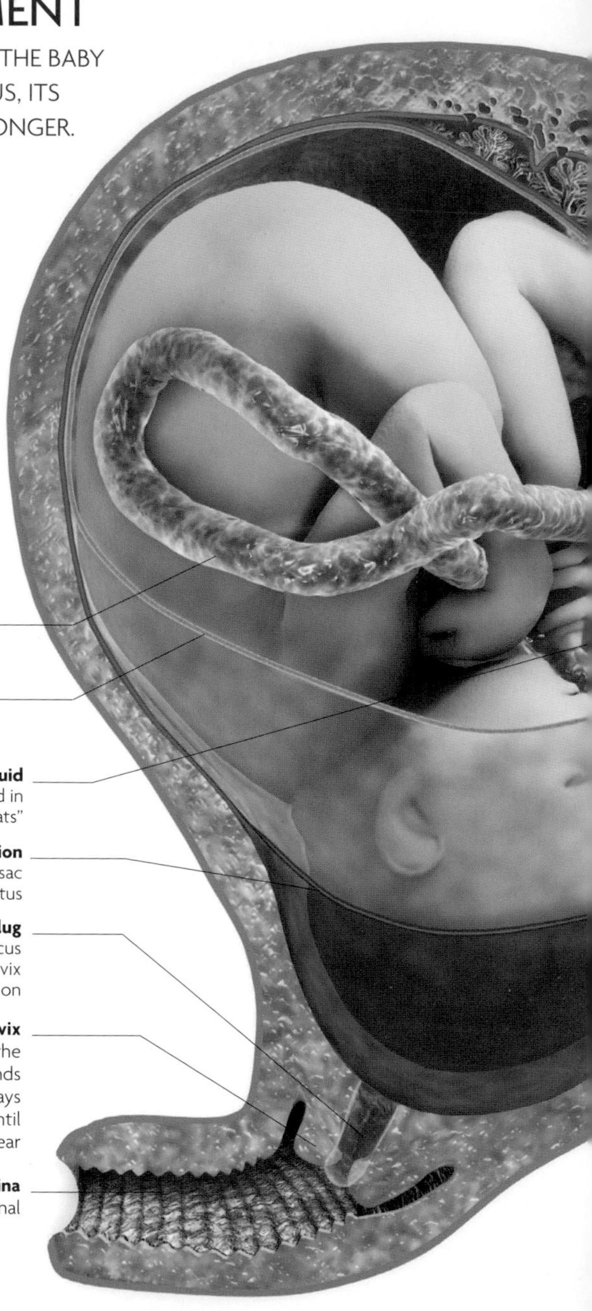
Umbilical cord
Immunological, nutritional, and hormonal link with the mother
Amnion
Strong, transparent sac within the chorion; it encloses amniotic fluid
Amniotic fluid
Shock-absorbing liquid in which the fetus "floats"
Chorion
Main protective sac around the fetus
Cervical plug
Plug of thick mucus that blocks the cervix to prevent infection
Cervix
Lower part of the uterus that extends into the vagina; it stays tightly closed until birth is near
Vagina
Birth canal

36 WEEKS
The fetus is now somewhat restricted by the uterus. The side of the placenta facing the fetus is smooth and circular in outline, with the umbilical cord attached at the center.

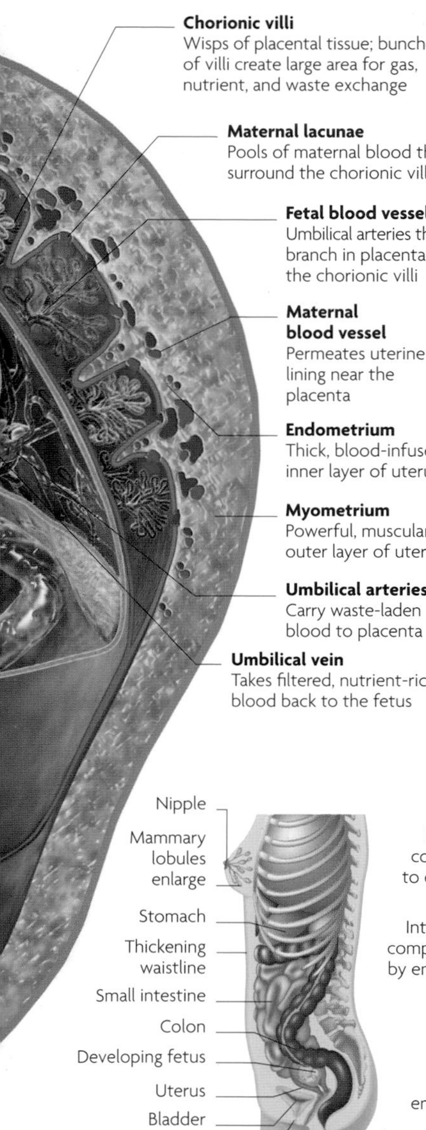
Chorionic villi
Wisps of placental tissue; bunches of villi create large area for gas, nutrient, and waste exchange
Maternal lacunae
Pools of maternal blood that surround the chorionic villi
Fetal blood vessels
Umbilical arteries that branch in placenta to the chorionic villi
Maternal blood vessel
Permeates uterine lining near the placenta
Endometrium
Thick, blood-infused inner layer of uterus
Myometrium
Powerful, muscular outer layer of uterus
Umbilical arteries
Carry waste-laden blood to placenta
Umbilical vein
Takes filtered, nutrient-rich blood back to the fetus

HOW THE PLACENTA WORKS

Oxygen, nutrients, and antibodies pass from the mother to the fetus in the umbilical veins; fetal waste passes to the mother in the umbilical arteries.

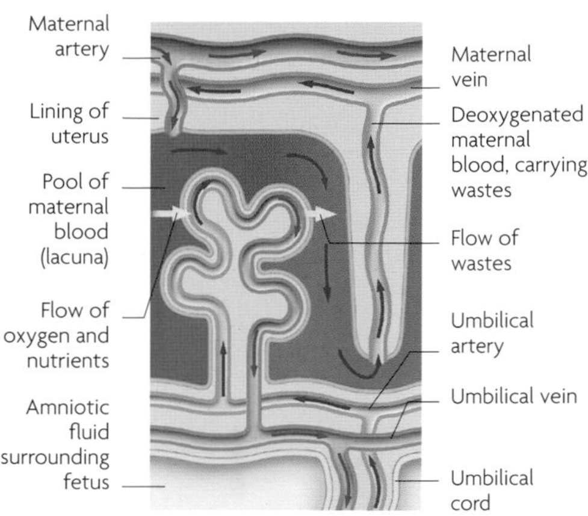
Maternal artery
Lining of uterus
Pool of maternal blood (lacuna)
Flow of oxygen and nutrients
Amniotic fluid surrounding fetus
Maternal vein
Deoxygenated maternal blood, carrying wastes
Flow of wastes
Umbilical artery
Umbilical vein
Umbilical cord

EXCHANGE OF OXYGEN AND NUTRIENTS

A thin barrier of cells in the chorion allows the exchange of gases, nutrients, and waste between mother and fetus.

CHANGES IN THE MOTHER

Pregnancy is divided into trimesters, each lasting about three calendar months. During this time, the mother's body changes to support the fetus and to prepare itself for childbirth and breast-feeding.

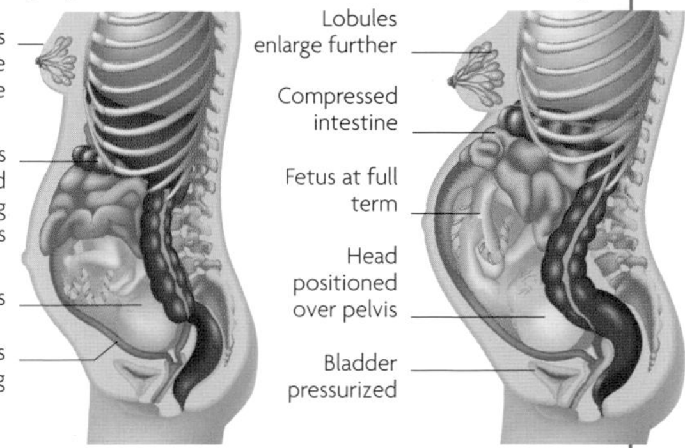
Nipple
Mammary lobules enlarge
Stomach
Thickening waistline
Small intestine
Colon
Developing fetus
Uterus
Bladder
Urethra
Breasts continue to enlarge
Intestines compressed by enlarging uterus
Fetus
Uterus enlarging
Lobules enlarge further
Compressed intestine
Fetus at full term
Head positioned over pelvis
Bladder pressurized

FIRST TRIMESTER

Breasts become tender and larger, with darkened areolas; nausea and vomiting are common.

SECOND TRIMESTER

Enlarging uterus shows; heart rate increases; forehead and cheek skin may temporarily darken.

THIRD TRIMESTER

Abdominal skin stretches; fatigue, back pain, heartburn, and some breathlessness may occur.

PREPARING FOR BIRTH

CHANGES DURING LATE PREGNANCY SIGNAL THE APPROACH OF CHILDBIRTH. THE HEAD OF THE FETUS DROPS LOWER INTO THE PELVIS; THE EXPECTANT MOTHER MAY EXPERIENCE WEIGHT LOSS; AND THERE MAY BE EARLY UTERINE CONTRACTIONS.

MULTIPLE PREGNANCY AND FETAL POSITIONS

The presence of more than one fetus in the uterus is called a multiple pregnancy. Twins occur in approximately one in 80 pregnancies, and triplets in about one in 8,000. After about 30 weeks, the most common fetal position is head down, facing the mother's back, with the neck flexed forward. Such a position eases the baby's passage through the birth canal. However, about 1 in 30 full-term deliveries is breech, in which the baby's buttocks emerge before the head.

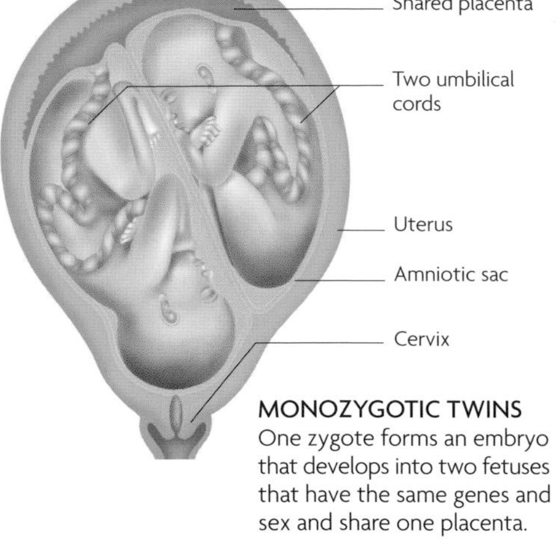
Shared placenta
Two umbilical cords
Uterus
Amniotic sac
Cervix

MONOZYGOTIC TWINS
One zygote forms an embryo that develops into two fetuses that have the same genes and sex and share one placenta. They are "identical" twins.

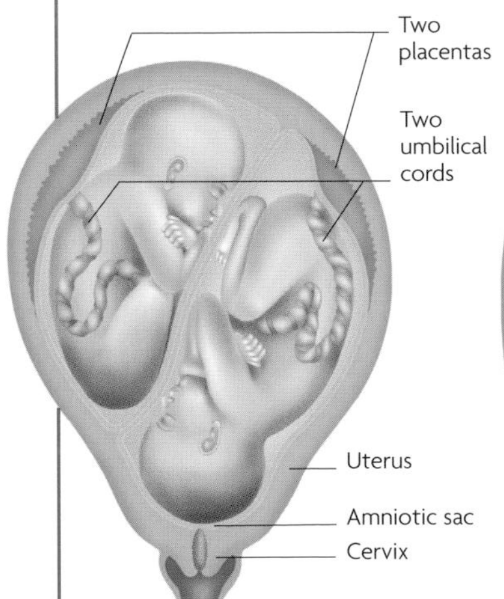
Two placentas
Two umbilical cords
Uterus
Amniotic sac
Cervix

DIZYGOTIC TWINS
Two zygotes develop separately, each with its own placenta. They may be different or the same sex. They are "fraternal twins" and are like any brothers and sisters.

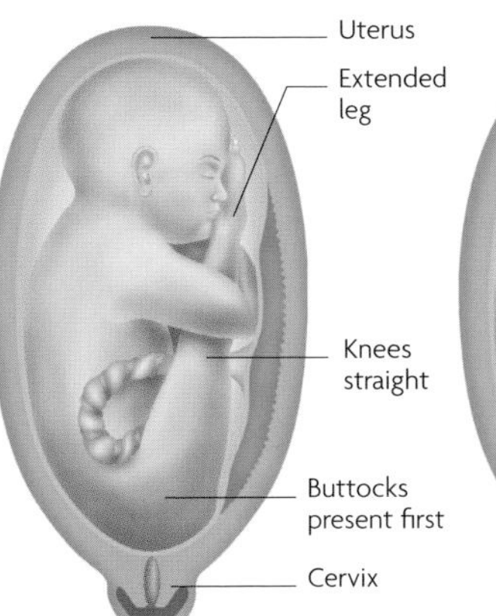
Uterus
Extended leg
Knees straight
Buttocks present first
Cervix

FRANK BREECH
The baby fails to turn head-down in the uterus. The hips are flexed and the legs are straight, extending alongside the body so that the feet are positioned beside the head.

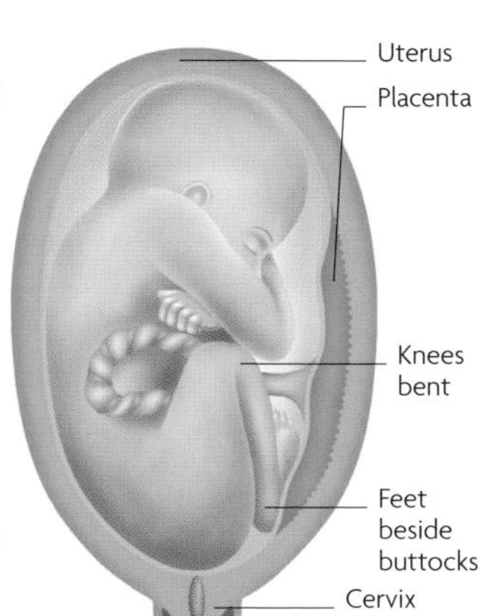
Uterus
Placenta
Knees bent
Feet beside buttocks
Cervix

COMPLETE BREECH
The legs are flexed at the hips and knees, so the feet are next to the buttocks. The incidence of breech delivery is much higher among premature babies.

CHANGES IN THE CERVIX

The cervix is the firm band of muscle and connective tissue that forms the neck-like structure at the bottom of the uterus. In late pregnancy, it softens in readiness for childbirth. Sporadic uterine tightenings, known as Braxton–Hicks contractions, help to thin the cervix so that it merges with the uterus's lower segment. These are usually painless and become noticeable only after the middle of pregnancy.

CERVIX SOFTENING

As labor nears, the cervix tissues lose their firm consistency. They become softer and more spongy, affected by natural substances in the blood called prostaglandins.

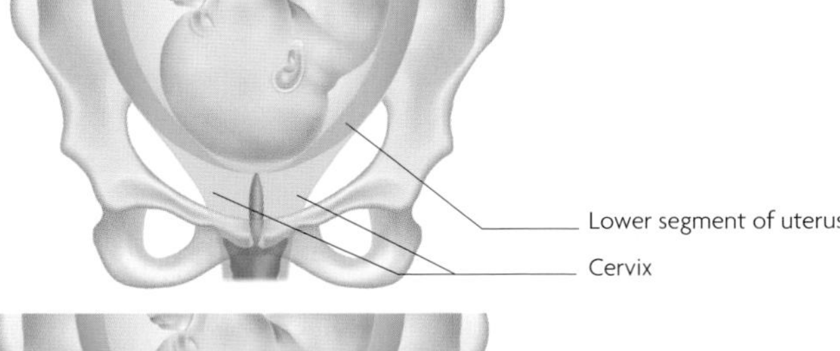
Lower segment of uterus
Cervix

CERVIX THINNING

The cervix becomes wider and thinner, and merges smoothly into the uterus wall above. The process of softening and thinning is known as effacement.

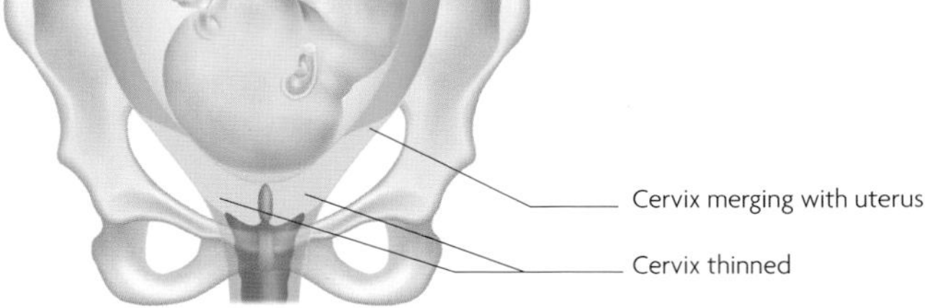
Cervix merging with uterus
Cervix thinned

CONTRACTIONS

The shortening of uterine muscles, with the eventual aim of expelling the fetus, are called contractions, which are regular and become steadily more frequent, more painful, and longer-lasting. The main area of contraction is in the muscles of the uterine fundus (upper uterus), which stretches, causing the lower uterus and cervix to thin.

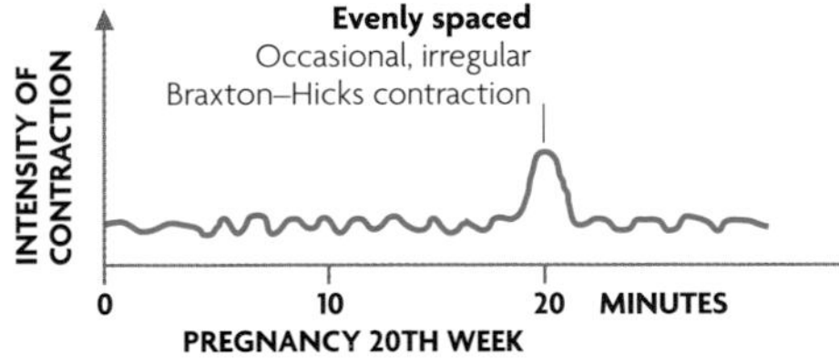
Evenly spaced
Occasional, irregular
Braxton–Hicks contraction
INTENSITY OF CONTRACTION
0
10
20
MINUTES
PREGNANCY 20TH WEEK

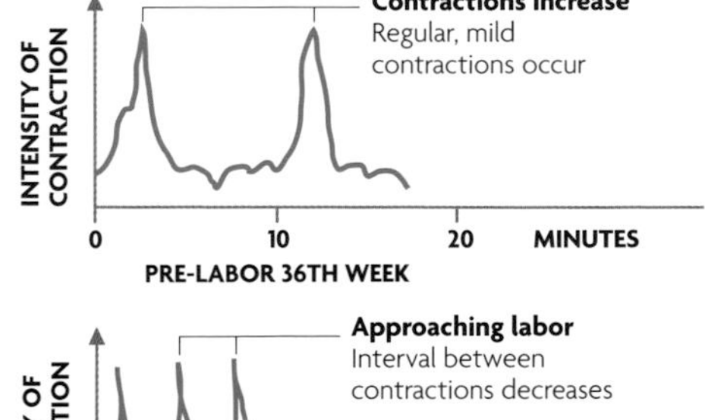
Contractions increase
Regular, mild
contractions occur
INTENSITY OF CONTRACTION
0
10
20
MINUTES
PRE-LABOR 36TH WEEK

Approaching labor
Interval between
contractions decreases
INTENSITY OF CONTRACTION
0
10
20
MINUTES
EARLY LABOR 40TH WEEK

PROGRESS OF CONTRACTIONS

Gentle, partial contractions occur through much of pregnancy. True contractions begin late in pregnancy. At first, they are occasional and relatively weak. As labor intensifies, they are more frequent and last longer, putting more downward pressure on the baby.

LABOR

LABOR USUALLY MEANS THE FULL PROCESS OF GIVING BIRTH. IT CAN BE DIVIDED INTO THREE PHASES OR STAGES: ONSET OF CONTRACTIONS TO FULL DILATION OF THE CERVIX; DELIVERY OF THE BABY; AND DELIVERY OF THE PLACENTA (AFTERBIRTH).

ENGAGEMENT

Toward the end of pregnancy, the part of the baby that will emerge first—usually the head—descends into the pelvic cavity. This is called engagement. Many women feel a sensation of dropping and "lightening" as it happens because the movement of the baby lowers the upper uterus, relieving the pressure on the diaphragm and making it easier for the mother to breathe. Engagement usually takes place at about 36 weeks during a first pregnancy and at the onset of labor during subsequent pregnancies.

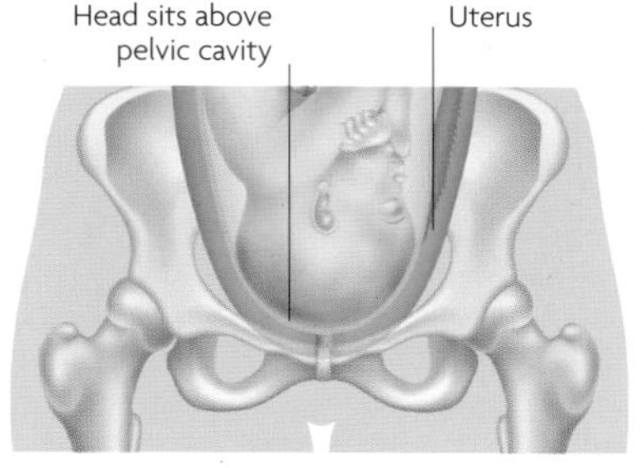
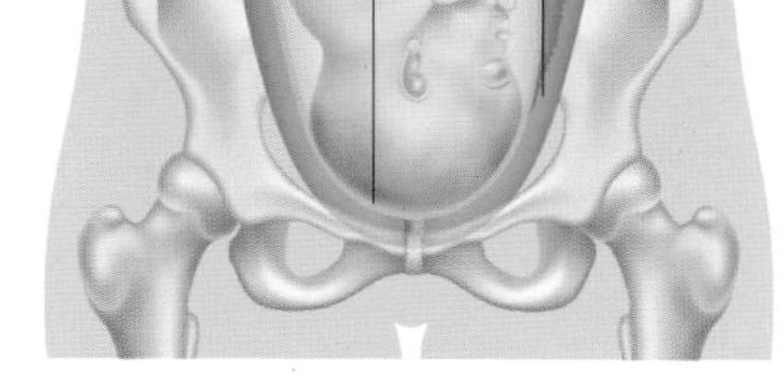
Head sits above pelvic cavity
Uterus

BEFORE THE HEAD ENGAGES
Before engagement, the top of the uterus reaches the breastbone. The baby's head has yet to pass through the inlet of the pelvis into the cavity.

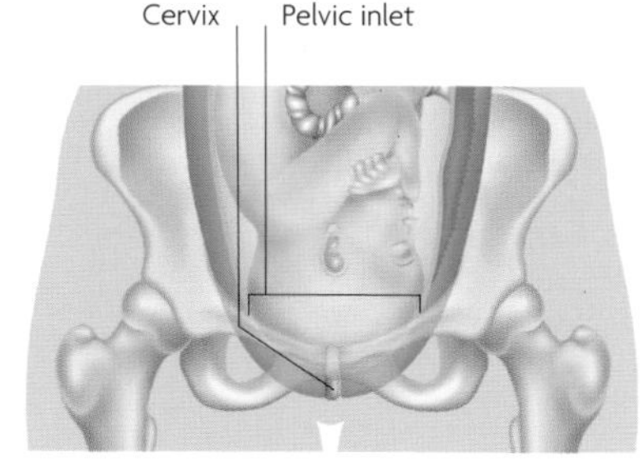
Cervix
Pelvic inlet

AFTER ENGAGEMENT
The baby's head descends into the pelvic cavity. The overall position of the uterus drops, and the baby's head rests against the uterine cervix.

CERVICAL DILATION

Labor begins with the onset of regular, painful contractions, which dilate the cervix. These occur mainly in the upper uterus, which shortens and tightens, pulling and stretching the lower uterus and cervix. For a first baby, the cervix dilates at about ½ in (1cm) per hour on average; progress is usually quicker for subsequent babies. In most women, the cervix is fully dilated when it opens to around 4 in (10 cm).

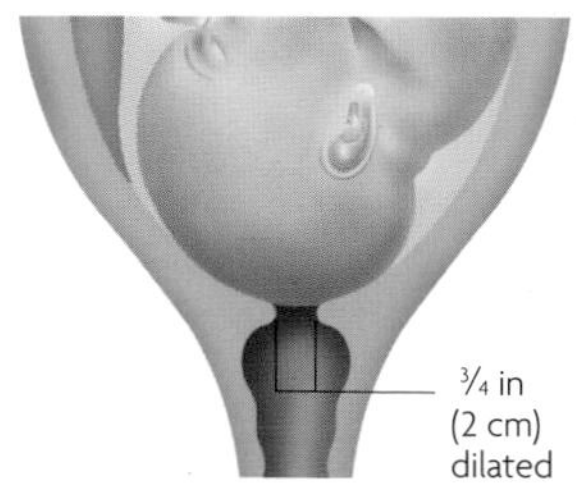
¾ in
(2 cm)
dilated

INITIAL DILATION

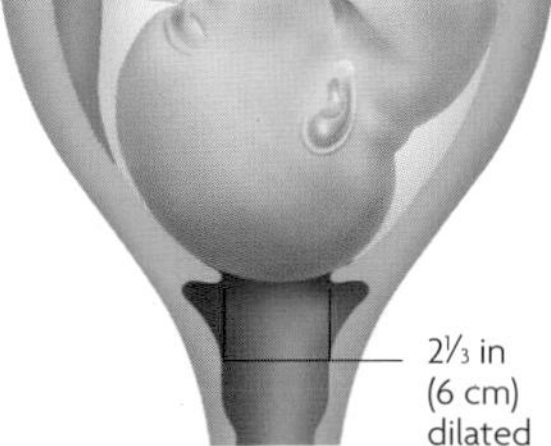
2⅓ in
(6 cm)
dilated

CERVIX WIDENS

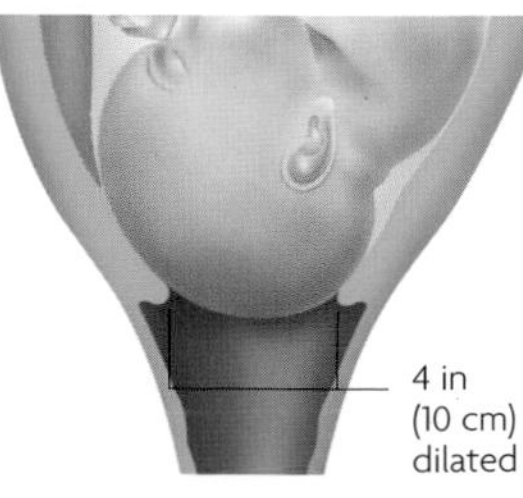
4 in
(10 cm)
dilated

FULLY DILATED

SIGNS OF EARLY LABOR

Every woman's personal experience of childbirth is different, but generally there are three particular signs that labor is starting. First there is a "show," followed by contractions, and finally the water breaks. Before labor begins (usually less than 3 days), the mucous plug in the cervix, which has been acting as a seal during pregnancy, is passed as a blood-stained or brownish discharge (the "show"). As the contractions of the uterus become stronger and more regular, the membranes that retain the amniotic fluid rupture (break), allowing the fluid (water) to leak out via the birth canal.

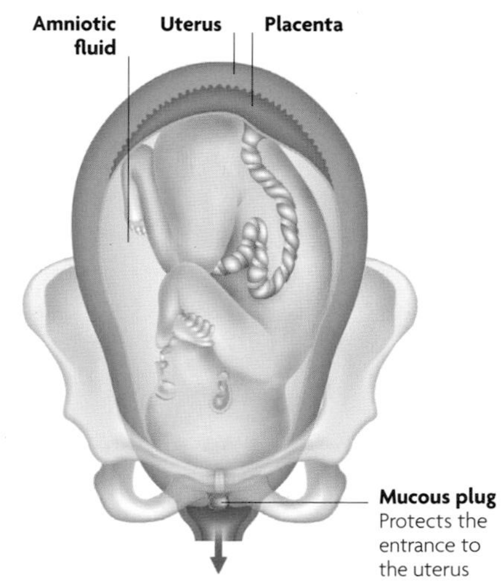
Amniotic fluid
Uterus
Placenta
Mucous plug
Protects the entrance to the uterus

1 THE "SHOW"
For most of the time during a pregnancy, the mucous plug in the cervix prevents microbes from entering the uterus. As the cervix widens slightly, the plug loosens and falls out.

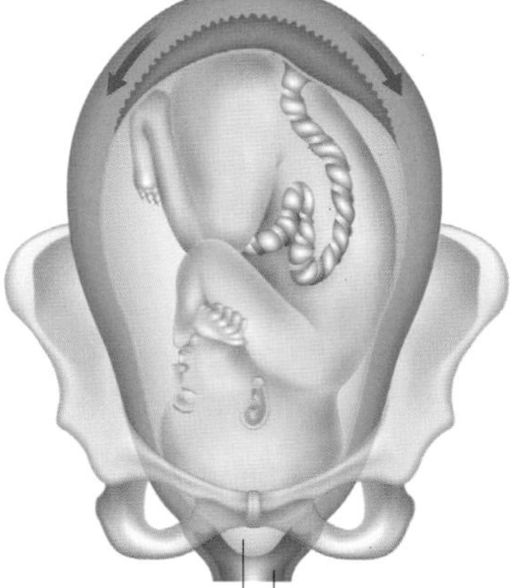

Fundus
Muscular wall contracts most forcefully here

Bulging sac
Pressure causes the membranes to bulge at the cervical opening

Dilating cervix
The cervix thins and its passageway widens (dilates)

2 CONTRACTIONS
Coordinated muscular contractions are generated in the upper part of the uterus, called the fundus. This helps gradually open, or dilate, the cervix.

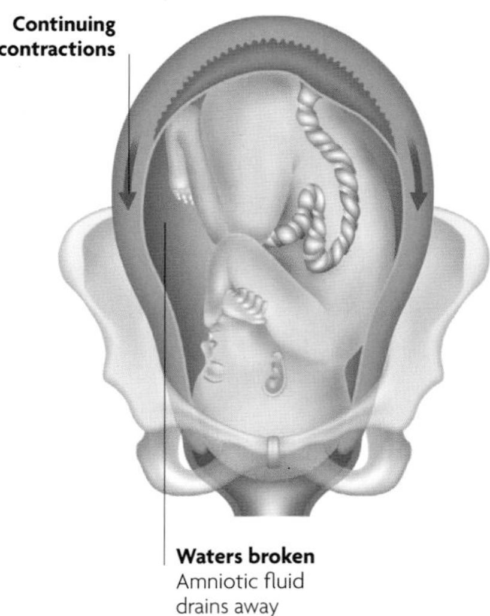
Continuing contractions
Waters broken
Amniotic fluid drains away

3 WATER BREAKS
The amniotic sac (membrane) around the baby ruptures, or breaks, allowing colorless amniotic fluid to pass out through the birth canal.

DELIVERY

THE CULMINATION OF PREGNANCY AND LABOR, DELIVERY OF THE BABY AND THE PLACENTA INVOLVES A COMPLEX SEQUENCE OF EVENTS THAT ULTIMATELY SEPARATES CHILD FROM MOTHER, ALLOWING THE START OF THEIR INDEPENDENT RELATIONSHIP.

THREE STAGES OF CHILDBIRTH

During the first stage, the cervix dilates and the water breaks (see p.269). The second stage, delivery, sees uterine contractions synchronize with shifts in the baby's position as it fits its large head into the birth canal and then travels along it to the outside world. In the third stage, the placenta, or "afterbirth," is delivered, often with the help of an obstetrician or a midwife gently pulling on the cord.

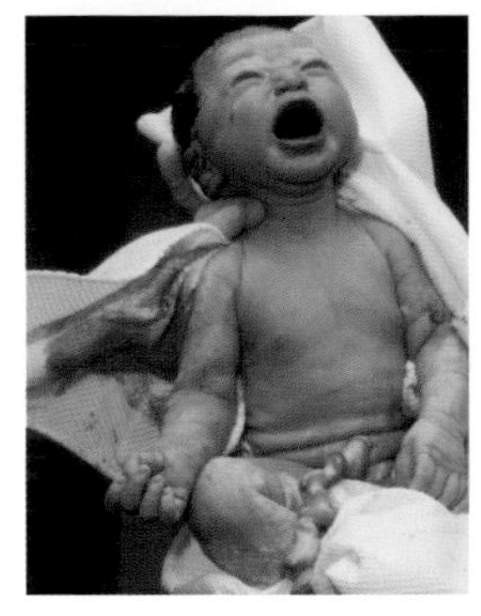

NORMAL DELIVERY
Newborn babies are usually covered with a combination of blood, mucus, and vernix (the greasy covering that protected the fetus in the uterus). This baby's umbilical cord has not yet been clamped and cut.

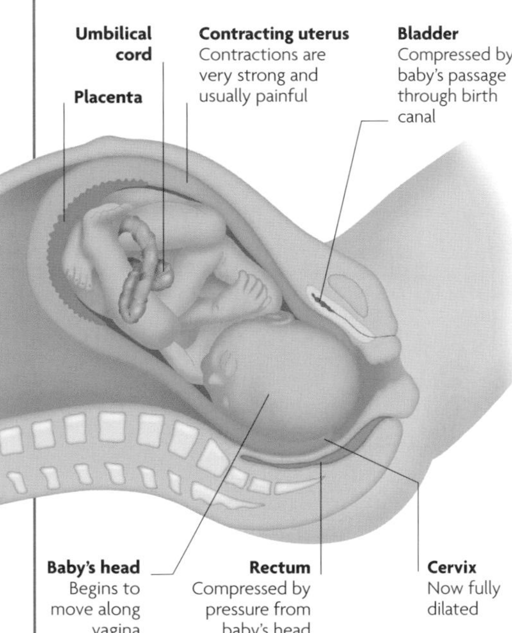
Umbilical cord
Placenta
Contracting uterus Contractions are very strong and usually painful
Bladder Compressed by baby's passage through birth canal
Baby's head Begins to move along vagina
Rectum Compressed by pressure from baby's head
Cervix Now fully dilated

1 DILATION OF THE CERVIX
When the cervix is fully dilated, the baby turns so that the widest part of its skull aligns with the widest part of the mother's pelvis. As the baby tucks in its chin, it starts moving out of the uterus.

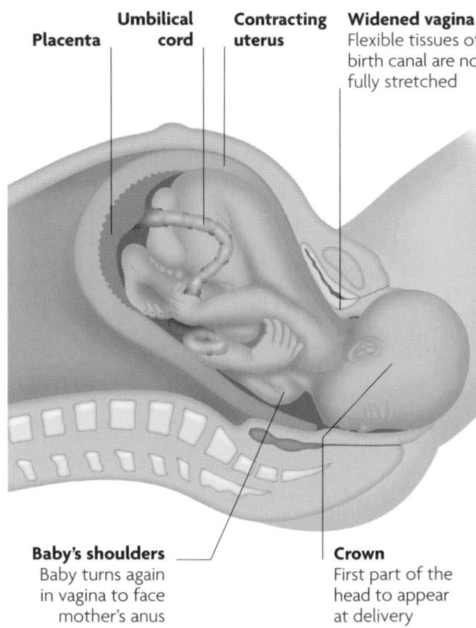
Placenta
Umbilical cord
Contracting uterus
Widened vagina Flexible tissues of birth canal are now fully stretched
Baby's shoulders Baby turns again in vagina to face mother's anus
Crown First part of the head to appear at delivery

2 DESCENT THROUGH THE BIRTH CANAL
The top of the baby's head appears ("crowning"). Usually, the baby faces the mother's anus, allowing the emerging head to negotiate the bend in the fully stretched vagina. Birth is usually imminent at this point.

PELVIC SHAPES

A woman's pelvis is adapted to child-bearing and delivery, but it varies greatly in shape. Some shapes make childbirth easier than others. The classic "female pelvis" (gynecoid) has a generous capacity, and usually results in few problems. A pelvis that is more like a man's (android) is less spacious and can cause difficulties with childbirth.

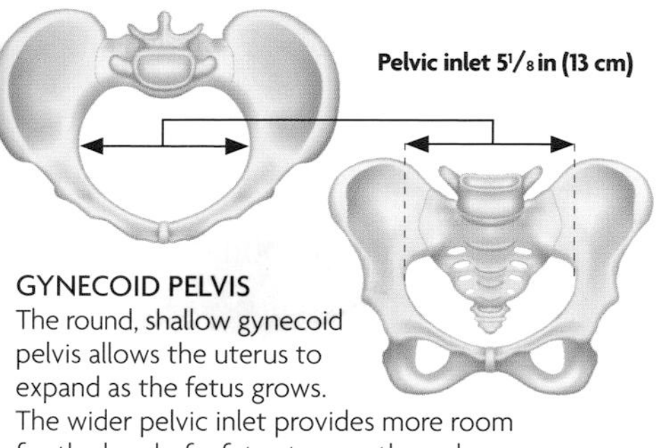
Pelvic inlet $5^{1}/_{8}$ in (13 cm)

GYNECOID PELVIS

The round, shallow gynecoid pelvis allows the uterus to expand as the fetus grows. The wider pelvic inlet provides more room for the head of a fetus to pass through.

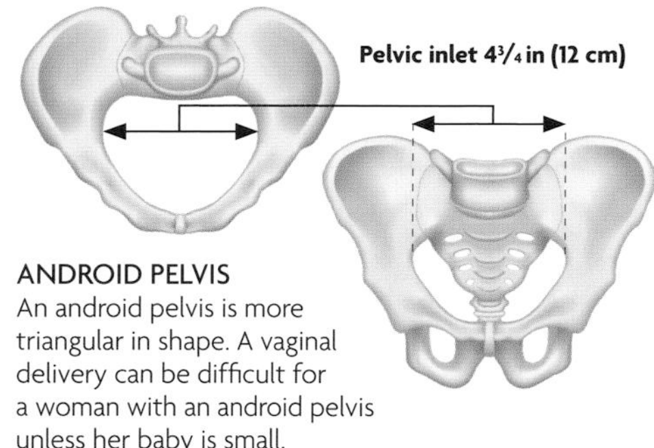
Pelvic inlet $4^{3}/_{4}$ in (12 cm)

ANDROID PELVIS

An android pelvis is more triangular in shape. A vaginal delivery can be difficult for a woman with an android pelvis unless her baby is small.

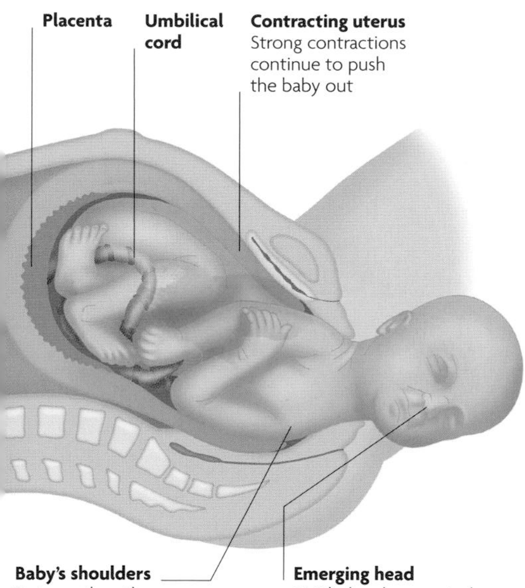

Placenta

Umbilical cord

Contracting uterus Strong contractions continue to push the baby out

Baby's shoulders First one then the other shoulder emerges

Emerging head Guided and supported by midwife as it emerges from mother's vagina

Detaching placenta Separates 5–15 minutes after delivery

Cord traction Pressing the lower abdomen provides traction as the cord is gently pulled and the placenta eased out

Rectum No longer compressed

Birth canal Starts to return to normal size

3 DELIVERY OF THE BABY

The doctor checks the cord is not around the baby's neck, and clears mucus from its nose and mouth. The baby rotates again so the shoulders can slip out easily, one shoulder quickly followed by the other.

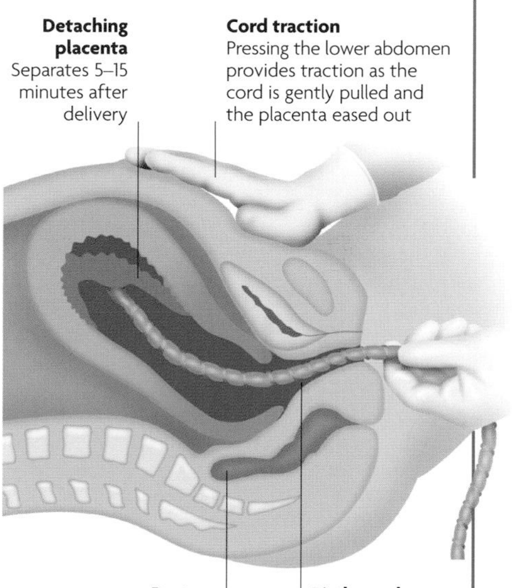

4 DELIVERY OF THE PLACENTA

The uterus mildly contracts soon after the baby is born, sealing any bleeding blood vessels. The placenta separates from the uterus and is eased out by gently pulling the cord while pressing on the lower abdomen.

AFTER THE BIRTH

OVER 40 WEEKS, THE FERTILIZED EGG HAS CHANGED FROM EMBRYO TO NEWBORN BABY. ALL ORGAN SYSTEMS ARE IN PLACE—SOME QUICKLY ADAPT TO LIFE WITHOUT AN UMBILICAL CORD, WHILE OTHERS DO NOT DEVELOP FULLY UNTIL ADOLESCENCE.

NEWBORN ANATOMY

Special features in a baby's anatomy help it grow and develop outside the uterus. Fontanelles allow the skull to expand as the brain grows; they become bone by the time the child is about six years old. Cartilage in the joints and at the end of long bones allows the skeleton to grow rapidly. In the fetus, the liver produced all the red blood cells, but this task is now taken over by the bone marrow.

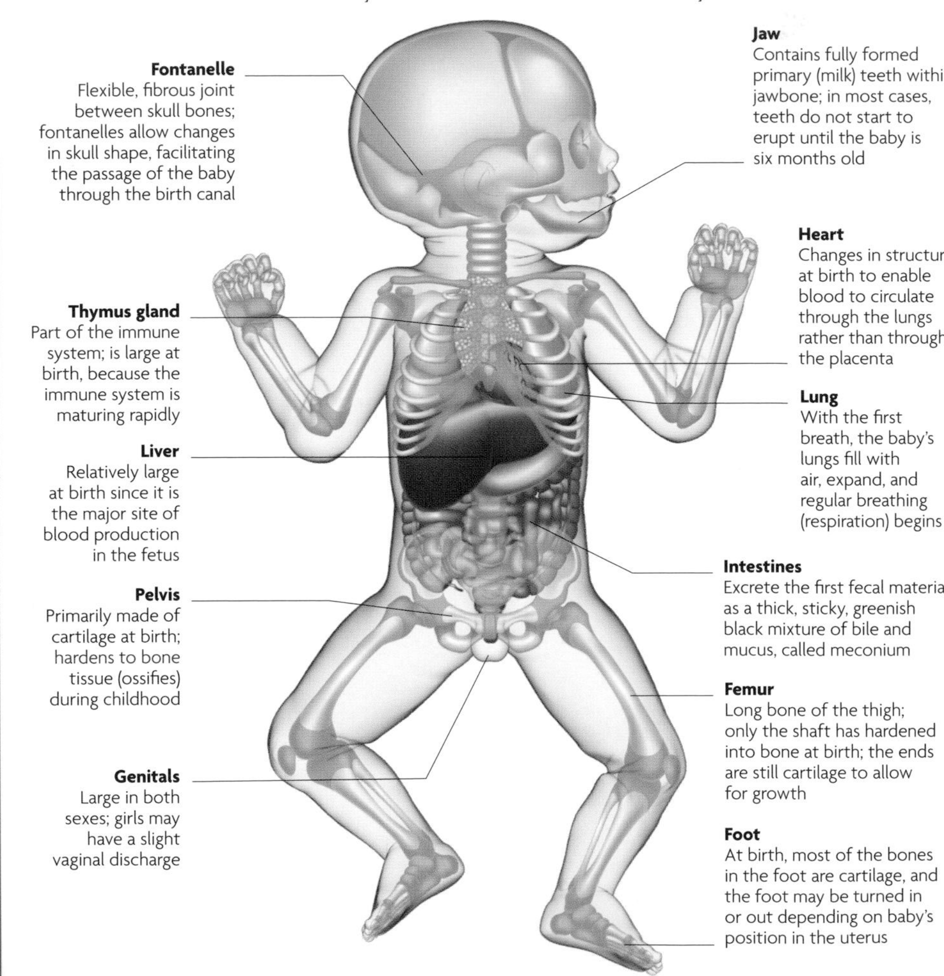

Fontanelle
Flexible, fibrous joint between skull bones; fontanelles allow changes in skull shape, facilitating the passage of the baby through the birth canal
Jaw
Contains fully formed primary (milk) teeth within jawbone; in most cases, teeth do not start to erupt until the baby is six months old
Thymus gland
Part of the immune system; is large at birth, because the immune system is maturing rapidly
Heart
Changes in structure at birth to enable blood to circulate through the lungs rather than through the placenta
Lung
With the first breath, the baby's lungs fill with air, expand, and regular breathing (respiration) begins
Liver
Relatively large at birth since it is the major site of blood production in the fetus
Intestines
Excrete the first fecal material as a thick, sticky, greenish black mixture of bile and mucus, called meconium
Pelvis
Primarily made of cartilage at birth; hardens to bone tissue (ossifies) during childhood
Femur
Long bone of the thigh; only the shaft has hardened into bone at birth; the ends are still cartilage to allow for growth
Genitals
Large in both sexes; girls may have a slight vaginal discharge
Foot
At birth, most of the bones in the foot are cartilage, and the foot may be turned in or out depending on baby's position in the uterus

CHANGES IN THE MOTHER

Many physiological changes take place in the mother after birth, for which her body has prepared during pregnancy. The process of enhancing breast tissue in anticipation of breast-feeding begins early in pregnancy: the breasts enlarge visibly, and the alveoli in each of the milk-producing glands (lobules) swell and multiply. From three months into the pregnancy, the breasts can produce colostrum, a fluid rich in antibodies (which help protect a newborn from allergies and respiratory and gastrointestinal infections), water, protein, and minerals. After the birth, colostrum supplies a breast-fed baby with nutrition until the mother's milk begins to flow several days later. Soon after birth, the uterus begins to shrink to its prepregnancy size–a process that is helped by breast-feeding.

UTERUS SHRINKS

After delivery of the baby and the placenta in the second and third stages of labor, hormones in the mother's body cause her uterus and vagina to shrink back to their normal size and position in her body.

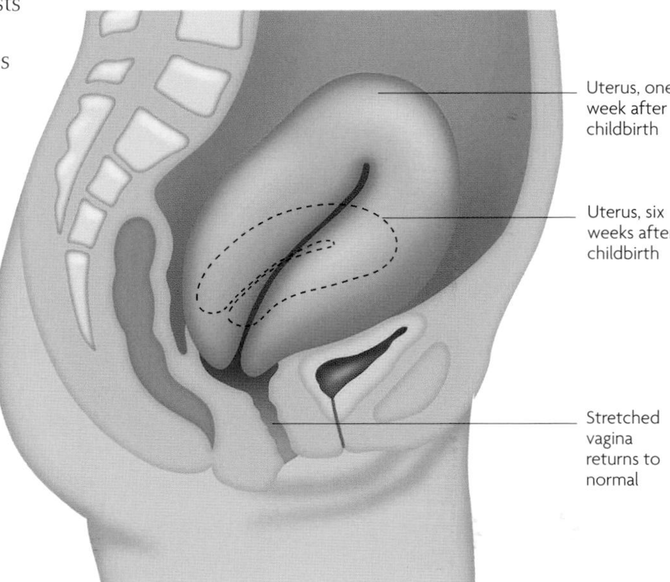
Uterus, one week after childbirth
Uterus, six weeks after childbirth
Stretched vagina returns to normal

LACTATION

During pregnancy, lobules (milk-producing glands) increase in size and number in preparation for breast-feeding the baby. By the end of the first trimester, they can produce colostrum, the yellow fluid that provides antibodies to protect against allergies and gastrointestinal and respiratory infections in the newborn.

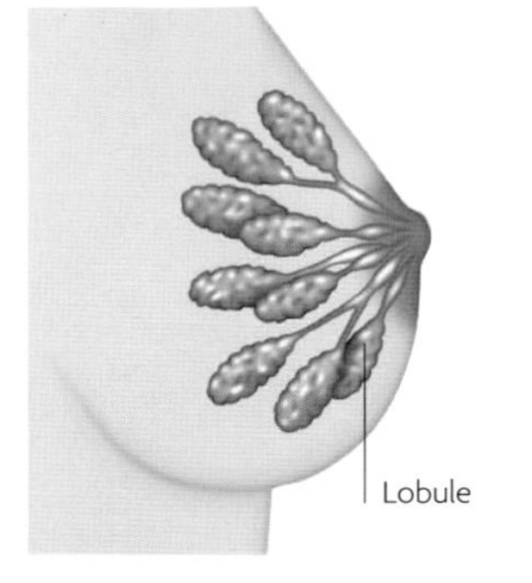
Lobule

BEFORE PREGNANCY

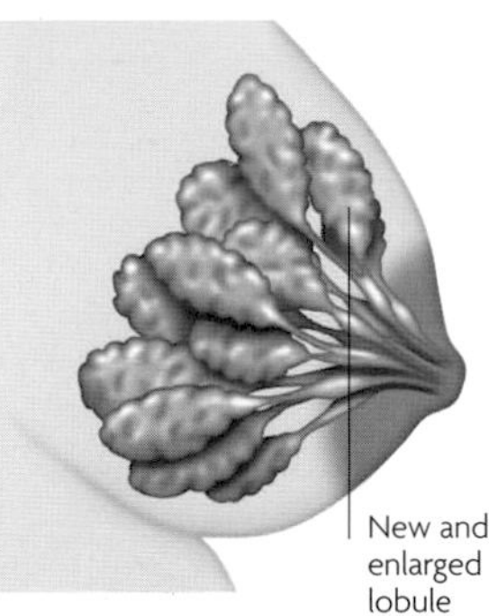
New and enlarged lobule

DURING PREGNANCY AND LACTATION

CIRCULATION IN THE UTERUS

As the placenta provides oxygen and nutrients, the fetal circulation has anatomical variations ("shunts") that bypass the not-yet-functioning liver and lungs. The ductus venosus shunts incoming blood through the liver to the right atrium, which shunts it through a gap, the foramen ovale, to the left atrium (mostly bypassing the right ventricle) and onward to the body. Any blood that enters the right ventricle passes into the pulmonary artery but is shunted into the aorta by the ductus arteriosus, thus bypassing the lungs.

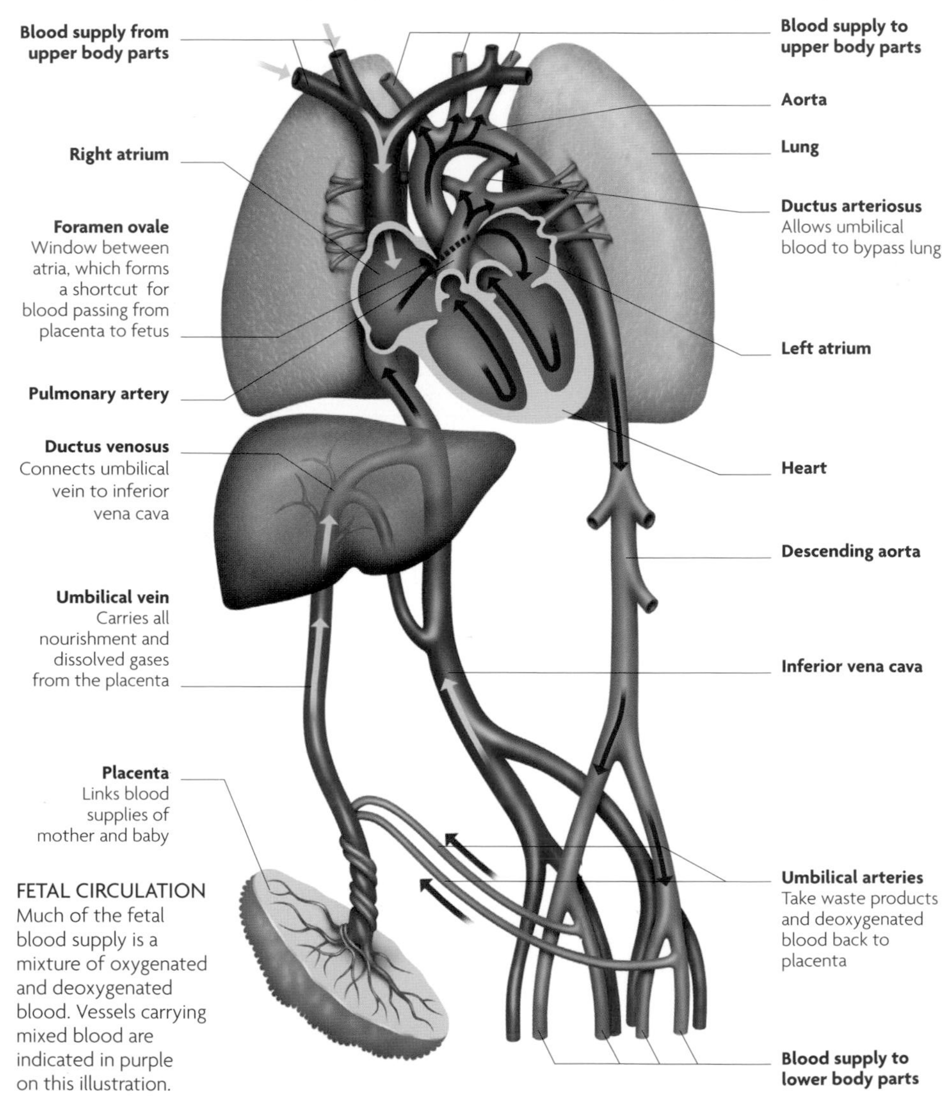
Blood supply from upper body parts
Right atrium
Foramen ovale
Window between atria, which forms a shortcut for blood passing from placenta to fetus
Pulmonary artery
Ductus venosus
Connects umbilical vein to inferior vena cava
Umbilical vein
Carries all nourishment and dissolved gases from the placenta
Placenta
Links blood supplies of mother and baby
Blood supply to upper body parts
Aorta
Lung
Ductus arteriosus
Allows umbilical blood to bypass lungs
Left atrium
Heart
Descending aorta
Inferior vena cava
Umbilical arteries
Take waste products and deoxygenated blood back to placenta
Blood supply to lower body parts

FETAL CIRCULATION

Much of the fetal blood supply is a mixture of oxygenated and deoxygenated blood. Vessels carrying mixed blood are indicated in purple on this illustration.

CIRCULATION AT BIRTH

At birth, the baby takes its first breaths and the umbilical cord is clamped. This forces the circulatory system into a monumental response: to convert itself immediately to obtain its oxygen supply via the lungs. Blood is sent to the lungs to retrieve oxygen, and the pressure of this blood returning from the lungs into the left atrium forces shut the foramen ovale between the two atria, thus establishing normal circulation. The ductus arteriosus, the ductus venosus, and the umbilical vein and arteries close up and become ligaments.

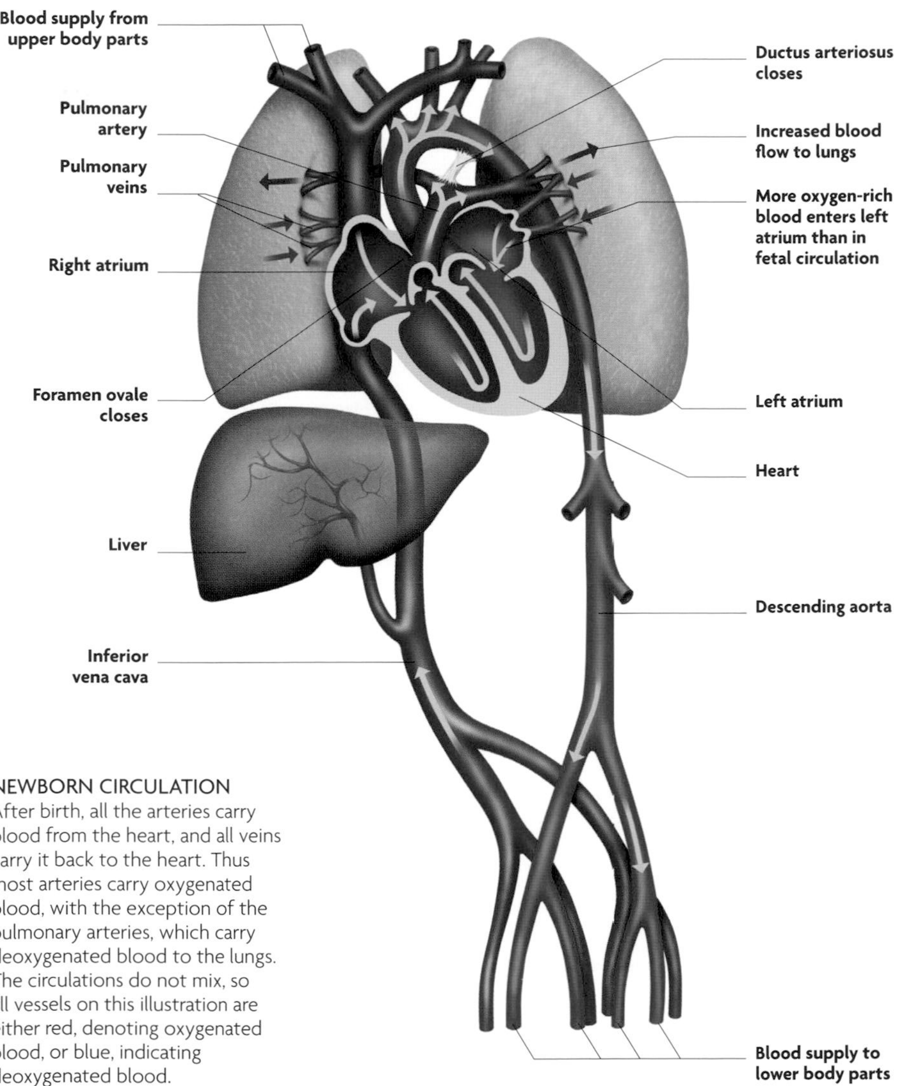
Blood supply from upper body parts
Pulmonary artery
Pulmonary veins
Right atrium
Foramen ovale closes
Liver
Inferior vena cava
Ductus arteriosus closes
Increased blood flow to lungs
More oxygen-rich blood enters left atrium than in fetal circulation
Left atrium
Heart
Descending aorta
Blood supply to lower body parts

NEWBORN CIRCULATION

After birth, all the arteries carry blood from the heart, and all veins carry it back to the heart. Thus most arteries carry oxygenated blood, with the exception of the pulmonary arteries, which carry deoxygenated blood to the lungs. The circulations do not mix, so all vessels on this illustration are either red, denoting oxygenated blood, or blue, indicating deoxygenated blood.

GROWTH AND DEVELOPMENT

YOUNG CHILDREN DEVELOP BASIC PHYSICAL SKILLS AND THEN BECOME MORE AGILE, WITH INCREASED INTELLECTUAL ABILITIES. PHYSICAL GROWTH RATE IS RAPID DURING INFANCY, AND THEN IS FAIRLY STEADY UNTIL IT SPEEDS UP AGAIN AT PUBERTY.

BONE GROWTH

Body growth depends on the increasing size of the skeleton. The long leg bones provide most of the increase in height. Many long bones develop from cartilage precursors, by a sequence of changes (ossification) that starts before birth at primary centers in the bone shafts. After birth, secondary centers develop near the bone ends. Growth ceases once ossification is complete, at 18–20 years of age.

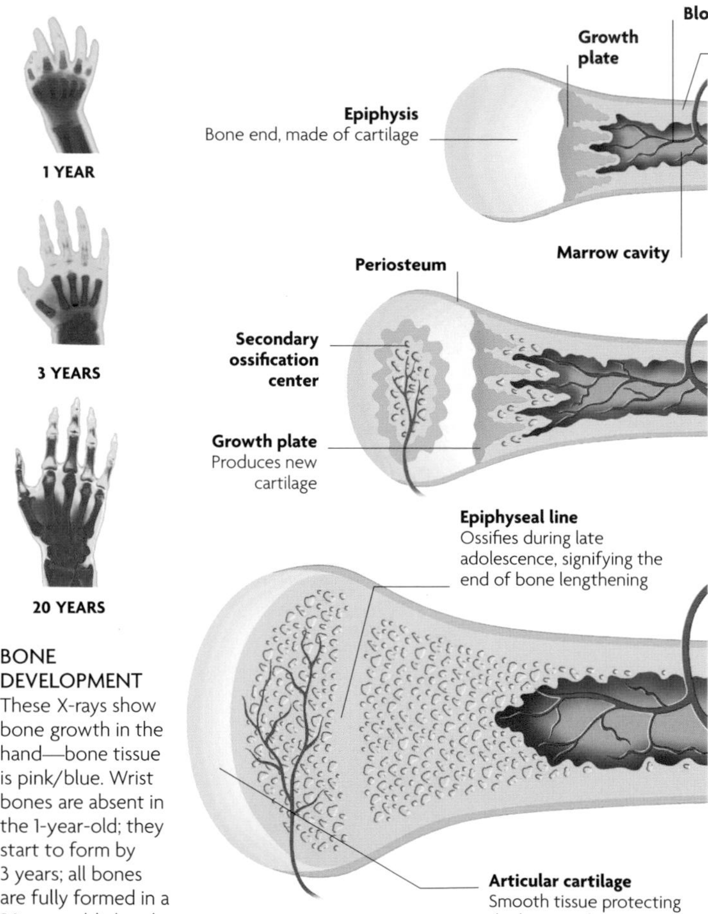
1 YEAR
3 YEARS
20 YEARS
Blood vessel
Growth plate
Shaft
Epiphysis
Bone end, made of cartilage
Marrow cavity
Periosteum
Secondary ossification center
Growth plate
Produces new cartilage
Epiphyseal line
Ossifies during late adolescence, signifying the end of bone lengthening
Articular cartilage
Smooth tissue protecting the bone end

BONE DEVELOPMENT

These X-rays show bone growth in the hand—bone tissue is pink/blue. Wrist bones are absent in the 1-year-old; they start to form by 3 years; all bones are fully formed in a 20-year-old's hand.

LONG BONE OF A NEWBORN

The shaft turns to hard bone from the primary ossification center, and has a marrow cavity. The head is all cartilage, and is relatively soft.

LONG BONE OF A CHILD

A secondary ossification centre inside the end begins to change the surrounding cartilage to hardened, mineralized bone. An elongating growth area (growth plate) forms between the shaft and the end.

LONG BONE OF AN ADULT

By about 18–20 years, all zones have hardened into true bone, with the epiphyseal growth plate represented by a line of dense bony tissue. The only remaining cartilage is smooth and slippery articular cartilage, which covers the head of the end inside the joint.

CHANGING PROPORTIONS

A newborn's head is relatively large, being wider than the shoulders and representing about a quarter of the baby's total height; the legs are about three-eighths of this height. As the child grows, the arms and legs "catch up." At two years, the head is about a sixth of the total height. When final adult size is reached during adolescence, the head is only about an eighth of the body length, and the legs one-half.

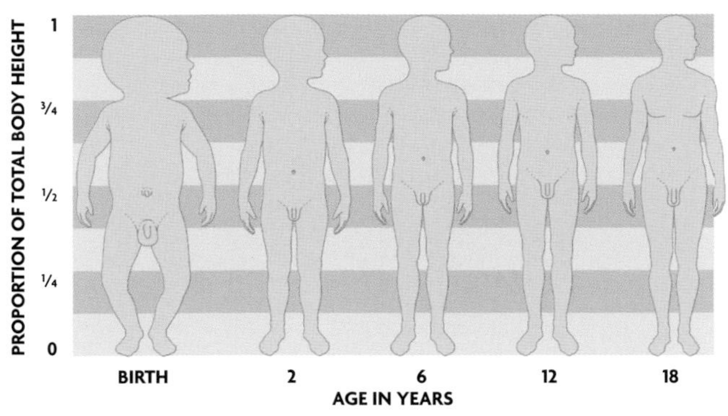
PROPORTION OF TOTAL BODY HEIGHT
1
3/4
1/2
1/4
0
BIRTH
2
6
12
18
AGE IN YEARS

HEAD–BODY PROPORTIONS
If the body's height at different ages is superimposed onto a grid, the changes in head–body proportions that take place from birth to adulthood are clearly shown. The overall growth trend is for the head to lead, growing first and fastest. Then the other regions of the body catch up: first the torso, followed by the arms, and finally the legs.

SKULL AND BRAIN

At birth, the brain is a quarter of its adult size. It has almost its full complement of neurons, but they have yet not made many interconnections. Gaps (fontanelles) and seams (sutures) between the skull bones allow for expansion. By two years, the brain is four-fifths of its adult size, and neurons are forging links into networks.

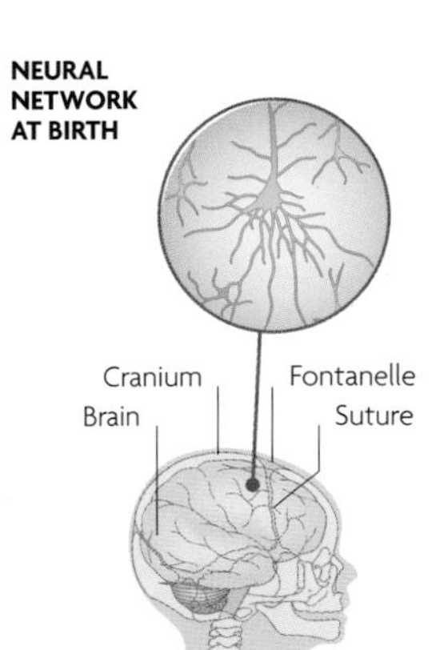
NEURAL NETWORK AT BIRTH
Cranium
Brain
Fontanelle
Suture

BIRTH
The cranium and brain are huge compared to the small facial bones. The neurons make limited links.

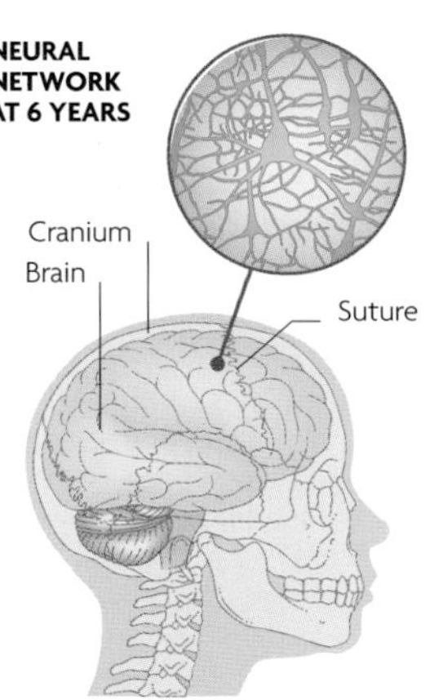
NEURAL NETWORK AT 6 YEARS
Cranium
Brain
Suture

SIX YEARS
Cranial bones are fusing at the sutures. Neurons rapidly extend their projections and their links.

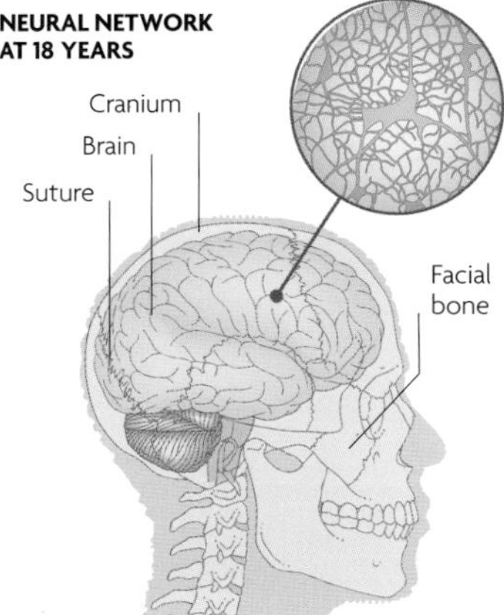
NEURAL NETWORK AT 18 YEARS
Cranium
Brain
Suture
Facial bone

ADULT
The cranium is solidly fused, the brain is full-sized, and new neural links are made less often.

DENTAL DEVELOPMENT

The first set of teeth, known as the primary or deciduous dentition, erupts through the gums in a set order, from about six months into the third year. In general, apart from the canines, the teeth appear from the front to the back. However, the exact times and order vary between individuals, and occasionally a baby is born with one or more teeth. Primary teeth loosen and fall out as the adult, or permanent, dentition erupts through the gums. This usually starts at about six years of age. The set of 32 permanent teeth is complete once the third molars (known as wisdom teeth) appear in the late teens or early twenties. In some people, however, the third molars never make an appearance above the gum.

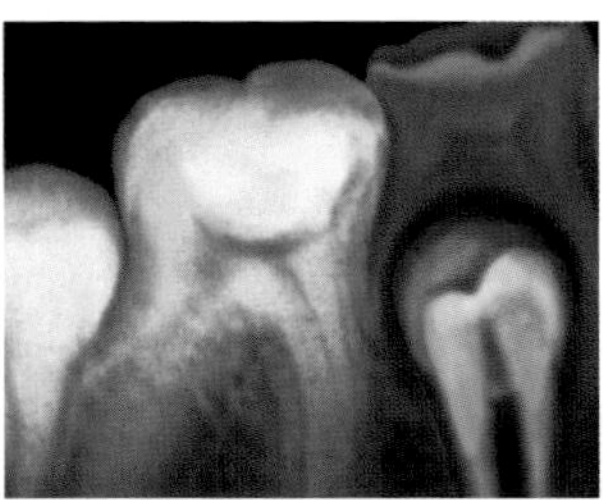

TOOTH ERUPTION
In this color-enhanced X-ray, a permanent, or adult, tooth (green) is shown erupting under a child's milk, or deciduous, teeth.

STAGES OF DEVELOPMENT

Babies are born able to see, hear, and perform reflex actions, such as grasping, urination, and defecation. Gradually, the infant learns to bring these reflexes under conscious control. As the eyes develop the ability to focus clearly, the baby watches his or her hands, and learns how conscious movements formulated in the brain result in actual physical movement. During early childhood, these basic motor skills are refined further. The child also gains a range of social developmental skills, such as smiling, to elicit a response from those nearby. For most children, development takes place in a fairly

0 1

MOTOR SKILLS

Basic motor coordination starts by "trial and error." An infant learns to associate a movement pattern with its mental intention to make the movement. Muscles gradually become coordinated as the brain learns to combine patterns of movements by reinforcing and linking the neural pathways that control them.

Can lift head to 45°
Can walk without help
Can bear weight on legs
Can roll over
Can stand by hoisting up own weight
Can sit unsupported
Can crawl

VISION AND MANUAL DEXTERITY

A new baby can focus clearly on objects up to a yard away. After six months, items several yards away are clear. The eyes are more coordinated, rather than occasionally squinting. Hand–eye coordination soon develops as the baby watches its fingers and senses what they touch.

Holds hands together
Plays with feet
Reaches out for a rattle
Can pick up a small object
Can grasp an object with finger and thumb

SOCIAL AND LANGUAGE SKILLS

After a few weeks, a baby starts to turn toward sounds. Language develops from listening and associating sounds with objects, and by practicing first words. In the second year, a child learns words at an astonishing rate. Social skills develop in tandem with language skills.

Smiles spontaneously
Squeals
Says "dada" and "mama" to parents

0 2 4 6 8 10 12

predictable sequence: for example, standing must occur before walking. However, there is great variation in the ages at which stages are reached; acquiring a skill early does not always mean the skill will improve later. Some babies and children miss stages and go straight on to the next ones.

NEONATAL GRASP
A newborn's grasp, when its palm is touched, is one of the primitive reflexes, which disappear in a few months.

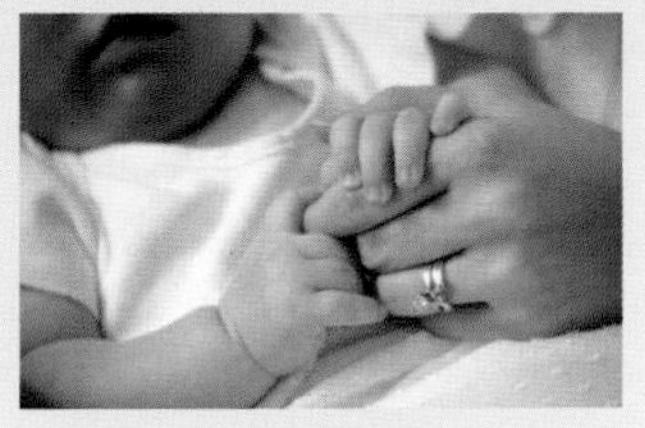

AGE IN YEARS

2 3 4 5

Can hop on one leg

Can walk upstairs without help

Can balance on one foot for a second

Can pedal a tricycle

Can kick a ball

Can catch a bounced ball

Likes to scribble

Can copy a circle

Can draw a straight line

Can copy a square

Can draw a rudimentary likeness of a person

Can drink from a cup

Stays dry at night

Stays dry in the day

Knows first and last names

Can put two words together

Can dress without help

Starts to learn single words

Can talk in full sentences

14 16 18 20 22 24 26 28 30 32 34 36 38 40 42 44 46 48 50 52 54 56 58 60

AGE IN MONTHS

PUBERTY

AT PUBERTY, A NUMBER OF HORMONAL CHANGES STIMULATE PHYSICAL GROWTH AND THE DEVELOPMENT OF THE SEX ORGANS. IN BOTH SEXES, EMOTIONAL, BEHAVIORAL, AND PSYCHOLOGICAL CHANGES ALSO OCCUR.

MALE PUBERTY

In boys, the physical changes of puberty start later than in girls, around age 12 or 13. Most show signs of development by age 14, and complete the changes of puberty by age 17 or 18. The testes and penis get bigger first, then hair grows in the pubic area and armpits. Muscles increase in bulk, and some breast tissue might also develop. The hormone testosterone causes cartilage in the voice box to grow larger and thicker, which results in the vocal cords getting longer and thicker. This causes the cords to vibrate at a lower frequency, so the voice becomes deeper. Finally, facial hair appears, which may be accompanied by acne. Boys are more likely than girls to experience problems with perspiration and oily skin. The sign of sexual maturation for boys is ejaculation. Although they are capable of having an erection from birth, boys only produce sperm when testosterone begins circulating in their bodies. It is then that they are able to ejaculate for the first time.

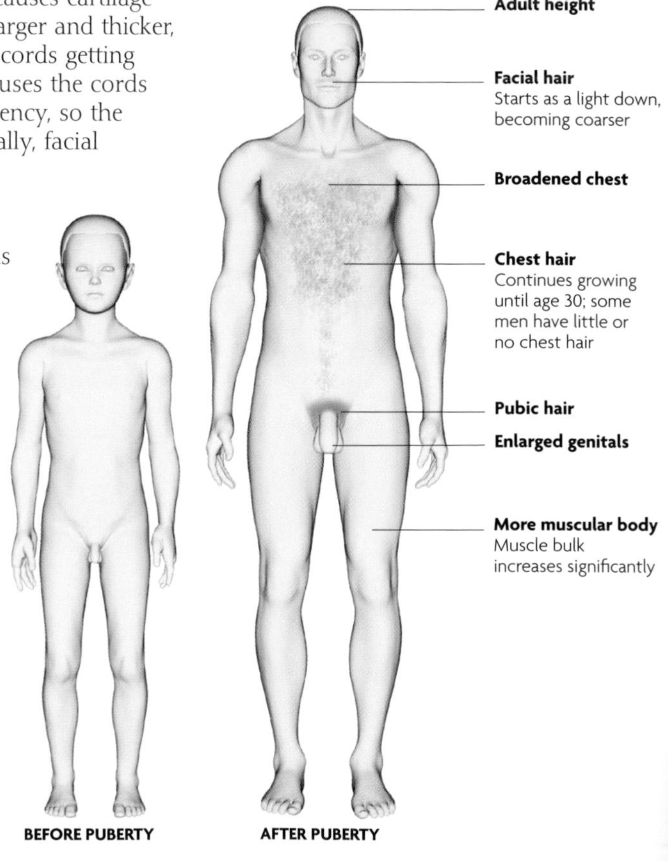

Adult height
Facial hair
Starts as a light down, becoming coarser
Broadened chest
Chest hair
Continues growing until age 30; some men have little or no chest hair
Pubic hair
Enlarged genitals
More muscular body
Muscle bulk increases significantly
BEFORE PUBERTY
AFTER PUBERTY

CHANGES IN THE BODY

Boys start growing later than girls. Once they begin growing, however, they grow faster and for a longer period, thus attaining a greater adult height. At age 14 or 15, the average boy is taller, heavier, and stronger than an average girl and is still growing.

SPERM PRODUCTION

Sperm develop in the seminiferous tubules of the testes. Sperm cells gradually move away from the supporting cells and mature as they pass through the seminiferous tubule and epididymis. The process takes about 74 days.

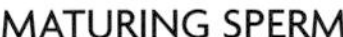

MATURING SPERM
This cross section through several seminiferous tubules in the testis shows maturing sperm with tails in the center (blue).

MALE HORMONE CONTROL

Hormone production is often regulated by feedback (see p.139), when the amount of a substance controls how much of it is made. The testes, hypothalamus, and pituitary gland control production of sperm and male hormones in this way. Gonadotropin-releasing hormone (GnRH) from the hypothalamus stimulates the pituitary to control testis function via follicle-stimulating hormone (FSH) and luteinizing hormone (LH). High levels of testosterone act on the pituitary to slow the release of LH and FSH.

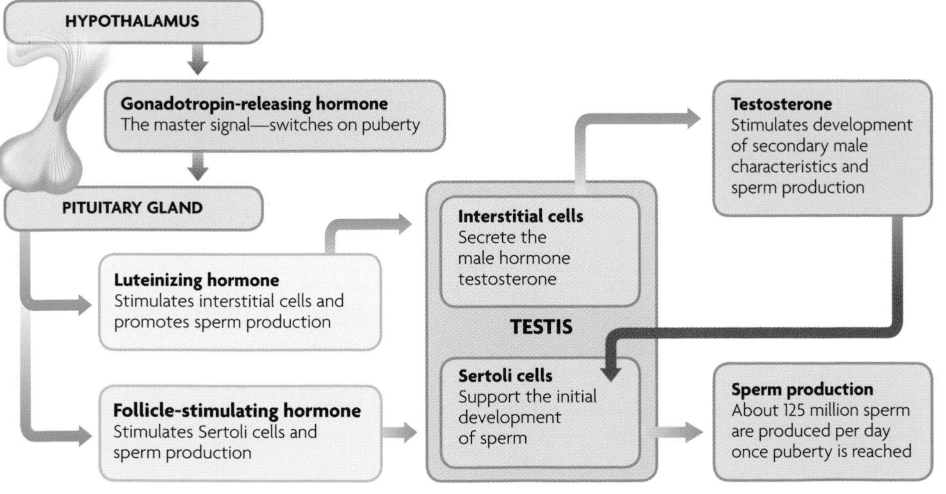
HYPOTHALAMUS
Gonadotropin-releasing hormone
The master signal—switches on puberty
PITUITARY GLAND
Luteinizing hormone
Stimulates interstitial cells and promotes sperm production
Follicle-stimulating hormone
Stimulates Sertoli cells and sperm production
Interstitial cells
Secrete the male hormone testosterone
TESTIS
Sertoli cells
Support the initial development of sperm
Testosterone
Stimulates development of secondary male characteristics and sperm production
Sperm production
About 125 million sperm are produced per day once puberty is reached

FEMALE PUBERTY

Changes to the female body are caused by the hormones estrogen and progesterone. The first sign of puberty is the development of breasts, which starts around 10–11 years. Then hair grows in the armpits and pubic area. Leg hair thickens, and body shape changes as body fat increases. Hair and skin become oily, which may cause acne. Periods tend to begin at 12 to 13 years. Girls may experience mood swings and irritability.

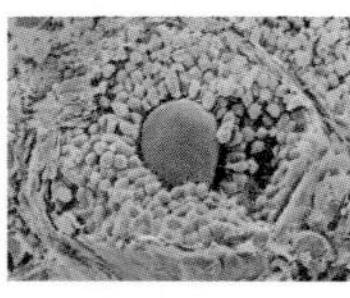

FOLLICLES IN AN OVARY
At puberty, a girl's ovaries start to form mature follicles (collections of blue cells), each with a ripe egg (red).

CHANGES IN THE BODY
Growth rate peaks at about age 12, when girls grow up to $3^{1}/_{2}$ in (9 cm) a year. Growth usually stops by the age of 16.

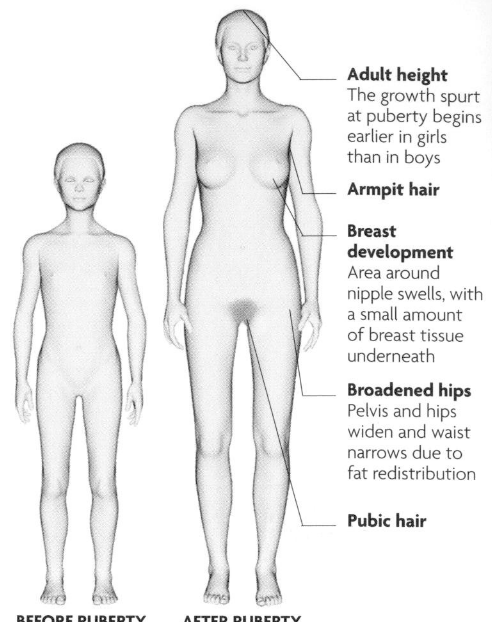
Adult height The growth spurt at puberty begins earlier in girls than in boys
Armpit hair
Breast development Area around nipple swells, with a small amount of breast tissue underneath
Broadened hips Pelvis and hips widen and waist narrows due to fat redistribution
Pubic hair
BEFORE PUBERTY
AFTER PUBERTY

FEMALE HORMONE CONTROL

In the menstrual cycle, the hypothalamus releases gonadotropin-releasing hormone (GnRH) to trigger the pituitary to secrete luteinizing hormone (LH) and follicle-stimulating hormone (FSH). These hormones control the activity of the ovaries and female hormones, and also send feedback (see p.139) to the hypothalamus and the pituitary gland.

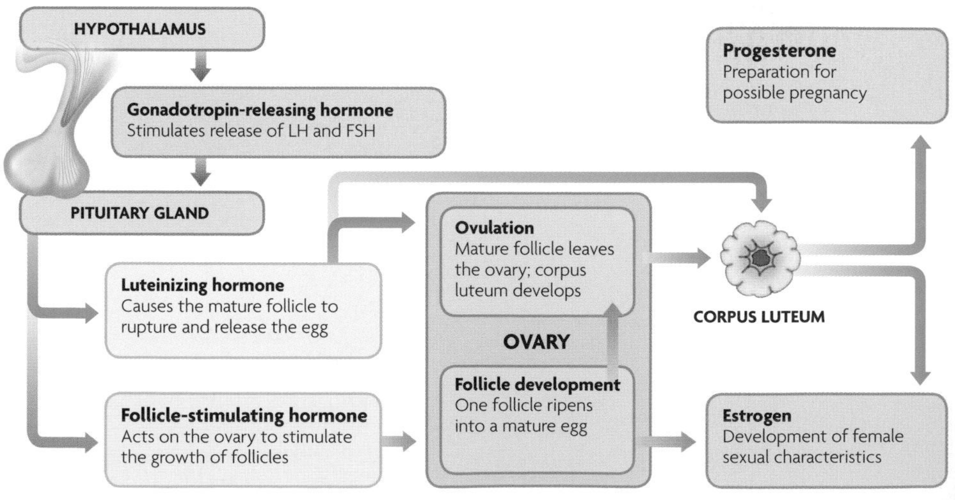
HYPOTHALAMUS
Gonadotropin-releasing hormone Stimulates release of LH and FSH
PITUITARY GLAND
Luteinizing hormone Causes the mature follicle to rupture and release the egg
Follicle-stimulating hormone Acts on the ovary to stimulate the growth of follicles
Ovulation Mature follicle leaves the ovary; corpus luteum develops
OVARY
Follicle development One follicle ripens into a mature egg
CORPUS LUTEUM
Progesterone Preparation for possible pregnancy
Estrogen Development of female sexual characteristics

THE MENSTRUAL CYCLE

For a few days each month, the lining of the uterus is shed and blood passes out through the vagina. The lining thickens again to prepare for the implantation of a fertilized egg. This is the menstrual cycle. It starts when the pituitary gland releases FSH (see opposite), which stimulates egg follicles in the ovary. The follicles secrete estradiol, a form of estrogen. This triggers the release of LH, which matures the egg and weakens the follicle wall, allowing the release of the mature egg (ovum). Whether the right or left ovary ovulates is entirely random. If fertilized, the embryo is implanted into the uterine wall, and signals its presence by releasing human chorionic gonadotropin (HCG), the hormone measured in pregnancy tests. This signal maintains the corpus luteum and enables it to continue producing progesterone. In the absence of a pregnancy and without HCG, the corpus luteum dies and progesterone levels fall. Progesterone withdrawal leads to menstrual bleeding and, as FSH levels rise, a new crop of follicles is formed–the cycle begins again.

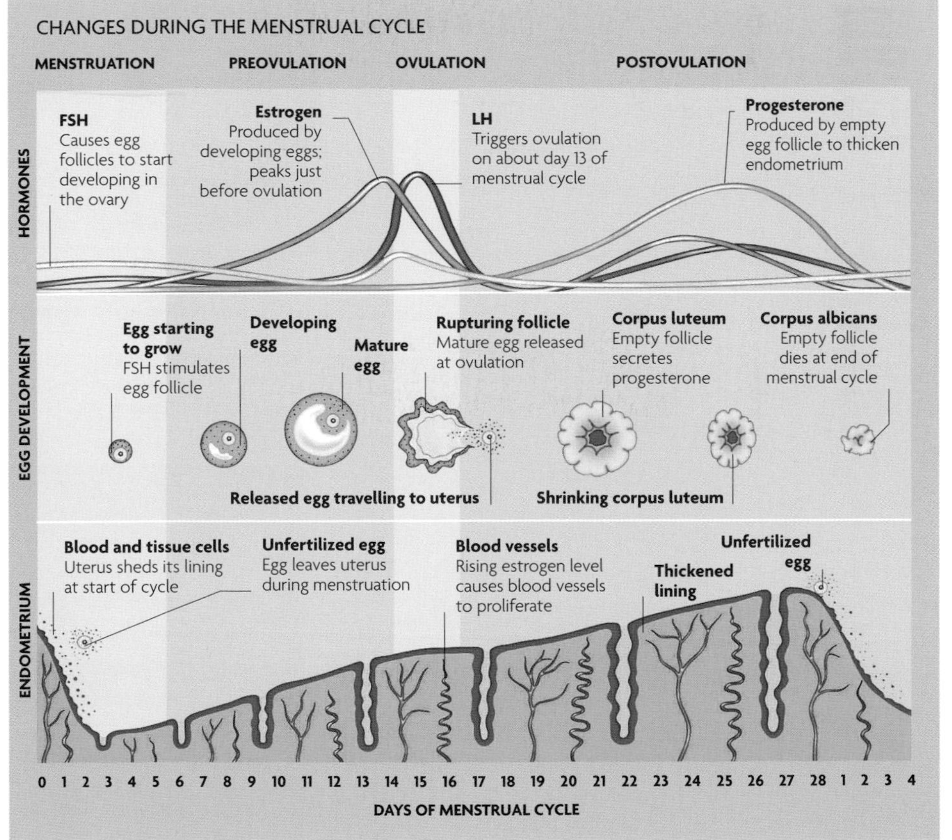
CHANGES DURING THE MENSTRUAL CYCLE
MENSTRUATION
PREOVULATION
OVULATION
POSTOVULATION
HORMONES
FSH
Causes egg follicles to start developing in the ovary
Estrogen
Produced by developing eggs; peaks just before ovulation
LH
Triggers ovulation on about day 13 of menstrual cycle
Progesterone
Produced by empty egg follicle to thicken endometrium
EGG DEVELOPMENT
Egg starting to grow
FSH stimulates egg follicle
Developing egg
Mature egg
Rupturing follicle
Mature egg released at ovulation
Corpus luteum
Empty follicle secretes progesterone
Corpus albicans
Empty follicle dies at end of menstrual cycle
Released egg travelling to uterus
Shrinking corpus luteum
ENDOMETRIUM
Blood and tissue cells
Uterus sheds its lining at start of cycle
Unfertilized egg
Egg leaves uterus during menstruation
Blood vessels
Rising estrogen level causes blood vessels to proliferate
Thickened lining
Unfertilized egg
0 1 2 3 4 5 6 7 8 9 10 11 12 13 14 15 16 17 18 19 20 21 22 23 24 25 26 27 28 1 2 3 4
DAYS OF MENSTRUAL CYCLE

AGING

THE BRAIN, MUSCLES, JOINTS, EYES, AND OTHER ORGANS ALL DECLINE WITH AGE, BUT CHANGES ARE USUALLY SMALL UNTIL AFTER THE AGE OF 60. GENETICS AND LIFESTYLE ARE MAJOR CONTRIBUTORS TO A PERSON'S LIFESPAN.

CELLULAR DETERIORATION

Cells divide a fixed number of times and then stop functioning properly. Connective tissue becomes increasingly stiff, making the organs, blood vessels, and airways more rigid. Changes in cell membranes impede the delivery of oxygen and nutrients and the removal of carbon dioxide and wastes, causing an increase in pigments and fatty substances inside cells. How quickly a person's cells deteriorate, and therefore how long he or she lives, is a balance between how fast things go wrong with cells and how efficiently the body functions to prevent damage from building up.

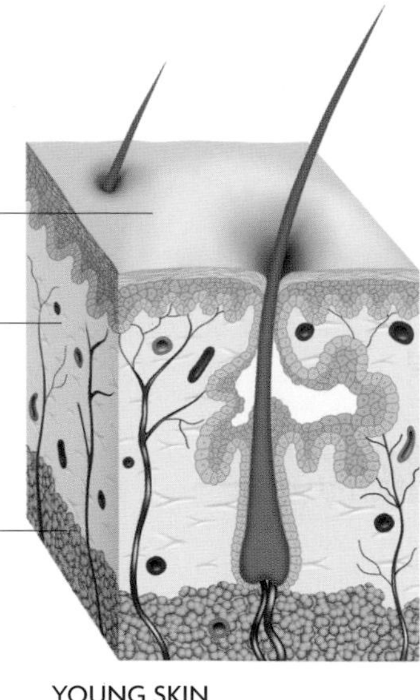
Epidermis
Thick layer confers strength
Dermis
Well supplied with elastic and collagen fibers
Fatty layer
Gives strong support to upper skin layers

YOUNG SKIN
A thick top layer, many elastic and collagen fibers in the deeper layers, good layers of supporting fat, and plenty of sebaceous glands producing oil all help maintain the smoothness and suppleness of young skin.

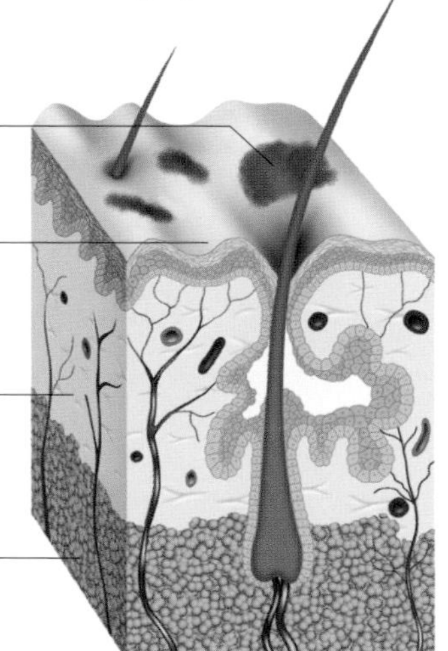
Liver spots
Pigmentation patches in areas exposed to sun
Wrinkles
Creased, sagging skin that gives a lined appearance
Dermis
Thinner; contains fewer collagen fibers, causing reduced elastic recoil
Fatty layer
Thicker relative to dermis, although may also be thinner with age

OLDER SKIN
A thinner outer layer, and fewer elastic fibers and collagen in the deeper layers, result in skin that appears loose, with deeper creases and wrinkles.

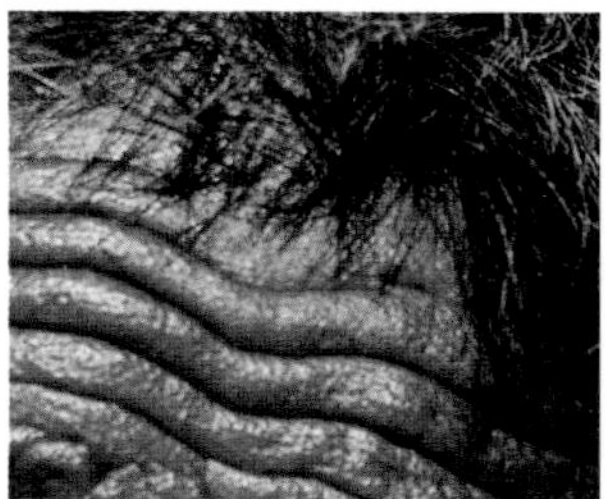

AGING GRACEFULLY
Skin wrinkling is one of the most visible signs of aging. Creased and sagging skin, seen as wrinkles, can be predetermined by genes.

NERVOUS SYSTEM

As people age, the brain and nervous system undergo changes, losing nerve cells. Messages are transmitted more slowly, and the senses may be affected. Reflexes may be lost, leading to problems with movement and safety. Waste products may collect in the brain tissue. Some slowing of memory and thinking occurs.

HEARING

More than half of people over 60 have hearing difficulties. Problems are caused by changes in the cochlea in the inner ear (see p.117). At birth, there are about 15,000 hair cells in the inner ear, but they gradually reduce with age, and the body is unable to generate new cells.

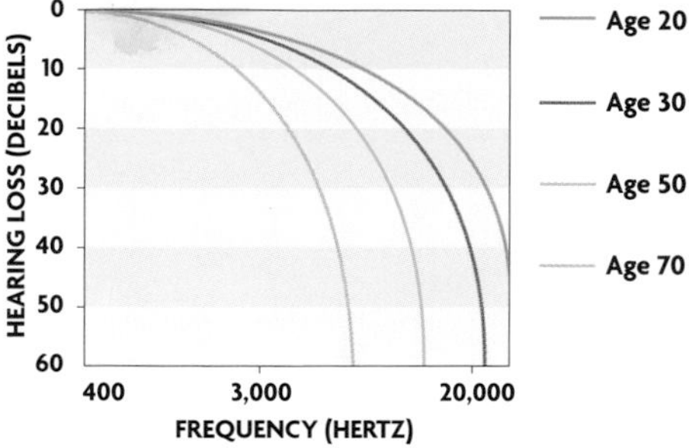
HEARING LOSS (DECIBELS)
0
10
20
30
40
50
60
400
3,000
20,000
FREQUENCY (HERTZ)
Age 20
Age 30
Age 50
Age 70

HIGH-FREQUENCY DROP-OFF

Aging usually causes a loss of sensitivity to sounds; they may become dull or distorted so that speech is difficult to follow. The first sign is often difficulty hearing high-frequency sounds. Hearing aids may enhance the ability to understand speech.

VISION

Older people are susceptible to a number of visual disorders (see p.129). In a cataract, for example, the normally transparent lens of the eye becomes cloudy. Meanwhile, macular degeneration can affect the retina, causing detailed vision to deteriorate.

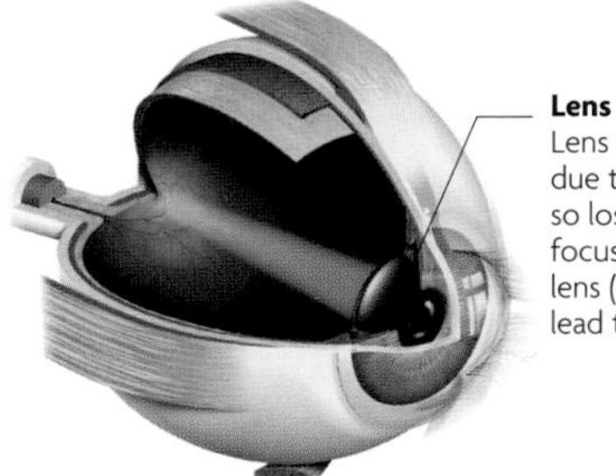
Lens degeneration
Lens usually stiffens due to presbyopia, so loses ability to focus; or a cloudy lens (cataract) can lead to poor vision

COMMON EYE PROBLEMS

A range of eye and visual disorders, particularly those involving the ability of the lens to focus, is more likely to occur with increasing age. Presbyopia, in which people lose the ability to adjust their eyes so they can see nearby objects, is almost universal.

MENOPAUSE

Menopause results from decreased production of sex hormones. Symptoms include hot flashes, insomnia, night sweats, and headaches. Falling estrogen levels can also cause depression. Menstruation may be irregular for several years up to menopause, which is complete once a woman has not had a period for one year. The average age for menopause in developed countries is 51 years.

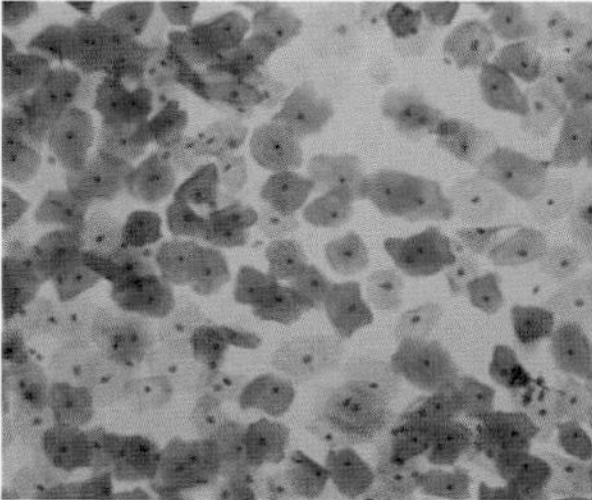

PREMENOPAUSAL CERVICAL SMEAR

Before menopause, the vaginal lining is thick and well lubricated. The cervical smear reveals large cells with small nuclei.

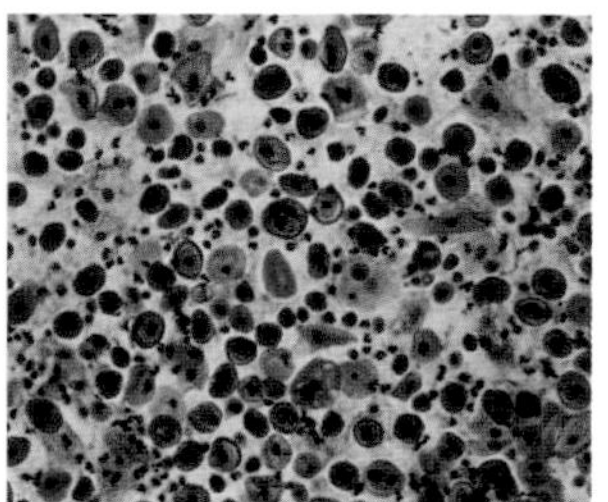

POSTMENOPAUSAL CERVICAL SMEAR

Declining estrogen levels cause the vaginal lining to thin. The smear reveals fewer cells (which clump together), with larger nuclei.

INHERITANCE

THE PASSING OF GENETIC INFORMATION FROM PARENT TO CHILD IS KNOWN AS INHERITANCE. THE INFORMATION IS CONTAINED IN CHEMICAL CODES CARRIED BY DEOXYRIBONUCLEIC ACID (DNA) IN THE SEX CELLS (EGGS AND SPERM).

INHERITANCE OF GENES

Everything that specifies a person is found in their genes. Each gene carries a "blueprint" to make a particular product, some of which affect appearance or biology–skin pigment, for instance. Other gene products combine to produce a complex trait, such as athletic ability. Simple features controlled by single genes are inherited in predictable patterns (see pp.290–93). However, complex traits are controlled by many genes.

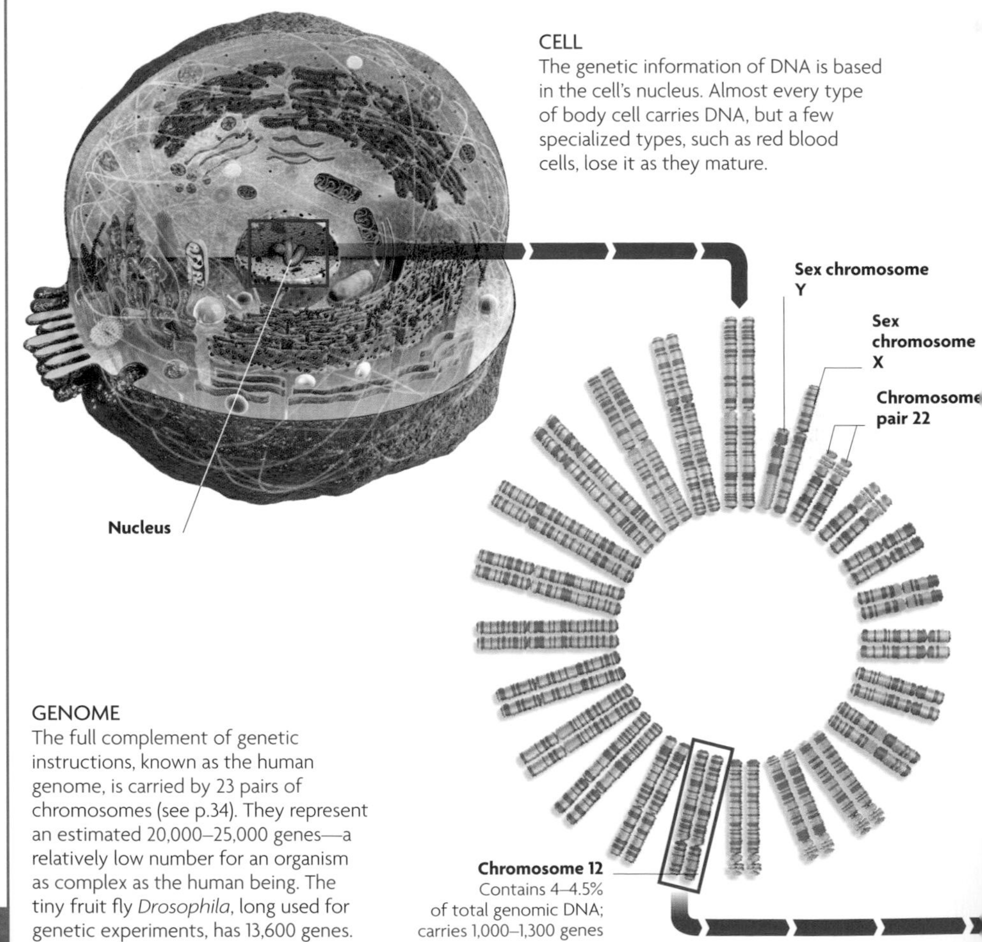
Sex chromosome Y
Sex chromosome X
Chromosome pair 22
Nucleus

CELL
The genetic information of DNA is based in the cell's nucleus. Almost every type of body cell carries DNA, but a few specialized types, such as red blood cells, lose it as they mature.

GENOME
The full complement of genetic instructions, known as the human genome, is carried by 23 pairs of chromosomes (see p.34). They represent an estimated 20,000–25,000 genes—a relatively low number for an organism as complex as the human being. The tiny fruit fly *Drosophila*, long used for genetic experiments, has 13,600 genes.

Chromosome 12
Contains 4–4.5% of total genomic DNA; carries 1,000–1,300 genes

SEQUENCING THE GENOME

The Human Genome Project was set up in 1990 to identify all the base pairs along all the strands of DNA in every chromosome of the genome. In 2003, a complete list of more than three billion base pairs was published. A major technique used in DNA sequencing is gel electrophoresis. DNA is extracted from cells, purified, and broken into smaller fragments of known length by chemicals known as restriction enzymes. The DNA fragments are separated out and stained with dye, showing up as dark stripes, like bar codes (see right). Computers can read these bar codes and reveal the sequences of base pairs.

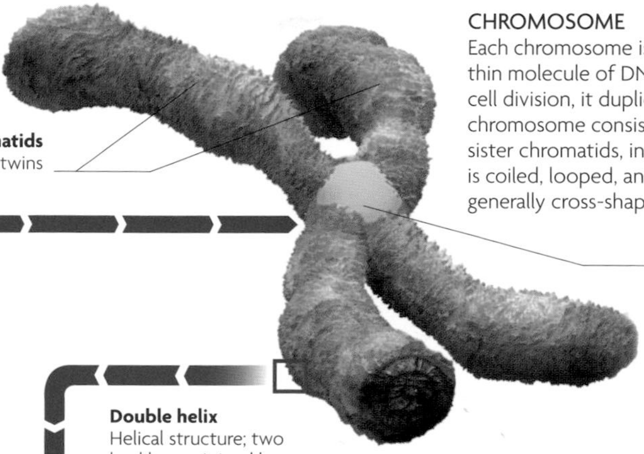

CHROMOSOME

Each chromosome is an immensely long, thin molecule of DNA. When ready for cell division, it duplicates itself as a double-chromosome consisting of two identical sister chromatids, in which the DNA is coiled, looped, and folded into one generally cross-shaped structure.

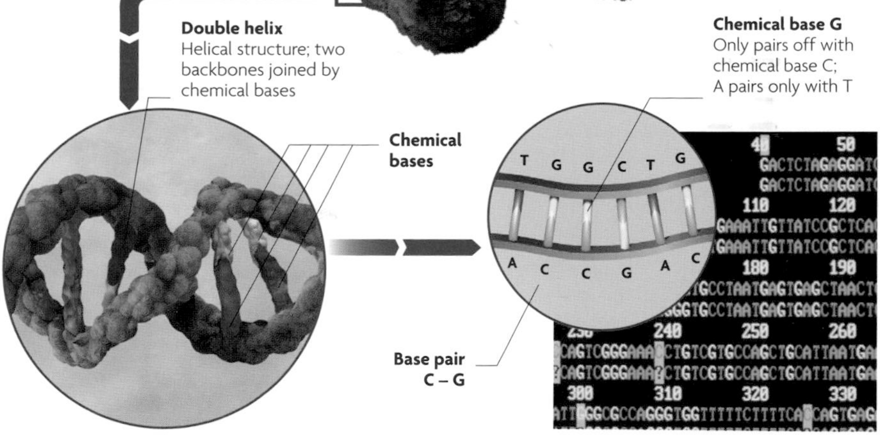

Chromatids
Replicated twins

Centromere
Links chromatids; attaches to spindle during cell division

Double helix
Helical structure; two backbones joined by chemical bases

Chemical bases

Chemical base G
Only pairs off with chemical base C; A pairs only with T

Base pair
C – G

GENETIC CODE

DNA consists of two spiral backbones joined by cross-rungs, which are pairs of chemical bases. The bases are adenine (A), thymine (T), guanine (G), and cytosine (C).

GENETIC SEQUENCE

The order of base pairs on DNA represents the coded genetic information. Using chemicals to identify the bases, DNA-sequencing machines can show the data on screen as lists of letters.

DNA REPLICATION

Apart from carrying genetic information in chemically coded form, as its sequences of base pairs, DNA has another key feature. It can make exact copies of itself, a process known as replication, by separating the two backbone strands and the bases attached to them, at the bonds between the base pairs. Then each strand acts as a template to build a complementary partner strand. DNA replication takes place before cell division (see right).

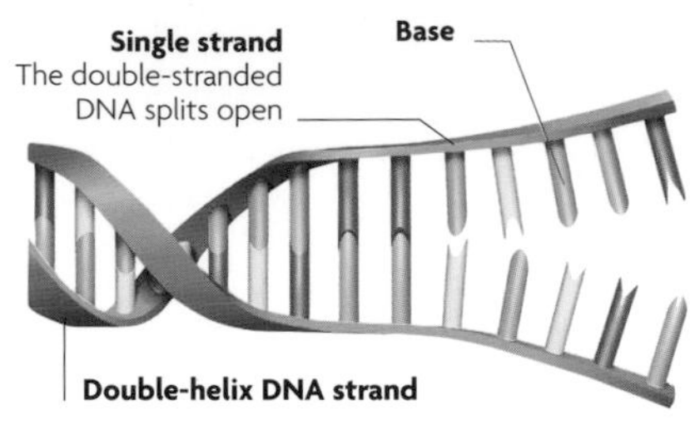
Single strand
The double-stranded DNA splits open
Base
Double-helix DNA strand

1 SEPARATION
The two strands of the double helix separate at the base pair links. Each base is exposed, ready to latch onto its partner in the newly constructed strand.

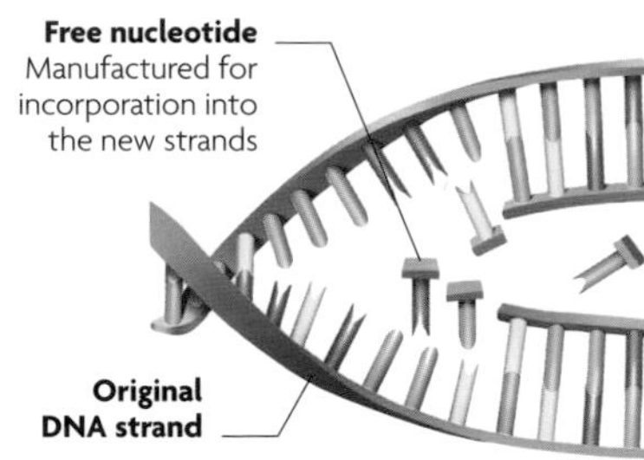
Free nucleotide
Manufactured for incorporation into the new strands
Original DNA strand

2 BASES JOIN
Free nucleotides, each one a base combined with a portion of DNA backbone, join to the two sets of exposed bases. This can only happen in the correct order, since A always pairs with T, and C with G.

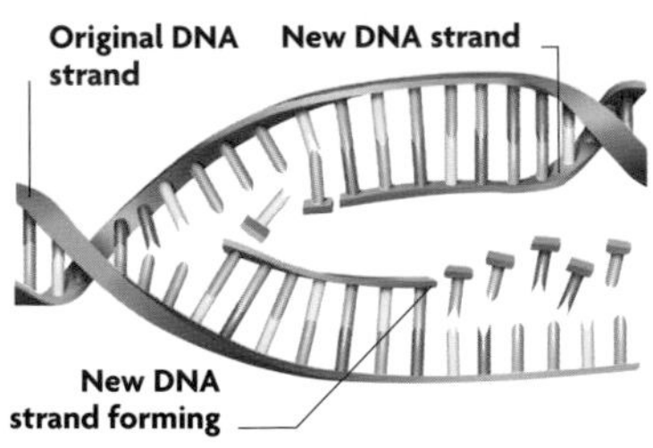
Original DNA strand
New DNA strand
New DNA strand forming

3 TWO STRANDS FORM
More nucleotides join, linked by a new backbone. Each strand now has a new "mirror-image" partner, giving two double helices, which are identical to each other and to the original.

MUTATIONS

DNA replication usually works well. However, factors such as radiation or certain chemicals may cause a fault, where one or more base pairs do not copy exactly. This change is a mutation. The new base sequence may produce a different protein, which could cause a problem in the body.

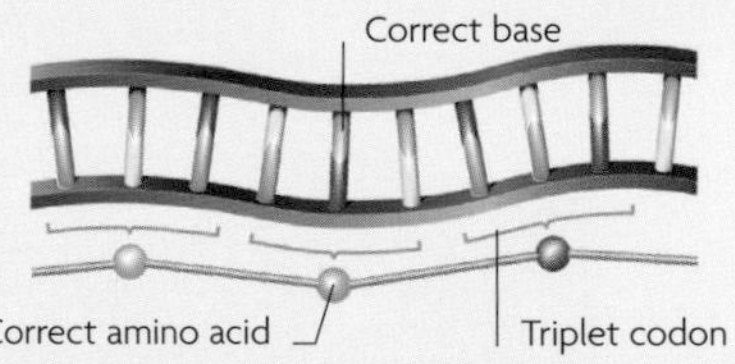
Correct base
Correct amino acid
Triplet codon

NORMAL GENE

Each set of three base pairs (a triplet codon) specifies which amino acid should be added to the series of amino acids that make the normal protein for that gene.

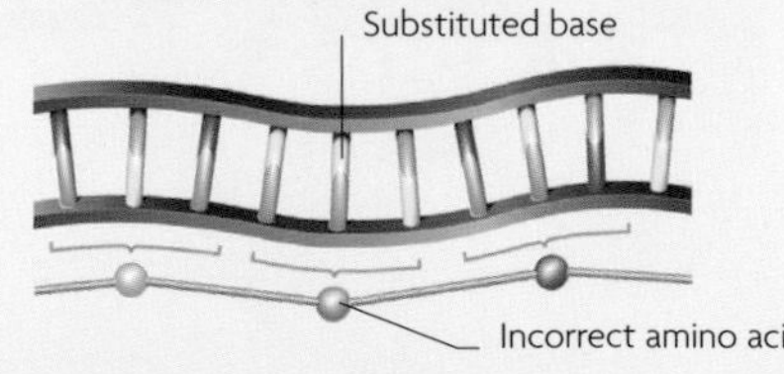
Substituted base
Incorrect amino acid

MUTATED GENE

In a point mutation, one base pair has become altered and substituted. A different amino acid may be specified, which will disrupt the protein's eventual shape and function.

MAKING NEW BODY CELLS

Cell division (mitosis) produces new cells for growth, maintenance, and repair. First, all the DNA replicates and the chromosomes are duplicated. These double-chromosomes form a line and then migrate away from each other as the cell splits in two.

1 PREPARATION DNA replicates and forms double-chromosomes. The nuclear membrane breaks down.

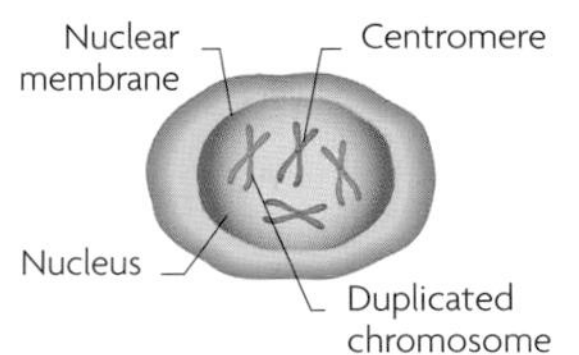
Nuclear membrane
Centromere
Nucleus
Duplicated chromosome

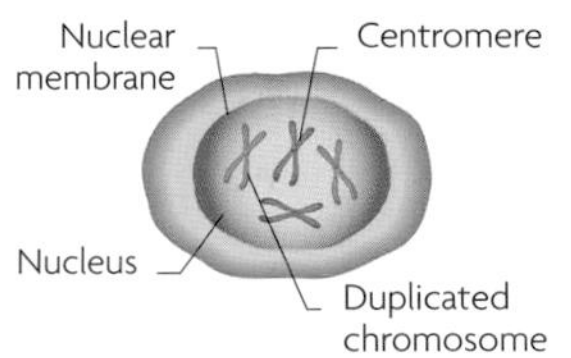

2 ALIGNMENT A spindle fiber holds the centromere of each double-chromosome as it lines up in the middle of the cell.

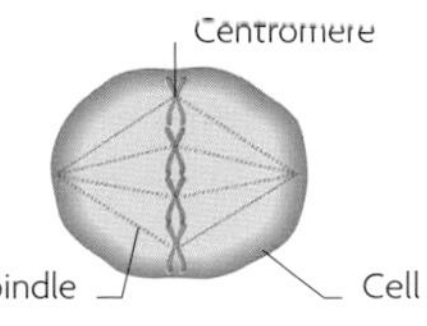
Centromere
Spindle
Cell

3 SEPARATION Each centromere splits so that single chromosomes move to each end of the cell.

Single chromosome
Spindle

4 SPLITTING As the spindle disappears, nuclear membranes form around the two groups of chromosomes.

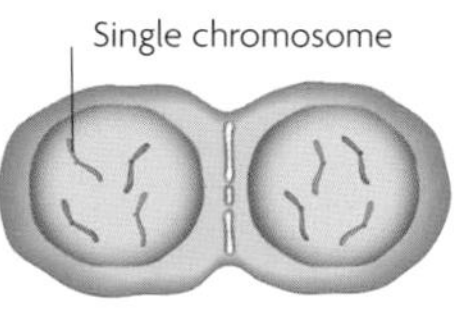
Single chromosome

5 OFFSPRING The cytoplasm divides and the cell splits into two. Each of the new cells has 23 pairs of chromosomes. Only two pairs are shown here for clarity.

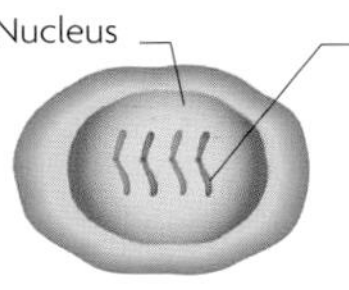
Nucleus
Chromosome

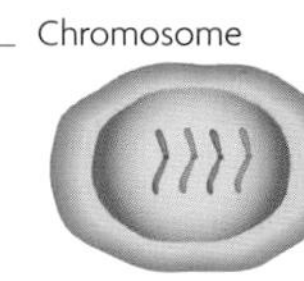

MAKING SEX CELLS

Sperm and egg cells divide by meiosis, into four sex cells (eggs or sperm) that have only one member of a chromosome pair. At fertilization, when egg and sperm unite, the full set (23 pairs) is restored and all subsequent cell divisions are by mitosis.

1 PREPARATION DNA strands replicate and coil up in the nucleus, forming X-shaped double-chromosomes.

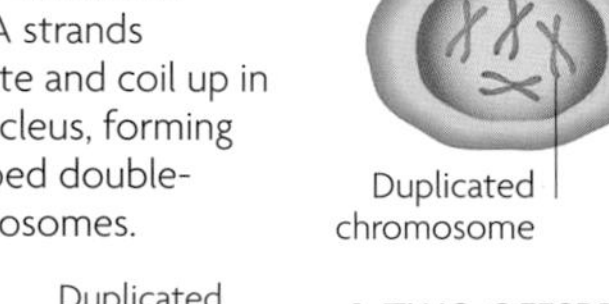
Duplicated chromosome

2 PAIRING The matching (homologous) pairs align, make contact, and exchange genetic material.

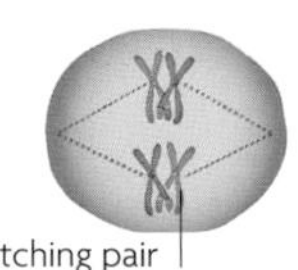
Matching pair of chromosomes

3 FIRST SEPARATION A thread-like spindle pulls one of each pair to each end as the cell splits.

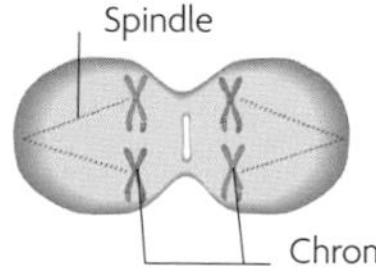
Spindle
Chromosome pair separates

4 TWO OFFSPRING Each cell has one double-chromosome of each pair, as a random choice during separation.

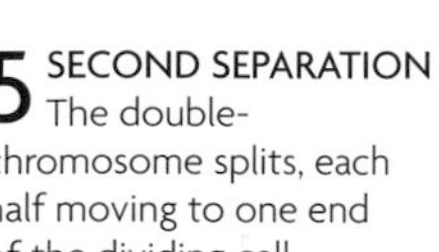
Duplicated chromosomes

5 SECOND SEPARATION The double-chromosome splits, each half moving to one end of the dividing cell.

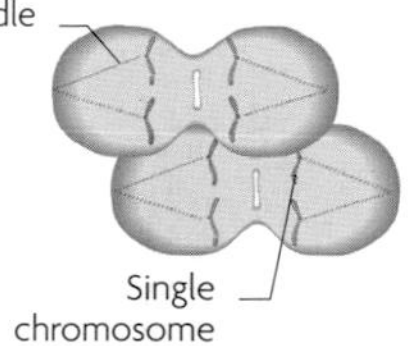
Spindle
Single chromosome

6 FOUR OFFSPRING The four sex cells differ from each other and the parent cell in their genetic composition.

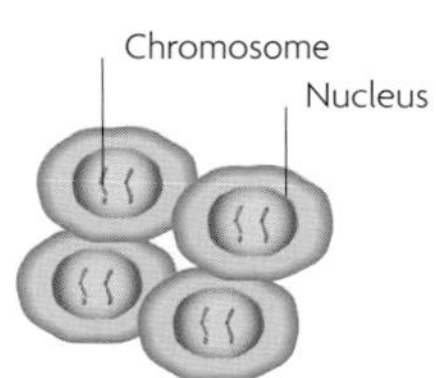
Chromosome
Nucleus

PATTERNS OF INHERITANCE

GENES ARE PASSED FROM ONE GENERATION TO THE NEXT, IN A VAST SEQUENCE OF INHERITANCE. THEY ARE RESHUFFLED AT EACH STAGE SO THAT OFFSPRING ARE UNIQUE, BUT THERE ARE PATTERNS IN THE MODE OF INHERITANCE.

VERSIONS OF GENES

Each cell in a body contains a double set of genetic material, in the form of 23 pairs of chromosomes. One chromosome of each pair, and the genes on it, come from the mother. The other chromosome is from the father. So there are, in effect, two versions of every gene in the set—one maternal and one paternal. These versions of genes are called alleles. Inheritance patterns vary depending on how these two versions interact, because they may be identical or slightly different.

TWO BY TWO
Chromosome pairs have the same sets of genes. But the individual allele on one chromosome may differ slightly from its equivalent allele on the other chromosome.

GENERATIONAL SEQUENCE

The two versions of the genes (alleles) are mixed, or reshuffled, as they are inherited at each generation. In effect, a child inherits one-quarter of its total genes from each grandparent. The child's inherited features strongly resemble a mixture of those from his or her parents, but the features from the grandparents appear to be less marked.

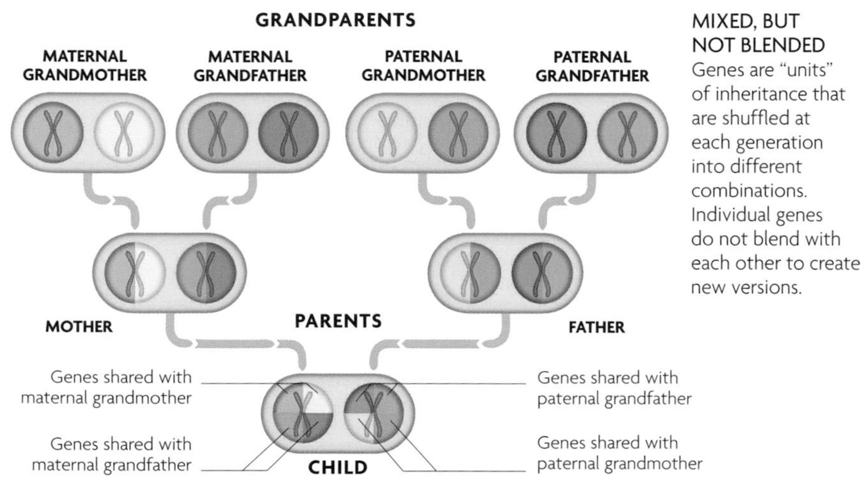
GRANDPARENTS
MATERNAL GRANDMOTHER
MATERNAL GRANDFATHER
PATERNAL GRANDMOTHER
PATERNAL GRANDFATHER
MOTHER
PARENTS
FATHER
Genes shared with maternal grandmother
Genes shared with maternal grandfather
Genes shared with paternal grandfather
Genes shared with paternal grandmother
CHILD

MIXED, BUT NOT BLENDED

Genes are "units" of inheritance that are shuffled at each generation into different combinations. Individual genes do not blend with each other to create new versions.

INHERITANCE OF GENDER

Gender depends on which sex chromosome—an X or a Y—is inherited. Females have two X chromosomes; males have an X and a smaller Y, with male genes. A woman's egg cells all contain an X, whereas half a man's sperm cells contain an X and the other half a Y. Thus, the gender of offspring is always determined by the father.

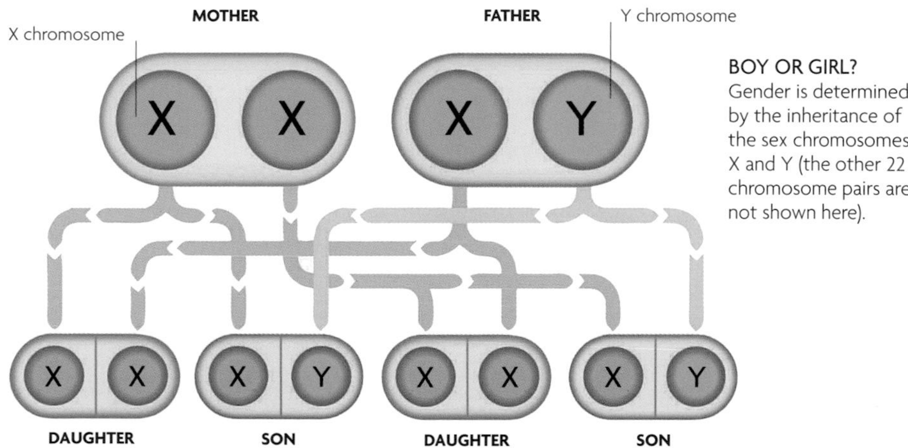
MOTHER
FATHER
X chromosome
Y chromosome
X
X
X
Y
X
X
X
Y
X
X
X
Y
DAUGHTER
SON
DAUGHTER
SON

BOY OR GIRL?

Gender is determined by the inheritance of the sex chromosomes, X and Y (the other 22 chromosome pairs are not shown here).

RECESSIVE AND DOMINANT GENES

Each gene in a cell exists in two versions, one inherited from each parent. In some cases these gene versions, or alleles, are different, and produce slightly different results. One allele may be dominant and "overpower" the other, which is recessive. An example is eye color, although this is not as simple as depicted below.

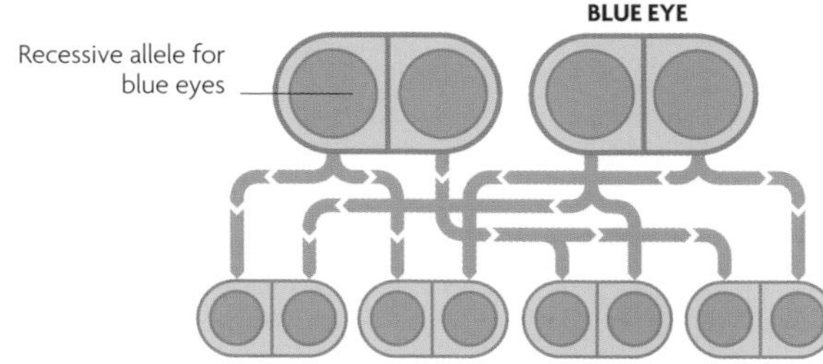
BLUE EYE
Recessive allele for blue eyes
ALL INDIVIDUALS HAVE BLUE EYES

RECESSIVE AND RECESSIVE

Each parent has two alleles for eye color. Here, both parents have only "blue" alleles. When both alleles are the same, the individual is said to be "homozygous." Their children can only inherit "blue" alleles, so all have blue eyes.

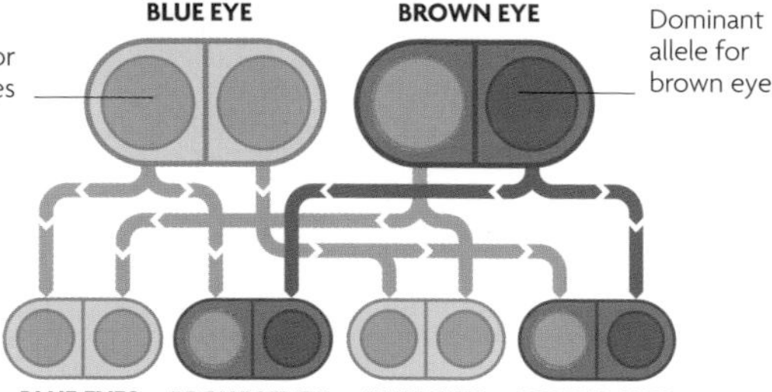
BLUE EYE
BROWN EYE
Recessive allele for blue eyes
Dominant allele for brown eyes
BLUE EYES
BROWN EYES
BLUE EYES
BROWN EYES

RECESSIVE AND MIXED

One parent has two "blue" alleles; the other, one "blue" and one "brown" allele. "Brown" is dominant and takes over when it occurs with "blue". So the chance is 1 in 2 that each offspring has brown eyes.

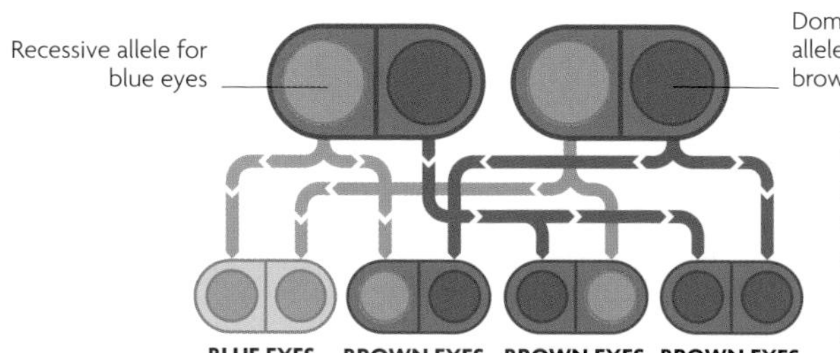
Recessive allele for blue eyes
Dominant allele for brown eyes
BLUE EYES
BROWN EYES
BROWN EYES
BROWN EYES

MIXED AND MIXED

Each parent has a "brown" and a "blue" allele. Individuals who carry two different alleles are "heterozygous." Only one of the four possible combinations leads to offspring with blue eyes.

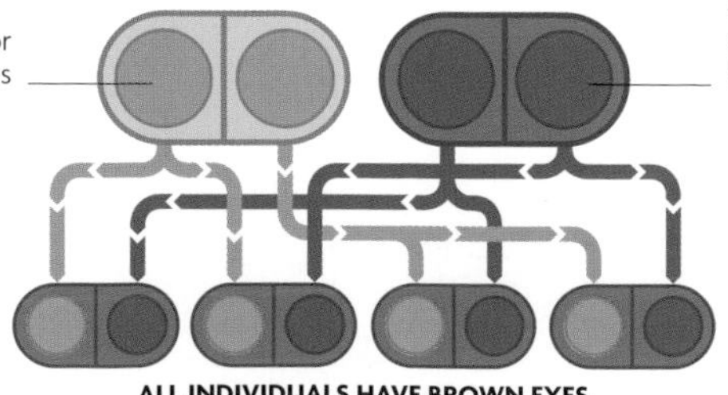
Recessive allele for blue eyes
Dominant allele for brown eyes
ALL INDIVIDUALS HAVE BROWN EYES

DOMINANT AND RECESSIVE

One parent has two "blue" alleles; the other, two "brown" alleles. The four possible combinations all produce offspring with brown eyes, but all four still carry "blue" alleles.

SEX-LINKED INHERITANCE

The pattern of inheritance changes when alleles for a body feature are carried on the sex chromosomes. If an allele on a man's X chromosome does not have its equal on the Y chromosome, or vice versa, only one allele can determine the feature. For example, the problem allele for color-impaired vision is on the X chromosome.

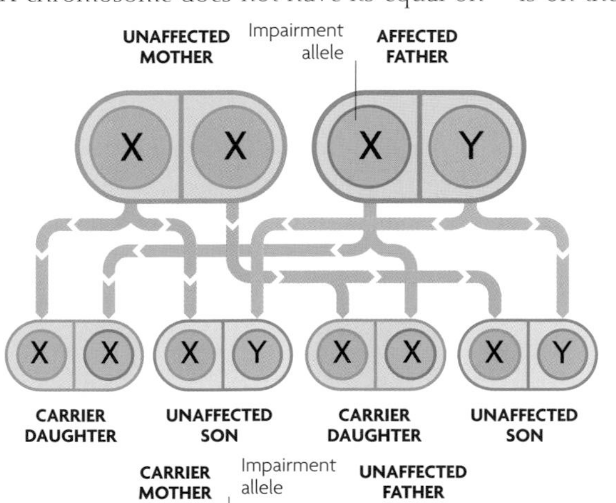
UNAFFECTED MOTHER
Impairment allele
AFFECTED FATHER
X
X
X
Y
X X
X Y
X X
X Y
CARRIER DAUGHTER
UNAFFECTED SON
CARRIER DAUGHTER
UNAFFECTED SON

COLOR-BLIND FATHER AND UNAFFECTED MOTHER
Sex chromosomes combine in four possible ways, governed by chance. Here, any daughter will inherit the color-impairment allele, and will be a carrier, but she also has the normal allele on her other X chromosome, to give normal vision. No sons can be affected, nor can they be carriers.

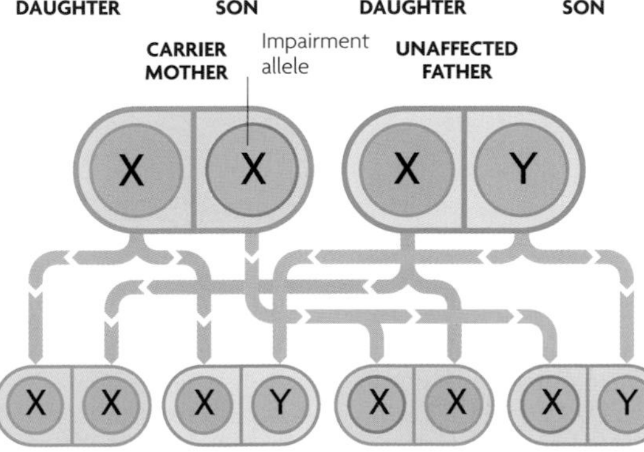
CARRIER MOTHER
Impairment allele
UNAFFECTED FATHER
X
X
X
Y
X X
X Y
X X
X Y
UNAFFECTED DAUGHTER
UNAFFECTED SON
CARRIER DAUGHTER
AFFECTED SON

CARRIER MOTHER AND UNAFFECTED FATHER
The four possible combinations give a one-in-four-chance each for unaffected sons and daughters. There is also a one-in-four chance that an offspring will be a carrier daughter or son who has inherited the color-impairment allele. He has no second X chromosome and therefore no normal allele, so the result is impaired color vision.

MULTIPLE-GENE INHERITANCE

Some body traits follow clear single-gene inheritance patterns. However, the situation becomes more complex in two ways. First, there may not be only two alleles of a gene with a simple dominant–recessive interaction between them. There may be three alleles or more in existence in the general population, although each person can have only two of them. An example is the blood group system, with alleles for A, B, and O. Second, a trait may be influenced by more than one gene. These two situations mean that a trait can be governed by multiple genes, and for each of these genes, by multiple alleles of the gene—added to which, the genes may interact in different ways, according to which alleles are present in each of them. In such cases, the numbers of possible combinations multiply, consequently making multi-gene inheritance exceptionally difficult to unravel.

MALE REPRODUCTIVE DISORDERS

DISORDERS AFFECTING THE EXTERNAL PARTS OF THE MALE REPRODUCTIVE TRACT ARE USUALLY APPARENT AT AN EARLY STAGE; THOSE AFFECTING INTERNAL PARTS, SUCH AS THE PROSTATE GLAND, MAY NOT BE NOTICED UNTIL LATER, WHEN SUCCESSFUL TREATMENT MAY BE HARDER TO ACHIEVE.

PROSTATE DISORDERS

Conditions that affect the prostate gland range from inflammation and benign enlargement to serious disorders such as cancer. Prostate disorders are very common and tend to occur in the middle and later years of a man's life. Prostate cancer, although potentially life-threatening, tends to occur most commonly in elderly men, in whom it often grows slowly and may not cause symptoms. New screening techniques are detecting the condition in much younger men, who do need treatment. Enlargement of the prostate is extremely common and is considered part of the aging process; most men over age 50 have it to some degree. If the enlarged gland constricts the urethra, it can cause distressing urinary symptoms, including frequent urination, delay in starting to urinate, weak flow, dribbling, and a feeling of incomplete bladder emptying. Prostatitis (see below) is a common condition, often caused by infection.

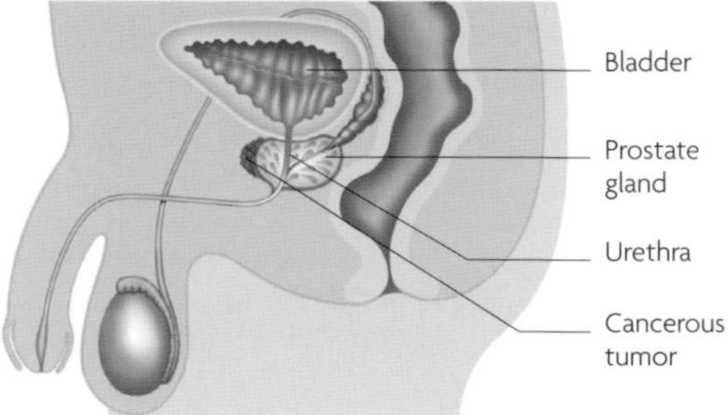
Bladder
Prostate gland
Urethra
Cancerous tumor

PROSTATE CANCER
A cancerous tumor of this size on the prostate gland is unlikely to cause immediate problems, but as it grows it may press on the urethra, and may spread to other parts of the body.

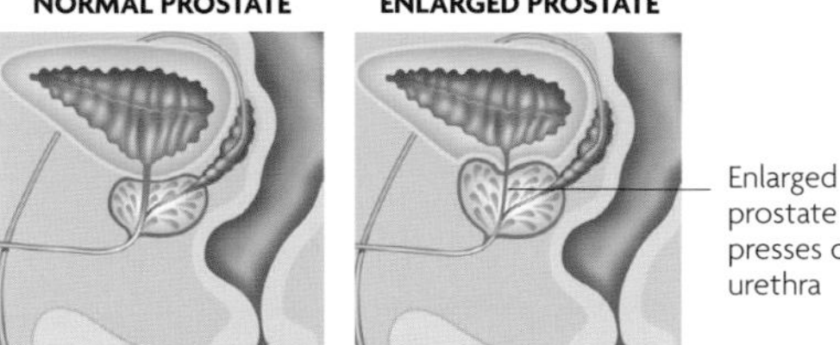
NORMAL PROSTATE
ENLARGED PROSTATE
Enlarged prostate presses on urethra

ENLARGED PROSTATE
A normal prostate gland fits snugly around the urethra and abuts the bladder; enlargement can squash the urethra.

PROSTATITIS

Inflammation of the prostate gland, or prostatitis, can be acute or chronic. The acute type is less common; severe symptoms such as fever and pain in the lower back come on suddenly, but usually clear up quickly. Chronic prostatitis features longstanding but often mild symptoms that are difficult to treat, such as groin and penis pain, pain on ejaculation, blood in semen, and painful urination. Possibly caused by a bacterial infection from the urinary tract, both types are most common in men between 30 and 50 years old.

CAUSATIVE BACTERIUM
This electron micrograph shows the bacterium *Enterococcus faecalis*, implicated in prostatitis. It is a normal, harmless inhabitant of the human gut.

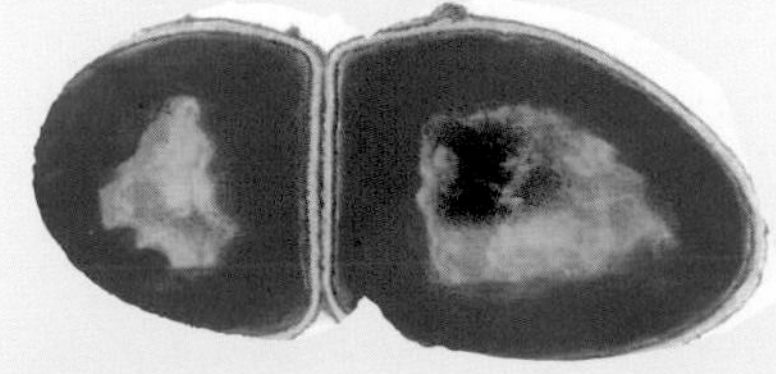

TESTICULAR CANCER

Cancer of the testis is one of the most commonly occurring cancers in men between the ages of 20 and 40. Although it is easily curable if discovered early, the cancer can spread to lymph nodes and to other parts of the body if not treated. Symptoms of testicular cancer include a hard, painless lump in the testis; a change in the size and appearance of the testis; or a dull ache in the scrotum. There are three different types of testicular tumour (germ cell tumor, stromal cell tumor, and secondary testicular tumor), all of which develop in the sperm-producing cells of the testis. Because early treatment of the cancer is vital and has a very high cure rate, all men should regularly examine their testes; any swellings or changes in the scrotal skin should be reported immediately. Soft lumps or painful swellings are likely to be caused by a cyst or infection, but should still be checked.

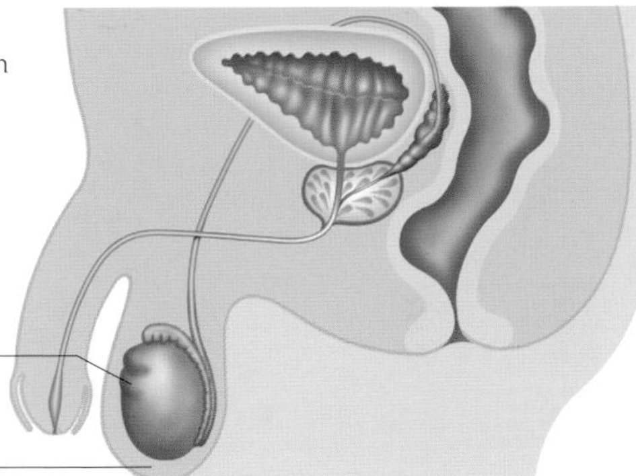
Tumor
Tiny growth on the testis
Scrotum

TUMOR ON TESTIS
A tumor of this size on the outer wall of the testis would be clearly felt through the thin outer skin and layers of the scrotum.

HYDROCELE

Each testis is surrounded by a double-layered membrane, which under normal conditions contains a small amount of fluid. In a hydrocele, an excessive amount of fluid forms, causing the testis to appear swollen. The condition occurs most frequently in infants and elderly people. The cause of hydrocele is not usually known, although infection, inflammation, or injury to the testis are possible triggers. A hydrocele does not usually cause any pain, but may result in a dragging sensation due to the increased size and weight of the scrotum. In younger sufferers, the condition often gets better without the need for treatment. However, if the condition is causing discomfort, the hydrocele may be surgically removed or, for those who are not fit enough for surgery, the fluid may be drained from the area using a needle and syringe.

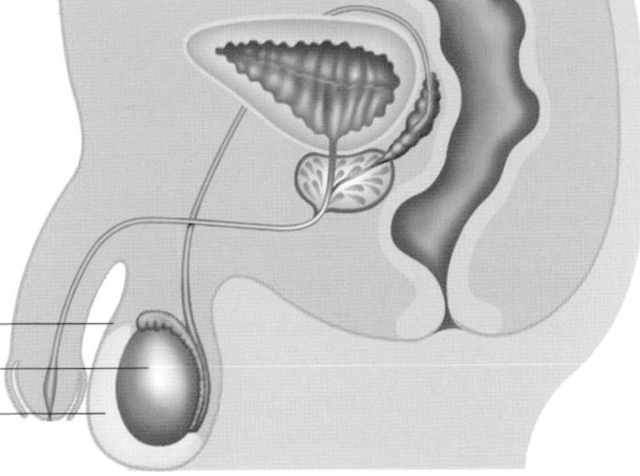
Scrotum
Testis
Fluid
Fluid accumulates around the testis

SWOLLEN TESTIS
A hydrocele is the result of excess fluid filling the double-layered membrane that surrounds the testis; it causes the scrotum to appear swollen.

FEMALE REPRODUCTIVE DISORDERS

MANY FEMALE REPRODUCTIVE DISORDERS ARE HARMLESS, AND SOME ARE EVEN SYMPTOMLESS. HOWEVER, EXCESSIVE HORMONAL FLUCTUATIONS AND PHYSIOLOGICAL STRESSES CAN LEAD TO MORE SERIOUS DISORDERS, INCLUDING VARIOUS TYPES OF CANCER.

BREAST LUMPS

A breast lump is a solid or swollen area that can be felt or seen in the tissue of the breast. General lumpiness is common as breasts change shape during puberty, pregnancy, and prior to menstruation. Nonspecific lumpiness usually relates to the hormonal fluctuations of the menstrual cycle.

A single lump may be an overgrown lobule, and a more defined one may be a cyst. Only a small percentage of lumps are a symptom of breast cancer. All women should familiarize themselves with the shape of their breasts during the menstrual cycle, so that they can look and feel for abnormal changes, and immediately report them to their doctor. From the age of 50, women should have regular mammograms.

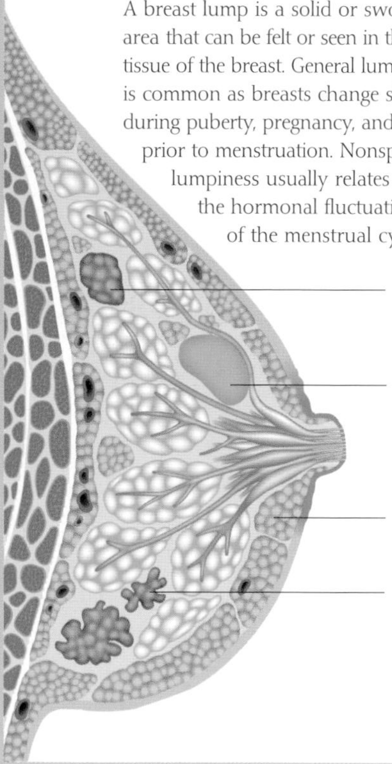

Fibroadenoma A common noncancerous breast lump

Cyst One or more fluid-filled sacs within breast

Fatty tissue

Nonspecific lumpiness Usually related to menstruation; often called fibrocystic disease

BREAST CANCER

Cancer of the breast is the most common female cancer. The risk increases with age, doubling every 10 years. The causes are unclear, but risk factors have been identified. Women with higher exposure to estrogen—for example, through having an early puberty, late menopause, or no children—have a higher risk. Age is significant, with many more cases occurring over the age of 50. Faulty genes are also a known cause. A breast lump, usually painless, is often the first sign of breast cancer.

ENDOMETRIOSIS

Endometriosis is a common condition, affecting many women of childbearing age. It can cause debilitating pain and very heavy periods; in severe cases, the condition can lead to fertility problems. The endometrium, the lining of the uterus, is shed approximately once every month as part of the menstrual cycle. Endometriosis causes small pieces of the uterine lining to become attached to other nearby organs, such as the ovaries or large intestine. These pieces of tissue respond to hormonal changes and bleed during menstruation. Since the blood cannot leave the body through the vagina, its normal exit, it irritates nearby tissues, causing pain and eventually forming scars. The cause of the disorder is unknown.

FIBROIDS

Fibroids are very common, occurring in about one-third of women of childbearing age. They can occur singly or in groups, and can be the size of a pea or as large as a grapefruit. Small fibroids are unlikely to cause any problems. Larger ones may result in prolonged and heavy menstrual bleeding, and increasingly severe period pain. Large fibroids can distort the uterus, which may cause infertility, or put pressure on other organs, such as the bladder or rectum.

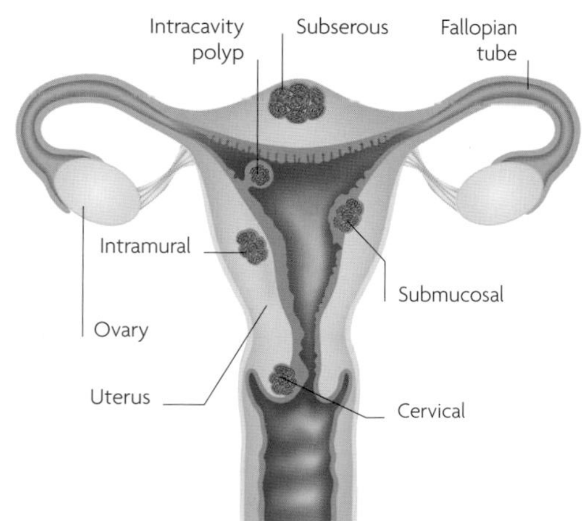

Intracavity polyp
Subserous
Fallopian tube
Intramural
Submucosal
Ovary
Uterus
Cervical

SITES AND TYPES OF FIBROIDS

Fibroids can occur in any part of the uterus wall and are named according to their site—for example, in the cervix—or in the tissues they occupy, such as submucosal fibroids.

PROLAPSED UTERUS

Prolapse of the uterus is more likely to occur after menopause, when low estrogen levels affect the ability of the ligaments to retain the uterus. Childbirth, obesity, and straining while coughing or opening the bowels are contributing factors. The uterus protrudes down into the vagina, and in severe cases may reach as far as the vulva. Symptoms may include a feeling of fullness in the vagina, pain in the lower back, and difficulty urinating or passing feces.

Pelvic brim
Cervix
Levator ani muscle
Deep transverse perineal muscle
Obturator internus muscle

NORMAL UTERUS

The uterus is kept in place by muscles and ligaments. Regular Kegel exercises are important to maintain their strength and avoid prolapse.

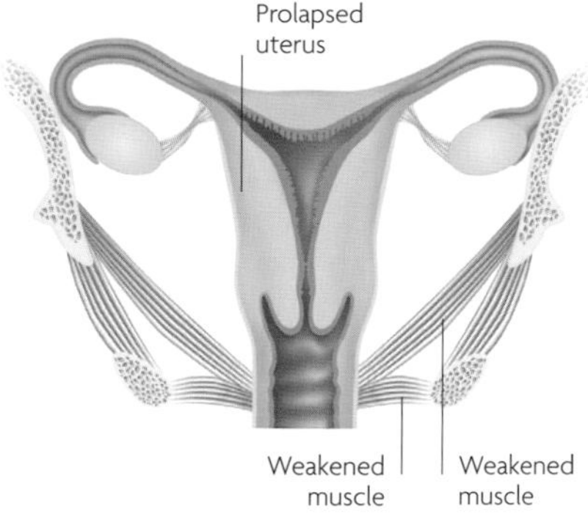

Prolapsed uterus
Weakened muscle
Weakened muscle

PROLAPSED UTERUS

In this case of uterine prolapse, the uterus has slipped down into the vagina because of weakened muscles. The wall of the vagina may also prolapse.

SEXUALLY TRANSMITTED INFECTIONS

SEXUALLY TRANSMITTED INFECTIONS (STIs), ALSO KNOWN AS SEXUALLY TRANSMITTED DISEASES (STDs), ARE INFECTIONS THAT ARE PASSED FROM PERSON TO PERSON BY SEXUAL ACTIVITY. GENITAL, ANAL, AND ORAL SEX CAN ALL PASS AN INFECTION ON TO ANOTHER PERSON.

GONORRHEA

Although gonorrhea tends to be more prevalent among males, it can also affect women. The main sites of infection are the urethra and, in women, the cervix. The symptoms often do not appear, but if they do they commonly include a discharge of pus from the penis or vagina and pain on urination. Women may also experience lower abdominal pain and irregular vaginal bleeding. Occasionally, the infection spreads to other parts of the body, such as the joints (via the bloodstream). If the disease is left untreated, it can cause infertility in women.

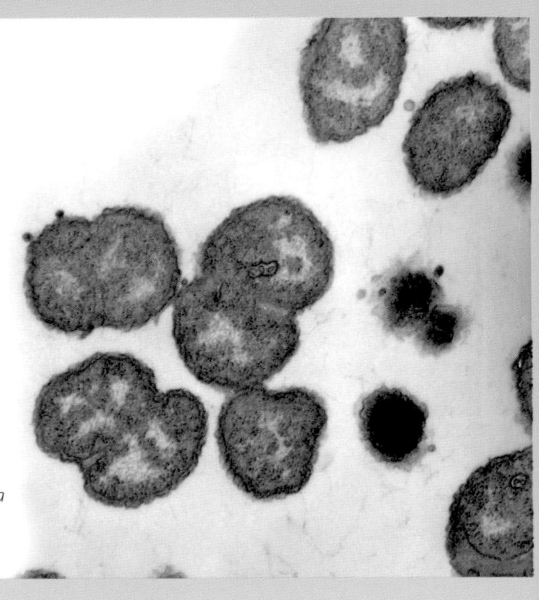

GONORRHEA BACTERIA
An electron micrograph of *Neisseria gonorrhoeae*, which is responsible for the STI gonorrhea.

PELVIC INFLAMMATORY DISEASE (PID)

PID is a common cause of pelvic pain in young women; other possible symptoms are fever, heavy or prolonged periods, and pain during sexual intercourse. Sometimes, there are no symptoms. Usually, PID is the result of an STI such as chlamydial infection or gonorrhea. Infection after childbirth or a pregnancy termination are also possible causes.

The inflammation starts in the vagina and spreads to the uterus and fallopian tubes. In severe cases, the ovaries are also infected. Left untreated, PID can lead to damage in the fallopian tubes, which may cause infertility and an increased risk of ectopic pregnancy (see p.302).

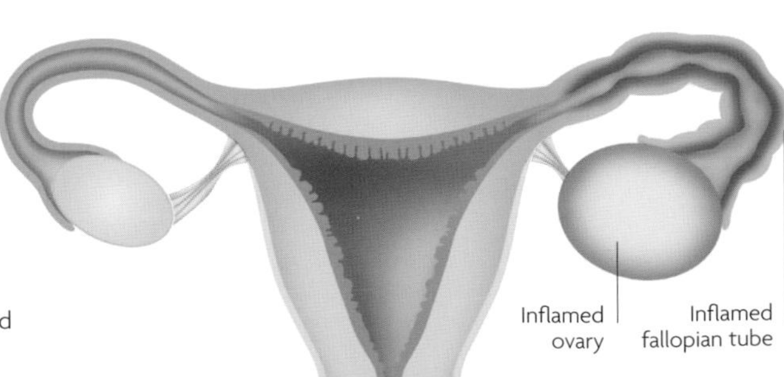
Inflamed ovary
Inflamed fallopian tube

INFECTED PARTS
The fallopian tube and ovary on the right of the image are inflamed and swollen as a result of PID.

NONGONOCOCCAL URETHRITIS

Nongonococcal urethritis (NGU) is one of the most common STIs affecting men. Typically, it features inflammation of the urethra, with or without a discharge of pus; inflammation and soreness at the end of the penis; and pain on urinating, particularly when the urine is concentrated first thing in the morning. In about half of all cases, the agent responsible is *Chlamydia trachomatis*, a bacterium that can also infect women, leading to chlamydial infection. Other possible causes of NGU include the bacterium *Ureaplasma urealyticum*; the protozoan *Trichomonas vaginalis*; the fungus *Candida albicans*; the genital warts virus (human papillomavirus, HPV); and the genital herpes viruses. It is important for both partners to seek treatment to avoid reinfecting each another. To prevent STIs, sexually active people should limit their sexual partners, and use a condom for penetrative sex.

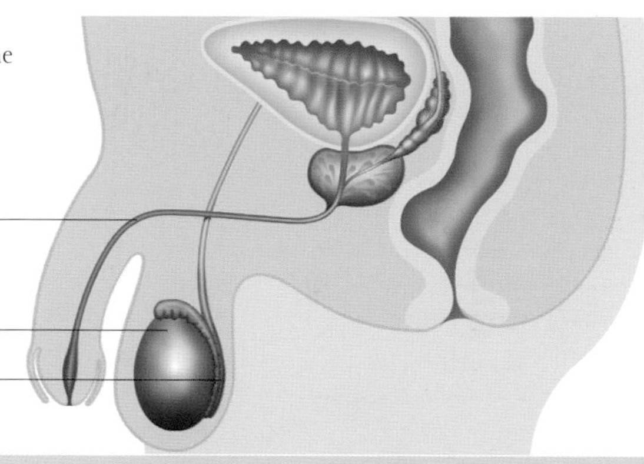
Urethra
Inflammation of urethra causes pain on passing urine
Testis
May become swollen if infection spreads
Epididymis
Sometimes also becomes inflamed

SYMPTOMS OF NGU
Inflammation of the urethra causes pain and soreness at the external opening on the penis, and painful urination. If the infection spreads, the epididymides and the testes may also become swollen.

SYPHILIS

Syphilis can affect both men and women. It is caused by *Treponema pallidum*, a bacterium that enters the body via the genitals. It first affects the organs of reproduction, and spreads to other parts of the body. An infectious sore (chancer) appears on the penis or vagina, lymph nodes swell, and then a rash and wartlike patches develop on the skin. With no treatment, it can proceed to a final, possibly fatal, stage characterized by personality changes, mental illness, and nervous system disorders. Today, the disease rarely progresses to this stage.

CHLAMYDIAL INFECTION

Chlamydial infection is a very common STI and occurs only in women. It is caused by *Chlamydia trachomatis*, which inflames the reproductive organs, and causes symptoms including vaginal discharge, a frequent urge to urinate, lower abdominal pain, and pain during intercourse. Chlamydial infection can lead to PID (see opposite), if left untreated, and may then cause infertility.

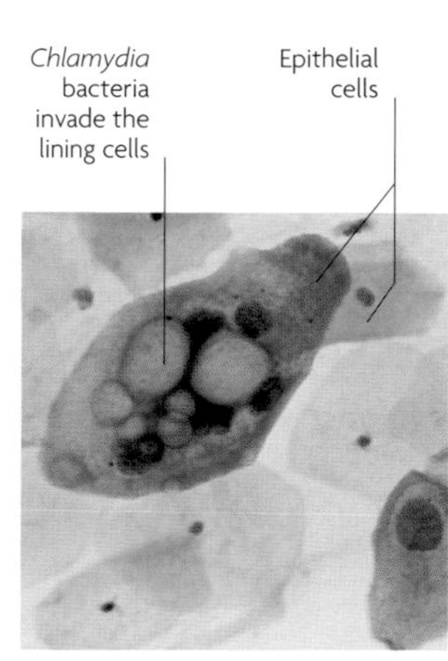
Chlamydia bacteria invade the lining cells
Epithelial cells

BACTERIA IN CERVICAL SMEAR
This micrograph (x400) of a cervical smear shows *Chlamydia trachomatis* bacteria (pink cells within large blue cell).

INFERTILITY DISORDERS

IF A COUPLE IS UNABLE TO CONCEIVE AFTER A YEAR OF HAVING UNPROTECTED SEX, ONE OR BOTH PARTNERS MAY HAVE A FERTILITY PROBLEM. THE LIKELIHOOD OF FERTILITY DISORDERS INCREASES WHEN COUPLES WAIT UNTIL THEY REACH THEIR 30s OR 40s TO START A FAMILY.

DAMAGED FALLOPIAN TUBE

The fallopian tube may become blocked as a result of endometriosis (see p.296), in which fragments of the uterine lining (endometrium) become embedded in the tube tissue. Pelvic inflammatory disease (see p.298), which is often caused by a sexually transmitted infection such as chlamydia (see p.299), may go unnoticed at the time of infection, but scarring due to the inflammation can cause problems with fertility later. An intrauterine contraceptive device can increase the risk of PID developing. Usually, only one tube is affected, which means that a woman's chance of conceiving is halved.

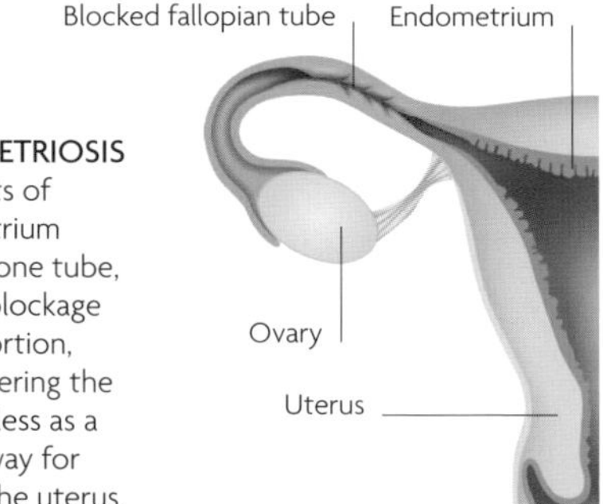
Blocked fallopian tube
Endometrium
Ovary
Uterus

ENDOMETRIOSIS Fragments of endometrium lodge in one tube, causing blockage and distortion, and rendering the tube useless as a passageway for eggs to the uterus.

OVULATION PROBLEMS

Any deviation from the normal ovulation pattern can cause problems with fertility. The precise problem can range from complete absence of egg release to infrequent release. Factors that can lead to ovulation problems include pituitary and thyroid gland disorders, polycystic ovary syndrome, long-term use of oral contraceptives, being very over- or underweight, stress, excessive exercise, and premature menopause.

CERVICAL PROBLEMS

The cervix, or neck of the womb, produces mucus that is usually thick; just before ovulation, when the level of estrogen increases, the mucus becomes less viscous to allow sperm to penetrate. If estrogen levels are low or if there is infection within the reproductive tract, the mucus may remain thick and impregnable to sperm. Another problem that may make the cervix inhospitable is that sometimes a woman's immune system forms antibodies to her partner's sperm, which will then damage or kill the sperm in the cervix. Polyps, fibroids (see p.297), narrowing (stenosis), and distortion are other problems of the cervix that may lead to infertility.

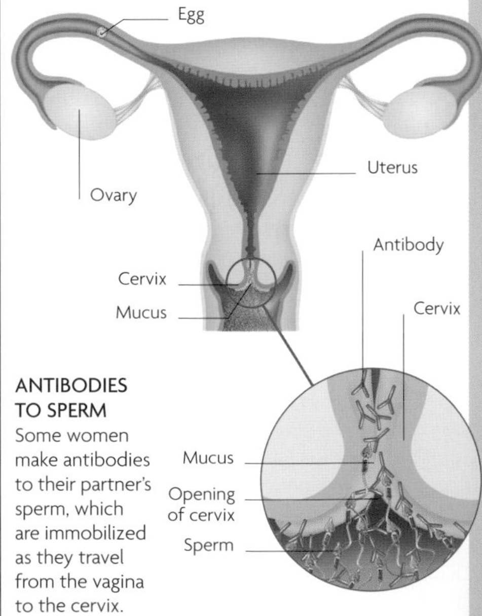
Egg
Ovary
Uterus
Cervix
Mucus
Antibody
Cervix
Mucus
Opening of cervix
Sperm

ANTIBODIES TO SPERM Some women make antibodies to their partner's sperm, which are immobilized as they travel from the vagina to the cervix.

PROBLEMS WITH SPERM PRODUCTION

A man may produce sperm in low quantities, or his sperm may be deformed or unable to swim properly. All these problems reduce the likelihood that his sperm can contribute to conception. Huge numbers of sperm must be produced in order for fertilization to occur; men in whom this does not happen have a low sperm count. Microscopic examination can reveal this problem and can also look at the size, shape, and movement (motility) of individual sperm. Problems in any of these areas can cause reduced fertility. If only a small volume of semen is produced per ejaculation, fertility may also be reduced.

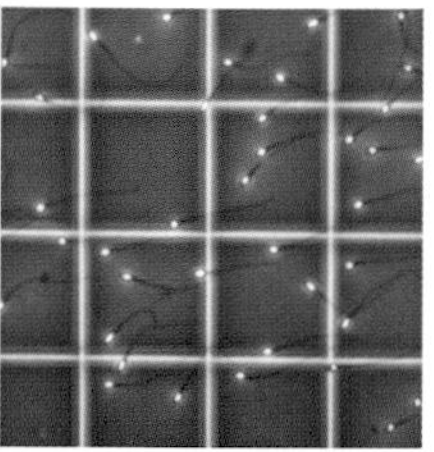

NORMAL SPERM COUNT

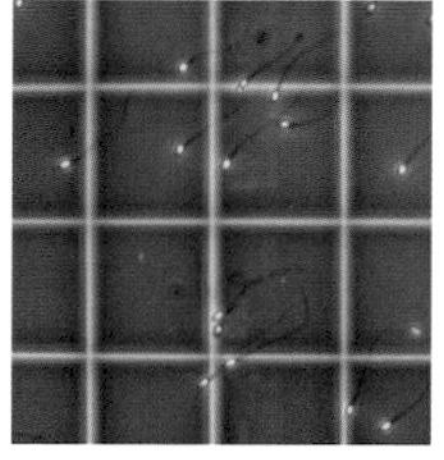

LOW SPERM COUNT

EJACULATION PROBLEMS

A number of ejaculation problems prevent sperm from arriving in the vagina by the normal means, making fertilization impossible. The most common is erectile dysfunction (the difficulty in achieving or maintaining an erection). This condition may be a result of diabetes mellitus (see pp.142–43), a spinal cord disease, impaired blood flow, certain drugs, or psychological problems. Another problem, retrograde ejaculation, causes semen to flow back into the bladder because of faulty valves; this can be a complication of surgery for partial or complete removal of the prostate gland. Various treatments are available that can help reduce erectile dysfunction, depending on the nature and cause of the problem.

DIFFICULT PASSAGE OF SPERM

Sperm has a long and tortuous journey from its source in the testis until it is ejaculated. Narrowing, blockage, or other distortion of any of the tubes, including the epididymis and vas deferens, that make up this network can slow or completely block the passage of sperm. Causes of this problem are various, but infection of the male reproductive system is most likely. Some sexually transmitted infections (STIs, see p.298), most notably gonorrhea, can cause inflammation of the tubes, which leaves scar tissue that can distort their structure and affect their sperm-carrying ability.

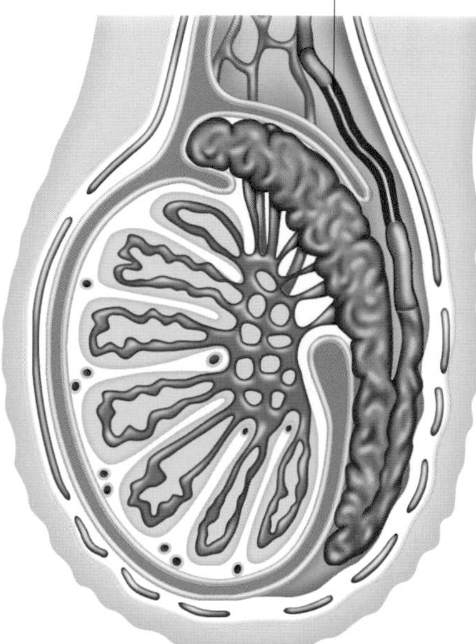
Narrowed lumen of vas deferens

INFLAMED VAS DEFERENS

Damage to the vas deferens, one of the tubes that transports the sperm, can prevent or slow down its passage. Infection, usually by a sexually transmitted organism, can be responsible for such damage.

PREGNANCY AND LABOR DISORDERS

PROBLEMS CAN ARISE IN NORMALLY HEALTHY WOMEN DURING PREGNANCY AND LABOR, WHICH MAY ENDANGER BOTH THE MOTHER'S AND THE BABY'S HEALTH. FEW DISORDERS OF PREGNANCY AND LABOR HAVE ANY PERMANENT PHYSICAL EFFECT ON EITHER MOTHER OR BABY.

ECTOPIC PREGNANCY

About 1 percent of pregnancies are ectopic; they are more common in women under the age of 30. The fertilized egg does not implant in the uterine lining, which is the normal place, but develops in one of the fallopian tubes or, more rarely, in another area altogether. The embryo does not develop normally, so the pregnancy usually fails. The embryo must be surgically removed to avoid rupture of the fallopian tube and to prevent internal bleeding.

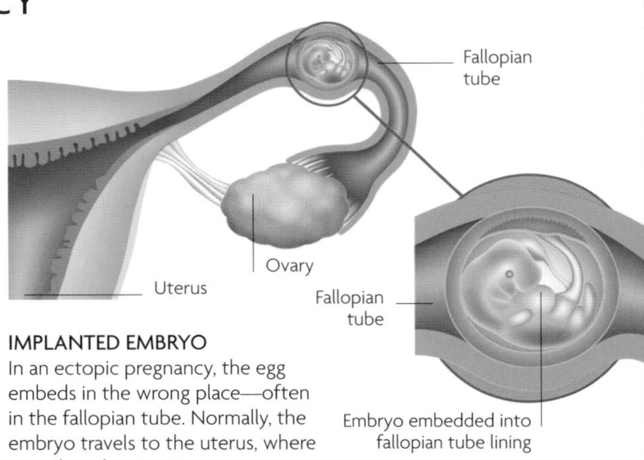
Fallopian tube
Ovary
Uterus
Fallopian tube
Embryo embedded into fallopian tube lining

IMPLANTED EMBRYO
In an ectopic pregnancy, the egg embeds in the wrong place—often in the fallopian tube. Normally, the embryo travels to the uterus, where growth and maturation occur.

PLACENTAL PROBLEMS

Two main problems can affect the placenta: placenta previa, in which the placenta covers the opening to the cervix; and placental abruption, in which the placenta separates from the uterine wall. The degree of severity in placenta previa depends on how much of the cervix is covered. If completely covered, the condition is serious. Placental abruption often comes on suddenly, and can threaten the fetus because essential blood supplies are compromised. Both conditions can cause vaginal bleeding, but in less severe cases, symptoms may go unnoticed.

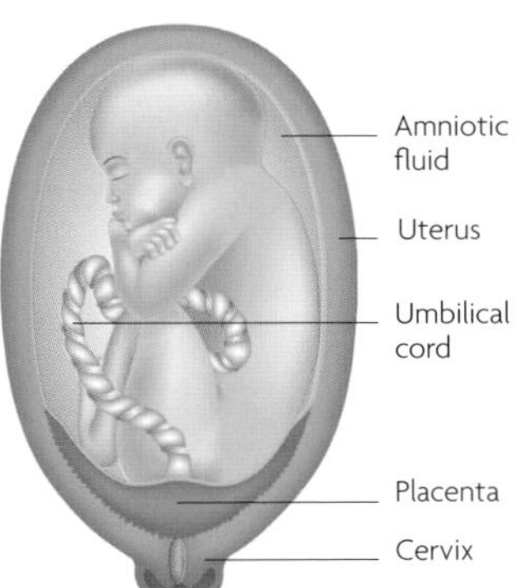
Amniotic fluid
Uterus
Umbilical cord
Placenta
Cervix

PLACENTA PREVIA
In complete placenta previa, as shown here, the placenta entirely covers the cervix. In a less severe form, the placenta only partially obstructs the exit from the uterus.

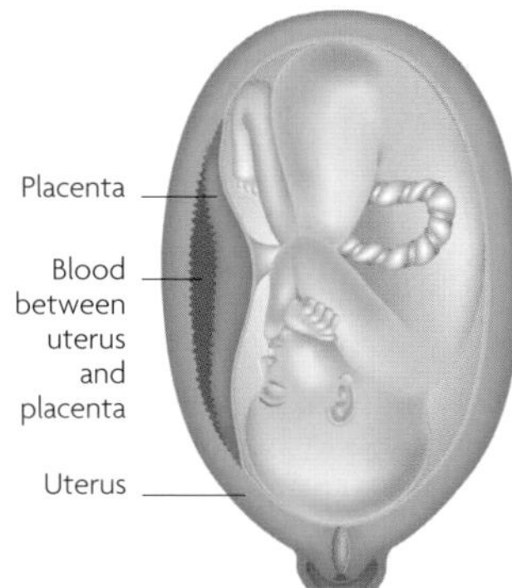
Placenta
Blood between uterus and placenta
Uterus

PLACENTAL ABRUPTION
Premature separation of the placenta from the uterus may be concealed, as shown here, in which case blood collects between the uterus and placenta.

MISCARRIAGE

Miscarriage, or spontaneous abortion, is the unintended end of a pregnancy before week 24. It is very common, occurring in 25 percent of all pregnancies. Most miscarriages occur in the first 14 weeks of pregnancy; over half of them are due to a genetic or fetal abnormality. Later miscarriages have various causes, ranging from physical problems with the cervix or uterus to severe infection. Smoking, alcohol, or drug abuse may also be factors. If three or more occur consecutively, it is known as recurrent miscarriage.

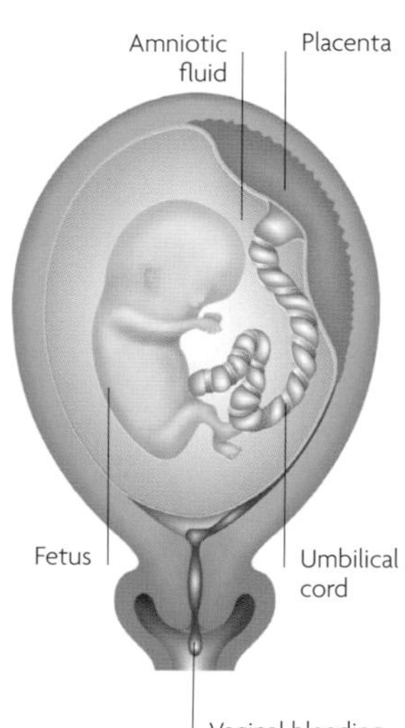
Amniotic fluid
Placenta
Fetus
Umbilical cord
Vaginal bleeding

THREATENED MISCARRIAGE
The fetus remains alive and the cervix is closed, although there is some blood loss. It may proceed to full miscarriage, when the fetus dies, or a successful birth.

PRETERM LABOR

Most pregnancies last for about 40 weeks, but delivery during the final three weeks is considered to be full term. Labor that occurs before 37 weeks is known as preterm and results in a premature baby. Premature labor rarely causes maternal problems, but the earlier the birth, the greater the problems encountered by the baby. The cause is not always known, but multiple births and urinary tract infection are known to be trigger factors. Sometimes, premature labor can be halted or delayed, giving the baby more time in the womb.

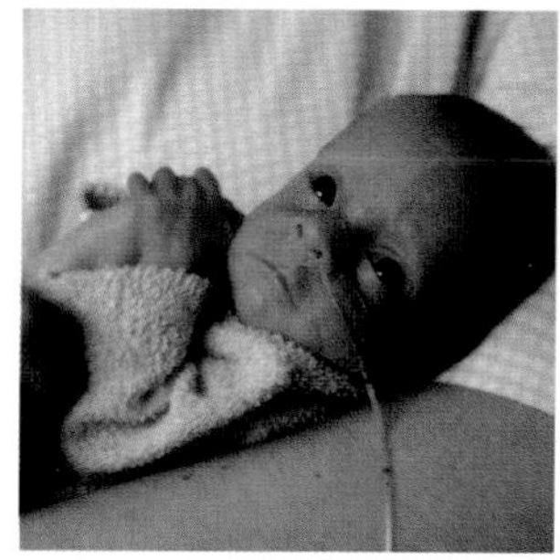

PREMATURE BABY
This premature baby is being fed through a nasogastric tube because his sucking reflex has yet to develop and his swallowing ability is poor. Other features are his tiny size, wrinkled and yellow skin, and disproportionately large eyes.

ABNORMAL PRESENTATION

Eighty percent of babies adopt the normal delivery position for birth, with the head down and facing toward the mother's back. The baby usually achieves this position by about week 36. Other babies are in a position that may cause problems during labor. Breech (see p.266) and occipitoposterior positions (see right) are the most common of these abnormal presentations.

In a breech birth, the baby's buttocks present first. Some presentations may allow the umbilical cord to drop through the birth canal and cause fetal distress. The cervix and vagina are more vulnerable to tears if the presentation is abnormal.

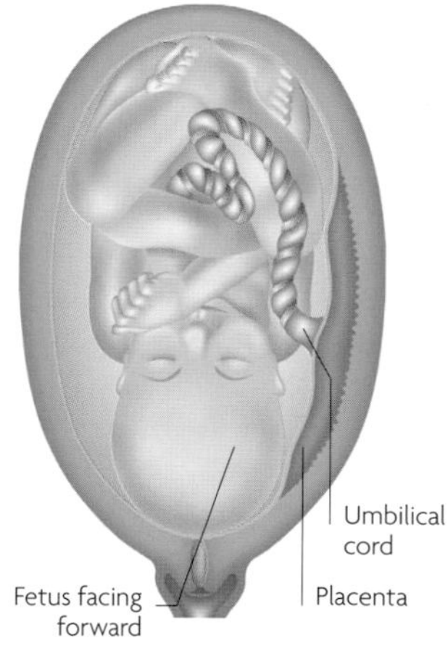
Umbilical cord
Fetus facing forward
Placenta

OCCIPITOPOSTERIOR POSITION
Although the baby's head is facing down, as is normal, the baby is turned 180° toward the front, instead of facing the mother's back.

INHERITED DISORDERS

INHERITED DISORDERS ARE CAUSED BY DEFECTIVE GENES OR ABNORMAL CHROMOSOMES. IN CHROMOSOME DISORDERS, THERE IS A PROBLEM IN THE NUMBER OR STRUCTURE OF CHROMOSOMES, WHEREAS IN GENE DISORDERS, THERE IS A FAULT IN ONE OR MORE GENES.

CHROMOSOME DISORDERS

Two-thirds of chromosome disorders are numerical—egg or sperm cells have either too many or too few chromosomes. In many cases, they result in a miscarriage. In a few exceptions, the fetus survives. The most common is Down syndrome, in which there is an extra chromosome 21. Abnormalities in the sex chromosomes have a less severe effect on the embryo, and there may not be any obvious signs of a problem. A girl with an extra X chromosome or a boy with an extra Y chromosome will probably go unnoticed. However, a boy who is born with an extra X chromosome (XXY) will have Klinefelter's syndrome, which becomes apparent at puberty when secondary sexual characteristics fail to develop. A girl who is born with only one X chromosome will have Turner's syndrome.

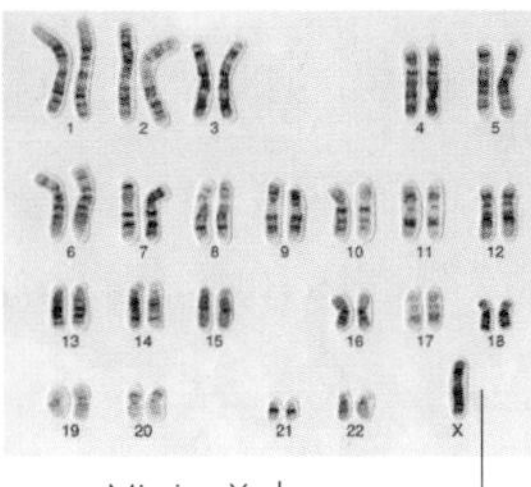
1
2
3
4
5
6
7
8
9
10
11
12
13
14
15
16
17
18
19
20
21
22
X
Missing X chromosome

TURNER'S SYNDROME
This chromosome set from a female shows only one X chromosome. She has Turner's syndrome, and will be short in stature and probably infertile.

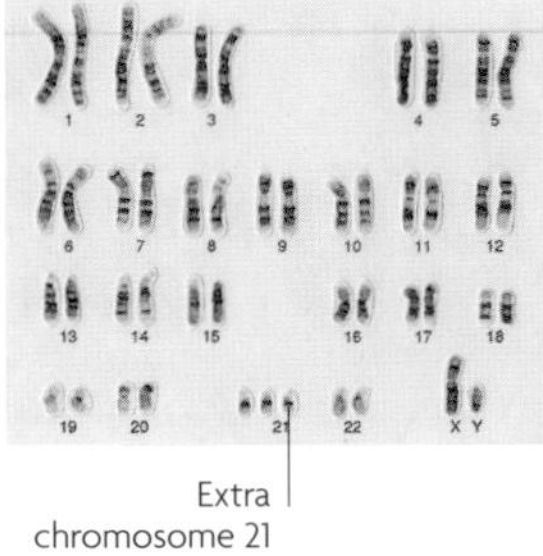
1
2
3
4
5
6
7
8
9
10
11
12
13
14
15
16
17
18
19
20
21
22
X Y
Extra chromosome 21

DOWN SYNDROME
This chromosome set from a male shows an extra chromosome 21. He has Down syndrome, and will have a distinct physical appearance and learning difficulties.

CYSTIC FIBROSIS

Cystic fibrosis is a gene disorder in which mucus glands produce abnormally thick secretions that cause repeated lung infections and problems digesting food. Weight gain is reduced, growth is slow, and life expectancy is shortened. Cystic fibrosis is caused by an abnormal gene that has to be received from each parent. Prenatal genetic testing and genetic counseling will be offered to parents of one affected child if they want to have more.

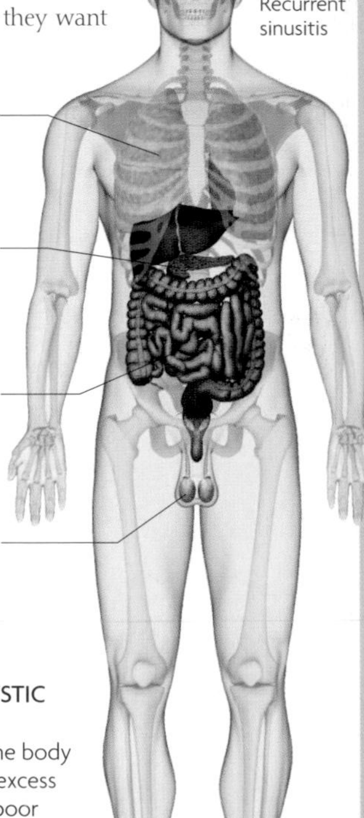
Sinuses
Recurrent sinusitis
Lungs
Lung infections, constant cough, and breathlessness
Pancreas
Lack of enzymes means digestion is inefficient
Intestines
Poor absorption of nutrients and intestinal blockage
Testes
Infertility as vas deferens and epididymis fail to develop properly

EFFECTS OF CYSTIC FIBROSIS
Many parts of the body are affected by excess mucus, causing poor health and slow physical development with intermittent episodes of serious illness.

CANCER

CANCER IS NOT A SINGLE DISEASE, BUT A LARGE GROUP OF DISORDERS WITH DIFFERENT SYMPTOMS. NEARLY ALL CANCERS HAVE THE SAME BASIC CAUSE: CELLS MULTIPLY UNCONTROLLABLY BECAUSE THE NORMAL REGULATION OF THEIR DIVISION HAS BEEN DAMAGED

CANCEROUS (MALIGNANT) TUMORS

A cancerous (malignant) tumor is a mass of abnormal cells that divide excessively quickly and do not carry out the normal functions of their tissue. These cells are often irregular in size and shape, and bear little resemblance to the normal cells from which they arose. This irregular appearance is often used to diagnose cancer during microscopic examination of a small sample of tissue taken from a tumor. The tumor gradually enlarges, crowding out normal cells, pressing on nerves, and infiltrating blood and lymph vessels. It is important to distinguish a malignant tumor from a nonmalignant one, because cancerous cells can spread to other parts of the body.

MALIGNANT TUMOR GROWTH

A cancerous tumor grows and spreads by forcing its way between other cells and infiltrating the tissues, eventually interfering with their function.

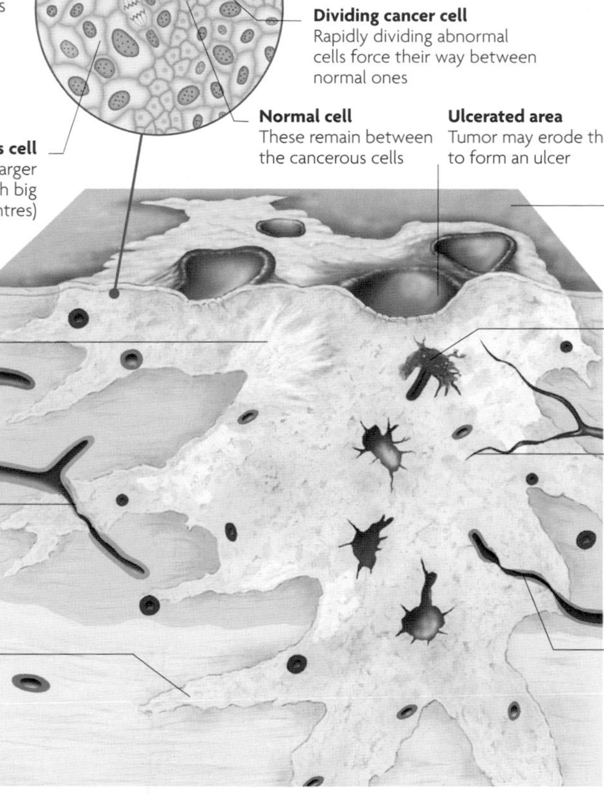
Dividing cancer cell
Rapidly dividing abnormal cells force their way between normal ones
Normal cell
These remain between the cancerous cells
Ulcerated area
Tumor may erode the epithelial layer to form an ulcer
Cancerous cell
These are often larger than normal cells, with big nuclei (control centres)
Epithelial layer
Covers and lines tissues and organs; tumors often form in this layer;
Calcium deposits
Hard deposits of calcium minerals may build up in tumors
Bleeding
Cancerous cells disrupt and breach tiny blood vessels
Nerve
Pressure on the nerves may cause the tumor to become painful to surrounding tissues
Blood vessel
Blood circulation is one major route for the spread of cancerous cells
Tumor outgrowth
Cancerous cells form outgrowths that infiltrate surrounding tissues
Lymph vessel
Like blood vessels, lymphatic vessels provide a route for cancer cells to spread

GLOSSARY

Terms in ***bold italics*** refer to other entries that appear in the glossary.

A

Accommodation
The process by which the eyes adjust to focus on nearby or distant objects.

Adenoids
Clusters of lymphoid ***tissue*** on each side of the back of the upper part of the throat.

Adipose tissue
Tissue made of specialized cells that store fatty (***lipid***) substances for energy, for protective "padding," and to provide heat insulation.

Allele
Form or version of a ***gene***. For example, the gene for eye color has blue and brown alleles.

Alveolus (pl. alveoli)
One of many tiny air sacs at the ends of the airways in the lungs. Gases diffuse in and out of blood through the alveolar walls.

Amino acid
One of about 20 kinds of building-block subunits of ***protein***.

Antibody
A soluble ***protein*** that attaches to body invaders, such as ***bacteria***, and helps destroy them.

Aorta
The central and largest ***artery*** of the body. It arises from the left ***ventricle*** of the heart and supplies oxygenated blood to all other arteries except the pulmonary artery.

Aortic valve
A triple-cusped valve at the origin of the ***aorta*** that allows blood to leave the left ***ventricle*** of the heart but prevents backward flow.

Appendix
The wormlike, dead-ended structure attached to the large intestine. The appendix has no known function in humans.

Aqueous humor
The fluid filling the front chamber of the eye, between the back of the ***cornea*** and front of the iris and ***lens***.

Arteriole
A small terminal branch of an ***artery*** leading to even smaller ***capillaries***, which link to the ***veins***.

Artery
An elastic, muscular-walled tube that transports blood away from the heart to other body parts.

Atrium (pl. atria)
One of two thin-walled, upper chambers of the heart.

Autonomic nervous system (ANS)
The portion of the nervous system controling unconscious functions such as heartbeat and breathing.

Axon
The long, fiberlike process of a ***nerve*** cell that conducts nerve impulses to or from the cell body.

B

Bacterium (bacteria pl.)
A type of microorganism with one cell. Only a few of the many species of bacteria cause disease.

Base
In ***nucleic acids*** (***DNA***, ***RNA***), a nitrogen-containing chemical unit or nitrogenous base (adenine, thymine, guanine, cytosine, uracil), the order of which carries genetic information.

Bile
A greenish brown fluid secreted by the ***liver*** that is concentrated and stored in the ***gallbladder***; released following food intake to help the digestion of fats.

Biliary system
The network of ***bile*** vessels formed by the ducts from the ***liver*** and the ***gallbladder***, and the gallbladder itself.

Bone marrow
Fatty ***tissue*** within bone cavities that may be red or yellow. Red bone marrow produces ***red blood cells***.

Brainstem
The lower part of the brain; houses the centers that control vital functions, such as breathing and the heartbeat.

Bronchus (pl. bronchi)
One of the larger air tubes in the lungs. Each lung has a main bronchus that branches into smaller and smaller airways.

C

Capillary
One of the numerous tiny blood vessels that link the smallest ***arteries*** and smallest ***veins***.

Cardiac
Relating to the heart.

Cartilage
Type of connective ***tissue*** that is tough and resilient, and often flexible; forms some structural parts, such as the ear and nose, and lines bone ends inside joints.

Central nervous system (CNS)
The brain and spinal cord; receives and analyzes sensory data, and initiates a response.

Cerebellum
A region of the brain located behind the ***brainstem***. It is concerned with balance, posture, and the control of fine movement.

Cerebrospinal fluid
A watery fluid that bathes the brain and spinal cord.

Cerebrum
The largest part of the brain; made up of two cerebral hemispheres. It contains the ***nerve*** centers for thought, personality, the senses, and voluntary movement.

Chromosome
A threadlike structure, present in all nucleated body cells, that carries the genetic code for the formation of the body. During cell division, chromosomes coil into "X" shapes. A normal human body has 23 pairs of chromosomes.

Cochlea
The coiled structure in the inner ear that contains the *organ* of Corti, which converts sound vibrations into *nerve* impulses for transmission to the brain.

Collagen
The body's most important structural *protein*, present in bones, *tendons*, *ligaments*, and other connective *tissues*.

Colon
The part of the large intestine that extends from the cecum to the rectum. Its main function is to conserve water by absorbing it from the bowel contents.

Cornea
The transparent dome at the front of the eyeball that is the eye's main focusing *lens*.

Coronary
A term meaning "crown." Refers to the arteries that encircle and supply the heart with blood.

Corpus callosum
The wide, fan-shaped band consisting of about 20 million *nerve* fibers that connects the two hemispheres of the *cerebrum*.

Cortex
Outer layer in various *organs*, such as the cerebral cortex (brain), *renal* cortex (*kidney*), and adrenal cortex (hormone-producing gland on top of the kidney).

Cranial nerves
The 12 pairs of *nerves* emerging from the brain and *brainstem*. They include the nerves for smell, sight, eye movement, facial movement and sensation, hearing, taste, and head movement.

Cytoplasm
Watery or jellylike fluid that fills the bulk of a cell; it contains many *organelles*.

D

Dermis
The thick inner layer of skin, made of connective *tissue*; contains structures such as sweat glands.

Diaphragm
The dome-shaped muscular sheet that separates the chest from the abdomen. When the muscle contracts, the dome flattens, increasing chest volume and drawing air into the lungs.

Diastole
The period in the heartbeat cycle when all four chambers are relaxed and the heart is filling with blood. See *systole*.

Digestive system
The mouth, *pharynx*, *esophagus*, stomach, and intestines. Associated *organs* are the *pancreas*, *liver*, and *gallbladder*.

DNA (Deoxyribonucleic acid)
A chemical with a double-helix structure that carries genetic information in the sequence of its paired subunits (*bases*); packaged into *chromosomes*.

Duodenum
The C-shaped first part of the small intestine, into which the stomach empties. Ducts from the *gallbladder*, the *liver*, and the *pancreas* all enter the duodenum.

E

Eardrum
The membrane separating the outer ear from the *middle ear* that vibrates in response to sound.

Embryo
The developing baby from conception until the eighth week of pregnancy. See *fetus*.

Endocrine gland
A gland that produces *hormones* (chemical messenger substances) that are released directly into the bloodstream rather than along tubes or ducts.

Endorphin
A morphinelike substance produced naturally by the body in times of pain and stress, and also activated during exercise.

Enzyme
A *protein* that accelerates chemical reactions within cells.

Epidermis
The outer layer of the skin; its box-shaped cells become flatter and scalier toward the surface.

Epiglottis
A leaflike flap of *cartilage* located at the entrance of the *larynx*, which covers the opening of the airways during swallowing and helps prevent food or liquid from entering the windpipe (*trachea*).

Epithelium
Specialized covering or lining *tissue* that forms sheets and layers around and within many *organs* and other tissues.

Esophagus
The muscular tube, also known as the food tube, that connects the *pharynx* with the stomach.

Estrogen
A *sex hormone* that prepares the uterine lining for an implanted, fertilized egg and stimulates the development of a female's secondary sexual characteristics.

Eustachian tube
Tube connecting the back of the nose to the *middle ear* cavity; allows air pressure to equalize on either side of the eardrum.

F

Fallopian tube
One of the two tubes along which an *ovum* travels to the *uterus*, after release from an *ovary*; its fingerlike projections help sweep the ovum into the tube.

Fertilization
The union of a sperm and an egg, after sexual intercourse or artificial insemination, or in a test tube.

Fetus
The developing baby from about the eighth week after *fertilization* until the time of birth. See *embryo*.

G

Gallbladder
The fig-shaped bag lying under the *liver*, into which *bile* secreted by the liver passes to be stored.

Gastric juice
Liquid produced by the stomach lining that contains hydrochloric acid and digestive *enzymes*.

Gastrointestinal tract
The muscular tube that extends from the mouth, through the *pharynx*, *esophagus*, stomach, and intestines, to the rectum. Also known as the digestive tract.

Gene
A distinct section of a *chromosome* that is the basic unit of inheritance. Each gene consists of a segment of deoxyribonucleic acid (*DNA*) containing the code that governs the production of a specific *protein*.

Genome
The full set of *genes*, or hereditary information, for a living organism; the human genome consists of 20,000–25,000 genes.

Gray matter
The darker-colored regions of the brain and spinal cord that comprise mainly *neuron* cell bodies as opposed to their projecting fibers, which form *white matter*.

Hair follicle
A pit on the surface of the skin from which hair grows.

Heart valve
One of four structures in the heart that allow the passage of blood in one direction only.

Hemoglobin
The *protein* in *red blood cells* that combines with oxygen to carry the gas from the lungs and distribute it around the body.

Hepatic
Concerning the *liver*.

Hepatocyte
A type of *liver* cell with many functions, including making *bile*.

Hippocampus
A structure in the *limbic system* in the brain concerned with learning and long-term memory.

Hormone
A chemical released by the *endocrine glands* and some *tissues*. Hormones act on specific receptor sites in other parts of the body.

Hypothalamus
A small structure located at the base of the brain, where the nervous and hormonal systems of the body interact. It is linked to the *thalamus* above and the *pituitary gland* below.

Ileum
The final segment of the small intestine, where most absorption of nutrients takes place.

Kidney
One of two bean-shaped organs in the back of the abdominal cavity that filter blood and remove wastes, particularly *urea*.

Killer T cells
White blood cells that can destroy damaged, infected, or malignant body cells by using proteins called lymphokines.

Larynx
The structure in the neck at the top of the *trachea*, known as the voice box, that contains the *vocal cords*.

Lens
The internal lens of the eye, also called the crystalline lens; it fine-focuses vision by adjusting its curvature. The outer lens is called the *cornea*.

Ligament
A band of *tissue* consisting of *collagen*–a tough, fibrous, elastic *protein*. Ligaments support bones, mainly in and around joints.

Limbic system
A collection of structures in the brain that plays an important role in the automatic (involuntary) body functions, instinctive behavior, emotions, and the sense of smell.

Lipid
Fatty or oily substance, insoluble in water, with varied roles in the body, including formation of *adipose tissue*, cell membranes (phospholipid), and steroid *hormones*.

Liver
The large *organ* in the upper right abdomen that performs vital chemical functions, including detoxification of poisons and conversion of waste products to *urea*.

Lobe
A rounded projection or subdivision forming part of a larger structure such as the brain, lung, or *liver*.

Lymphatic system
An extensive network of transparent lymph vessels and *lymph nodes*. It returns excess *tissue* fluid to the circulation and combats infections and cancer cells.

Lymph node
A small, oval gland packed with *white blood cells* that acts as a barrier to the spread of infection. Nodes occur in series along lymph vessels.

Lymphocyte
White blood cell that is part of the immune system; it protects against *virus* infections and cancer.

M

Medulla
The inner part of an *organ*, such as the *kidneys* or adrenal glands. Also refers to the part of the *brainstem* lying immediately above the start of the spinal cord, just in front of the *cerebellum*.

Meninges
Three membrane layers around the brain and spinal cord: the pia mater on the inside, the arachnoid and the dura mater next to the skull.

Meniscus
A crescent-shaped, shock-absorbing pad of *cartilage* found in the knee and some other joints.

Menopause
The end of the reproductive period in women, when the *ovaries* have ceased their production of eggs and menstruation has stopped.

Metabolism
The sum of all the physical and chemical processes that take place in the body.

Middle ear
The air-filled cleft within the temporal bone between the ***eardrum*** and the outer wall of the inner ear; contains ***ossicles***. Also called the tympanic cavity.

Mitochondrion (pl. mitochondria)
A cell ***organelle*** containing genetic material; it is also involved in the production of energy for cell functions. It is covered with a double-layered membrane.

Mitral valve
The valve that lies between the left ***atrium*** and left ***ventricle*** of the heart.

Motor neuron
A ***nerve*** cell that carries the impulses to muscles that cause movement.

Mucous membrane
The soft, skinlike, mucus-secreting layer lining the tubes and cavities of the body.

Myocardium
The special muscle of the heart. The fibers form a network that can contract spontaneously.

Myofibril
Cylindrical element within muscle cells (fibers) consisting of thinner filaments that move to produce muscle contraction.

N

Nephron
The ***kidney's*** filtering unit, consisting of a filtration capsule (glomerulus) and a series of tubules, that reabsorbs or excretes water and wastes to control fluid balance.

Nerve
Bundle of threadlike projections from individual ***neurons*** (nerve cells), held together by a fibrous sheath. Nerves carry electrical impulses to and from the brain and spinal cord and other body parts.

Neuron
A single ***nerve*** cell with long projections, the function of which is to transmit electrical impulses.

Nociceptor
A ***nerve*** ending responding to painful stimuli.

Nucleic acid
Deoxyribonucleic acid (*DNA*) or ribonucleic acid (*RNA*): chains of ***nucleotides***, with genetic information in the order of the nucleotide bases.

Nucleotide
Building-block subunit of a ***nucleic acid*** (*DNA, RNA*), consisting of a sugar, phosphate, and a nitrogen-containing ***base***.

Nucleus (pl. nuclei)
Control center of a cell, containing the genetic material *DNA*.

Olfactory nerve
One of two ***nerves*** of smell that run from the olfactory bulb in the roof of the nose directly into the underside of the brain.

Optic nerve
One of the two ***nerves*** of vision. Each one has about one million nerve fibers running from the ***retina*** to the brain, carrying visual stimuli.

Organ
Discrete body part or structure with a vital function: for example, the heart, ***liver***, brain, or ***spleen***.

Organelle
A tiny structure inside a cell that has a specific role. The ***nucleus***, ***mitochondrion***, and ribosomes are examples.

Ossicle
One of three tiny bones (the incus, malleus, and stapes) of the ***middle ear*** that convey vibrations from the ***eardrum*** to the inner ear.

Ossification
The process of formation, renewal, and repair of bone. Most bones in the body develop from ***cartilage***.

Osteon
The rod-shaped unit, also called a Haversian system, that is the building block of cortical bone.

Ovary
One of two structures lying at the end of the ***fallopian tubes*** on each side of the ***uterus***. They store ovarian follicles, release the mature ***ova***, and produce the female ***sex hormones*** (***estrogen*** and ***progesterone***).

Ovulation
The release of an ***ovum*** from a mature follicle in the ***ovary*** about midway through the menstrual cycle; if not fertilized, the egg is shed during menstruation.

Ovum (pl. ova)
The egg cell; if ***fertilization*** occurs, the ovum may implant in the ***uterus*** and develop into an ***embryo***.

P

Pancreas
A gland behind the stomach that secretes digestive ***enzymes*** and also ***hormones*** that regulate blood glucose levels.

Parasympathetic nervous system
One of the two divisions of the ***autonomic nervous system***. It maintains and restores energy–for example, by slowing the rate and strength of the heartbeat.

Parathyroid glands
Two pairs of yellowish ***endocrine glands***, located behind the thyroid gland, that help control the level of calcium in the blood.

Parotid glands
The large pair of salivary glands situated, one on each side, above the angles of the jaw just below and in front of the ears.

Pelvis
The basinlike ring of bones to which the lower end of the ***spine*** is attached and with which the thigh bones articulate. The term also refers to the general lower abdominal area.

Pericardium
The layers of membrane surrounding the heart. The outer fibrous sac encloses the heart and the roots of the major blood vessels

emerging from it. The inner layer attaches to the heart wall.

Periosteum
The tough ***tissue*** that coats all bone surfaces except joints and from which new bone can be formed; contains blood and lymphatic vessels and ***nerves***.

Peripheral nervous system
All the ***nerves*** that fan out from the brain and spinal cord, linking these parts with the rest of the body. The system consists of ***cranial nerves*** and ***spinal nerves***.

Peristalsis
A coordinated succession of contractions and relaxations of the muscular wall of a tubular structure, such as the intestines, that moves the contents along.

Peritoneum
The double-layered membrane that lines the inner wall of the abdomen. The peritoneum covers and partly supports the abdominal ***organs***. It also secretes a fluid that lubricates the movement of the intestines.

Phagocyte
A ***white blood cell*** or similar cell that surrounds and engulfs unwanted matter, such as invading microbes and cellular debris.

Pharynx
The passage leading down from the back of the nose and the mouth to the ***esophagus***; it consists of the nasopharynx, the oropharynx, and the laryngopharynx.

Pituitary gland
A pea-sized gland hanging from the underside of the brain. The pituitary secretes ***hormones*** that control many other glands in the body, and is regulated by the ***hypothalamus***.

Placenta
The disk-shaped ***organ*** that forms in the ***uterus*** during pregnancy. It links the blood supplies of the mother and ***fetus*** via the ***umbilical cord*** and nourishes the growing ***fetus***.

Plasma
The fluid part of the blood from which all cells have been removed; it is mostly water, but contains some ***proteins***, salts, and various nutrients, including glucose.

Platelet
Tiny fragment of a type of large cell manufactured in bone marrow and known as a megakaryocyte. Platelets are vital for blood clotting.

Pleura
A double-layered membrane, the inner layer of which covers the lung while the outer layer lines the chest cavity. A layer of fluid lubricates and enables movement between the two.

Progesterone
A female ***sex hormone*** secreted by the ***ovaries*** and ***placenta*** that allows the ***uterus*** to receive and retain a fertilized egg.

Prostaglandins
A group of fatty acids, made in the body, that have various functions and influence some hormones.

Prostate gland
A male accessory sex gland situated at the base of the bladder and opening into the ***urethra***. It secretes some of the fluid in semen.

Protein
Huge molecule composed of chains of ***amino acids***; the basis of many structural materials (keratin, ***collagen***), ***enzymes***, and ***antibodies***.

Pulmonary artery
The large ***artery*** that conveys deoxygenated blood from the right ***ventricle*** of the heart to the lungs to be reoxygenated.

R

Red blood cells
Biconcave, disk-shaped cells, without ***nuclei***, that contain ***hemoglobin***. There are 4–5 million red cells in $^1/_{500}$ pint (1 milliliter) of blood.

Renal
Relating to the ***kidneys***.

Respiration
1. Body movements of breathing. 2. Gas exchange of oxygen for carbon dioxide in the lungs. 3. Similar gas exchange in the ***tissues*** (cellular respiration). 4. Breakdown of molecules such as glucose to release their energy for cellular functions.

Retina
A light-sensitive layer lining the inside of the back of the eye; it converts optical images to ***nerve*** impulses, which travel to the brain via the ***optic nerve***.

RNA
Ribonucleic acid, a substance present in cells; different forms carry out various functions, including the manufacture of ***proteins***.

S

Saliva
A watery fluid secreted into the mouth by the salivary glands to aid tasting, chewing, and digestion.

Sex hormones
Steroid substances, including ***testosterone*** in males and ***estrogen*** and ***progesterone*** in females, that bring about the development of sexual characteristics. Sex hormones also regulate sperm and egg cell production and the menstrual cycle.

Sphincter
A muscle ring, or local thickening of the muscle coat, surrounding an opening in the body, such as the anus or the ***urethra***.

Spinal nerves
The 31 pairs of combined motor and sensory ***nerves*** that emerge from and enter the spinal cord.

Spine
The column of 33 ringlike bones, called ***vertebrae***, that divides into seven cervical vertebrae, 12 thoracic vertebrae, five lumbar vertebrae, and the fused vertebrae of the sacrum and coccyx.

Spleen
A lymphatic ***organ***, situated on the upper left of the abdomen, that destroys worn-out ***red blood cells***, filters out impurities from the blood, and helps fight infection.

Stem cell
Generalized type of cell, usually fast-dividing, with the potential to

become many different kinds of specialized cells.

Sympathetic nervous system
One of the two divisions of the ***autonomic nervous system***. It prepares the body for action–for example, by constricting blood vessels in the intestines and skin.

Synapse
The junction between two ***nerve*** cells, or between a ***nerve*** cell and a muscle fiber or a gland. Chemical messengers are passed across a synapse to produce a response in a target cell.

Synovial fluid
Thin, slippery, lubricating fluid within a joint.

Synovial joint
A mobile joint with a membrane that produces a lubricating fluid.

Systole
The period in the heartbeat cycle during which first the ***atria*** and then the ***ventricles*** contract to force blood out of the heart. See ***diastole***.

T

Taste bud
A spherical nest of receptor cells found mainly on the tongue; each bud responds most strongly to a sweet, salty, sour, or bitter flavor.

Tendon
A strong band of ***collagen*** fibers that joins muscle to bone and transmits the pull caused by muscle contraction.

Testis (pl. testes)
One of a pair of the sperm- and ***hormone***-producing sex glands in the scrotum.

Testosterone
The principal male ***sex hormone;*** produced in the ***testis*** and in small amounts in the adrenal gland on top of the kidney, and in the ***ovary***.

Thalamus
A mass of ***gray matter*** found deep in the brain, on top of the ***brainstem***. The thalamus receives and processes sensory information.

Thorax
The part of the trunk between the neck and the abdomen that contains the heart and the lungs.

Tissue
Body structure made of similar cells that perform one main function; types include muscle and connective tissues.

Tonsils
Oval masses of lymphoid tissue on the back of the throat. They help protect against childhood infections by attacking microorganisms that enter through the nose and mouth.

Trachea
A muscular tube lined with ***mucous membrane*** and reinforced by about 20 rings of ***cartilage***.

Umbilical cord
The structure that connects the ***placenta*** to the ***fetus***. It provides the immunological, nutritional, and hormonal link with the mother.

Urea
A waste product of the breakdown of ***proteins***; the nitrogen-containing component of urine.

Ureter
Tube through which urine passes from each ***kidney*** to the bladder.

Urethra
The tube that carries urine from the bladder to the exterior; much longer in the male than in the female.

Urinary tract
The system that forms and excretes urine; made up of the ***kidneys***, ***ureters***, bladder, and ***urethra***.

Uterus
A hollow muscular structure in which the ***fetus*** grows until birth.

Vagina
The muscular passage from the ***uterus*** to the outside of the body; it stretches during sexual intercourse and childbirth.

Vagus nerves
The tenth pair of ***cranial nerves***; helps control automatic functions such as heartbeat and digestion.

Vas deferens
One of a pair of tubes that lead from the ***testes***; each tube carries sperm, which mix with fluid before entering the ***urethra***

Vein
A thin-walled blood vessel that returns blood at low pressure from body ***organs*** and ***tissues*** to the heart.

Vena cava
One of the two large ***veins***, the superior and inferior vena cavae, that empty into the right ***atrium*** of the heart.

Ventricle
A chamber or compartment, usually fluid-filled. Examples include two ***cardiac*** ventricles in the heart.

Vertebra (pl. vertebrae)
One of the 33 bones of the vertebral column (***spine***).

Virus
The tiniest form of infectious microorganism (germ). It takes over a cell to produce copies of itself.

Vocal cords
One of two sheets of ***mucous membrane*** stretched across the inside of the ***larynx*** that vibrate to produce voice sounds when air passes between them.

W–Z

White blood cell
Any of the colorless blood cells that play a role in the immune system.

White matter
Nerve tissue in the brain and spinal cord formed mainly of the projecting fibers, or ***axons***, of ***neurons*** (nerve cells).

Zygote
The single cell produced when an ***ovum*** is fertilized by a sperm; it contains the genetic material (***DNA***) for a new person.

INDEX

Page numbers in **bold** type indicate main references.

A

B

N

O

P

R

ACKNOWLEDGMENTS

Dorling Kindersley would like to thank several people for their help in the preparation of this book. Anna Barlow contributed valuable comments on the cardiovascular system. Peter Laws assisted with visualization, and additional design work was done by Mark Lloyd and Louise Waller. Three-dimensional illustrations were created from a model supplied by Zygote Media Group, Inc. Ben Hoare, Peter Frances, and Ed Wilson all provided editorial assistance, Katie John did the proof reading, and Hilary Bird provided the index. Marianne Markham and Andrea Bagg contributed to the initial development work.

The Concise Human Body Book Picture Credits

The publisher would like to thank the following for their kind permission to reproduce their photographs:

(Key: a-above; b-below/bottom; c-center; f-far; l-left; r-right; t-top)

5 Science Photo Library: Sovereign, ISM. **6 Science Photo Library. 10-11 Science Photo Library:** Francois Paquet-Durand. **12 Alamy Images:** Rob Walls (br). **Science Photo Library:** CNRI (bl). **Wellcome Library, London:** K. Hodivala-Dilke & M. Stone (cra). **13 Alamy Images:** Chad Ehlers (bl). **Getty Images:** Science Faction / L. Steinmark - CMSP (br). **Science Photo Library:** CNRI (bc); Wellcome Dept. of Cognitive Neurology (cr); Zephyr (tr). **Wellcome Library, London:** Mark Lythgoe & Chloe Hutton (tc). **18 Science Photo Library. 20 Still Pictures:** PHONE Labat J.M. / F. Rouquette (bl); Volker Steger (tr). **21 Science Photo Library:** CNRI (tl). **22 Science Photo Library:** Richard Wehr / Custom Medical Stock Photo; Adam Hart-Smith (crb). **23 Science Photo Library:** Adam Hart-Davis (cr). **25 Corbis:** Visuals Unlimited (tl). **27 Science Photo Library:** Volker Steger (cla). **28 Alamy Images:** Phototake Inc. (tc). **Science Photo Library:** Jean-Claude Revy, ISM (cb). **29 Science Photo Library:** Professors P. Motta & T. Naguro (tl) (tc) (tr). **30 Science Photo Library:** Lawrence Livermore Laboratory (clb). **33 Alamy Images:** Bjanka Kadic (tr). **34 Science Photo Library:** L. Willatt, East Anglian Regional Genetics Service (tl). **35 Science Photo Library:** Alain Pol, ISM (tl). **36 Alamy Images:** Phototake Inc. (tr). **Science Photo Library:** Nancy Kedersha / UCLA (tc). Still Pictures: Ed Reschke (br). **37 Corbis:** Visuals Unlimited (br). **Science Photo Library:** Innerspace Imaging (tl); Claude Nuridsany & Marie Perennou (bc). **Still Pictures:** Ed Reschke (bl) (tc). **Wellcome Library, London:** David Gregory & Debbie Marshall (tr). **38–61 Wellcome Library, London:** Professor Alan Boyde (sidebars). **43 Science Photo Library:** Biophoto Associates (bl). **Wellcome Library, London:** M.I. Walker (cra). **44 DK Images:** Philip Dowell / Courtesy of The Natural History Museum, London (clb). Wellcome Library, London: (bl). **46 Science Photo Library:** GJLP (bl). **47 Science Photo Library:** Eye of Science (tc). **48 Science Photo Library:** Simon Brown (tl). **50 Science Photo Library:** Anatomical Travelogue (tr). **59 Science Photo Library:** CNRI (cla). **60 Science Photo Library:** Princess Margaret Rose Orthopaedic Hospital (cl). **61 Science Photo Library:** CNRI (cl). **62–75 Science Photo Library:** Eye of Science (sidebars). **67 Science Photo Library:** (crb/Smooth) (br/Cardiac). **Wellcome Library, London:** M.I. Walker (bc/ Striated). **68 Still Pictures:** Ed Reschke (tl). **70 Science Photo Library:** Steve Gschmeissner (cla). **76–129 Science Photo Library:** Nancy Kedersha (sidebars). **80 Wellcome Library, London:** Dr. Jonathan Clarke (clb). **81 Science Photo Library:** Steve Gschmeissner (br). **83 Wellcome Library, London:** (br); Prof. Peter Brophy (bc). **86 Alamy Images:** allOver Photography (bl). **Science Photo Library:** Zephyr (br). **Still Pictures:** Alfred Pasieka (tr). **91 Science Photo Library:** Zephyr (clb). **92 Science Photo Library:** Bo Veisland, MI&I (br). **99 Alamy Images:** Phototake Inc, (t). **104 Science Photo Library:** CNRI (tr). **111 Corbis:** Visuals Unlimited (b). **112 Science Photo Library:** Eye of Science (tr). **113 Science Photo Library:** Pascal Goetgheluck (t). **114 Alamy Images:** Phototake Inc. (br). **118 Science Photo Library:** Susumu Nishinaga (tr). **121 Louise Thomas:** (fcra) (fcr). **123 Wellcome Library, London:** (cla) (cra). **125 Science Photo Library:** Alfred Pasieka (c). **126 Alamy Images:** Medical-on-Line (br); Phototake Inc. (bc). 1 28 **Science Photo Library:** Bo Veisland (clb). **130–143 Wellcome Library, London:** University of Edinburgh (sidebars). **137 Science Photo Library:** Steve Gschmeissner (br); Manfred Kage (bl). **141 Wellcome Library, London:** (br). **144–159 Wellcome Library, London:** EM Unit / Royal Free Med. School (sidebars). **151 Science Photo Library:** Manfred Kage (tr). **152 Science Photo Library:** (cr); Professors P.M. Motta & G. Macchiarelli (bl). **153 Wellcome Library, London:** M.I. Walker (tr). **156 Science Photo Library:** CNRI (crb). **157 Science Photo Library:** BSIP VEM (cra). **160–175 Science Photo Library:** GJLP (sidebars). **167 Alamy Images:** Phototake Inc. (tc). **169 Science Photo Library:** Zephyr (t). **171 Science Photo Library:** CNRI (cl) (c); Dr. Gary Settles (cr). **172 Science Photo Library:** Dr. Gary Settles (cra). **176–189 Science Photo Library:** Steve Gschmeissner (sidebars). **178 Science Photo Library:** Sheila Terry (tl). **182 Alamy Images:** Phototake Inc. (bl). **Science Photo Library:** J.C. Revy (cl)**. 184 Science Photo Library:** Steve Gschmeissner (bl); Prof. P. Motta / Dept. of Anatomy / University, "La Sapienza," Rome (cr). **185 Science Photo Library:** Prof. P. Motta / Dept. of Anatomy / University "La Sapienza," Rome (tr). **186 DK Images:** Steve Gorton (tr); Jules Selmes and Debi Treloar (cl). **Science Photo Library:** Alfred Pasieka (bc). **187 DK Images:** Susanna Price (cr); Jules Selmes and Debi Treloar (cl) (c). **188 Alamy Images:** Medical-on-Line (bl). Mediscan: (cl). **Wellcome Library, London:** (cr) (br). **190–209 Science Photo Library:** Francis Leroy, Biocosmos (sidebars). **196 Science Photo Library:** CNRI (bl). **199 Alamy Images:** Phototake Inc. (tr) (tl). **205 Alamy Images:** Scott Camazine (br). **206 Science Photo Library:** Eye of Science (tr). **207 Science Photo Library:** David Scharf (cl) (bl). **210–239 Science Photo Library:** Eye of Science (sidebars). **215 Science Photo Library:** Steve Gschmeissner (crb). **216 Science Photo Library:** Eye of Science (tr). **217 Science Photo Library:** CNRI (tr). **221 Science Photo Library:** Prof. P. Motta / Dept. of Anatomy / University "La Sapienza," Rome (bl). **224 Science Photo Library:** Professors P. Motta & F. Carpino / University "La Sapienza," Rome (br). **230 Science Photo Library:** Dr. T. Blundell, Dept. of Crystallography, Birkbeck College (bl). **232 Corbis:** Frank Lane Picture Agency (tr). **235 Science Photo Library:** P. Hawtin, University of Southampton (tr). **244 Wellcome Library, London:** David Gregory & Debbie Marshall (br). **248 Science Photo Library:** Professor P.M. Motta et Al (br). **249 Science Photo Library:** Steve Gschmeissner (bl). **250–305 Science Photo Library:** Susumu Nishinaga (sidebars). **254 Science Photo Library:** Steve Gschmeissner (clb). **255 Science Photo Library:** Parviz M. Pour (b). **256 Wellcome Library, London:** Yorgos Nikas (bl). **257 Alamy Images:** Phototake Inc. (tl). **258 Science Photo Library:** Professors P.M. Motta & J. Van Blerkom (tr). **270 Science Photo Library:** Keith / Custom Medical Stock Photo (cra). **276 Science Photo Library:** (l). **278 Science Photo Library:** BSIP VEM (bl). **279 Corbis:** Jose Luis Pelaez, Inc. (tr). **281 Science Photo Library:** Susumu Nishinaga (t). **282 Science Photo Library:** Professor P.M. Motta, G. Macchiarelli, S.A, Nottola (cla). **284 Alamy Images:** Chad Ehlers (br). **285 Science Photo Library:** (bc) (br). **287 Corbis:** Andrew Brookes (tr). **Science Photo Library:** Philippe Plailly (br). **294 Science Photo Library:** CNRI (br). **298 Mediscan:** CDC (t). **299 Science Photo Library:** (br). **301 Science Photo Library:** James King-Holmes (fcl) (cl). **303 Alamy Images:** Janine Wiedel Photolibrary (tr). **304 Wellcome Library, London:** Wessex Reg. Genetics Centre (clb) (bl)